Fungi from Different Substrates

Fungi from Different Substrates

Editors

J.K. Misra
SLM Bharatiya Vidya Bhavan Girls Degree College
Vineet Khand, Gomti Nagar
Lucknow
India

J.P. Tewari
Department of Agricultural, Food, and Nutritional Science
University of Alberta
Alberta
Canada
and
Lung, Allergy, Sleep Centers of America
224 W. Exchange Street, Ste. 380
Akron, Ohio 44302
U.S.A.

S.K. Deshmukh
Piramal Enterprises Limited
Department of Natural Products
Goregaon (East), Mumbai
India

Csaba Vágvölgyi
Department of Microbiology
Faculty of Science and Informatics
University of Szeged
Szeged
Hungary

CRC Press is an imprint of the
Taylor & Francis Group, an **informa** business
A SCIENCE PUBLISHERS BOOK

Cover illustrations reproduced by kind courtesy of Dr. Kim Fleming, Dr. Hoai Xuan Truong and Dr. Shubha Pandey

CRC Press
Taylor & Francis Group
6000 Broken Sound Parkway NW, Suite 300
Boca Raton, FL 33487-2742

First issued in paperback 2020

ISBN-13: 978-1-4822-0960-0 (hbk)
ISBN-13: 978-0-367-73942-3 (pbk)

Visit the Taylor & Francis Web site at
http://www.taylorandfrancis.com

and the CRC Press Web site at
http://www.crcpress.com

Preface

Fungi are achlorophilous heterotrophs and thrive in or on any animate or inanimate object in nature. They flourish in any environment—terrestrial or aquatic, freshwater or marine. Saprophytically, they proliferate on a variety of materials of plant, animal and human origin, found in any ecosystem on the earth. Thus, the range and variety of substrata on which fungi can be found growing are enormous. The biodegrading activity of fungi is immense without which humans will drown in an ocean of leaf litter. These fascinating organisms also are known to grow parasitically on plants, animals and human beings causing various diseases. The symbiotic or mutualistic relationships of fungi with many other entities are also of great significance such as their mycorrhizal associations with various plant roots, endophytic associations with almost all known plants on earth, but for a few, and their associations with algal forms in case of lichens. In a nut shell, fungi can be found on anything and everything on earth in one way or the other. This emphasizes their importance in our life as they can be seen or isolated from a tiny soil particle to a big stone piece in a museum including many objects submerged in water consisting of decomposing plant or animal materials. Various substrata for fungi have been looked for these organisms by different mycologists all over the world. In this book, fungi observed on some of the important substrata have been presented and their significance discussed.

The book encompasses 17 chapters on fungi from different substrates including fossilized leaves. In addition to non-living substrates, association of pathogenic fungi with important plants like Wheat, Rice, Soybean and Brassicas have also been discussed and reviewed in different chapters. Some animal and human pathogens have also been covered. Besides the taxonomic information, some ecological aspects like distribution and substrate/host preferences have also been dealt with in some chapters, where appropriate. Fungi-like organisms, the myxomycetes, are discussed in Chapter 6. Thus, the book has covered a variety of substrates, their fungal colonizers, and ecology as well. Updated information, for all these aspects have been discussed and presented. How and up to what extent, the fungi associated with various materials deteriorate them, have also been elaborated in chapters dealing with dairy products and materials of cultural heritage significance.

Thus, the book has attempted to present information for as many substrates as possible—living, senescing, and decomposing, with widely approached and updated information. However, these are not complete in a sense, as all information available cannot be put in one book.

The editors are grateful to the contributors for their efforts in preparing and providing the chapters for the book.

Dr. J.K. Misra is thankful to the Management (Justice V. K. Mehrotra, Chairman and Brig. R.N. Misra, Secretary) and the Principal (Dr. Rachna Mishra) of SLM Bharatiya Vidya Bhavan Girls Degree College, Lucknow, for always encouraging him to undertake such academic ventures by providing all possible logistic supports through the institution. S.K. Deshmukh thankfully acknowledges the support and encouragement received from Piramal Enterprises Limited, Mumbai, while Csaba Vágvölgyi also gratefully acknowledges the editorial activity support received in the frame of TÁMOP 4.2.4. A/2-11-1-2012-0001 project.

J.K. Misra
J.P. Tewari
S.K. Deshmukh
Csaba Vágvölgyi

Contents

CHAPTER 1

Fungal Genera from Fossilized-leaf Surfaces

S.K.M. Tripathi

ABSTRACT

A variety of known fossilized fungal remains are described here. These are germinating spores, mycelia and fruiting bodies belonging to two families, Meliaceae and Microthyriaceae. Members of Microthyriaceae are ectoparasites on leaves of higher plants of tropical to subtropical zones growing particularly in areas with high humidity. Although, the earliest record of microthyriaceous fruiting body (ascocarp) is from early Cretaceous, these are frequently found in Tertiary sequences all over the world. The ascocarps contain asci that are surrounded by protective tissues. The ascocarps may be in the form of closed, globose structures or flask-shaped bodies with an opening known as the ostiole or saucer shaped open structures. The size of fossil fruiting bodies generally ranges between 80 and 160 μm. The fossil ascocarps are classified under an artificial system, grouping them with Fungi Imperfecti. Characters taken into consideration for the classification of fossil ascocarps are the shape and margin of the fruiting body, nature of ostiole and central part of the fruiting body and the presence or absence of pores in the individual cells. Fossil microthyriaceous fruiting bodies are quite helpful in interpreting palaeoclimatic conditions.

Introduction

Fruiting bodies (ascocarps) of epiphyllous fungi were amongst the first fungal groups that were unquestionably identified in microfossil

Birbal Sahni Institute of Palaeobotany, 53, University Road, Lucknow 226007, India.
Email: suryatripathi.2009@rediffmail.com

assemblages. These belong to the family Microthyriaceae (Ascomycetes). Numerous types of ascocarps of epiphyllous fungi and fungal hyphae growing on leaf surfaces have been described from Cretaceous and Tertiary strata throughout the world. Fossil records indicate that the morphological diversity of microthyriaceous fruiting bodies considerably increased during mid-Tertiary (Cookson, 1947; Elsik, 1978; Ramanujam, 1982; Kalgutkar and Jansonius, 2000; Saxena and Tripathi, 2011). In most cases, distinctive morphological features of fossil ascocarps facilitate their comparison with extant counterparts. Microthyriaceous fungi have scutate fruit bodies called thyriothecia (singular thyriothecium) or ascocarps. Generally, these possess radiating rows of mycelial cells giving an appearance of tissues arranged in a radial fashion. The ascocarps contain asci that are surrounded by or enclosed within a protective tissue. The ascocarps may be in the form of closed, globose structures or flask-shaped bodies with an opening known as ostiole or saucer shaped open structures. The size of fossil fruiting bodies generally ranges between 80 and 160 µm. Dilcher (1965, 1973) made comprehensive studies on epiphyllous fungi and documented numerous fungal remains associated with angiospermic leaves of the Eocene age.

This chapter incorporates the methodology and description of fossil ascocarps known from all over the world.

Methodology

Fossil fungal remains are found along with palynofossils. Methods applied for extraction of both of these entities from rock samples are therefore, the same. Depending upon the mineral contents, rock samples are treated with different acids. Crushed shale samples are kept in 40% hydrofluoric acid for 3–4 days and after washing with water, are sieved with 400 mesh (38 µm) screen. Carbonaceous shale samples are treated with commercial nitric acid for 24 hours. Rock samples having calcareous contents are first kept in concentrated hydrochloric acid for 12 hours and then treated with hydrofluoric acid. Lignite samples are kept in concentrated nitric acid for 24–36 hours. After acid treatment, the macerated residues are treated with a solution of potassium hydroxide (5–15%) for 2–5 minutes. Water-free residues are mixed with a few drops of polyvinyl alcohol, spread uniformly over the cover glass, dried in an oven for about 30 minutes and mounted in Canada balsam for observations.

Classification of Fossil Ascocarps

Modern microthyriaceous fungi are classified on the basis of the mode of dehiscence of their fruiting bodies. Dehiscence may be either through a regular or irregular cracking or by formation of a central pore (ostiole). Other characters taken into consideration for distinguishing different taxa are that of mycelium, asci and ascospores. Mycelia and spores are seldom found associated with fossil fruiting bodies making it difficult to relate these with extant genera. However, several workers attempted to classify and formally describe the fossil thyriothecia (Edwards, 1922; Rosendahl, 1943; Cookson, 1947; Rao, 1958; Dilcher, 1965; Venkatachala and Kar, 1969; Jain and Gupta, 1970; Elsik, 1978). Fossil species of this fungal group are classified, under the artificial system, grouping them with Fungi Imperfecti. Considering features like, shape and margin of the fruiting body, characters associated with the ostiole, the presence or absence of pores in the individual cells and the nature of the central part of the fruiting body, a comprehensive scheme of classification was proposed by Elsik (1978) for fossil ascocarps.

Classification of multicellular fruiting bodies is primarily based on porate or aporate individual cells. Forms without pores in individual cells are divided into Radiate and Non-radiate forms which may be ostiolate or non-ostiolate. Radiate forms are further subdivided into genera having smooth, fimbriate or spinose margins. Radiate fruiting bodies with smooth to fimbriate margins are further divided on the basis of the presence, absence or nature of ostiole.

Essentials of this scheme are given in Table 1.1.

Table 1.1. Basic classification scheme proposed for fossil ascocarps.

<table>
<tr><td colspan="2" rowspan="2">Ascocarp cells aporate</td><td colspan="2">Ascocarp cells porate</td></tr>
<tr><td>Body radiate (Callimothallus)</td><td>Body non-radiate (Ratnagiriathyrites) (Microthyriella)</td></tr>
<tr><td colspan="3">Body radiate, margin smooth or irregular</td><td rowspan="2">Body radiate, margin with projecting spines (Parmathyrites)
Body radiate, eccentric, margin smooth (Kutchiathyrites)</td></tr>
<tr><td>Ostiole distinct
1. Body made up of intertwined thin hyphae; ostiole margin thickened (Plochmopeltinites)
2. Ostiole bordered with single/double walled cells (Trichothyrites)</td><td>Ostiole indistinct
1. Central cells simple (Paramicrothallites)
2. Central cells thick-walled (Microthyriacites)</td><td>Non-ostiolate
1. Central cells modified, provided with star-shaped opening (Asterothyrites)
2. Central cells unmodified (Phragmothyrites)</td></tr>
<tr><td colspan="4">Ascocarp cells aporate, body non-radiate</td></tr>
<tr><td colspan="3">Ostiolate</td><td rowspan="2">Non-ostiolate (Trichopeltinites)</td></tr>
<tr><td colspan="2">Irregular ostiole, body fan-shaped (Brefeldiellites)</td><td>Elongated dehiscence, body multi fan-shaped (Euthythyrites)</td></tr>
</table>

Salient features of different genera along with their sketch diagrams are given below.

Table 1.2. Diagnostic features and diagrammatic sketches of fossil ascocarps (fruiting bodies).

Taxa and Diagnostic Features	Illustrations
Asterina Léveillé, 1845 **MycoBank No.**: MB 409 **Type Species:** *Asterina melastomatis* Léveillé, 1845 *Asterina eocenica* Dilcher, 1965 MycoBank No.: MB 326723 **Characteristic features:** Fruiting bodies round, radiate, consists of prosenchymatous cells, and 45–225 µm in diameter. Central cells isodiametric, marginal cells elongate and frequently bifurcating. Fruiting bodyies astomate and split open radially at maturity, exposing radially arranged ascospores. Spores two-celled, cells unequal in size, and echinate. **Age:** Early Eocene	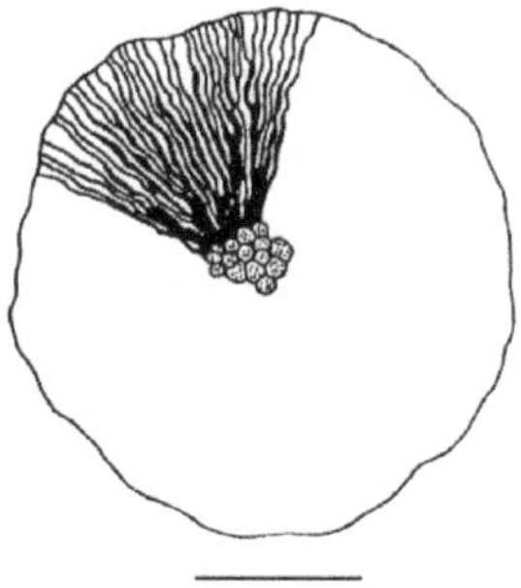 *Asterina eocenica* (Bar = 50 µm)
Asterothyrites Cookson, 1947 emend. Kalgutkar and Jansonius, 2000 **MycoBank No.**: MB 21026 **Type Species:** *Asterothyrites minutus* Cookson 1947 (designated by Jansonius & Hills, 1976: card 186) ***Asterothyrites edvensis*** (Rao and Ramanujam) Kalgutkar and Jansonius, 2000 **Characteristic features:** Ascomata circular with radially arranged hyphae which are laterally interconnected to form a pseudoparenchymatous tissue. Cells isodiametric, squarish or elongate rectangular. Ascomata ostiolate. Ostiole simple and small or large with irregular or regular outline. **Age:** Miocene	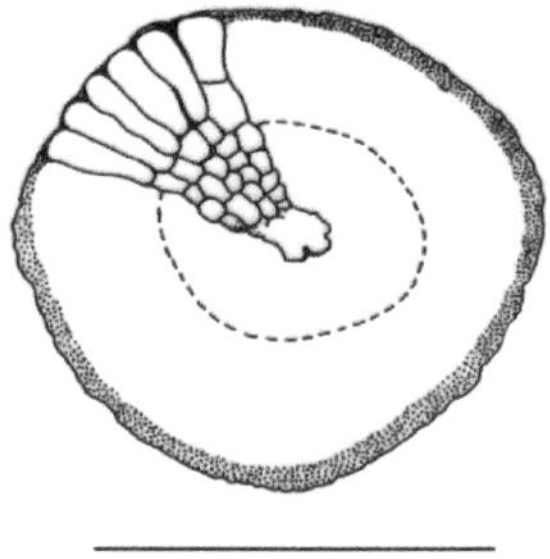 *Asterothyrites edvensis* (Bar = 50 µm)
Callimothallus Dilcher, 1965 ex Jansonius and Hills, 1977 **MycoBank No.**: MB 21042 **Type species:** *Callimothallus pertusus* Dilcher 1965 **Characteristic features:** No free hyphae. Stroma round, radiate, 50–250 µm in size, no central dehiscence (non-ostiolate) and individual cells may possess a single pore. Spores undetermined. **Age:** Early Eocene	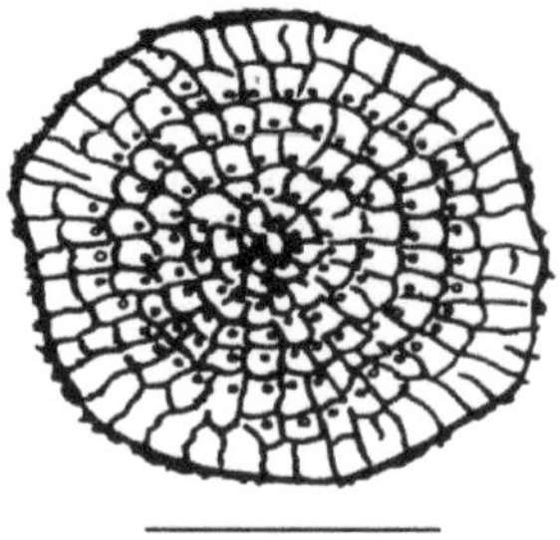 *Callimothallus pertusus* (Bar = 50 µm)

Table 1.2. contd....

Table 1.2. contd.

Taxa and Diagnostic Features	Illustrations
Cucurbitariaceites Kar et al., 1972 **MycoBank No.**: MB 21067 **Type species:** *Cucurbitariaceites bellus* Kar et al., 1972 **Characteristic features:** Pseudoperithecia subcircular to circular and 40–120 µm in size. Peripheral part dark brown and central part translucent, characterized by presence of a polygonal area formed by the interconnection of basal parts of asci. Asci ± cylindrical, sometimes swollen tipped, and always originate from upper surface of stroma. **Age:** Palaeocene	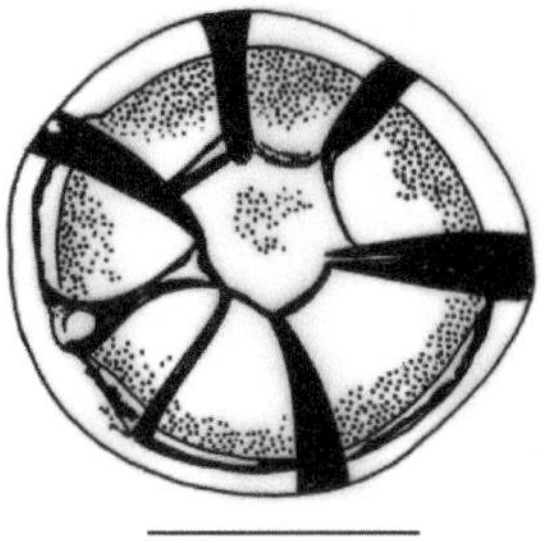 *Cucurbitariaceites bellus* (Bar = 50 µm)
Diplodites Teterevnikova-Babaian and Taslakhchian, 1973 ex Kalgutkar et al., 1993 **MycoBank No.**: MB 532862 **Type species:** *Diplodites sweetii* Kalgutkar et al., 1993 **Characteristic features:** Hyphae intercellular, septate, branched, smooth and thick-walled. Pycnidia superficial or immersed, with no definite orientation, shape and size variable, globose to subglobose, ovate-oblong or pyriform, dark, thick-walled and wall tissue pseudoparenchymatous. Pycnidia generally ostiolate, solitary or aggregated in small groups and stromate. Stroma dark brown or black, composed of thick-walled cells and uniloculate. Conidia uniseptate or aseptate and both kinds occurring in the same pycnidium. **Age:** Late cretaceous	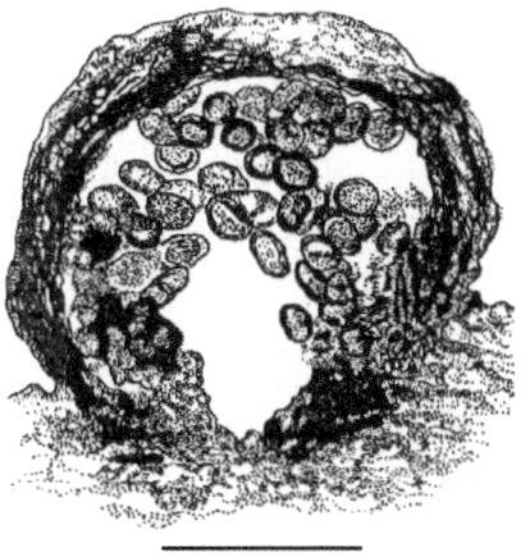 *Diplodites sweetii* (Bar = 30 µm)
Euthythyrites Cookson, 1947 **MycoBank No.**: MB 21098 **Type species:** *Euthythyrites oleinites* Cookson, 1947 *Euthythyrites bifidud* Kar et al., 2010 **Characteristic features:** Ascomata linear, elliptical to oblong, ends rounded or flattened, lateral margins uneven, dehiscence by a longitudinal slit, cells radiating from mid-vertical line, hyphopodiate, and hyphopodia small. **Age:** Miocene	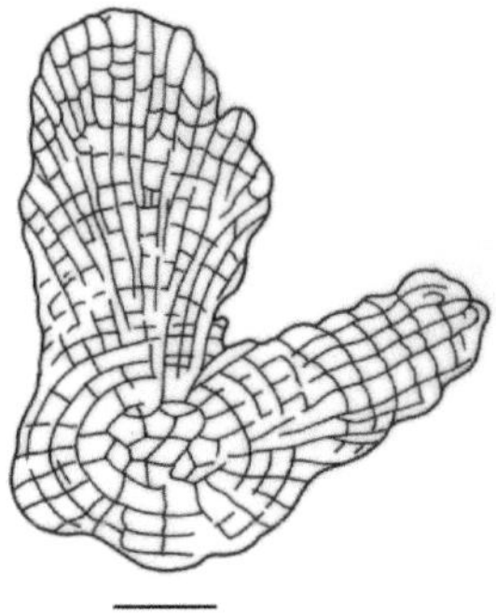 *Euthythyrites bifidud* (Bar = 10 µm)

Table 1.2. contd....

Table 1.2. contd.

Taxa and Diagnostic Features	Illustrations
Haplopeltis Theissen, 1914 **MycoBank No.**: MB 2234 **Type species:** *Haplopeltis bakeriana* (Rehm) Theissen, 1914 *Haplopeltis neyveliensis* Reddy et al., 1982 **MycoBank No.**: MB 519809 **Characteristic features:** Ascomata rounded, pseudoparenchymatous, brownish, 75–145 μm in diameter, conspicuously and ostiolate. Ostiole rounded, elevated, 8–15 μm in diameter, bordered by 2–4 layers of small, angular, dark brown, thick-walled cells, and ostiole border 5–8 μm thick. Fruiting body cells non-radiating, angular, thin-walled, margin firm, entire and composed of flattened cells. **Age:** Miocene	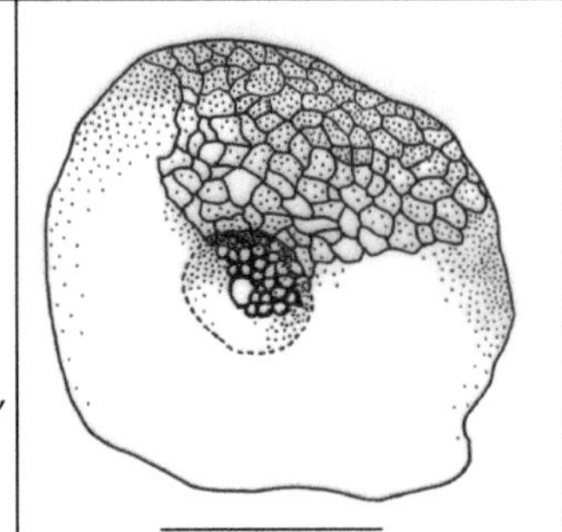 *Haplopeltis neyveliensis* (Bar = 50 μm)
Kalviwadithyrites Rao, 2003 **MycoBank No.**: MB 519802 **Type species:** *Kalviwadithyrites saxenae* Rao, 2003 **Characteristic features:** Cleistothecia subcircular to circular in shape, dimidiate and non-ostiolate. Fruiting bodies made up of two types of aporate cells. Marginal cells larger in size and rectangular to polygonal. Central cells 2 or 3 layered, squarish and isodiametric. **Age:** Miocene	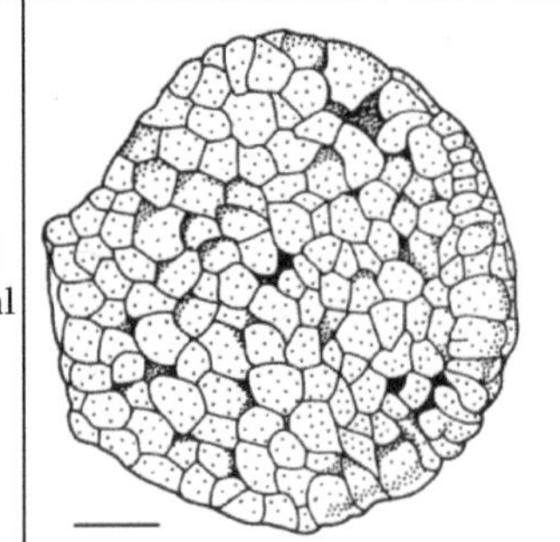 *Kalviwadithyrites saxenae* (Bar = 25 μm)
Koshalia Sarkar and Prasad, 2003 MycoBank No.: MB 519804 **Type species:** *Koshalia enigmata* Sarkar and Prasad, 2003 **MycoBank No.**: MB 519805 **Characteristic features:** Thyriothecia subspherical, 90–150 μm in diameter, multicellular, 9–10 cells arranged in compact rings around an ostiole, marginal cells extremely large (35–45 x 65–85 μm) and inner cells small (8–15 x 10–20 μm). Individual cells radially arranged and interconnected to form a shield-shaped body. Ostiole centrally placed, circular, 6–10 μm across and surrounded by 3–4 dark cells. **Age:** Early Eocene	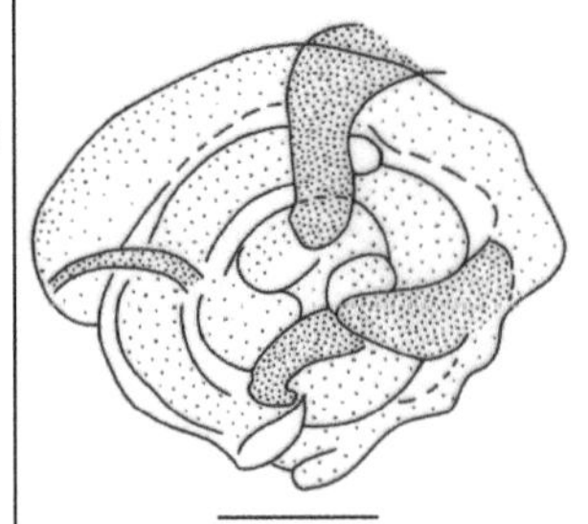 *Koshalia enigmata* (Bar = 20 μm)

Table 1.2. contd....

Table 1.2. contd.

Taxa and Diagnostic Features	**Illustrations**
Kutchiathyrites Kar, 1979 **MycoBank No.:** MB 21145 **Type species:** *Kutchiathyrites eccentricus* Kar 1979 **Characteristic features:** Ascostromata eccentric, without free hyphae, dimidiate, nonostiolate, radially arranged hyphae thick, dark and diverging from one another. Transverse hyphae comparatively thinner, interconnecting radial ones to form squarish, pseudoparenchymatous cells and individual cell aporate. **Age:** Oligocene	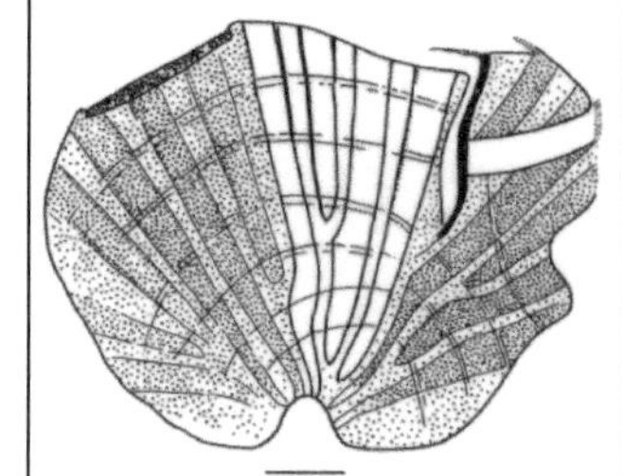 *Kutchiathyrites eccentricus* (Bar = 10 μm)
Genus: ***Lithopolyporales*** Kar et al., 2003 **MycoBank No.:** MB 28754 **Type species:** *Lithopolyporales zeerabadensis* Kar et al. 2003 **Characteristic features:** Fruiting bodies (basidiocarp) macroscopic, tough-textured, sessile and minute hyphal strands forming the network could be seen in section. Tiny dot-like spores, presumably the basidiospores, found scattered within the body. **Age:** Late Cretaceous	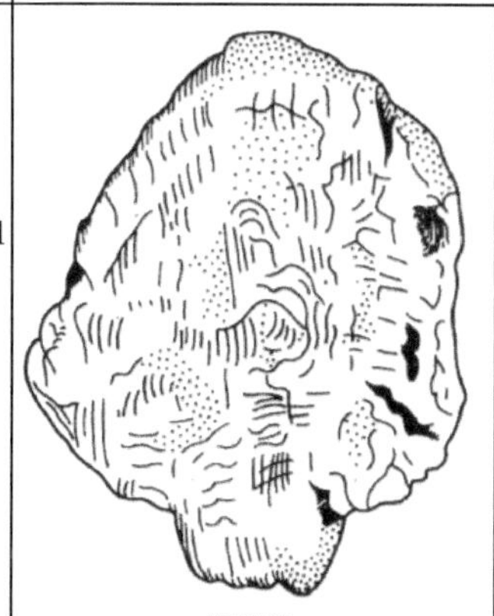 *Lithopolyporales zeerabadensis* (Bar = 25 μm)
Meliolinites Selkirk, 1975 **MycoBank No.:** MB 21162 **Type Species:** *Meliolinites spinksii* (Dilcher) Selkirk, 1975 **Characteristic features:** Mycelium and spores constituting fossil fungal colonies with general characteristics of members of Meliolaceae. Mycelial setae absent. Information regarding perithecial structure and nature of perithecial appendages uncertain or lacking. **Age:** Early Eocene to Miocene	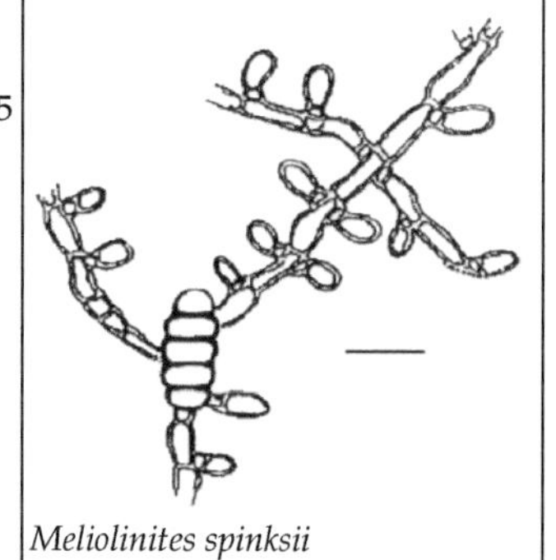 *Meliolinites spinksii* (Bar = 10 μm)

Table 1.2. contd....

Table 1.2. contd.

Taxa and Diagnostic Features	Illustrations
Microthyriacites Cookson, 1947 **MycoBank No.:** MB 21166 **Type Species:** *Microthyriacites grandis* Cookson, 1947 *Microthyriacites cooksoniae* Rao, 1958 **Characteristic features:** Ascomata composed of two parts, a central part of squarish to hexagonal thick walled cells with little or no radial pattern and a broad zone of hyphae composed of more elongate cells that are interconnected to form a pseudoparenchymatous fabric with a distinct radial pattern. **Age:** Early Eocene to Miocene	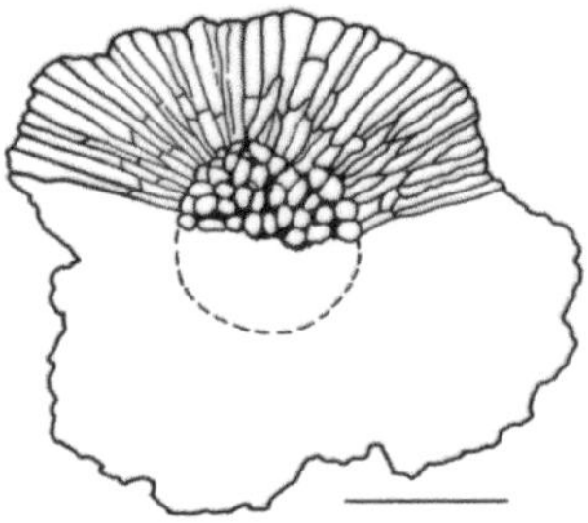 *Microthyriacites cooksoniae* (Bar = 50 µm)
Genus: ***Microthyriella*** Höhnel, 1909 **MycoBank No.:** MB 3202 **Type Species:** *Microthyriella rickii* (Rehm) Höhnel, 1909 *Microthyriella diporata* Rao and Ramanujam, 1976 **MycoBank No.:** MB 317745 **Characteristic features:** Free mycelium lacking. Ascomata flattened, irregular in shape and highly variable in size (50–150 µm). Ascomata cells pentagonal to hexagonal, irregularly arranged, porate, pores randomly disposed and mostly two per cell. **Age:** Miocene	 *Microthyriella diporata* (Bar = 30 µm)
Netothyrites Misra et al., 1996 **MycoBank No.:** MB 519790 **Type Species:** *Netothyrites vertistriatus* Misra et al. 1996 **Characteristic features:** Fruiting bodies pitcher-shaped with distinct collar, hollow neck and main body with closed reticulated bottom. Proximal opening (? ostiole) distinct, bordered with dark, multicellular cells forming a distinct collar. Main body hangs down from the collar, with a distinct neck in between. Side walls of neck and main body bear number of longitudinal ribs which run down parallel or anastomose to form reticulum. Bottoms of main body densely reticulated and closed. **Age:** Palaeocene	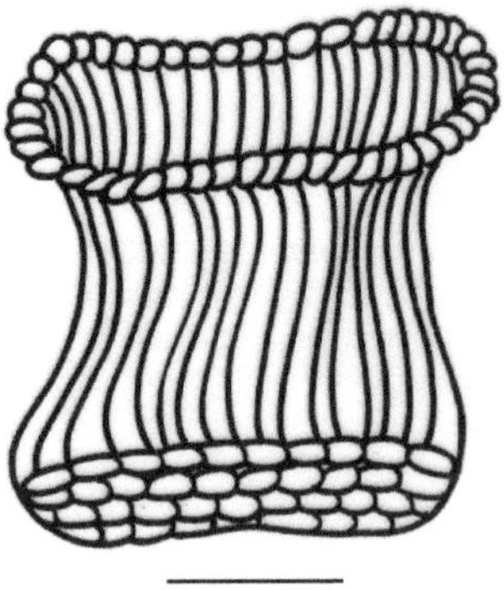 *Netothyrites vertistriatus* (Bar = 20 µm)

Table 1.2. contd....

Table 1.2. contd.

Taxa and Diagnostic Features	Illustrations
Palaeocercospora Mitra and Banerjee, 2000 **MycoBank No.**: MB 28426 **Type Species:** *Palaeocercospora siwalikensis* Mitra and Banerjee, 2000 **Characteristic features:** The hyphomycetous epiphyllous fungi showing distinct, well developed circular stroma (16.8 to 37.8 µm in diameter) consists of deep brown compactly arranged hyphal cells. Conidiophores fasciculate, in fascicles of 7–20 divergent stalks emerging from the stromata which are straight to flexuous, simple, slightly thick walled, smooth and pluriseptate. Conspicuous spore scars present at the point of geniculations of conidiophores. Conidiophores long (42–92.4 µm), slender (width 4.2–6.3 µm), tip somewhat pointed and conidia absent. **Age:** Middle Miocene	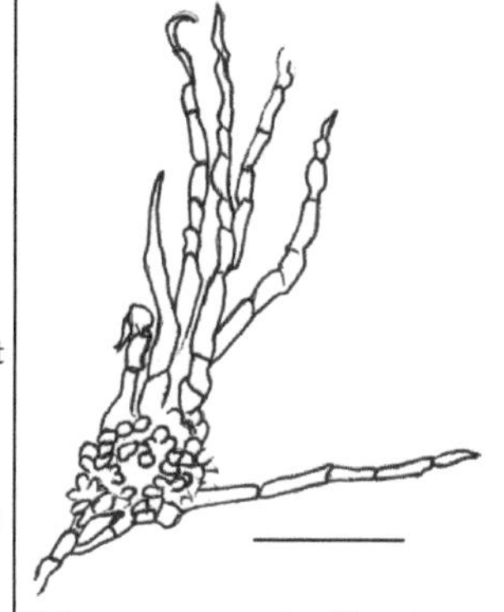 *Palaeocercospora siwalikensis* (Bar = 20 µm)
Palaeocolletotrichum Mitra and Banerjee, 2000 **MycoBank No.**: MB 28427 **Type Species:** *Palaeocolletotrichum graminioides* Mitra and Banerjee, 2000 **Characteristic features:** The coelomycetous epiphyllous fungi occurring on cuticular layers show numerous setae scattered singly or in groups of 3–6 both on veins and inter-venal regions. Faint outline of hyphal mass of acervular structure observed in the subepidermal position. Individual setae 84–126 µm long, stiff, pointed, moderately thick (4–5.5 µm) and dark brown in colour project from the leaf surfaces. The setae with a bulbous base of 7–8.4 µm in diameter, septate and the number of septa varying from 5–8. **Age:** Middle Miocene	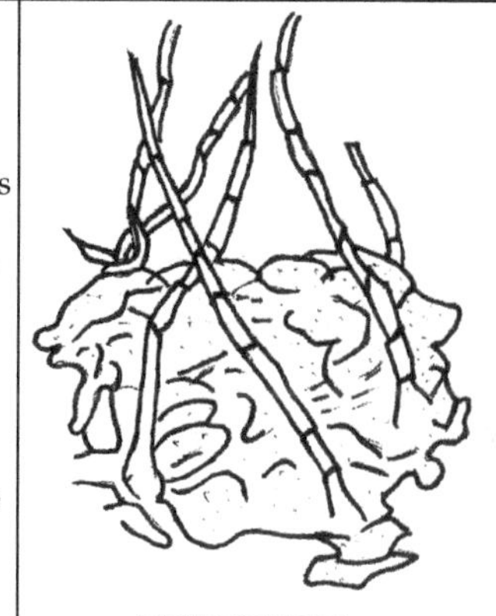 *Palaeocolletotrichum graminioides* (Bar = 25 µm)
Palaeophoma Singhai, 1974 **MycoBank No.**: MB 21213 **Type Species:** *Palaeophoma intertrappea* Singhai, 1974 **Characteristic features:** Pycnidium brown and spherical, measuring 224 x 200 µm, thick-walled (12–32 µm) and pseudoparenchymatous. Ostiole not seen. Conidia unicelled, hyaline, bent or curved or lunate or spherical, thin walled, smooth, and measuring 5–8 x 2–4 µm. **Age:** Late Cretaceous	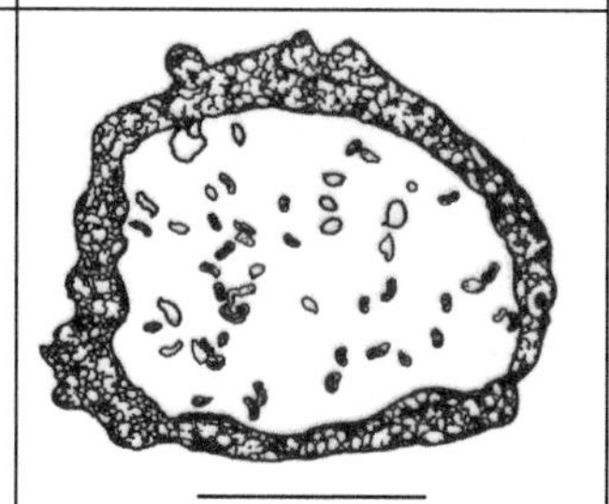 *Palaeophoma intertrappea* (Bar = 100 µm)

Table 1.2. contd....

Table 1.2. contd.

Taxa and Diagnostic Features	Illustrations
Palaeosordaria Sahni and Rao, 1943 **MycoBank No.**: MB 21216 **Type Species:** *Palaeosordaria lagena* Sahni and Rao, 1943 **Characteristic features:** Perithecia black, flask-shaped; body smooth and spherical, about 140 µm in diameter, external surface reticulate; neck tapering, about 180 µm long, with traces of short hairs round the tip, wall composed of one layer of cells, mycelium septate. **Age:** Palaeocene	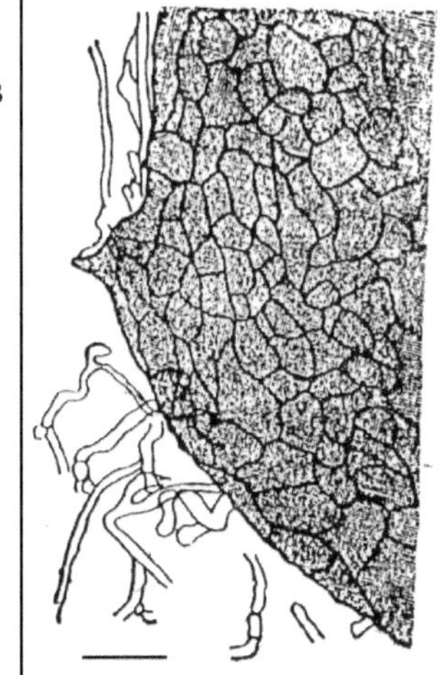 *Palaeosordaria lagena* (Bar = 20 µm)
Genus: *Parmathyrites* Jain and Gupta, 1970 **MycoBank No.**: MB 21223 **Type Species:** *Parmathyrites indicus* Jain and Gupta, 1970 **Characteristic features:** Ascomata ostiolate, flattened, circular, one layer thick, hyphae radially arranged, interconnected and forming pseudoparenchymatous non-porate cells. Outer peripheral cells prominent with thickened radial walls, spines peripheral, spine sheath present or absent and ascospores unknown. Ostiole distinct, not surrounded by specialized cells. **Age:** Early Miocene	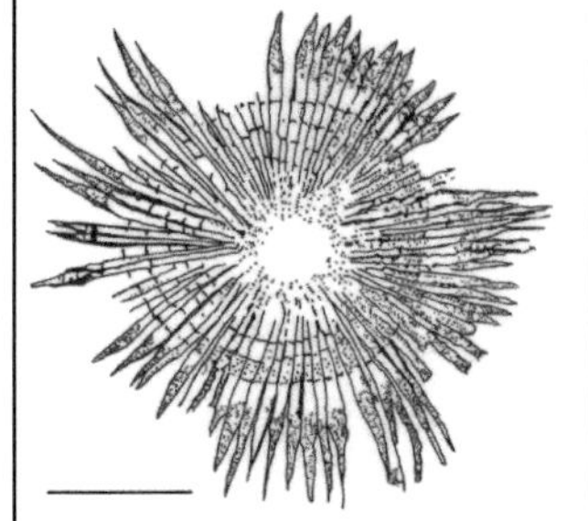 *Parmathyrites indicus* (Bar = 50 µm)
Phragmothyrites Edwards, 1922 emend. Kalgutkar and Jansonius, 2000 **MycoBank No.**: MB 21244 **Type Species:** *Phragmothyrites eocaenicus* Edwards 1922 Characteristic features: Ascomata subcircular to circular, astomate and with radially arranged hyphae that interconnect laterally to form a pseudo-parenchymatous tissue. Hyphal cells uniform in shape or size or showing various development in different regions of the ascoma, e.g., elongated rectangular or isodiametric. Central cell or cells in the immediate central area, may be cubical, hexagonal or subcircular. **Age:** Eocene	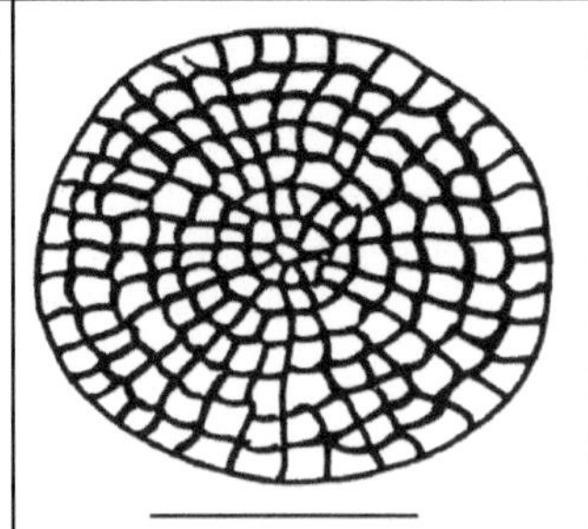 *Phragmothyrites eocaenicus* (Bar = 60 µm)

Table 1.2. contd....

Table 1.2. contd.

Taxa and Diagnostic Features	Illustrations
Plochmopeltinites Cookson, 1947 **MycoBank No.:** MB 21254 **Type Species:** *Plochmopeltinites masonii* Cookson, 1947 *Plochmopeltinites cooksoniae* Ramanujam and Rao, 1973 **Characteristic features:** Ascomata epiphyllous, discoid to irregular in shape and ostiolate. Ostiole rounded to irregular, centric or slightly sub-centric, border slightly elevated and made up of two or three layers of dark brown, thick-walled and angular cells. Covering membrane of ascomata plectenchymatous, comprising sinuous, irregularly branched and intertwining hyphae showing a distinct but locally distorted radial pattern. Many hyphal branches end blindly against adjacent hyphae, hyphal branching more pronounced near margin. Cells in individual hyphae narrow, elongated of variable size. Margin of fruiting body wavy and locally discontinuous. Free hyphae present, extending locally beyond the margins of ascomata and associated with marginal cells. **Age:** Late Miocene	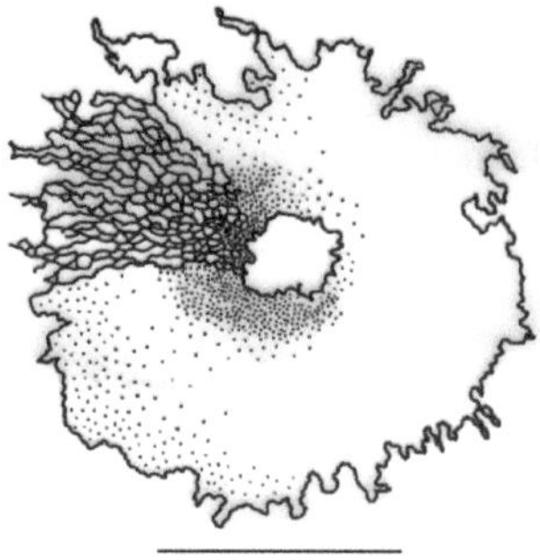 *Plochmopeltinites cooksoniae* (Bar = 50 μm)
Ratnagiriathyrites Saxena and Misra, 1990 MycoBank No.: MB 28615 **Type Species:** *Ratnagiriathyrites hexagonalis* Saxena and Misra, 1990 **Characteristic features:** Ascostromata subcircular or irregular in shape, dark brown, non-ostiolate, cells not arranged radially, individual cells porate. Pores generally distributed throughout stromata. Cells hexagonal and bigger towards periphery than in the central region. Margin thick and wavy. **Age:** Miocen	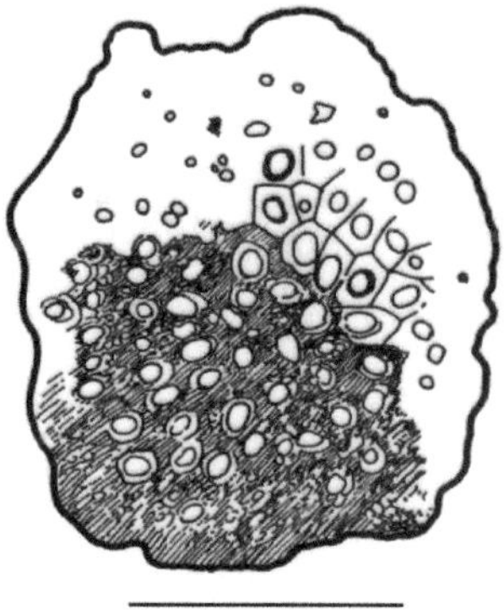 *Ratnagiriathyrites hexagonalis* (Bar = 50 μm)

Table 1.2. contd....

Table 1.2. contd.

Taxa and Diagnostic Features	Illustrations
Trichopeltinites Cookson, 1947 **MycoBank No.:** MB 21340 **Type Species:** *Trichopeltinites pulcher* Cookson, 1947 Trichopeltinites folius Kar et al., 2010 **Characteristic features:** Ascomata developed as thickened areas of the thallus and dehiscing by an irregular ostiole. **Age:** Miocene	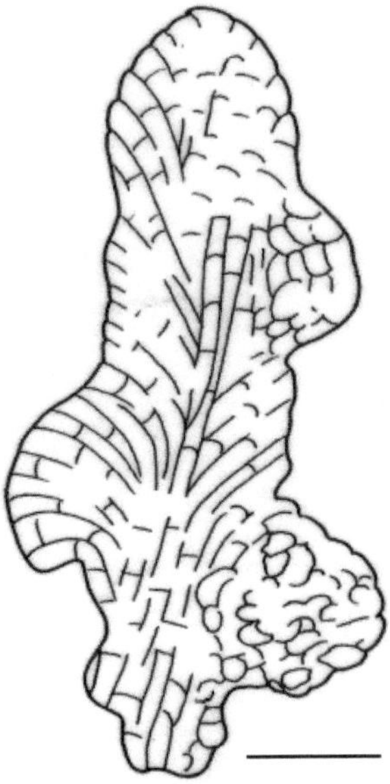 *Trichopeltinites folius* (Bar = 10 μm)
Trichothyrites Rosendahl, 1943 **MycoBank No.**: MB 21342 **Type Species:** *Trichothyrites pleistocaenicus* Rosendahl, 1943 ***Trichothyrites padappakkarensis*** (Jain and Gupta) Kalgutkar and Jansonius, 2000 **Characteristic features:** Thyriothecia disc- or saucer-shaped, lobate and smooth possessing definite upper and lower walls of radiate rows of almost square cells (3–8 x 3–8 μm). Cell walls of upper layer of thyriothecium generally more strongly thickened than those of the lower layer. Thyriothecia 70 μm to 200 μm in diameter, bearing on upper wall an erect ostiolar collar (papilla) made up of 2–6 tiers of small, thick-walled quadrilateral cells. Upper most tier (ostiolar margin) of cells may have short prolongations (setae) in some cases. **Age:** Miocene	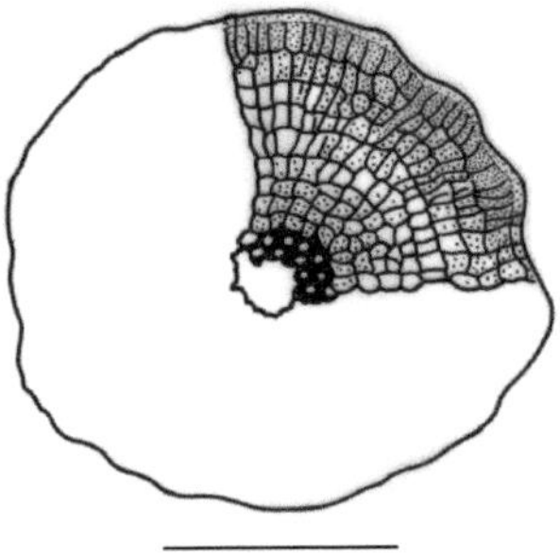 *Trichothyrites padappakkarensis* (Bar = 100 μm)

Stratigraphical Implications of Fossil Ascocarp Genera

The earliest undisputed microthyriaceous fungus, *Stomiopltites* is reported from Lower Cretaceous of Wealden, Isle of Wright (Alvin and Muir, 1970). It is a dome shaped ostiolate ascocarp belonging to the family Microthyriaceae. A specimen referred to as *Phragmothyrites* was described by Singh (1971) from Late Albian of Alberta. Records of fossil Microthyriaceous fungi provide ample evidence that they occur in major parts of the Cenozoic but due to taxonomic confusion, stratigraphic applications of different genera and species are still obscured. However, Tripathi (2009) and Saxena and Tripathi (2011) attempted to appraise the stratigraphic distribution of Indian fossil fruiting bodies. The basics of fossil ascocarp distribution in Indian Tertiary sediments are summarized in Table 1.3.

Table 1.3. Stratigraphic distribution of fossil ascocarp genera in Indian Tertiary sediments.

Taxa	**Palaeocene**	**Eocene**	**Oligocene**	**Miocene**	**Pliocene**
Callimothallus Dilcher	—	—	—	—	—
Cucurbitariaceites Kar et al.	—	—	—	—	—
Phragmothyrites Edwards	—	—	—	—	
Microthyriacites Cookson		—	—	—	
Kutchiathyrites Kar		—	—	—	
Kalviwadithyrites Rao				—	
Parmathyrites Jain & Gupta				—	
Plochmopeltinites Cookson				—	
Ratnagiriathyrites Saxena & Misra				—	
Trichopeltinites Cookson				—	
Trichothyrites Rosendahl				—	
Asterothyrites Cookson				—	
Euthythyrites Cookson				—	

Palaeoclimatic Significance of Fossil Epiphyllous Fungi

Most of the fossil fungal fruiting bodies discussed in this chapter morphologically resemble with those of extant members of the family Microthyriaceae. These are ectoparasites on leaves of higher plants of tropical to subtropical zones growing particularly in areas with high humidity. Microthyriaceous fungi grow best in rain forests and rain forest margins (Dilcher, 1965; Ramanujam, 1982; Prasad, 1986; Kalgutkar and Jansonius, 2000). Their presence is therefore, generally indicative of a wet tropical climate with heavy precipitation. However, the ecological interpretations based on epiphyllous fungi should be made with caution because some of these are reported to occur in wider latitudinal ranges (Dilcher, 1965; Selkirk, 1975). Palaeoenvironmental interpretations based on complete palynological assemblage and, if possible, also coordinated with studies on megafossils may provide more accurate information about the palaeoclimate.

General Remarks and Futuristic Approach

Most of the fungi are found in close association with specific plants and animals and, if found in a fossil state are indicative of similar kind of situations during the geological past. Fossil fungi, therefore, may provide useful information about the palaeoecology, past habitats and their hosts. In this regard fossil epiphyllous fungi can be more reliable for palaeoclimatic interpretations because these can be related to the modern fungi with more accuracy. Occurrence of these fossils reflects moist and humid climate of tropical to subtropical belts. Environmental interpretations based on the presence of microthyriaceae may, however, sometimes be hampered due to the incorrect identification of the material. For example, the red alga *Caloglossa leprieurii*, generally found on grasses of brackish water marshes may be confused with *Trichopeltinites* due to a morphological resemblance. Similarly, marine green alga *Ulvella lens* also resembles the fructifications of Microthyriaceae. Their presence in dispersed fossil assemblage should, therefore, be ascertained before deciphering the past climate. Studies particularly focusing on host fungus relationship are also of great significance in attempting palaeoenvironmental interpretations.

References

Alvin, K.L. and Muir, M.D. 1970. An epiphyllous fungus from the Lower Cretaceous. Biological Journal of Linnean Society, 2: 55–59.

Cookson, I.C. 1947. Fossil fungi from Tertiary deposits in the southern hemisphere. Part I. Proceedings of the Linnean Society, New South Wales, 72: 207–214.

Dilcher, D.L. 1965. Epiphyllous fungi from Eocene deposits in Western Tennessee, USA. Palaeontographica Abt. B, 116: 1–54.

Dilcher, D.L. 1973. A revision of the Eocene flora of southeastern North America. Palaeobotanist, 20: 7–18.

Edwards, W.N. 1922. An Eocene microthyriaceous fungus from Mull, Scotland. Transactions of the British Mycological Society, 8: 66–72.

Elsik, W.C. 1978. Classification and geologic history of the microthyriaceous fungi. Proceedings of the IV International Palynological Conference, Lucknow (1976–77), 1: 331–342.

Jain, K.P. and Gupta, R.C. 1970. Some fungal remains from the Tertiaries of Kerala Coast. Palaeobotanist, 18(2): 177–182.
Jansonius, J. and Hills, L.V. 1976. Genera file of fossil spores. Special Publication, Department of Geology, University of Calgary, Canada, 1–3287.
Jansonius, J. and Hills, L.V. 1977. Genera file of fossil spores-supplement. Special Publication, Department of Geology, University of Calgary, Canada, 3288–3431.
Kalgutkar, R.M. and Jansonius, J. 2000. Synopsis of fungal spores, mycelia and fructifications. AASP Contribution Series, 39: 1–423.
Kar, R.K. 1979. Palynological fossils from the Oligocene sediments and their biostratigraphy in the District of Kutch, Western India. Palaeobotanist, 26(1): 16–49.
Kar, R.K., Sharma, N., Agarwal, A. and Kar, R. 2003. Occurrence of fossil wood rotters (Polyporales) from Lameta Formation (Maastrichtian), India. Current Science, 85: 37–40.
Kar, R.K., Singh, R.Y. and Sah, S.C.D. 1972. On some algal and fungal remains from Tura Formation of Garo Hills, Assam. Palaeobotanist, 19(2): 146–154.
Léveillé, J.H. 1845. Descriptions des champignons du Muséum du Paris. Annaes de Sciencias Naturaes, 5: 111–304.
Misra, C.M., Swamy, S.N., Prasad, B., Pundeer, B.S., Rawat, R.S. and Singh, K. 1996. *Netothyrites* gen. nov. a fungal fossil fruit-body from the Paleocene sediments of India. Geoscience Journal, 17(1): 17–23.
Mitra, S. and Banerjee, M. 2000. On the occurrence of epiphyllous deuteromycetous fossil fungi *Palaeocercospora siwalikensis* gen. et sp. nov. and *Palaeocolletotrichum graminioides* gen. et sp. nov. from Neogene sediments of Darjeeling Foothills, Eastern Himalaya. Journal of Mycopathological Research, 37(2): 7–11.
Prasad, M.N.V. 1986. Fungal remains from the Holocene peat deposits of Tripura state, North-eastern India. Pollen et Spores, 28(3-4): 365–390.
Ramanujam, C.G.K. 1982. Recent advances in the study of fossil fungi. pp. 287–301. *In:* Bharadwaj, D.C. (ed.). Recent Advances in Cryptogamic Botany 2, Palaeobotanical Society, Lucknow.
Rao, A.R. 1958. Fungal remains from some Tertiary deposits of India. Palaeobotanist, 7(1): 43–46.
Rao, M.R. 2003. *Kalviwadithyrites,* a new fungal fruiting body from Sindhudurg Formation (Miocene) of Maharashtra, India. Palaeobotanist, 52(1-3): 117–119.
Reddy, P.R., Ramanujam, C.G.K. and Srisailam, K. 1982. Fungal fructifications from Neyveli lignite, Tamil Nadu—their stratigraphic and palaeoclimatic significance. Records of the Geological Survey of India, 114(5): 112–122.
Rosendahl, C.O. 1943. Some fossil fungi from Minnisota. Bulletin of the Torrey Botanical Club, 70: 126–138.
Sahni, B. and Rao, H.S. 1943. A silicified flora from the Intertrappean cherts at Sausar in the Deccan. Proceedings of the National Academy of Sciences, India, 13(1): 36–75.
Sarkar, S. and Prasad, V. 2003. *Koshalia,* an Incertae sedis fossil from the Subathu Formation (Late Ypresian) Himachal Pradesh, India. Palaeobotanist, 52: 113–116.
Saxena, R.K. and Misra, N.K. 1990. Palynological investigation of the Ratnagiri Beds of Sindhu Durg District, Maharashtra. Palaeobotanist, 38: 263–276.
Saxena, R.K. and Tripathi, S.K.M. 2011. Indian Fossil Fungi. Palaeobotanist, 60(1): 1–208.
Selkirk, D.R. 1975. Tertiary fossil fungi from Kiandra, New South Wales. Proceedings of the Linnean Society, New South Wales, 100: 70–94.
Singh, C. 1971. Lower Cretaceous microfloras of the Peace River area, north Western Alberta. Bulletin of the Research Council of Alberta, 28: 1–542.
Singhai, L.C. 1974. Fossil fungi from the Deccan Intertrappean Beds of Madhya Pradesh, India. Journal of Biological Sciences, 17: 92–102.
Teterevnikova-Babaian, D.N. and Taslakhchian, M.G. 1973. New data on fossil fungal spores in Armenia. Academy of Sciences of the USSR, Mycology and Phycology, 4: 159–164.

Theissen, F. 1914. Trichopeltaceae n. fam. Hemisphaerialium. Zentralblatt für Bakteriologie und Parasitenkunde, 93: 625–640.

Tripathi, S.K.M. 2009. Fungi from palaeoenvironments: their role in environmental interpretations. pp. 1–27. *In*: Misra, J.K. and Deshmukh, S.K. (eds.). Science Publishers, Enfield, NH, USA.

Venkatachala, B.S. and Kar, R.K. 1969. Palynology of the Tertiary sediments in Kutch-2. Epiphyllous fungal remains from the borehole no. 14. Palaeobotanist, 17(2): 179–183.

CHAPTER 2

Wood-Inhabiting Fungi

Ivan V. Zmitrovich,[1,*] *Solomon P. Wasser*[2] and *Daniel Ţura*[3]

ABSTRACT

The chapter summarizes the current developments on wood-inhabiting fungi. Taxonomically, wood-decomposers are represented by some groups of Basidiomycota, especially belonging to the order Polyporales (class Agaricomycetes). Composition of wood-inhabiting fungi established on generic and higher levels are given. Morphologically parallel series of sporogeneous structures of wood-inhabiting fungi are presented as a biomorphic system. Enzymatic systems of fungal wood-decomposers highlight white-rot and brown-rot fungi. It is emphasized that both groups are capable of oxidizing C-C components of wood polymers, but their targets are diverse. The mycogeographical aspects of wood-inhabiting fungi have been overviewed. The trophic aspects as well as substrate groupings of these fungi and their distributional patterns in the forest ecosystems of the Northern Hemisphere have also been discussed.

Introduction

Woody vegetation predominates in moist and cold climates, but is scarce in arid ones. This vegetation represents a main environment softening component of terrestrial ecosystems and shows maximum biodiversity of a zonal bioms. Such types of vegetation in terrestrial environments

[1] V.L. Komarov Botanical Institute, Russian Academy of Sciences 2, Professor Popov St., St Petersburg, 197376, Russia.
[2] Institute of Evolution and Department of Evolutionary & Environmental Biology, Faculty of Science & Science Education, University of Haifa, Mt Carmel, Haifa 31905, Israel.
[3] Aloha Medicinals Inc., Carson City, Nevada 89706, USA.
* Corresponding author

are closely related to woody plant-fungal interactions, e.g., mycorrhizal symbiosis, wood decaying/decomposition and phyllophane fungal epi-endophytism.

The life history of a tree is connected, one way or the other, with diverse wood-inhabiting fungi. Fungal groups are connected spatially with wood at its various stages (from living trees to litter wood debris), and most of their representatives are capable of wood decomposition and utilization. However, some wood-inhabiting fungi are basically mycorrhiza-formers or mycoparasites.

Enzymatic systems of fungal wood-decomposers are adapted to the degradation of various components of wood, and the two main groups can be distinguished: the white-rot and brown-rot fungi. Both groups are capable of oxidizing C-C components of wood polymers. White-rot fungi target the lignin molecules, whereas the brown-rot fungi attack the celluloses and hemicelluloses.

The main diversity of wood-inhabiting fungi is represented by some groups of Ascomycota (with related anamorphic representatives), but the most adapted pool of wood decomposers is represented by some groups of Basidiomycota, especially belonging to the order Polyporales of class Agaricomycetes. All these fungi are characterized by having a wonderful parallel morphological adaptation for sporulation over various substrata such as small twigs, branches, trunks, stumps and fallen logs, and also amorphous wood-remnants within the forest litter.

It is hard to overestimate the role of wood-inhabiting fungi in forest ecosystems. In moist taiga environments these fungi are the key agents for enrichment of forest soil by humus-like compounds. The humic acids, released in the process of wood and litter decay in such forests, are powerful factors of relief formation in taiga landscapes. In mesophylic nemoral and moist tropical forests, the wood-inhabiting fungi are important agents for keeping the biomass balance, whereas in arid ecosystems their pathogenic role is important.

Pathogenic wood-inhabiting fungi produce heart-rots of many economically important trees; some species are connected with superficial necrosis or the colonization of total volume of the trunk. The control strategies of tree pathogens and timber fungi are, therefore, important issues of applied mycology.

Trophic and Topic aspects of Wood-Inhabiting Fungi

What are Wood-Inhabiting Fungi?

The term "wood-inhabiting fungi" is usually applied to the topic group, uniting those representatives of eumycetes (Ascomycota and Basidiomycota) whose sporulations are associated with wood under various conditions. If the vegetative mycelium of a fungus is associated with wood, whereas fungal nutrition is connected to wood degradation, such a fungus may be called a "wood-destroying fungus".

In some cases, the mycelium of a fungus spreads through forest litter to the humus-soil horizon and is capable of forming ectomycorrhizae, whereas the fruitbodies are obligately or facultatively associated with woody substrates. These fungi coincide with the tophic grouping of wood-inhabiting fungi, but their trophic affiliations are diverse. The situation is similar to some obligate parasitic wood-destroying fungi that develop their spores/fruiting bodies over woody substrates.

The lichenized Ascomycota (lichens), growing abundantly over wood of various types, are traditionally not considered as being part of wood-inhabiting fungi. However, lichenized Basidiomycota, whose dependence on algae is rather facultative and whose mycelium is capable of penetrating and destroying woody substrata, are traditionally considered as being representatives of wood-inhabiting fungi.

Biochemical Aspects of Wood Decay

Wood-decay is the most common mode of life of wood-inhabiting fungi. Historically, this mode descendents to Devonian transformation of plant ecomorphs from plagiotropic to orthotropic ones through the processes of sprouts lignification as result of some transformations of plant secondary metabolism (Ragan and Chapman, 1978; Karatygin, 1993). According to Ragan and Chapman (1978), the close relatives of Devonian plants with associated symbiotic fungi did not exclude a horizontal plant-fungus gene transfer, particularly expressed by genes related to building and destroying of the lignin molecular composites.

A basic feature of wood organization, as shown in Fig. 2.1, is the presence of vessels with secondarily thickened walls. The fibrils of cellulose and hemicelluloses (Fig. 2.1a) compose a fibrillar core of this structure, whereas the lingo-cellulose complex composes their amorphous matrix (Fig. 2.1b, c).

Fungi are adapted to destroying and assimilating such a specific substrate; this process occurs in two ways: 1) the development of an enzymatic device system for polysaccharides biodegradation and 2) the development of an enzymatic system for lignin and polysaccharide oxidation.

The main polysaccharides of cell walls of woody plants and the corresponding hydrolytic enzymes that hydrolyze internal glycosidic bonds have been tabulated below (Table 2.1). The result of this hydrolysis process is the disruption of microfibrillar structure of the wood.

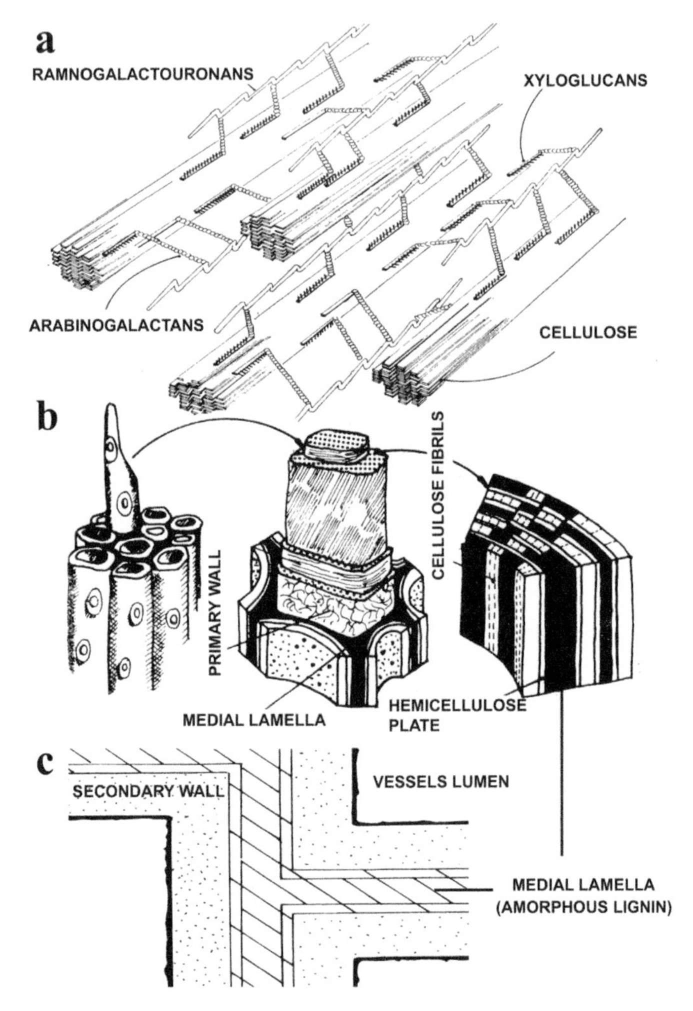

Figure 2.1. The structure and polymer composition of the xylem (according to Kirk 1983): (a) polysaccharide core, (b) the wall structure of xylem vessels; (c) schematic section crossing xylem tissue emphasizing secondary wall and medial lamella.

However, such a biodegradation of polysaccharide fibrillar core of the wood is rather rare. Particularly, this way of wood decomposition is a characteristic of cambial biotrophs that are associated with bark exfoliation. After exfoliation, sapwood can be attacked by highly specific hydrolizers produced by the so-called "blue staining fungi"—ascomycetes for example, genera like *Ophiostoma* and *Ceratocystis*, whose xylanase and pectinase activities are the reasons for changes in wood surface spectral characteristics. The other pool of hydrolytically active fungi is associated with wood

Table 2.1. The hydrolytic biodegradation of the wood.

Fractions of polysaccharides of wood cell wall	Hydrolytic enzymes of wood-inhabiting fungi	Products of degradation	Example species, authors
Cellulose	1,4-β-D-glucan cellobiohydrolases	cellobiose, cellooligomers, D-glucose	*Postia placenta* M.J. Larsen & Lombard, *Hypocrea jecorina* Berk. & Broome; *Phanerochaete chrysosporium* Burds. (Cowling, 1961; Aro et al., 2005)
Hemicelluloses	1,4-β-D-xylan xylohydrolase, β-xylosidase, acetyl xylan esterase; xylan-α-1,2-glucuronosidase; feruloyl esterase; α-L-arabinofuranosidase; endo-1,4-β-mannosidase; xyloglucan hydrolase	D-xylose; glucuronic acid; ferulic acid; L-arabinose; D-glucose; galactose	*Magnaporthe grisea* (T.T. Hebert) M.E. Barr (Polizeli et al., 2005; Gamauf et al., 2007)
Pectins	rhamnogalactouronan hydrolase, rhamnogalactouronan lyase; endo-β-1,6-galactanase, exogalactanase; L-arabinofuranosidase	D-galactouronic acid; D-galactose L-arabinose	*Ceratobasidium* spp. (González García et al., 2006); *Tremella aurantialba* Bandoni & M. Zang (Jing et al., 2007)

debris of forest litter, especially within pectin-rich remnants, such as fallen fruits and old herb stems. This group includes many members of the order Helotiales. But in most of the cases, the capacity of wood fungi to hydrolyze polysaccharides is combined with the capacity to oxidize C–C links both in polysaccharide and lignin-containing composites (= multicomponent substances: cellulose + lignin + pectines, etc.). Thus, depending on the targets of oxidation—polysaccharides or lignocelluloses, two types of wood decomposition are distinguishable: brown-rot and white-rot.

Brown-rot. The main targets of brown-rot wood-inhabiting fungi are cellulose and hemicelluloses; the lignin is subjected to slight modification via demethylation (Wright, 1985; Eriksson et al., 1980, 1990). The wood loses its fibrillar structure, and becomes fragile and cracks into a red-brownish mass due to lignin modification.

Koenigs (1972) demonstrated that such wood decomposers are capable of producing a huge amount of H_2O_2. This substance is generated by extracellular enzymatic systems (peroxidases). The mechanism of attacking the C–C-links of crystalline cellulose and similar composites is known as the Fenton reaction:

$$H_2O_2 + Fe^{2+} + H^+ \rightarrow H_2O + Fe^{3+} + HO\cdot$$

$$\text{-RCH (OH)-} + 2HO\cdot \rightarrow \text{-RHO} + CO_2 + H_2O$$

In order to avoid the destruction of hyphal wall and to act on lignified parts of the secondary cell wall, the OH-radicals should be produced at a distance from the hypha, and the fungal reductants should be stable enough to diffuse before they react to reduce Fe (III) and oxygen to Fe (II) and peroxide (Fig. 2.2). The production of OH-radicals takes place in several ways following different systems including secretion of hydroquinones, cellobiose dehydrogenases, low-molecular-weight glycopeptides and phenolate chelators (Gamauf et al., 2007).

The brown-rot fungi are adapted to rapid xylolysis, and are capable of causing heart-rots and decomposition of stumps and logs, presumably of conifers, which are characterized by highly resistant lignocellulose complexes. The wood cracks into polygons or chips, and this lignin-rich material is a predecessor of humus in forest soils (Fig. 2.3).

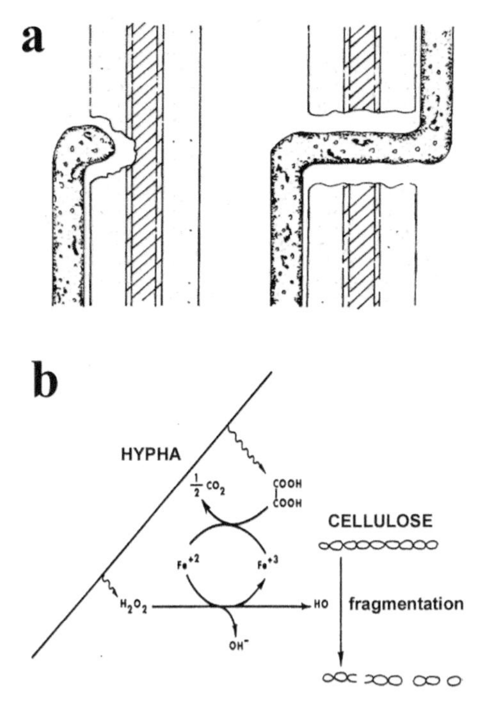

Figure 2.2. Scheme showing the vessel wall penetration by hypha of brown-rot fungus (a) and free radical production by such a hypha (b) according to Wright (1985).

Figure 2.3. A pine stump decomposed by brown-rot fungus *Neolentinus lepideus* (Fr.) Redhead & Ginns: on the foreground the large wood chip is appreciable whereas decomposed cubical material is converted into forest litter.

Under xerophylic conditions, in exposed fallen logs and decorticated stands, the brown-rot fungi demonstrate rather weak activity causing the so-called "dry-rot". The typical dry-rot producers are representatives of the genera *Dacrymyces* and *Gloeophyllum*. However, when the process of water evaporation outside the wood is hampered by forest shade, by ground conditions or the abundance of fallen wet logs, the rot stays active, and in many cases is accompanied by self-moisturizing of wood due to metabolic water. Such an active—wet brown-rot is characteristic of *Serpula*, *Tapinella*, and many species of *Antrodia* and *Fomitopsis* (Fig. 2.4).

White-rot. The white-rot fungi are adapted to deep degradation of lignin and partial decomposition of polysaccharides. The lignin is a highly inert biopolymer with a high molecular weight, which is composed of many stacked polyphenol moieties. The simple type structure of this polymer is presented in Fig. 2.5.

White-rot fungi decompose this resistant biopolymer through many steps of the oxidative process, involving peroxidases and laccases (phenol oxidases), which act non-specifically by generating lignin free radicals and then undergo spontaneous cleavage reactions (Rabinovich et al., 2001; Gamauf et al., 2007).

The laccases, represented in wood-fungi by Lignin Peroxidases (LiPs), Manganese Peroxidases (MnPs), and Versatile Peroxidases (VP) have a high

Figure 2.4. Key brown-rot fungi: (a) dry-rot producer *Gloeophyllum sepiarium* (Wulfen) P. Karst.; (b) superficial dry-rot producer *Dacrymyces punctiformis* Neuhoff; (c) heart dry-rot producer *Phaeolus schweinitzii* (Fr.) Pat.; (d) wet brown-rot producer *Tapinella atrotomentosa* (Batsch) Šutara; (e) mixed brown-rot producer *Antrodia crassa* (P. Karst.) Ryvarden.

level of redox potential and thus are capable of oxidizing polyphenolic and other polycyclic aromatic composites. The active center of laccases is presented by iron-containing gem structure (Fig. 2.6).

These peroxidases become highly oxidized when H_2O_2 is reduced to H_2O, and a two-electron reaction allows two activated substrate units to be in a resting state once again before their reduction by peroxidase. The scheme of basic pathways of lignin oxidation by this system is presented in Fig. 2.7.

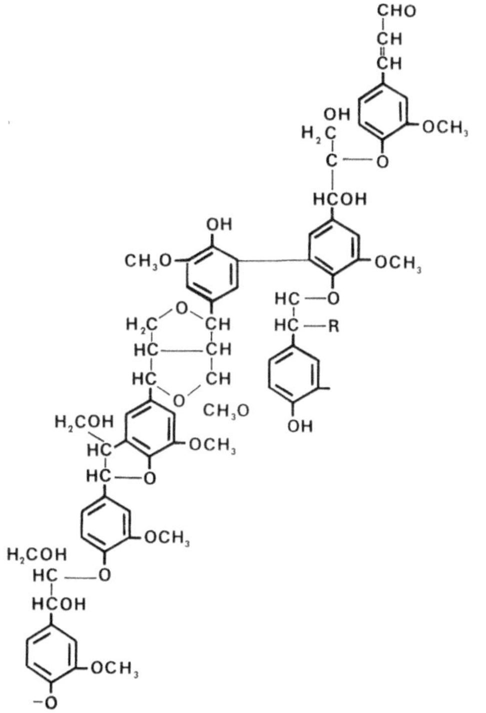

Figure 2.5. The type module of lignin molecular composite after Wright (1985).

There are two basic white-rot patterns:

1) Non-selective delignification attacks mainly hardwood and degrades cellulose, lignin, and hemicellulose simultaneously. The vessel walls are degraded progressively from the lumen towards the middle lamella. The wood remnants are represented by discolored lignin derivatives and breaking cellulose threads of white—stramineous colors. This pattern is characteristic to most white-rot producers, both for ascomycetes and basidiomycetes.
2) Selective delignification attacks hard and soft woods. In this case lignin and hemicelluloses are primarily attacked, and then cellulose. The wood remnants, in many cases, keep their regular crystalline structure or are penetrated by regular pockets. Discoloration occurs in many cases up to white or stramineous colors, but in some cases may be produced to an unusual red or green coloration due to polyphenol remnants. This pattern is a characteristic of some basidiomycetes only, e.g., *Pleurotus* spp., *Phanerochaete* spp., *Ceriporiopsis subvermispora* (Pilát) Gilb. & Ryvarden, and such white-rot fungi may be a good source for biotechnological appications (Schmidt, 2006).

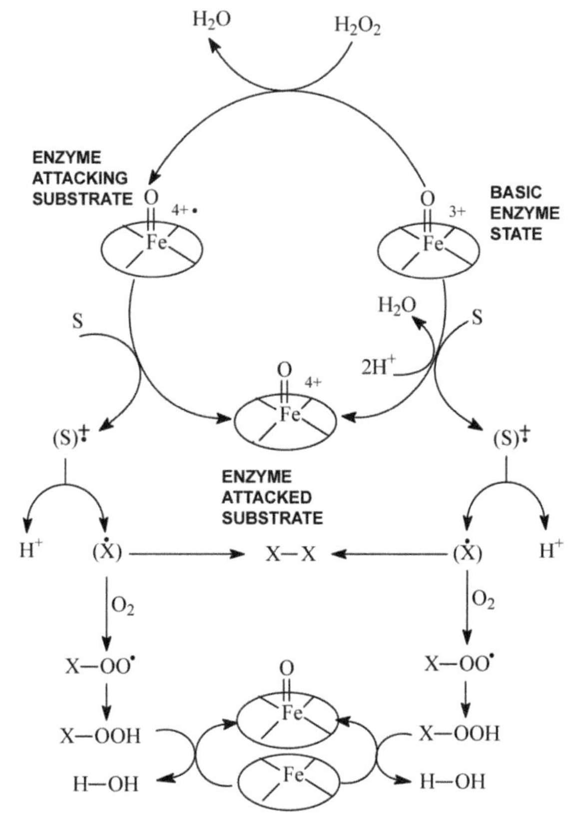

Figure 2.6. The principle structure of the active center of laccases and the mechanism of free-radical production and enzyme regeneration, according to Aisenstadt and Bogolytzin (2009).

White-rot fungi are represented by a huge pool of wood-decomposers, inhabiting a large range of ecological niches from living trees and shrubs to forest litter (Fig. 2.8).

According to Nobles (1958), white-rot fungi represent a derived and homogeneous group of xylotrophic fungi, which originated from brown-rotters via an elaborate system of oxidative enzymes. On the contrary, Gilbertson (1980) states that brown-rot fungi are phylogenetically younger than white-rotters and have multiple and independent origins. Indeed, remnants of gymnosperm predecessors (*Callyxylon* spp.) have characteristics similar to white-rot (Stubblefield et al., 1985). Gilbertson's view also supports this by revealing silent laccase genes in brown-rotters genome (D'Souza et al., 1996). The oxidative (not hydrolytic!) nature of cellulose degradation by brown-rotters may be interpreted as a secondary strategy of rapid colonization of wood substrates.

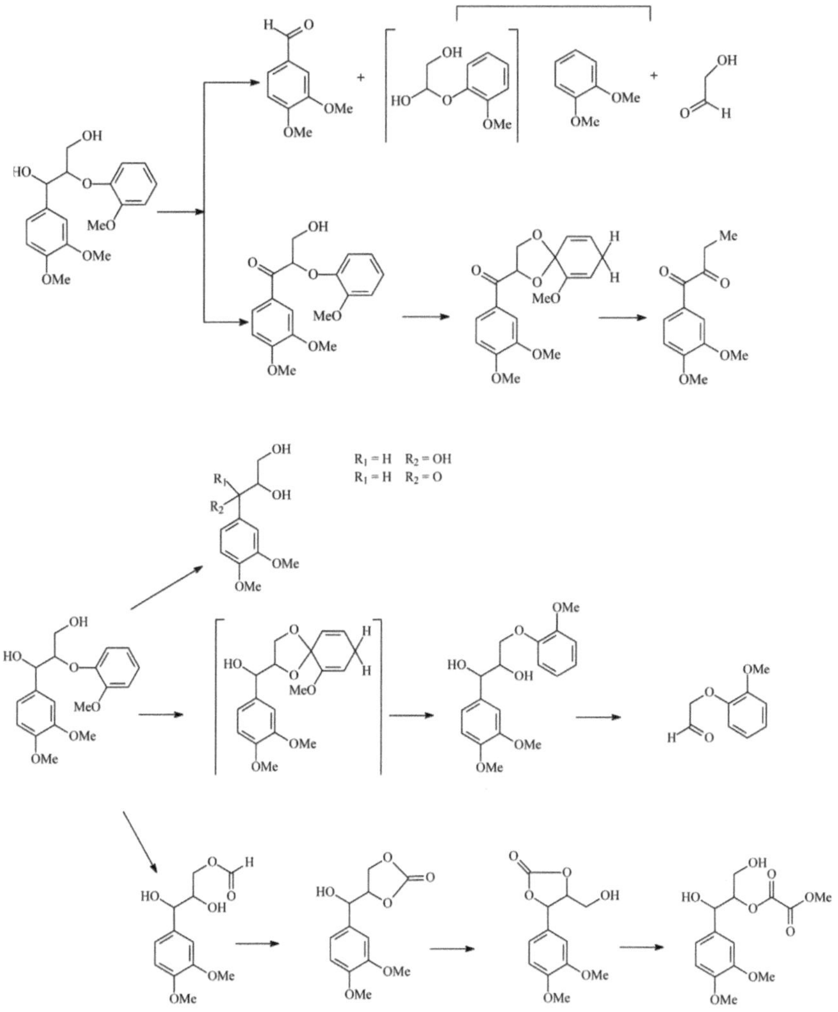

Figure 2.7. The basic ways of oxidation of lignin modules by laccase-generated radicals, as it is shown in Fig. 6 (according to Aisenstadt and Bogolytzin 2009).

C. Trophic Differentiation

According to a basic terminological revision of fungal modes of nutrition, presented by Cooke and Whipps (1980), there are five main modes of nutrition of plant-associated fungi (Fig. 2.9): facultative biotrophs, obligate biotrophs, facultative necrotrophs, obligate saprotrophs and obligate necrotrophs.

Biotrophs and necrotrophs cover an old uncertain category of "parasites". The latter reflects a mode of life rather than a mode of nutrition; therefore, it should be excluded from the trophic category.

Figure 2.8. Key niches representatives of white-rot fungi: (a) *Phellinus tremulae* (Bondartsev) Bondartsev & P.N. Borisov on living aspen tree; (b) *Fomes fomentarius* (L.) J.J. Kickx on fallen birch tree; (c) *Peniophora rufomarginata* (Pers.) Litsch. in Keissler on fallen lime branch; (d) *Merulius tremellosus* Schrad. on debris of alder tree; (e) *Piloderma olivaceum* (Parmasto) Hjortstam on debris of pine tree and forest litter.

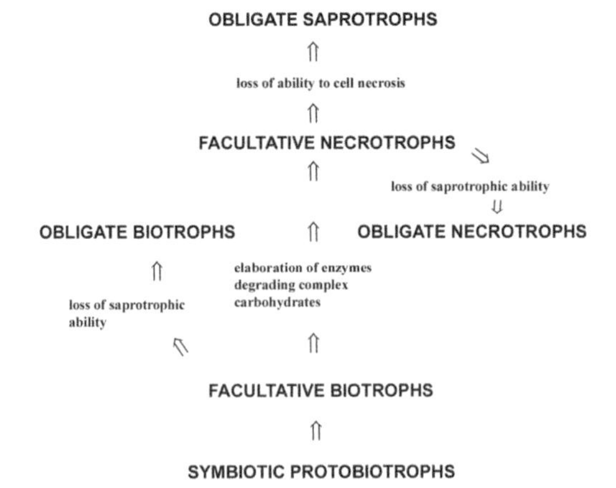

Figure 2.9. The evolution of nutritional modes of fungi according to Cooke and Whipps (1980). The wood-inhabiting fungi belong in their predominating mass to diapason from facultative necrotrophs to obligate necrotrophs and obligate saprotrophs.

True biotrophs are associated with grasses and leaves of woody plants. Their mycelium is adapted to weak exploitation of the host protoplast via elaboration of the so-called "interaction zone", appressoria and haustoria. Wood-inhabiting fungi, in most cases, are devoid of these specialized structures because wood cells are furnished with secondarily thickened walls and usually do not have a protoplast. However, within the Tremellales and some Pucciniomycetes, there are fungi combining xylosaprotrophic ability with a biotrophic (mycoparasitic) one. For example, the *Tremella* representatives are able to infest only cells of fungal hymenium (so-called "intrahymenial parasites"), or transform the whole host fructification (Fig. 2.10a) or occupy xylotroph-attacked wood and vegetative mycelium

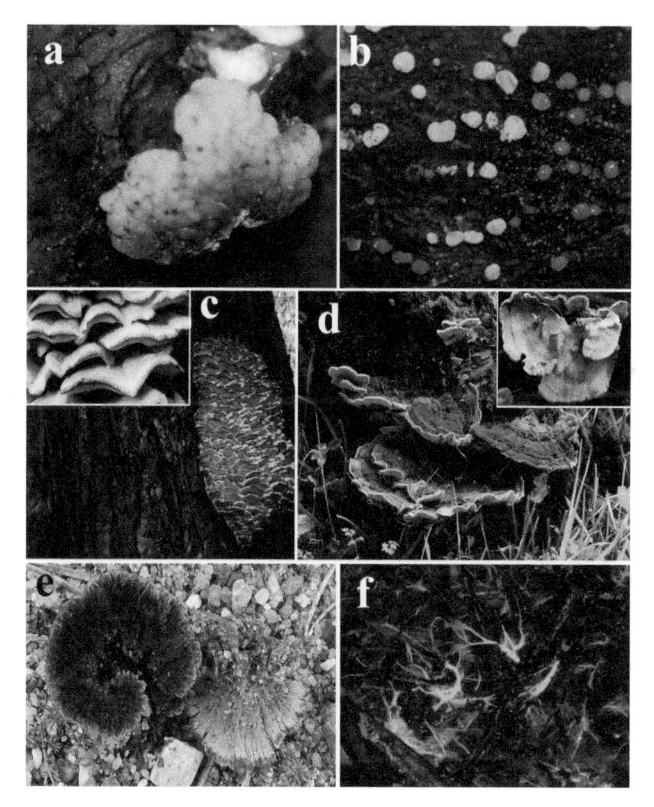

Figure 2.10. Representatives of main trophic groupings of wood-inhabiting fungi: (a) facultative biotrophs—mycoparasites [*Tremella encephala* Willd., transformed fruitbody of host xylotrophic fungus *Stereum sanguinolentum* (Alb. & Schwein.) Fr.]; (b) facultative necrotrophs [*Tubercularia vulgaris* Tode—anamorph of *Nectria cinnabarina* (Tode) Fr.]; (c) pathogenic saprotrophs [*Climacodon septentrionalis* (Fr.) P. Karst. on living maple tree]; (d) non-pathogenic saprotrophs [*Cerrena unicolor* (Bull). Murrill on birch stump]; (e) facultative humus saprotrophs/mycorrhiza-formers (*Thelephora terrestris* Ehrh.—with epiphytic pileate fruitbody); (f) ibid. [*Piloderma bicolor* (Peck) Jülich—with rhizomoid intramatrical to epiphytic fruitbody].

of xylosaprotrophs, penetrating them by haustoria or similar structures (Olive, 1946; Bandoni, 1961, 1987; Zugmaier et al., 1994; Torkelsen, 1997). In a similar way, some *Hypocreopsis* species (ascomycetes from Hypocreales order) protrude wood-destroying mycelium of basidiomycetes. Some wood-inhabiting fungi parasitize scale insects on bark, as *Septobasidium* (Pucciniomycetes) or *Myriangium* (Dothideomycetes). However, a predominant mass of wood-inhabiting fungi lie in diapason from facultative necrotrophs to obligate necrotrophs and obligate saprotrophs according to Cooke and Whipps' classification.

1. **Facultative necrotrophs**. These fungi are localized in the cambial zone of trees and shrub shoots. Having an enzymatic complex, which influences plant cell walls, these fungi kill living cells and exhaust the products of protoplast degradation; however, being adapted to a certain degree of xylolysis (carbohydrate hydrolysis, rarely white-rot patterns), they continue to superficially decay wood. As a result, the bark exfoliates and sprouts die. Some species produce cancer spots. Taxonomically, facultative wood necrotrophs belong to Ascomycota (mostly to orders Diaporthales, Xylariales, Hypocreales and their anamorphs). The most important representatives are *Nectria* (and their *Tubercularia*-anamorphs; Fig. 2.10b), *Diaporthe, Diatrype, Biscogniauxia, Hypoxylon*, some *Valsa* and *Diplodia*-species.
2. **Obligate necrotrophs**. Unlike facultative necrotrophs, these fungi are not capable of asimilating the necrotic tissues and are, therefore, adapted to destroy cells and protoplasts. Their wood-degrading capacity is weak. This group may be considered as a specialized derivative of the previous one. The activity of obligate necrotrophs results in the death of sprouts. As a rule, the target sprouts in a terminal phase. Taxonomically, obligate wood necrotrophs belong to the anamorphic genera of Ascomycota (*Phoma, Camarosporium, Phomopsis, Ottia, Cytospora*, and *Cucurbitaria*).
3. **Obligate saprotrophs.** This type of fungi has the capacity of saprotrophic utilization of the cell wall skeleton of the wood substitute, and has the ability to interact with living plant cells. Probably, in some cases, some haustoria-like structures may be observed near living plant cells (as it was reported in cases of *Stereum* and *Chondrostereum* by Davydkina, 1980), but in general, the mycelium is adapted to colonization and a distant oxidation of wood biopolymers. The predominat part of wood-inhabiting fungi belongs to obligate saprotrophs. Both brown-rot and white-rot fungi belong to this group. Taxonomically, certain representatives of Ascomycota (Xylariales, Pezizales, and Helotiales) belong to this group, but the basic pool is composed of wood-inhabiting Basidiomycota belonging to the orders Polyporales, Agaricales,

Hymenochaetales, Cantharellales, Auriculariales, Dacrymycetales, Tremellales, and Platygloeales. Because wood-inhabiting saprotrophs have a wider range, a finer arrangement of this group is necessary.

a. Pathogenic saprotrophs. These fungi colonize heart wood of living trees and shrubs or fresh decorticated branches, which are rich in water, mineral nutrients and oxygen. The nutrient source of fungi is carbohydrates or lignin of xylem vessels, together with a proper water level. After the tree dies, this type of fungi stop their activity (*Vuilleminia, Godronia, Chondrostereum, Climacodon, Oxyporus* spp., *Pholiota* spp., *Phellinus, Phaeolus, Laetiporus, Piptoporus, Fomes, Fomitopsis* spp., *Ganoderma* spp., *Inonotus* spp., *Lentinus* spp., and *Pleurotus* spp.; Fig. 2.10c). Some of these genera (e.g., *Phellinus* spp., *Inonotus* spp.), simultaneously die when their host tree dies, while some of them (e.g., *Chondrostereum, Ganoderma* spp., *Laetiporus, Piptoporus*, etc.) are capable of continuing their activity after the death of the host tree as non-pathogenic saprotrophs.
b. Non-pathogenic saprotrophs. These fungi form a giant saprotroph pool. They colonize standing dead trees, fallen trees, branches, stumps and woody remnants of lost structures within forest litter. As a result, these fungi are associated with the wood at all stages of their decay. Certain species occupy spatially and temporarily localized niches. Initial stages and niches are associated with the same range of "terminal pathogenic saprotrophs" taxa. Fallen logs are found infested by many decorticators (such as *Ceriporiopsis, Junghuhnia, Oxyporus, Cylindrobasidium, Stereum*, etc.), superficial and per-volume wood decomposers (*Antrodia, Fomitopsis, Gloeophyllum, Skeletocutis, Tyromyces, Postia, Lentinus, Ossicaulis, Xylaria, Rosellinia*, etc.), followed by wood debris decomposers (*Anomoporia, Athelia, Botryobasidium, Serpulomyces, Peziza* spp., *Stropharia* spp., *Coprinus* spp., *Lycoperdon* spp., etc.). The stumps and fallen logs are infested by some characteristic ("leaders") species such as *Fomitopsis, Cerrena, Ganoderma, Fistulina, Lentinus, Tapinella, Clavicorona, Hypholoma*, etc. (Fig. 2.10d).
c. Facultative humus saprotrophs/mycorrhiza-formers. The litter wood debris is connected to humus soil horizon by a continuous range of lignocellulose composites (Table 2.2).

The mycelium of some wood-inhabiting basidiomycetes penetrates these horizons throughout. In some cases, they form ectomycorrhizal covers. The capacity of ectomycorrhiza formation is known for such genera of litter-wood decomposers as *Coltricia, Thelephora, Tomentella, Tylospora, Piloderma, Amphinema*, and *Byssocorticium* (Erland and Taylor, 1999; Zmitrovich, 2008; Fig. 2.10e, f). According to studies by Chen et al.

Table 2.2. The quality and rates of litter decomposition in boreal-nemoral forests (after Heal and Dighton, 1985).

	Mosses	Herbaceous plants	Angiosperm leaves	Coniferous needles	Wood
Cellulose (%)	16–35	20–37	6–22	20–31	36–63
Lignin (%)	7–36	3–30	9–42	20–58	17–35
C : N ratio	13–50	29–160	21–71	63–327	294–327
Decay (% year^{-1})	20	30–70	40–60	3–50	1–90

(2001), the genes for lignolytic enzymes, normally associated with white-rot fungi, are widespread in a broad taxonomical range of ectomycorrhizal mushrooms (*Hydnellum, Bankera, Ramaria, Amanita, Cortinarius, Rozites, Tricholoma, Paxillus, Tylopilus, Xerocomus, Chroogomphus*, etc.). Of course, these ectomycorrhiza-formers are not part of the wood-inhabiting fungi group. According to Bon (1991), these fungi may be considered as facultative mycorrihza-formers. However, their origin is connected to litter-inhabiting lignotrophic basidiomycetes in many phylogenetic lines (Hibbett and Donoghue, 1995; Binder and Hibbett, 2006; James et al., 2006).

D. Topic Groupings (Topic-Sic!)

Tropic (Sic!) groupings are important to be discussed here because of two main reasons: 1) some ectomycorrhizal mushrooms and mycoparasites are able to develop their fructification on wood and are, therefore, wood-inhabiting and, 2) the xylotrophic fungi are distributed over the wood showing some regularities.

Topic (Sic!) aspects of wood-inhabiting fungi have been discussed by Isikov and Konoplya (2004) and Arefiev (2010). According to Isikov and Konoplya (2004), the primary arrangement of topic (Sic!) groupings may be on the basis of topology of shoots (of I, II branching order, and stem). Arefiev (2010) highlighted the relationships of fungus with wood bark and distinguished transcortical vs. decortical species.

According to wood debris and its structural decomposition in time, the general topic groupings of wood-inhabiting fungi may be overviewed as follows:

1. **Wood debris inhabiting fungi.** The grouping is associated with buried wood, rotten fallen branches incorporated into litter and rotten stumps. The fruit bodies of these fungi are, as a rule, stipitate or negatively geotropic. Several characteristic representatives are: *Peziza* spp., *Cudonia, Chlorociboria, Calocera, Macrotyphula, Stropharia, Hypholoma*, etc. (Fig. 2.11a).

Figure 2.11. Main topic groupings of wood-inhabiting basidiomycetes: (a) wood debris inhabiting fungi [*Stropharia hornemannii* (Fr.) S. Lundell & Nannf. on stump debris]; (b) corticolous fungi [*Coprinellus disseminatus* (Pers.) J.E. Lange on bark lignose deposits]; (c) transcortical fungi [*Datronia mollis* (Sommerf.) Donk—the fruit bodies developing through small bark perforations]; (d) decortical fungi [*Bjerkandera adusta* (Willd.) P. Karst. on decorticated lime stump].

2. **Corticolous fungi.** The nutritional source of these fungi lies outside of cortex substratum. As a rule, they are connected to ectomycorrhiza —litter decomposition [*Paxillus involutus* (Batsch) Fr.; *Tylopilus felleus* (Bull.) P. Karst.], or to bark-surface lignocellulose deposites [*Coprinellus disseminatus*—Fig. 2.11b], or to the epiphytic algae/protonema of mosses (*Mycena pseudocorticola* Kühner). Some fungi are also parasites on scale insects (*Septobasidium, Myriangium, Podonectria*), but traditionally these groupings are not mentioned when wood-inhabiting fungi are discussed. Their fruiting bodies, as a rule, are negatively geotropic.
3. **Transcortical fungi.** These necrotrophic and xylosaprotrophic fungi are found infesting the shoots via natural bark perforations. Further mycelium development is connected to bark undergrowth, whereas

the sporulations develop mostly in the areas of penetration. Two main subgroupings can be distinguished here depending on shoot order association:

a. Crown fungi (*Valsa, Mycosphaerella, Phoma, Dothistroma, Diaporthe,* etc.);
b. Stem fungi (*Oxyporus, Datronia, Basidioradulum, Hyphoderma, Cylindrobasidium, Radulomyces,* etc.; Fig. 2.11c).

4. **Decortical fungi.** This type of fungi attacks the wood after sufficient bark disruptions (frost cracks and keels and sores, insects or fungal exfoliation). The mycelium colonizes the general wood mass from alburnum to heart areas. In principle, all active tinder fungi belong to this group: *Fomitopsis, Antrodia, Gloeophyllum, Bjerkandera,* etc. (Fig. 2.11d), together with many corticioid fungal species and wood-inhabiting ascomycetes.

Biodiversity

A. Taxonomical Composition

Wood colonization by higher (multicellular) fungi are heterochronous and polyphyletic. As a result, all basic radiations of ascomycetes and basidiomycetes contain elements adapted to wood (Table 2.3). The worldwide these fungi are estimated to be 30,000 species.

Such a huge radiation of higher fungi, as was viewed by Hibbett et al. (2007), is predominated by wood-inhabiting genera and species. Within Ascomycota, such groups are represented by the orders: Helotiales (class Leotiomycetes), Pezizales (class Pezizomycetes), Diaporthales and Xylariales (class Sordariomycetes). Within Basidiomycota, the dominant order of wood-inhabiting fungi is Polyporales (class Agaricomycetes), which is represented, almost exclusively, by xylotrophic taxa. The other gross taxa belong to Agaricales, Hymenochaetales and Russulales.

The groups with necrotrophic activities (mainly various Sordariomycetes from the old order "Sphaeriales") are characterized by the absence of biochemical mechanisms responsible for the oxidation of wood composites; however, as a rule all saprotrophic lineages of Ascomycota and Basidiomycota are characterized by peroxidase and laccase activities. The nodes of brown-rot producers within Basidiomycota are scarce and species-powered, and as a rule, they are well delimited taxonomically (order Dacrymycetales in Dacrymycetes, order Gloeophyllales and Boletales in Agaricomycetes, family Fomitopsidaceae in Polyporales). However, within Polyporaceae, the brown-rot taxa are distributed rather stochastically, and some genera (e.g., *Grifola*) may be heterogeneous on the basis of rot produced.

Table 2.3. The taxonomical composition of wood-inhabiting fungal groups (Zmitrovich et al., 2007, with addition).

Phyla, classes	**Orders**	**Key genera of wood-inhabiting fungi**
Ascomycota Caval.-Sm.		
Dothideomycetes O.E. Erikss. & Winka	Capnodiales Woron.	**Mycosphaerella* Johans., **Metacapnodium* Speg.
	Dothideales Lindau	**Dothidea* Fr., **Polwrightia* Sacc.
	Pleosporales Luttr. ex M.E. Barr	**Astrosphaeriella* Sydow, **Caryospora* de Not., **Cucurbitaria* Gray, **Fenestella* Tul., **Herpotrichia* Fuckel, **Karstenula* Speg., **Massaria* de Not., **Massarina* Sacc., **Melanomma* Nitschke, **Melomastia* Nitschke, **Otthia* Nitschke, **Pteridiospora* Penzig & Sacc., **Teichospora* Fuckel, **Thaxteria* Sacc., **Thyridaria* Sacc., **Trematosphaeria* Fuckel, **Pleomassaria* Speg.
	Hysteriales Lindau	*Hysterium* DC. ex Mérat
Leotiomycetes O.E. Erikss. & Winka	Cyttariales Luttr. ex Gamundí	*Cyttaria* Berk.
	Helotiales Nannf.	*Arachnopeziza* Fuckel, *Ascocorticium* Bref., *Ascocorticiellum* Jülich & B. de Vries, *Ascocoryne* Groves & D.E. Wilson, *Ascotremella* Seaver, *Bisporella* Sacc., *Bulgaria* Fr., *Bulgariella* P. Karst., *Cenangium* Fr., *Chloencoelia* Dixon, *Chlorociboria* Seaver ex Ramamurthi, Korf & Batra, *Cistella* Quél., *Claussenomyces* Kirschst., *Crocicreas* Fr., *Cudoniella* Sacc., *Dasyscyphella* Tranzschel, *Dematioscypha* Svrček, *Dencoeliopsis* Korf, *Discocainia* J. Reid & Funk, *Encoelia* (Fr.) P. Karst., *Eriopezia* (Sacc.) Rehm, *Godronia* Moug. & Lév., *Gorgoniceps* P. Karst., *Gremmeniella* M. Morelet, *Hamatocanthoscypha* Svrček, *Holwaya* Sacc., *Hyaloscypha* Boud., *Hymenoscyphus* Gray, *Incrupila* Raitv., *Ionomidotis* E.J. Durand, *Lachnellula* P. Karst., *Lachnum* Retz., *Lasiobelonium* Ellis & Everh., *Mollisia* (Fr.) P. Karst., *Neobulgaria* Petr., *Neodasyscypha* Spooner, *Patinellaria* P. Karst., *Perrotia* Boud., *Phaeohelotium* Kanouse, *Proliferodiscus* J.H. Haines & Dumont, *Tympanis* Tode, *Velutarina* Korf, *Rutstroemia* P. Karst., *Unguicularia* Höhn.
	Rhytismatales M.E. Barr ex Minter	*Propolis* (Fr.) Corda
Orbiliomycetes O.E. Erikss. & Winka	Orbiliales Baral et al.	*Hyalinia* Boud., *Orbilia* Fr.

Table 2.3. contd....

Table 2.3. contd.

Phyla, classes	Orders	Key genera of wood-inhabiting fungi
Pezizomycetes O.E. Erikss. & Winka	Pezizales J. Schröt.	*Balsamia* Vittad., *Cheilymenia* Boud., *Discina* (Fr.) Fr., *Gyromitra* Fr., *Helvella* L., *Humaria* Fuckel, *Karstenella* Harmaja, *Microstoma* (Fr.) Kanouse, *Peziza* Fr., *Pithya* Fuckel, *Plectania* Fuckel, *Rhizina* Fr., *Scutellinia* (Cooke) Lambotte, *Urnula* Fr.
Sordariomycetes O.E. Erikss. & Winka	Coronophorales Nannf.	*Calyculosphaeria* Fitzp., **Coronophora* Fuckel, **Bertia* de Not., **Nitschkia* Otth, **Tympanopsis* Stärb.
	Hypocreales Lindau	*Arachnocrea* Z. Moravec, *Calonectria* de Not., *Gibberella* Sacc., *Hypocreopsis* P. Karst., *Nectria* (Fr.) Fr., *Podostroma* P. Karst.
	Boliniales P.F. Cannon	**Camarops* P. Karst., **Bolinia* (Nitschke) Sacc.
	Calosphaeriales M.E. Barr	**Calosphaeria* Tul. & C. Tul.
	Chaetosphaeriales Huhndorf, A. N. Mill. & F.A. Fernández	**Chaetosphaeria* Tul. & C. Tul., *Zignoëlla* Sacc.
	Coniochaetales Huhndorf, A. N. Mill. & F.A. Fernández	**Coniochaeta* (Sacc.) Massee
	Diaporthales Nannf.	*Apioporthe* Höhn., *Diaporthe* Nitschke, *Calosporella* J. Schröt., *Caudospora* Starb., *Cryptodiaporthe* Petrak, *Cryptospora* Tul. & C. Tul., *Cryptosporella* Sacc., *Endothia* Fr., *Hercospora* Fr., *Melanconiella* Sacc., *Prosthecium* Fr., *Pseudovalsa* Ces. & de Not., *Sillia* P. Karst., *Valsa* Fr.
	Ophiostomatales Benny & Kimbr.	**Ceratocystis* Ellis & Halst., **Ophiostoma* Syd. & P. Syd.
	Sordariales Chadef. ex D. Hawksw. & O. E. Erikss.	**Bombardia* Fr., **Camarops* P. Karst., *Chaetomium* Kunze ex Fr., **Lasiosphaeria* Ces. & de Not.

	Xylariales Nannf.	*Biscogniauxia* Kuntze, *Daldinia* Ces. & de Not., *Diatrype* Fr., *Diatrypella* (Ces. & de Not.) Cooke, *Entoleuca* Syd., *Entonaema* Möller, *Eutypa* Tul. & C. Tul., *Eutypella* (Nitschke) Sacc., *Hypoxylon* Bull., *Lopadostoma* (Nitschke) Traverso, *Nemania* Gray, *Rosellinia* de Not., *Ustulina* Tul., *Xylaria* Hill ex Schrank
	Trichosphaeriales M.E. Barr	**Trichosphaeria* Fuckel
	Triblidiales O.E. Erikss.	*Pseudographis* Nyl., *Triblidium* Rebentisch & Pers.
Basidiomycota R.T. Moore		
Pucciniomycetes Bauer et al.	Platygloeales R.T. Moore	**Eocronartium* G.F. Atk., *Platygloea* J. Schröt.
Tremellomycetes Dowled	Tremellales Fr.	*Tremella* Pers.
Dacrymycetes Dowled	Dacrymycetales Henn.	*Calocera* (Fr.) Fr., *Cerinomyces* G.W. Martin, *Dacrymyces* Nees, *Ditiola* Fr., *Femsjonia* Fr.
Agaricomycetes Dowled	Agaricales Underw.	*Armillaria* (Vahl) P. Kumm, *Arrhenia* Fr., *Calyptella* Quél., *Campanella* Henn., *Caripia* Kuntze, *Cellypha* W.B. Cooke, *Chaetocalathus* Singer, *Cheimonophyllum* Singer, *Chondrostereum* Pouzar, *Chromocyphella* De Toni & Levi, *Claudopus* Gillet, *Clitopilus* (Fr. ex Rabenh.) P. Kumm., *Collybia* (Fr.) Staude, *Coprinus* Pers., *Coronicium* J. Erikss. & Ryvarden, *Cotylidia* P. Karst., *Crepidotus* (Fr.) Staude, *Crinipellis* Pat., *Cylindrobasidium* Jülich, *Deflexula* Corner, *Episphaeria* Donk, *Favolaschia* (Pat.) Pat., *Filoboletus* Henn., *Fistulina* Bull., *Flagelloscypha* Donk, *Flammulina* P. Karst., *Gymnopilus* P. Karst., *Hemipholiota* (Singer) Bon, *Henningsomyces* Kuntze, *Hohenbuehelia* Schulzer, *Hypholoma* (Fr.) P. Kumm., *Hypsizygus* Singer, *Kuehneromyces* Singer & A.H. Sm., *Lachnella* Fr., *Lampteromyces* Singer, *Lentinula* Earle, *Lycoperdon* P. Micheli, *Merismodes* Earle, *Mucronella* Fr., *Mycena* (Pers.) Roussel, *Nothopanus* Singer, *Omphalina* Quél., *Ossicaulis* Redhead & Ginns, *Panellus* P. Karst., *Pellidiscus* Donk, *Phaeosolenia* Speg., *Pholiota* (Fr.) P. Kumm., *Pleurocybella* Singer, *Pleurotus* (Fr.) P. Kumm., *Pluteus* Fr., *Podoscypha* Pat., *Porotheleum* Fr., *Radulomyces* M.P. Christ., *Rectipilus* Agerer, *Resupinatus* Nees ex Gray, *Rhodotus* Maire, *Rimbachia* Pat., *Schizophyllum* Fr., *Stigmatolemma* Kalchbr., *Strobilurus* Singer, *Stromatocyphella* W.B. Cooke, *Stropharia* (Fr.) Quél., *Tricholomopsis* Singer, *Trogia* Fr., *Typhula* (Pers.) Fr., *Volvariella* (Fr.) P. Kumm.

Table 2.3. contd....

Table 2.3. contd.

Phyla, classes	Orders	Key genera of wood-inhabiting fungi
	Atheliales Jülich	*Amylocorticium* Pouzar, *Amylocorticiellum* Spirin & Zmitr., *Athelia* Pers., *Ceraceomyces* Jülich, *Piloderma* Jülich, *Serpulomyces* (Zmitr.) Zmitr., *Tylospora* Donk
	Auriculariales J. Schröt.	*Auricularia* Bull. ex Juss., *Basidiodendron* Rick, *Bourdotia* (Bres.) Bres. & Torrend, *Craterocolla* Bref., *Ductifera* Lloyd, *Eichleriella* Bres., *Exidia* Fr., *Exidiopsis* (Bref.) Möller., *Heterochaete* Pat., *Pseudohydnum* P. Karst., *Stypella* Möller, *Tremellostereum* Ryvarden
	Boletales E.-J. Gilbert	*Bondarcevomyces* Parmasto, *Coniophora* DC., *Gyrodontium* Pat., *Jaapia* Bres., *Leucogyrophana* Pouzar, *Meiorganum* R. Heim, *Paxillus* Fr., *Serpula* (Pers.) Gray, *Tapinella* E.-J. Gilbert, *Tylopilus* P. Karst.
	Cantharellales Gäum.	*Botryobasidium* Donk, *Ceratobasidium* D.P. Rogers, *Clavulicium* Boidin, *Oliveonia* Donk, *Scotomyces* Jülich, *Suillosporium* Pouzar, **Thanatephorus* Donk, *Tulasnella* J. Schröt.
	Corticiales K.H. Larss.	*Corticium* Pers., *Cytidia* Quél., *Punctularia* Pat., *Vuilleminia* Maire
	Gloeophyllales Thorn	*Boreostereum* Parmasto, *Donkioporia* Kotl. & Pouzar, *Gloeophyllum* P. Karst., *Neolentinus* Redhead & Ginns, *Veluticeps* Cooke
	Gomphales Jülich	*Hydnocristella* R.H. Petersen, *Kavinia* Pilát, *Lentaria* Corner, *Ramaricium* J. Erikss.
	Hymenochaetales Oberw.	*Asterodon* Pat., *Basidioradulum* Nobles, *Cyclomyces* Kunze ex Fr., *Fibricium* J. Erikss., *Hydnochaete* Bres., *Hymenochaete* Lév., *Hyphodontia* J. Erikss., *Inonotopsis* Parmasto, *Inonotus* P. Karst., *Leucophellinus* Bondartsev & Singer, *Oxyporus* (Bourdot & Galzin) Donk, *Phellinidium* (Kotl.) Fiasson & Niemelä, *Phellinus* Quél., *Phylloporia* Murrill, *Pyrrhoderma* Imazeki, *Repetobasidium* J. Erikss., *Resinicium* Parmasto, *Schizopora* Velen., *Sidera* Miettinen & K. H. Larss., *Stipitochaete* Ryvarden, *Subulicium* Hjortstam & Ryvarden, *Trichaptum* Murrill, *Tubulicrinis* Donk
	Polyporales Gäum.	*Abortiporus* Murrill, *Amauroderma* Murrill, *Amylocystis* Bondartsev & Singer, *Anomoporia* Pouzar, *Antrodia* P. Karst., *Antrodiella* Ryvarden & I. Johans., *Aurantiporus* Murrill, *Auriculariopsis* Maire, *Auriporia* Ryvarden, *Bjerkandera* P. Karst., *Bulbillomyces* Jülich, *Byssomerulius* Parmasto, *Ceriporia* Donk, *Cerrena* Gray, *Climacocystis* Kotl. & Pouzar, *Climacodon* P. Karst., *Cryptoporus* (Peck) Shear, *Dacryobolus* Fr., *Daedalea* Pers., *Daedaleopsis* J. Schröt., *Datronia* Donk, *Dichomitus* D.A. Reid, *Flaviporus* Murrill, *Fomes* (Fr.) Fr., *Fomitopsis* P. Karst., *Ganoderma* P. Karst., *Gloeoporus* Mont., *Grammothele* Berk. & M.A. Curtis, *Grifola* Gray, *Haddowia* Steyaert, *Hapalopilus* P. Karst., *Haploporus* Bondartsev & Singer, *Hexagonia* Fr., *Humphreya* Steyaert, *Hydnophlebia* Parmasto, *Hymenogramme*

		Mont. & Berk., *Hyphoderma* Wallr., *Hyphodermella* J. Erikss. & Ryvarden, *Hypochnicium* J. Erikss., *Intextomyces* J. Erikss. & Ryvarden, *Ischnoderma* P. Karst., *Irpex* Fr., *Jahnoporus* Nuss, *Junghuhnia* Corda, *Laetiporus* Murrill, *Lentinus* Fr., *Leptoporus* Quél., *Lopharia* Kalchbr. & MacOwan, *Loweomyces* (Kotl. & Pouzar) Jülich, *Megasporoporia* Ryvarden & J.E. Wright, *Meripilus* P. Karst., *Microporus* P. Beauv., *Mycorrhaphium* Maas Geest., *Nigroporus* Murrill, *Oligoporus* Bref., *Pachykytospora* Kotl. & Pouzar, *Panus* Fr., *Parmastomyces* Kotl. & Pouzar, *Perenniporia* Murrill, *Phaeolus* Pat., *Phanerochaete* P. Karst., *Phlebia* Fr., *Phlebiella* P. Karst., *Phlebiopsis* Jülich, *Piptoporus* P. Karst., *Polyporus* P. Micheli, *Porogramme* (Pat.) Pat., *Postia* Fr., *Pycnoporellus* Murrill, *Pyrofomes* Kotl. & Pouzar, *Scopuloides* (Massee) Höhn. & Litsch., *Skeletocutis* Kotl. & Pouzar, *Sparassis* Fr., *Spongipellis* Pat., *Steccherinum* Gray, *Trametes* Fr., *Theleporus* Fr., *Tyromyces* P. Karst., *Wolfiporia* Ryvarden & Gilb., *Xenasma* Donk
	Russulales Kreisel ex Kirk et al.	*Acanthobasidium* Oberw., *Acanthophysellum* Parmasto, *Aleurodiscus* Rabenh. ex J. Schröt., *Amylonotus* Ryvarden, *Amylosporomyces* S.S. Rattan, *Asterostroma* Massee, *Auriscalpium* Gray, *Boidinia* Stalpers & Hjortstam, *Bondarzewia* Singer, *Clavicorona* Doty, *Conferticium* Hallenb., *Dentipellis* Donk, *Dentipratulum* Domański, *Dichopleuropus* D.A. Reid, *Dichostereum* Pilát, *Echinodontium* Ellis & Everh., *Gloeocystidiellum* Donk, *Gloeodontia* Boidin, *Gloiodon* P. Karst., *Gloiothele* Bres., *Hericium* Pers., *Heterobasidion* Bref., *Laurilia* Pouzar, *Laxitextum* Lentz, *Lentinellus* P. Karst., *Megalocystidium* Jülich, *Pseudoxenasma* K.H. Larss. & Hjortstam, *Scytinostroma* Donk, *Scytinostromella* Parmasto, *Stecchericium* D.A. Reid, *Stereofomes* Rick, *Stereum* Hill ex Pers., *Vararia* P. Karst., *Wrightoporia* Pouzar, *Xylobolus* P. Karst.
	Sebacinales M. Weiß et al.	**Sebacina* Tul.
	Thelephorales Corner ex Oberw.	*Amaurodon* J. Schröt., *Pseudotomentella* Svrček, *Thelephora* Ehrh. ex Willd., *Tomentella* Pers. ex Pat., *Tomentellopsis* Hjortstam, *Tomentellago* Hjortstam & Ryvarden
	Trechisporales K.H. Larss.	*Litschauerella* Oberw., *Subulicystidium* Parmasto, *Trechispora* P. Karst.

Note. The genera, devoid of oxidative enzymes, are marked by an asterisk, the genera, associated with a brown-rot are underlined, and the rest of the genera are associated with white-rot.

It is obvious that colonization of wood, wood remnants and wood surfaces was realized on several levels of organization of ascomycetes and basidiomycetes, and these adaptations as well as certain convergence and unifications were overbuilded on various organization types and biochemical pathways. The most dramatic changes occur in the external mycelial structures—the stromata and fruiting bodies.

Biomorphic Diversity

The aerial mycelia of wood-inhabiting fungi develop in a rather limited range of parameters such as the presence or absence of bark, lacunes on wood material, wood remnants exposure and forest insolation regimes. Most fungal wood-inhabiting fructifications are mechanically layed out above ground; therefore, their main morphogenetic tendency is to develop a positive geotropism and prostrate growth form.

As a result of adaptation to wood overgrowth, the sporulation structures of ascomycetes and basidiomycetes have a convergent similarity (Bondartseva, 2001), producing a wide biomorphic diversity. The overview of their biomorphic diversity is given below.

α. Ascomycetous cycle of forms (hyphae with one to multinucleate segments; septum with a central pore without doliolum; meiospores formed endogenously in a stichal mode; meiotangia—asci; the arrangement of aerial mycelium mostly cladomian, with axial and pleuridial differentiation, organized as [a] stromata with many carpocenters, [b] single ascocarps, and [c] clustered ascocarps).

1. Ascolocular series (obligate stromatic; conceptacles originating as locules into the stroma; in some cases the interlocular stromatic context disintegrates and conceptacles transform into pseudothecia; sporulation passive).
    a. Elsinoid group (stroma pulvinate with stochastically dispersed locules—the so-called "myriothecium")
        - Elsinoid biomorph (stroma immersed into host tissue): *Myriangium* and some others.
    b. Dothideoid group (stroma globose to pulvinate, with singular or multiplicate pseudothecia; mostly subepidermal):
        - Dothideoid biomorph—*Dothidea, Mycosphaerella, Cucurbitaria* and some others.
    c. Hysterioid group (stroma elongated—with fusoid or cylindrical outline, revealed by linear fissure; subepidermal):
        - Hysterioid biomorph—*Hysterium* and some others.
2. Pyrenocarpous series (conceptacles differentiated as individuals—perithecia—surrounded by a wall of stromatic origin: pitcher-like,

pear-like or subglobose, in some cases with developed excretory channel; integrated by common stroma or disintegrated; sporulation active).

A. Hypocreoid infraseries (stromatic wall not carbonized, of *textura globulosa—epidermoidea*, light or bright-colored; stromata fleshy or subceraceous).

Hypocreoid group (stroma pulvinate to prostrate)

- Creopioid biomorph (stroma gelatinous, pulvinate): *Creopus* and some others.
- Hypocreoid biomorph (stroma non-gelatinized, pulvinate to prostrate, persisting): *Hypocrea, Hypocreopsis, Hypomyces, Gibberella* and some others.
- Nectrioid (stroma non-gelatinized, pulvinate, transforming into perithecia clusters).

B. Xylarioid infraseries (stromatic wall carbonized, of *textura angularis—porrecta*, blackish-colored; stromata of hard consistency).

a. Xylarioid group (perithecia submerged into stroma).

- Hypoxyloid biomorph (perithecia copious, stroma superficial, globose or pulvinate): *Hypoxylon, Daldinia* and some others.
- Ustulinoid (perithecia copious, stroma superficial, crustose): *Ustulina, Hypoxylon* pr. p.
- Xylospheroid biomorph (perithecia copious, stroma superficial, erect—clavate or staghorn-like): *Xylosphaera, Thamnomyces* and some others.
- Nummularioid biomorph (stroma submerged; ectostroma prostrate, with singular or copious perithecia).
  - Nummularioid (strict) biomorph (perithecia copious): *Biscogniauxia* and some others
  - Rosellinioid biomorph (stroma monoperithecial): *Rosellinia, Phyllachora* and some others.

b. Diatrypoid group (perithecia deeply immersed, with long necks).

- Diatrypoid biomorph (ectostroma enlarged, perithecial necks not clustered): *Diatrype, Diaporthe, Glomerella, Eutypa, Cryptospora, Endothia* and some others.
- Valsoid biomorph (ectostroma local, perithecial necks clustered): *Valsa, *Cytospora* and some others.

C. Sphaerioid infraseries (perithecia disintegrated—singular or grouped over common subiculum).

a. Sordarioid group (perithecia pear-like, mazaedium present or absent).

- Sordarioid biomorph (perithecia without hyphal appendages and mazaedium): *Sordaria, Nectria* and some others.
- Chaetomioid biomorph (perithecia with hyphal appendages; apical apparatus transformed into mazaedium): *Chaetomium*, **Chaetomella* and some others.

b. Sphaerioid group (perithecia subglobose, mazaedium as a rule present).

- Sphaerioid biomorph (perithecium without neck): *Lasiosphaeria, Podospora,* **Rabenhorstia* and some others.
- Ophiostomoid biomorph (perithecium having an elongated neck): *Ophiostoma, Ceratocystis,* **Ceratopycnis* and some others.

3. Discomycete series (ascocarp open, differentiated into hymenium and excipulum (so-called apothecium): cupulate, discoid, pulvinate with prominent or reduced stipe; sporulation active).

a. Pezizoid group (apothecia more than 1.5 cm across, fleshy-ceraceous, with reduced stipe, symmetrically or asymmetrically cupulate to discoid).

- Pezizoid biomorph (apothecia symmetrical, cupulate, sometimes lobate): *Peziza* and some others.
- Discinoid biomorph (apothecia symmetrical, prostrate plate-like to turned out plate-like): *Discina, Rhizina, Plicaria* and some others....
- Gyromitroid biomorph (apothecia asymmetrical, cerebriform, with irregularly-folded hymenial surface and increscent margin): *Gyromitra*.
- Helvelloid biomorph (apothecia asymmetrical, irregularly-lobed, even or folded, with free margin): *Helvella* and some others.
- Humarioid biomorph (apothecia cupulate, sessile or substipitate, excipulum/margin pubescent): *Humaria, Trichophaea, Rutstroemia* pr. P and some others.
- Scutellinioid biomorph (apothecia discoid-cupulate, sessile, excipulum and margin prominently ciliate): *Scutellinia* and some others.

b. Sclerotinioid group (ascocarp less than 1.5 cm across, ceraceous, stipitate, cupulate to discoid).

- Ciborioid biomorph (sclerotium absent, excipulum even): *Ciboria, Rutstroemia, Chlorociboria, Hymenoscyphus, Cudoniella, Cyathicula* and some others.
- Dasyscyphoid biomorph (sclerotium absent, excipulum pubescent): *Dasyscyphus, Lachnellula* and some others.

c. Geoglossoid group (ascocarps medium-sized to small, waxy to subgelatinous, differentiated into stipe and fertile head).
   - Cudonioid biomorph (fertile head pileate with inrolled margin, cerebriform): *Leotia lubrica, Cudonia* and some others.

d. Helotioid group (apotheciun less than 1.5 cm across, waxy to subgelatinous, discoid, substipitate).
   - Helotioid biomorph (apothecium waxy, excipulum even): *Helotium, Pezicula, Bisporella, Pezizella, Mollisia, Calycellina, Phialina, Pithya, *Jaczewskiella* and some others.
   - Hyaloscyphoid biomorph (apothecium waxy, excipulum pubescent): *Hyaloscypha, Hyalopeziza, Patellariopsis, Tapezia, Cenangium, *Pseudocenangium* and some others.
   - Ascocoryneoid biomorph (apothecium subgelatinous, excipulum even): *Ascocoryne* and some others.

e. Bulgarioid group (apothecia 1–12 cm across, gelatinous, turbinate to barrel-shaped):
   - Bulgarioid biomorph—*Bulgaria.*

f. Ascotremelloid group (apothecia 0.5–1.5 cm across, in brain-like to lobed clusters):
   - Ascotremelloid biomorph—*Ascotremella, *Coryne* and some others.

g. Scleroderrioid group (apothecia less than 1.5 cm across, bowl-shaped, with villous excipulum, of corneous consistency, clusterized):
   - Scleroderrioid biomorph—*Godronia, Tympanis* and some others.

h. Cyttarioid group (apothecia less than 1.5 across, bowl-shaped, immersed into fleshy-gelatinous globose to trametoid stroma):
   - Cyttarioid biomorph—*Cyttaria.*

i. Cryptodiscoid group (apothecia less than 1.5 cm across, pitcher-like, immersed into substratum):
   - Cryptodiscoid biomorph—*Cryptodiscus, Pyrenopeziza, Stictis* and some others.

j. Clitrioid group (apothecia small, lanceolate, with prominent excipulum, immersed into substrate):
   - Clitroid biomorph—*Propolis, Colpoma, Lophodermium* and some others.

k. Ascocorticioid group (excipulum reduced, hypothecium prostrate):
   - Ascocorticioid biomorph—*Ascocorticium, Ascocorticiellum.*

β. Basidiomycetous cycle of forms (hyphae of secondary mycelium dikaryotic to multicellular, mostly doliporous; meiospores exogeneous; meiotangia as basidia—auricularioid, tremelloid, dacrymycetoid, tulasnelloid heterobasidia or homobasidia; structure presumable hemicladome without prominent pleuridia; cladothalle as basidiocarps of various structure).

1. Clavarioid series (basidiocarps erect, without pileus differentiation).
   a. Clavarioid group (basidiocarps unbranched, non-gelatinous).
      - Clavariadelphoid biomorph (basidiocarps large, hollow, clavate): *Macrotyphula* and some others.
      - Typhuloid biomorph (basidiocarps small, monolith).
        - Strict typhuloid biomorph (emerging from sclerotiun): *Typhula*.
        - Pistillarioid (without sclerotium): *Pistillaria, Mucronella* and some others.
   b. Caloceroid group (basidiocarps erect, unbranched or with apical branching, gelatinous or viscous):
      - Caloceroid biomorph—*Calocera*.
   c. Ramarioid group (basidiocarps erect, more or less sympodially branching, non-gelatinous):
      - Ramarioid biomorph—*Ramaria, Lentaria*.
   d. Clavicoronoid group (basidiocarp erect, branched via scyphoid proliferation, non-gelatinose):
      - Clavicoronoid biomorph—*Clavicorona*.
2. Cantharelloid series (basidiocaps erect, funnel-shaped, pileus plectologically not differentiated from the stipe).
   a. Grifoloid group (basidiocarps erect, multipileate, with even or poroid hymenophore).
      - Amylarioid biomorph (hymenophore even): *Amylaria, Sparassis*.
      - Grifoloid biomorph (hymenophore poroid): *Grifola, Meripilus*.
   b. Polyporoid group (basidiocarps erect to laterally-attached, unipileate, funnel-shaped to tongued, with central to lateral stipe and cellar to poroid hymenophore):
      - Polyporoid biomorph—*Polyporus, Microporus, Coltricia, Coltriciella, Ganoderma* pr. p., *Ischnoderma* pr. p.
   c. Hericioid group (basidiocarps erect, ramose, with spinose hymenophore):
      - Hericioid—*Hericium*.
3. Corticioid series (basidiocarps positively geotropic to ageotropic with radial tendency to expansion of hyphal masses).
   a. Merulioid group (basidiocarps prostrate with free margin or in lateral forms semipileate; double-layered with loose pubescent

abhymenial stratum and gelatinized hymenophoral stratum; hymenophore even, wrinkled or tubulose, cornescent).

- Merulioid biomorph (basidiocarps prostrate of pileate, with wrinkled hymenophore): *Merulius, Serpula* and some others.
- Phlebioid biomorph (basidiocarp prostrate with concentrically folded to even hymenophore): *Phlebia, Punctularia* and some others.
- Chondrostereoid biomorph (basidiocarp prostrate to pileate with even hymenophore): *Chondrostereum, Auriculariopsis, Auricularia mesenterica, Gloeostereum.*
- Gloeoporoid biomorph (basidiocarps prostrate to pileate, with poroid hymenophore): *Gloeoporus, Skeletocutis, Gelatoporia* and some others.

b. Porioid group (basidiocarps prostrate, annual—perennial, with non-gelatinized poroid hymenophore).

- Fuscoporioid biomorph (hymenophore multilayered, of tough consistency): *Phellinus* pr. p., *Fuscoporia, Phellinidium, Rigidoporus crocatus* and some others.
- Fibroporioid biomorph (hymenophore as a single layer, of fibrous consistency): *Antrodia* pr. p., *Fibroporia, Trametes, Antrodiella* pr. p., *Kneiffiella* pr. p., *Diplomitoporus* and some others.
- Ceriporioid biomorph (hymenophore as a single layer, of soft-ceraceous consistency): *Ceriporiopsis, Ceriporia, Oligoporus, Postia* pr. p., *Parmastomyces, Protomerulius* and some others.
- Cristelloid biomorph (hymenophore as a single layer, not condensed, of soft film consistency): *Trechispora* pr. p., *Sistotrema* pr. p., *Porpomyces mucidus* and some others.

c. Stereoid group (basidiocarps of hard consistency, homogeneous, pileate to resupinate, with or without stipe: hymenophore basically even, often irregularly sculptured, one-layered to multilayered).

- Stereoid biomorph (basidiocarps prostrate, with a smooth hymonophore): *Stereum, Amylostereum, Hymenochaete, Lopharia, Boreostereum, Cystostereum* and some others.
- Sterelloid biomorph (basidiocarps prostrate with border-like margin and even—papillose hymenophore): *Cylindrobasidium, Peniophora* pr. p., *Xylobolus, Aleurodiscus* and some others.
- Podoscyphoid biomorph (basidiocarps with sublateral to central stipe, hymenophore one-layered, even): *Podoscypha, Cotylidia, Cyphellostereum* and some others.

- Arrhenioid biomorph (basidiocarps with lateral, sometimes rudimentary, stripe and venose-sublamellate hymenophore sculpture): *Arrhenia, Caripia* and some others.

d. Raduloid group (basidiocarps prostrate, with dull-tooched? to radulose hymenophore):
- Raduloid biomorph—*Radulomyces, Basidioradulum, Sistotrema* pr. p., *Dentocorticium* and some others.

e. Corticioid group (basidiocarps prostrate, loose to dense consistency, with even or papillose hymenophore).
- Peniophoroid biomorph (basidiocarps of hard consistency, with multilayered hymenophore): *Peniophora, Dendrophora, Duportella* and some others.
- Corticioid biomorph (basidiocarps of hard consistency, with single-layered even hymenophore): *Corticium, Exidiopsis, Acanthophysellum, Dendrothele* and some others.
- Hyphodermoid biomorph (basidiocarps homogeneous, of ceraceous consistency, with even or sculptured hymenophore): *Hyphoderma, Gloeocystidiellum, Metulodontia, Phanerochaete* and some others.
- Athelioid biomorph (basidiocarps two-layered—with loose subiculum and pellicular even hymenium): *Athelia, Byssocorticium, Piloderma* pr. p., *Coniophora, Leptochaete* and some others.

f. Odontoid group (basidiocarps prostrate to semipileate, with toothed hymenophore).
- Grandinioid biomorph (basidiocarps prostrate, teeth small, as farinaceous tinge): *Grandinia, Kneiffiella, Resinicium, Trechispora* pr. p., *Steccherinum* pr. p., *Lyomyces* and some others.
- Sarcodontoid biomorph (basidiocarps prostrate or semipileate, teeth long, cylindrical): *Sarcodontia, Mycoacia, Dentipellis, Kavinia.*
- Irpicoid biomorph (basidiocarps semipileate to resupinate with teeth and disrupted pores): *Irpex, Steccherinum, Trichaptum* and some others.

4. Hypochnoid series (basidiocarps ageotropic, the growth mostly prostrate, mucedinous):

- Hypochnoid biomorph—*Amaurodon, Amauromyces, Botryobasidium, Botryohypochnus, Byssocorticium* pr. p., *Ceratobasidium, Coniophora olivacea, Epithele, Hypochnella, Pseudotomentella, Sistotrema* pr. p., *Subulicystidium, Suillosporium, Thanatephorus* s. l., *Tomentella* pr. p., *Tomentellago, Tomentellopsis, Tylospora* and some others.

5. Tremelloid series (basidiocarps ageotropic, strongly gelatinized, with more or less radial growth, prostrate, cushion-like, hemispheric and often lobate).
   a. Tremelloid group (basidiocarps strongly lobate):
      - Tremelloid biomorph—*Tremella, Tremiscus* and some others.
   b. Exidioid group (basidiocarps wrinkled, but not strongly lobate—prostrate, cushion-like, turbinate, ear-shaped):
      - Exidioid biomorph—*Exidia, Auricularia* pr. p., *Craterocolla, Tremella encephala* and some others.
   c. Platygloeoid group (basidiocarps prostrate, even, often incrusting the substrate):
      - Platygloeoid biomorph—*Platygloea, Galzinia* and some others.
   d. Dacrymycetoid group (basidiocarps somewhat reduced—hemispheric or discoid with non-differentiated excipulum):
      - Dacrymycetoid—*Dacryomyces, Femsjonia, *Linodochium* and some others.
6. Tyromycetoid series (basidiocarps parageotropic—sessile, pileate; annual, soaked; hymenophore one layered, tubular, rarely toothed).
   a. Pseudohydnoid group (basidiocarps strongly gelatinized, with toothed hymenophore):
      - Pseudohydnoid biomorph—*Pseudohydnum* and some others....
   b. Tyromycetoid group (basidiocarps not gelatinized).
      - Climacodontoid biomorph (basidicarps tongue-shaped, in clusters, hymenophore spinose): *Climacodon.*
      - Fistulinoid biomoph (basidiocarps tongue-shaped, singular, hymenophore polycyphelloid): *Fistulina.*
      - Phaeoloid biomorph (basidiocarps large, fan-shaped to plate-like, with pseudostipe, clustered): *Laetiporus, Phaeolus, Bondarzewia.*
      - Tyromycetoid biomorph (basidiocarps small or medium-sized, semicircular, spathulate or kidney-shaped, singular or in small clusters; laterally attached or with prostrate base): *Tyromyces, Postia, Leptoporus, Hapalopilus, Bjerkandera, Piptoporus* and some others.
   c. Trametoid group (basidiocarps persisting, sessile, of tough consistency, with tubular or derivative hymenophore).
      - Trametoid biomorph (hymenophore poroid, context thicker than hymenophoral layer): *Trametes, Antrodia, Ischnoderma* and some others.

- Corioloid biomorph (hymenophore poroid, context more or less equal in thickness to hymenophoral layer): *Antrodiella, Pycnoporus, Diplomitoporus* and some others.
- Scenidioid biomorph (hymenophore poroid, cellar, or irpicoid, context thinner than hymenophore layer): *Datronia, Earliella, Hexagonia, Trichaptum* and some others.
- Daedaleoid biomorph (hymenophore labyrhintine to lamellate): *Daedalea, Daedaleopsis, Gloeophyllum, Lenzites, Cerrena* pr. p.

d. Fomitoid group (basidiocarps perennial, sessile, of hard consistency, with multilayered poroid hymenophore):
- Fomitoid biomorph—*Fomes, Fomitopsis, Ganoderma* pr. p., *Phellinus* pr. p., *Oxyporus* pr. p., *Rigidoporus* pr. p.

7. Agaricoid series (basidiocarps negatively geotropic, pileate, stipe differentiated from pileus; hymenophore mostly lamellate).

a. Pluteoid group (pileus expanding, lamellae free):
- Pluteoid biomorph—*Pluteus* and some others.

b. Mycenoid group (pileus conical, stipe thin and elongated; lamellae of various attachments; ring and other velum derivatives none):
- Mycenoid biomorph—*Mycena, Conocybe* pr. p., *Bolbitius* pr. p., *Entoloma* pr. p.

c. Armillarioid group (pileus expanding, lamellae slightly decurrent, with ring and other velum derivatives):
- Armillarioid biomorph—*Armillaria, Tubaria, Kuehneromyces* and some others.

d. Coprinoid group (pileus expanding from semiclosed to conical; lamellae autolytic, narrowly attached; ring persistent in many representatives):
- Coprinoid biomorph—*Coprinus, Coprinopsis* and some others.

e. Pleurotoid group (stipe excentic to short; pileus sublateral, expanding; lamellae decurrent; ring and derivatives absent in most of representatives).
- Lentinoid biomorph (stipe prominent; basidiocarps tough, surface hispid or squamulose, lamellae sinuose or not, ring absent in most of representatives): *Lentinus*.
- Panelloid biomorph (stipe prominent, basidiocarps fleshy, surface matt, lamellae sinuose or not, without ring): *Panellus, Lentinellus* pr. p., *Panus* pr. p.
- Pleurotoid biomorph (stipe prominent or not, basidiocarp fleshy, surface naked—cuticulate, sometimes gelatinized, without ring): *Sarcomyxa, Hypsizygus, Ossicaulis, Rhodotus, Pleurotus* pr. p., *Hohenbuehelia* pr. p.

- Crepidotoid biomorph (stipe short, basidiocarp fleshy, surface matt to cuticulate, without ring): *Crepidotus, Resupinatus, Lentinellus* pr. p.
- Schizophylloid biomorph (without stipe, lamellae repeatedly splitting; basidiocarps coriaceous, surface hispid to matt, no ring): *Schizophyllum*.

8. Cyphelloid series (basidiocarps positively-, negatively, or parageotropic—as cupulate or tubular bodies—singular or united by a common subiculum).
   a. Cyphelloid group (subiculum absent—basidiocarps solitary or clustered).
      - Cyphelloid biomorph (basidiocarp cupulate, with prominent stipe).
        - Strictly cyphelloid (abhymenial surface naked): *Cyphella, Chromocyphella, Cellypha* and some others.
        - Lachnelloid (abhymenial surface villose): *Lachnella, Merismodes* and some others.
      - Calatelloid biomorph (basidiocarp cupulate with short stipe): *Calathella, Campanella, Woldmaria* and some others.
      - Solenioid biomorph (basidiocarps tubular with prominent or short stipe): *Phaeosolenia, Henningsomyces* and some others.
   b. Porotheleoid group (cyphelloid basidiocarps are united by a common subiculum).
      - Porotheleoid biomorph (basidiocarps initially hemispheric, then tubular): *Porotheleum*.
      - Stigmatolemmoid biomorph (basidiocarps cupulate on minute stipes): *Stigmatolemma* and some others.

C. Chorionomical Notes

Many wood-inhabiting species have a circumglobal distribution (approx. 5,000 species). Other species (approx. 15,000 species) are connected to zonal biomes (boreal or nemoral forests, arid zones and rainy tropical forests). For detailed chorionomical reconstructions, mycogeography generally compels the student to integrate vascular plants into more complex associations, for example, the host-associate connections (Pirozynski, 1983; Rajchenberg, 1989; Zmitrovich et al., 2003). Other dimensions of mycotas specificity are connected to isolation processes in the Southern Hemisphere, where specific segments contain roughly 5,000 species of wood-inhabiting fungi.

On the whole, the following gross units may be distinguished in the wood-inhabiting mycota.

Cosmopolitan species. These species are distributed worldwide in warm climates. In order to recognize such species, at least three "control points" are needed: circumboreal, pantropical, and New Zealand. Within wood-inhabiting fungi, at least 5,000 species both from Ascomycota and Basidiomycota are listed. The New Zealand control point is well represented in the check-list of Buchanan and Ryvarden (2000) which is highly informative because the list includes all groups of wood-inhabiting fungi together with their biogeographical descriptions.

Examples of cosmopolitan species:

Bjerkandera adusta (Willd.) P. Karst.
Schizopora paradoxa (Schrad.) Donk
Trametes versicolor (L.) Lloyd
Gloeophyllum trabeum (Pers.) Murrill
Byssomerulius corium (Pers.) Parmasto
Annulohypoxylon multiforme (Fr.) Y.M. Ju et al.
Nectria cinnabarina (Tode) Fr.

Holarctic species. It is assumed that the comprising Ascomycota and Basidiomycota is the richest in species diversity. These fungi, distributed over large parts of Eurasian and North American land massifs and many other areas, are associated with humid climates, and are adapted to active wood decay. The exact number of species is currently controversial and is a challenge for future research (Ginns, 1998). As of now we may refer to basic modern "large-scale Mycotas", only:

Bondartsev (1971)—polypores: East Europe and Caucasia;
Bondartseva (1998)—polypores: East Europe, Urals, Siberia, Far East Russia;
Dennis (1978)—Ascomycota: West Europe;
Jülich and Stalpers (1980)—corticioid fungi: Europe, North America;
Gilbertson and Ryvarden (1986, 1987)—polypores: North America;
Ryvarden and Gilbertson (1993, 1994)—polypores: Europe;
Teng (1996)—all groups: non-tropical Central and East Asia;
Nordic macromycetes. Vol. 3 (1997)—former Aphyllophorales: North Europe;
Nordic macromycetes. Vol. 1 (2000)—discomycetous and stromatic Ascomycota (including wood-inhabiting fungi): North Europe;
Funga Nordica (2008)—agaricoid fungi (including wood-inhabiting fungi): North Europe;
Ghobad-Nejhad (2011)—former Aphyllophorales: Caucasia.

Examples of Holarctic species:
Phellinus lundellii Niemelä
Ph. nigricans (Fr.) P. Karst.

Lentinus suavissimus Fr.
Trametes suaveolens (L.) Fr.
Phlebia centrifuga P. Karst.
Tylospora fibrillosa (Burt) Donk
Piloderma croceum J. Erikss. & Hjortstam

Palearctic species. These fungi are known only from the Eurasian segment of the Holarctic. The history of migration of species comprising this union is explained by Vasilyeva and Stephenson (2010).

Examples of Palearctic species:
Heterobasidion abietinum Niemelä & Korhonen
Pachykytospora wasseri Zmitr., V. Malysheva & Spirin
Ganoderma carnosum Pat.
Lentinus martianoffianus Kalchbr.
Loweomyces sibiricus (Penzina & Ryvarden) Spirin
Peniophora laeta (Fr.) Donk
Biscogniauxia maritima L.N. Vasilyeva

American radiating species. Species are distributed from temporal areas of North America to American arids and tropics. Certain biogeographical notes on these species are given by Gilbertson and Ryvarden (1986) and Vasilyeva and Stephenson (2010).

Examples of American radiating species:
Coriolopsis byrsina (Mont.) Ryvarden
C. hostmannii (Berk.) Ryvarden
Hexagonia variegata Berk.
Trametes ectypa (Berk. & M.A. Curtis) Gilb. & Ryvarden
T. pavonia (Hook.) Ryvarden
T. supermodesta Ryvarden & Iturr.
Pogonomyces hydnoides (Sw.) Murrill

Bi-polar species. These species have both boreo-nemoral and sub-antarctic circumpolar distribution but are lacking in the tropics and arid areas of the Northern Hemisphere. Biogeographical notes on some species are given by Rajchenberg (1989).

Examples of bi-polar species:
Phellinus inermis (Ellis & Everh.) G. Cunn.
Polyporus melanopus Fr.
Rigidoporus undatus (Fr.) Donk
Antrodia stratosa (J.E. Wright & J.R. Deschamps) Rajch.
Fibroporia gossypia (Speg.) Parmasto
Fibroporia vaillantii (DC.) Parmasto
Fistulina hepatica (Schaeff.) With.

Pantropical species. These type of species are present in American, African and Asian tropics (radiating to subtropics). Biogeographically, these could be "omnivorous" Gondwanian derivatives. Some basic tropical wood Mycotas are treated by:

Fidalgo and Fidalgo (1966)—polypores: tropical Central America;
Fidalgo and Fidalgo (1968)—polypores: tropical South America;
David and Rajchenberg (1985)—polypores: tropical South America;
Roy and De (1996)—polypores: tropical South Asia;
Parmasto (1986)—former Aphyllophorales; *Lentinus*: tropical East Asia;
Imazeki et al. (1988)—all groups: East Asia (Japan);
Wu (1990)—corticioids: East Asia (Taiwan);
Núñez and Ryvarden (2001)—polypores: tropical East Asia;

Examples of Pantropical species:
Pycnoporus sanguineus (L.) Murrill
Earliella scabrosa (Pers.) Gilb. & Ryvarden
Leiotrametes menziesii (Berk.) Welti & Courtec
Trametes tephroleuca Berk.
Microporus vernicipes (Berk.) Kuntze
Phellinus merrillii (Murrill) Ryvarden
Cyclomyces tabacinus (Mont.) Pat.

Paleoaustral species. Such species are distributed basically in the Southern Hemisphere, but also occur in gravitating tropical areas of the Northern Hemisphere. Mycogeographical aspects of this union are discussed by Rajchenberg (1989). Some basic paleoaustral wood Mycotas are mentioned by:

Ryvarden and Johansen (1980)—polypores: East Africa;
Härkonen et al. (2003)—polypores, corticioids, some other groups: East Africa;
Hood (2003)—wood-inhabiting fungi: Australia.
Paleoaustral species are subdivided here as follows:

a) Gondwanic (e.g., Southern South America, Australia, New Zealand, North of India or East of Africa):

Examples of Gondwanic species:
Microporus xanthopus (Fr.) Kuntze
M. affinis (Blume & T. Nees) Kuntze
Hexagonia niam-niamensis P. Henn.
Macrohyporia dictyopora (Cooke) I. Johans. & Ryvarden
Postia dissecta (Lév.) Rajchenb.
Postia pelliculosa (Berk.) Rajchenb.
Biscogniauxia philippinensis (Ricker) Whalley et Læssøe

b) Subantarctic (Southern South America, Australia and vicinities):

   Examples of Subanctarctic species:
   Ceriporiopsis merulinus (Berk.) Rajchenb.
   Grifola sordulenta (Mont.) Singer
   Laetiporus portentosus (Berk.) Rajchenb.
   Polyporus maculatissimus Lloyd
   Ryvardenia campyla (Berk.) Rajchenb.
   R. cretacea (Lloyd) Rajchenb.

c) South East Asian:

   Examples of South East Asian species:
   Lentinus polychrous Lév.
   Phellinus fastuosus (Lév.) Ryvarden
   Coriolopsis aspera (Jungh.) Teng
   C. telfairii (Klotzsch) Ryvarden
   T. conchifera (Schwein.) Pilát
   T. orientalis (Yasuda) Imazeki
   T. pocas (Berk.) Ryvarden

Nothofagus-area species. These types of fungi are found in warm-temperate areas of South America and in some Pacific islands, and are more or less associated with the distribution area of *Nothofagus*. This area is climatically isolated. Mycogeographical aspects of this union are discussed by Rajchenberg (1989). Such wood Mycotas are treated by:
   Wright and Deschamps (1972, 1975)
   Rajchenberg (1989)

   Examples of Nothofagus-area species:
   Bondarzewia guaitecasensis (P. Henn.) J.E. Wright
   Phellinus crustosus (Speg.) A.M. Gottlieb, J.E. Wright & Moncalvo
   Phellinus andinus Plank & Ryvarden
   Polyporus gayanus Lév.
   Skeletocutis australis Rajchenb.
   Cyttaria darwinii Berk.
   Biscogniauxia nothofagi Whalley, Læssøe et Kile.

Ecology

Substrate Groupings

According to Kirk et al. (2001), a substrate is "a material on which an organism is growing or to which it is attached". Therefore, we can distinguish nutritive substratum and attachment substratum (Yurchenko, 2006). In most of the cases both substrate types represent a union, but in some cases, the sporulation and nutrient consumtion is spatially incongruent.

Table 2.4 shows some key substrata, colonized by the most important wood-decay fungi. It is necessary to note that many wood-destroying fungi are capable of forming a secondary colonization of non-lignified substrata, such as mosses protonemata or algal cells. These phenomena were exhaustively presented by Yurchenko (2001, 2006). As a rule, greater amount of fungal mycelia colonize core lignin-containing substrate, but surface-associated hyphae form appressoria in the zone of interactions with green epiphytic cells. In some *Athelia*-species these green-cells—mycelium associations are rather stable (Zmitrovich, 2008), so are also in some representatives of the so-called *Rickenella*-family (Larsson et al., 2006).

The other circle of secondary phenomena is connected to the colonization of herbaceous plants by wood-rot fungi. These plants contain H-lignin (hydrophenilous lignin) (Manskaya and Kodina, 1975; Zmitrovich, 2010) and are probably derived from woody predecessors (Church, 1919; Chadefaud, 1950; Takhtajan, 1950). The most prominent lignifying component of such plants is in xerophylizied forms, where parenchymatous living tissues are reduced. The fungi colonizing such a substrate are presumably omnivorous and xerotolerant.

Obviously, the capacity of a wide-range of biopolymer decomposition is present in many taxa, whereas a real substrate specialization has ecological control and is correlated to insolation niche occupied by fungi and their biomorphic status.

Microsuccessions

In nature, groupings of wood-inhabiting fungi colonizing dried and fallen wood are temporally localized. The wood decay in forest ecosystems passes through several stages (Renvall, 1995; Kotiranta and Niemelä, 1993; Lindgren, 2001; Spirin, 2002). As shown in Table 2.5, in boreal forests, where soil water evaporation is non-intensive, the process of wood humification ranges from 15 to 20 years (in arid climates and rainy tropical forests with intensive evaporation, the decay rate is low).

In incipient stages of wood degradation, there is no contact with the ground; therefore, decay develops rather slowly. As a result, in such stages, pathogenic saprotrophs with some other xerotolerant saprotrophs are predominant.

After roots and gross branches are destroyed, the logs become gradually immersed into the ground, where contact with capillary connected water and soil mycelium is present. This contact represents a key event in the destruction process of wood. In spruce forests, the pioneer groupings of wood fungi are substituted by decortical fungi and strong saprotrophic

Table 2.4. An overview of key species of wood-destroying fungi on most widespread substrates of basic world biomes

Key substrata	Key fungal pathogens/decomposers	Literature
Trees and shrubs		
Betula sect. *Albae*	*Inonotus obliquus* (Ach. ex Pers.) Pilát, *Fomes fomentarius*, *Piptoporus betulinus* (Bull.) P. Karst., *Phellinus laevigatus* (P. Karst.) Bourdot & Galzin, Ph. *nigricans* (Fr.) P. Karst. f *betulae* comb. ined., *Ph. lundellii* Niemelä; *Trichaptum biforme* (Fr.) Ryvarden, *Antrodiella faginea* Vampola & Pouzar, *Gloeoporus dichrous* (Fr.) Bres.	Arefiev (2010)
Populus tremula	*Phellinus tremulae*, *Inonotus rheades* (Pers.) Bondartsev & Singer, *Peniophora rufa* (Fr.) Boidin, *Punctularia strigosozonata* (Schwein.) P.H.B. Talbot	Ershov and Ezhov (2009)
Alnus incana	*Biscogniauxia nummularia* (Bull.) Kuntze, *Vuilleminia alni* Boidin, Lanq. et Gilles, *Phellinus nigricans* var. alni Zmitr. et V. Malysheva, *Ph. conchatus* (Pers.) Quél., *Ph. punctatus* (Fr.) Pilát, *Stereum rugosum* Pers., *Bisporella citrina* (Batsch) Korf et S.E. Carp., *Hyphodontia crustosa* (Pers.) J. Erikss.	Strid (1975); Zmitrovich (2012)
Salix spp.	*Trametes suaveolens* (Fr.) Fr., *Haploporus odorus* (Sommerf.) Bondartsev & Singer, *Phellinus conchatus*, *Ph. punctatus*, *Lentinus suavissimus* Fr.	Andersson et al. (2009)
Picea abies/obovata	*Phellinus chrysoloma* (Fr.) Donk, *Fomitopsis rosea* (Alb. & Schwein.) P. Karst., *Pycnoporellus fulgens* (Fr.) Donk	Andersson et al. (2009)
Pinus sylvestris	*Phellinus pini* (Brot) Bondartsev & Singer, *Fomitopsis pinicola* (Sw.) P. Karst., *Phlebiopsis gigantea* (Fr.) Jülich	Sinadsky (1983)
Larix sibirica	*Phaeolus schweinitzii* (Fr.) Pat., *Phellinus niemelaei* (M. Fisch.) Zmitr., Malysheva & Spirin	Ezhov et al. (2011); Spirin et al. (2006)
Juniperus communis	*Amylostereum laevigatum* (Fr.) Boidin	Davydkina (1980)
Quercus robur	*Laetiporus sulphureus* (Bull) Murrill, *Phellinus robustus* (P. Karst.) Bourdot & Galzin, *Inonotus dryophilus* (Berk.) Murrill, *I. nidus-pici* Pilát, *Xylobolus frustulatus* (Fr.) Boidin	Chamuris (1988); Larsen and Cobb-Poulle (1990); Spirin (2002); Ghobad-Nejhad and Kotiranta (2008)

Table 2.4. contd....

Table 2.4. contd.

Key substrata	Key fungal pathogens/decomposers	Literature
Tilia cordata	*Polyporus squamosus* (Huds.) Fr., *Spongipellis spumea* (Sowerby) Pat., *Neolentinus schaefferi* (Weinm.) Redhead & Ginns	Malysheva and Malysheva (2008)
Juglans regia	*Inonotus plorans* (Pat.) Bondartsev & Singer	Ghobad-Nejhad and Kotiranta (2008)
Ceratonia siliqua	*Ganoderma australe* (Fr.) Pat., *Phellinus torulosus* (Pers.) Bourdot & Galzin	Ţura et al. (2011)
Robinia pseudoacacia	*Phellinus robiniae* (Murrill) A. Ames	Larsen and Cobb-Poulle (1990)
Quercus virginiana	*Phellinus coffeatoporus* Kotl. & Pouzar, *Ph. grenadensis* (Murrill) Ryvarden	Larsen and Cobb-Poulle (1990)
Eucalyptus camadulensis	*Laetiporus gilbertsonii* Burds., *Ganoderma australe* (Fr.) Pat.	Burdsall and Banik (2001); Ţura et al. (2011)
Tamarix aphylla	*Inonotus tamaricis* (Pat.) Maire, *Phellinus torulosus*, *Ganoderma australe*, *Peniophora tamaricicola* Boidin & Malençon	Ghobad-Nejhad and Kotiranta (2008); Ţura et al. (2011).
Cupressus sempervirens	*Phellinus torulosus*	Ţura et al. (2011)
Casuarina cunninghamiana	*Inonotus ochroporus* (Van der Bijl) Pegler, *I. patouillardii* (Rick) Imazeki	Gottlieb et al. (2002)
Nothofagus dombei	*Phellinus crustosus* (Speg.) Gottlieb, Wright & Moncalvo, *Ph. andinopatagonicus* (J.E. Wright & J.R. Deschamps) Ryvarden	Larsen and Cobb-Poulle (1990); Gottlieb et al. (2002)
Brugiera gymnorhiza	*Trametes cingulata* Berk., T. flavida (Lév.) Zmitr., Wasser & Ezhov	Gilbert et al. (2008)
Rhizophora apiculata	*Trametes sanguinaria* (Klotzsch) Corner, *T. nivosa* (Berk.) Murrill	Gilbert et al. (2008)
Sonneratia alba	*Phellinus fastuosus* (Lév.) Ryvarden, *Inonotus luteoumbrinus* (Romell) Ryvarden, *Trametes cingulata*	Gilbert et al. (2008)
Avicennia germinans	*Phellinus swieteniae* (Murrill) S. Herrera & Bondartseva, *Trichaptum biforme* (Fr.) Ryvarden	Gilbert and Sousa (2002)
Small schrubs and semi-arboreous plants		

Calluna vulgaris	*Acanthobasidium norvegicum* (J. Erikss. & Ryvarden) Boidin, *Acanthophysium apricans* (Bourdot) G. Cunn., *Corticium macrosporopsis* Jülicj, *Hyphodontia hastata* (Litsch.) J. Erikss, *Phanerochaete ericina* (Bourdot) J. Erikss. & Ryvarden, *Ph. martelliana* (Bres.) J. Erikss. & Ryvarden, *Sistotrema dennisii* Malençon	Domański (1988, 1991, 1992); Yurchenko (2006)
Rubus idaeus	*Ceratobasidium cornigerum* (Bourdot) D.P. Rogers, *Peniophora cinerea* (Pers.) Cooke, *P. incarnata* (Pers.) P. Karst., *Acanthobasidium norvegicum*, *Corticium macrosporopsis*, *Phanerochaete tuberculata* (P. Karst.) Parmasto	Domański (1988, 1991, 1992); Yurchenko (2006)
Actinidia spp.	*Peniophora sphaerocystidiata* Burds. & Nakasone	Yurchenko (2006)
Lignified herbaceous and succulent plants		
Chamaerion angustifolium	*Ceratobasidium pseudocornigerum* M.P. Christ., *Peniophora cinerea*, *Sistotrema octosporum* (J. Schröt.) Hallenb	Yurchenko (2006)
Humulus lupulus	*Aleurodiscus cerussatus* (Bres.) Höhn. & Litsch.	Yurchenko (2006)
Juncus sp.	*Tomentella juncicola* Svrček	Domański (1992)
Carnegiea spp.	*Hyphoderma fouquieriae* Nakasone & Gilb., *Peniophora tamaricicola*, *Phanerochaete omnivorum* (Schear) Burds. & Nakasone	Nakasone and Gilbertson (1978)
Opuntia spp.	*Crustoderma opuntiae* Nakasone & Gilb., *Uncobasidium calongei* (Tellería) Hjortstam & Tellería	Nakasone and Gilbertson (1978); Yurchenko (2006)
Bryophyta	*Ceratobasidium bicorne* J. Erikss. & Ryvarden, *Sistotrema muscicola* (Pers.) S. Lundell, *Lindtneria leucobryophila* (Henn.) Jülich, *Athelia epiphylla* Pers., *Tomentella sublilacina* (Ellis & Holw.) Wakef., *Amphimena byssoides* (Pers.) J. Erikss., *Tubulicrinis subulatus* (Bourdot & Galzin) Donk	Eriksson and Ryvarden (1973); Eriksson et al. (1984); Domański (1988, 1991); Yurchenko (2001, 2006)
Chlorophycophyta (epiphytic)	*Athelia epiphylla*, *A. phycophila* Jülich, *A. andina* Jülich, *Resinicium bicolor* (Alb. et Schwein.) Parmasto, *Hyphoderma* spp., *Sistotrema sernanderi* (Litsch.) Donk, *Sistotremastrum suecicum* Litsch. ex J. Erikss., *Sidera lenis* (P. Karst.) Miettinen, *Hyphodontia rimosissima* (Peck) Gilb.	Jülich (1972); Eriksson et al. (1981, 1984); Eriksson and Ryvarden (1975, 1976); Yurchenko and Golubkov (2003); Yurchenko (2006); Zmitrovich (2008); Miettinen and Larsson (2011)

Table 2.4. contd....

Table 2.4. contd.

Key substrata	Key fungal pathogens/decomposers	Literature
Lichenized Ascomycota	*Athelia arachnoidea* (Berk.) Jülich, *A. epiphylla*, *A. salicum* Pers., *Botryobasidium candicans* J. Erikss., *Peniophora cinerea*, *Sistotrema brinkmannii* (Bres.) J. Erikss.	Eriksson et al. (1978); Parmasto (1998); Yurchenko and Golubkov (2003); Zmitrovich (2008)
Chitinous substrata	*Hyphoderma setigerum* (Fr.) Donk (pyrenomycete stromata, insect exoskeleton), *Peniophora cinerea* (pyrenomycete stromata), *Antrodiella pallescens* (Pilát) Niemelä & Miettinen, *Sistotrema brinkmannii*, *Peniophora incarnata*, *Phanerochaete laevis* (Fr.) J. Erikss. & Ryvarden (polypore basidiomata)	Yurchenko and Zmitrovich (2001); Miettinen et al. (2006); Yurchenko (2006)
Humus soil horizon	*Piloderma croceum*, *Tylospora fibrillosa*, *Amphinema byssoides*, *Byssocorticium* spp., *Tomentella* spp., *Tomentellastrum* spp., *Tomentellopsis* spp., *Conohypha terricola* (Burt) Jülich, *Echinotrema clanculare* Park.-Rhodes, *Sistotrema hypogaeum* Warcup & P.H.B. Talbot, *Waitea circinata* Warcup & P.H.B. Talbot, *Dacryobasidium coprophilum* (Wakef.) Jülich	Jülich (1984); Yurchenko (2006); Zmitrovich (2008)
Antropogenic Composites	*Coniophora marmorata* Desm., *Serpula lacrymans* (Wulfen) J. Schröt., *Leucogyrophana olivascens* (Berk. & M.A. Curtis) Ginns & Weresub	Bondartsev (1956); Jülich (1984); Yurchenko (2006)

Table 2.5. Humification of spruce wood in boreal forersts with microsuccessions of wood-inhabiting fungi (according to Spirin, 2002).

Stage	Description
Fallen log (0–2 years)	Fresh wood material has intact branches and bark and their mechanical properties as in living trees. Predominate pathogenic saprotrophs: *Fomitopsis pinicola, Heterobasidion* spp.
Origin of decomposition (2–10 years)	The bark partially falls off and the skeletal branches break up. The wood located at the bottom side of the log changes its mechanical properties. Pathogenic saprotrophs continue the growth. The core saprotrophs species occur: *Phellinidium ferrugineofuscum, Fomitopsis rosea* (Alb. & Schwein.) P. Karst.
Intensive decomposition (10–15 years)	In this stage only the bark located above remains, whereas the log merges into the ground. The wood strongly changes its mechanical properties (becames friable and stratified); on contact with the ground the humification process starts. The pioneer species complexes are substituted as follows: *Fomitopsis rosea* → *Skeletocutis odora* (Sacc.) Ginns and *Phlebia centrifuga* P. Karst.; *Heterobasidion* spp. → *Junghuhnia collabens* (Fr.) Ryvarden, *Dichostereum boreale* (Pouzar) Ginns & M.N.L. Lefebvre; *Fomitopsis pinicola* → *Pycnoporellus fulgens* (Fr.) Donk, *Phellinidium sulphurascens* (Pilát) Y.C. Dai
Full decomposition (15–20 years)	The bark falls or is humified. The wood softens completrely and changes into a red-brown color. Saprotrophic fungi groups are represented by many ephemerous hygrophilic species as *Postia* spp., *Leptoporus mollis* (Pers.) Quél., *Physisporinus* spp., *Asterodon ferruginosus* Pat.

wood-colonizers: for example, *Heterobasidion parviporum* is changed by *Junghuhnia collabens, Perenniporia* spp., or *Dichostereum boreale.* A primary decayer such as *Fomitopsis pinicola* as a rule is changed by *Amylocystis lapponica* (Romell) Bondartsev & Singer, *Fomitopsis rosea, Pycnoporellus fulgens* then, the latter are changed with a rather large suite of tertiary successors such as *Postia* spp., *Skeletocutis* spp., *Crustoderma* spp., or *Phlebia centrifuga,* fungi responsible for starting the humification process.

Simultaneously, terminal branches and branchlets are decayed by certain specific wood fungi such as [*Peniophora pini* (Schleich.) Boidin—pine, *P. rufa* (Fr.) Boidin, *Punctularia strigosozonata* (Schwein.) P.H.B. Talbot—aspen, *Vuilleminia alni* Boidin, Lanq. & Gilles—alder] and non-specific [*Hyphodontia* spp., *Hyphoderma* spp., *Byssomerulius corium* (Pers.) Parmasto, *Cylindrobasidium evolvens* (Fr.) Jülich].

In the stage of humification, numerous ephemerous hygrophilic wood-inhabiting fungi (*Anomoporia* spp., *Ceriporiopsis* spp., *Ceriporia* spp., *Physisporinus* spp., *Trechispora* spp. and other corticioids) colonize the rest of the cavernose wood mass. There are many threads and rhizomorphs-forming wood fungi. The fallen log remnants become amorphous and protruded by roots of young trees and shrubs. Often these are overgrown by moss groupings and form certain "soil bolsters".

As a result of wood humification, the virgin boreal forests are characterized by fractured microrelief formed by the so-called "wood remnants-soil complexes" (Smirnova, 2004) and rather gross amorphous wood debris layer reaching 15 cm in thickness (Fig. 2.12). This layer is the main producer of humic and fulvic acids (Dighton et al., 2005), that migrate into the water bodies and accumulate colloids-linked Ca^{2+}, Mg^{2+} and K^{+}. Therefore, the boreal soils are acidic, and dominated by oligotrophic moss communities.

In a placore nemoral forests, water does not reach the ground surface during the warm period. In this type of forests, the rate of wood destruction and humification is lower, though humus accumulates at a high rate, whereas humic acids production falls. Such pioneers as *Ganoderma applanatum*, *Neolentinus schaefferi*, *Hypsizygus ulmarius* (Pers.) Singer, *Pleurotus dryinus* (Pers.) P. Kumm., *Inonotus dryophilus* (oak), *Aurantiporus fissilis* (Berk. & M.A. Curtis) H. Jahn, *Phellinus* spp., and *Spongipellis* spp. start the decaying process of stands and stumps. Fallen logs and dry stumps are colonized by such key species as *Bjerkandera adusta*, *Hapalopilus croceus* (Pers.) Donk (oak) or *Fomes fomentarius*. Transcortical species of *Junghuhnia* spp., *Oxyporus* spp., and *Hypochnicium* spp. are leader species in the process of wood decortications. Many *Peniopora* species decompose fallen terminal branches. After partial decortication and branch degradation the rich

Figure 2.12. Subvirgin boreal forest (Eu-Piceetum abietis) with abundance of overmossed spruce fallen logs on various stages of decaying. The fruit bodies of the key destructor *Fomitopsis pinicola* is seen on the spruce stump (Veps Plateau).

wood-inhabiting fungi communities are revealed by marker fungi species as *Crepidotus* spp., *Pluteus* spp., *Phlebia* spp. and many other corticioids (see Malysheva and Malysheva, 2008).

In the south of nemoral zone, the role of pathogenic saprotrophs (e.g., *Ganoderma* spp., *Phellinus* spp.) in the wood decay process is increasing. The fallen wood decomposes very slowly, and the key role within non-pathogenic decayers is held by the genus *Trametes*, represented in the southern areas by more than 50 species (Zmitrovich et al., 2012).

Syntaxonomical Aspects

The wood-inhabiting fungi constitute one of the basic functional blocks of forest ecosystems. Their main functions are to destroy and humify the wood debris and optimize the mineral exchange. In oligotrophic environments of the taiga forest communities containing wood fungi and ectomycorrhiza-formers (often these functions are performed by the same species) determine the development of forest vegetation (Zmitrovich, 2011).

The fungal communities represent a specific part of biogeocoenoses which do not have a description and a conventional classification yet. There were some attempts to classify the fungal communities in connection with phytosphere of coenosis-forming trees and bonds of edaphotop (Jahn, 1966; Darimont, 1973; Šmarda, 1972; Bon, 1981); however, most authors traced fungal species complexes into phytosociological classifications (Jahn, 1986; Bujakiewicz, 1992; Richard, 2000). Recently, a new approach has been proposed by Zmitrovich (2011). The approach is based on a strong dependence of mycosynusia from environment-making role of dominating trees. Our basic knowledge on communities of wood-inhabiting, litter, and mycorrhizal fungi in boreal and sub-nemoral European forests has been summarized (Table 2.6).

Concluding Remarks

The progress achieved during the past decades in research on wood-inhabiting fungi highlights both, their importance in nature and their high potential in biotechnology. In nature, wood fungi are key agents involved in the decomposition of wood, soil humus formation and nutrient recycling. In terms of evolution, various wood fungi species belong to different groupings. Some fungi are restricted to colonizing one type of substrate and follow the distribution range of the substrate that they prefer [e.g., *Inonotus tamaricis* (Pat.) Maire], while other fungal species evolved differently. For example, fungal species able to colonize wood at a fast rate, occurring on a broad range of substrata including both living and dead hardwoods and

Table 2.6. Syntaxonomical aspects of wood inhabiting fungi in boreal and subnemoral forests of Europe (according to Zmitrovich, 2011).

Fungal community syntaxonomy	Corresponding element of phytocoenotic mosaics	Key fungal species*	Synonymy (authors)
[1]. *Pilodermato crocei-Amylocystidietum* prov.	*Picea abies* (*Rubo chamaemori-Piceetum; Vaccinio myrtilli-Piceetum; Melico nutantis-Piceetum*)	*Piloderma croceum*, *Amanita pantherina*, *Tylospora fibrillosa*, *Russula claroflava*, *Cantharellus tubae formis*, *Phellodon niger*, *Clitocybe ditopa*, *Cortinarius* sp. div., *Tylopilus felleus*, *Phellinus chrysoloma*, *Amylocystis lapponica*, *Postia caesia*, *P. lateritia*, *Climacocyatis borealis*, *Onnia leporina*, *Fomitopsis rosea*, *Trichaptum abietinum*, *Phellinus nigrolimitatus*, *Phlebia centrifuga*, *Hyphodontia breviseta*	*Calocerettum viscosae* Ricek (1967); *Tyromycetum caesii* Ricek (ibid.); *Clitocybo-Phellodontetum nigrae* Šmarda (1973)
[2]. *Pilodermato crocei-Amyloporietum* prov.	*Pinus sylvestris* (*Vaccinio myrtilli-Pinetum, Vaccinio vitis-idaea-Pinetum*)	*Piloderma croceum*, *Amanita fulva*, *Rozites caperata*, *Leccinum vulpinum*, *Fomitopsis pinicola*, *Antrodiella citrinella*, *Amyloporia xantha*, *Antrodia sinuosa*, *A. serialis*, *Postia fragilis*, *Trichaptum fuscoviolaceum*, *Tapinella atrotomentosa*, *Serpulomyces borealis*, *Phlebiella vaga*	*Calocerettum viscosae* Ricek (1967)
[3]. *Pilodermato crocei-Piptoporetum* prov.	*Betula pubescens* (*Vaccinio myrtilli-Pinetum, Vaccinio vitis-idaea-Pinetum; Rubo chamaemori-Piceetum; Vaccinio myrtilli-Piceetum; Melico nutantis-Piceetum*)	*Piloderma croceum*, *Russula emetica*, *Cantharellus cibarius*, *Leccinum variicolor*, *Boletus edulis*, *Hydnum rufescens*, *Phellinus nigricans* f. *betulae*, *Piptoporus betulinus*, *Antrodiella pallescens*, *Mycena galericulata*, *M. haematopoda*, *Trichaptum biforme*, *Exidia repanda*, *Hyphodontia aspera*	*Mycenetum galericulatae* Ricek (1967)

[4]. *Pilodermato crocei-Crepidotetum calolepidis* prov.	*Populus tremula* (*Vaccinio myrtilli-Piceetum; Melico nutantis-Piceetum*)	*Piloderma croceum, Leccinum aurantiacum, L. albostipitatum, Phellinus tremulae, Junghuhnia pseudozilingiana, Inonotus rheades, Oxyporus corticola, Antrodia mellita, A. pulvinascens, Crepidotus calolepis, Ceriporiopsis aneirina, Radulodon erikssonii, Punctularia strigosozonata*	*Phellinetum tremulae* Jahn (1966); *Crepidotetum calolepidis* Jahn (ibid.)
[5]. *Thelephoro terrestridis-Gloeophylletum* prov.	*Pinus sylvestris* (*Ledo-Pinetum; Pino-Polytrichetum; Vaccinio uliginosi-Pinetum; Betulo-ledetum; Betulo pubescentis-Vaccinietum uliginosi*)	*Thelephora terrestris, Russula paludosa, Suillus variegatus, Cantharellus aurora, Hypholoma polytrichi, Gloeophyllum sepiarium, Trametes velutina, Pycnoporus cinnabarinus, Phyllotopsis nidulans, Chaetodermella luna*	–
[6]. *Galereto-Chaetodermetum* prov.	*Pinus sylvestris* (*Sphagno magellanici-Ledetum*)	*Galerina sphagnorum, G. paludosa, G. tibiicystis, Exobasidium andromedae, Chaetodermella luna*	–
[7]. *Galereto-Pycnoporetum* prov.	*Betula pubescens* (*Sphagno magellanici-Ledetum*)	*Russula paludosa, Galerina* sp. div., *Exobasidium* sp. div., *Pycnoporus cinnabarinus*	–
[8]. *Bankero-Amyloporietum* prov.	*Pinus sylvestris* (*Cladonio stellaris-Pinetum*)	*Bankera fuligineoalba, Sarcodon scabrosus, S. lundellii, Boletus pinophilus, Ramaria* sp. div., *Hydnellum ferrugineum , Fomitopsis pinicola, Amyloporia xantha, Phlebiopsis gigantea, Dacrymyces tortus, Exidia saccharina, Calocera viscosa*	*Caloceretum viscosae* Ricek (1967)
[9]. *Suillo bovinis-Amylostereetum laevigatidis* prov.	*Juniperus communis*—young *Pinus sylvestris* on fire places (*Cladonio stellaris-Pinetum*)	*Suillus bovinus, S. luteus, Chalciporus piperatus, Thelephora terrestris, Coltricia perennis, Fomitopsis pinicola, Dacrymyces tortus, D. chrysocomus, Amylostereum laevigatum, Resinicium furfuraceum*	–

Table 2.6. contd....

Table 2.6. contd.

Fungal community syntaxonomy	Corresponding element of phytocoenotic mosaics	Key fungal species*	Synonymy (authors)
[10]. *Leccino versipelle-Trametetum ochraceae* prov.	*Betula pubescens* (*Cladonio stellaris-Pinetum*)	*Leccinum versipelle*, *Suillus granulatus*, *Thelephora terrestris*, *Ramaria* sp. div., *Fomitopsis pinicola*, *Piptoporus betulinus*, *Postia tephroleuca*, *Trametes ochracea*, *Skeletocutis amorpha*, *Resinicium bicolor*	–
[11]. *Cortinario uliginosi-Cytidietum* prov.	*Salix* spp. *frutic.* (*Salicetum triandrae-viminalis*)	*Cortinarius uliginosus*, *Inocybe glabripes*, *I. lacera*, *Laccaria laccata*, *Leccinum scabrum*, *Phellinus punctatus*, *Physisporinus vitreus*, *Cytidia salicina*, *Peniophora violaceolivida*, *Tremella mesenterica*, *Exidia recisa*	*Tremelletum mesentericae* Darimont (1973)
[12]. *Cortinario uliginosi-Trametetum suaveolentis* prov.	*Salix fragilis* (*Salicetum triandrae-viminalis*)	*Cortinarius uliginosus*, *Inocybe* sp. div., *Laccaria laccata*, *Paxillus involutus*, *Phellinus igniarius* s.str., *Trametes suaveolens*	*Fometum igniarii* Pirk (1952)
[13]. *Lactario lilacini-Phellinetum* prov.	*Alnus glutinosa*, *A. incana*, *Salix caprea*, *Betula pubescens* (*Salici pentandrae-Betuletum pubescentis*; *Urtico dioicae-Alnetum glutinosae*)	*Lactarius lilacinus*, *L. mitissimus*, *Laccaria laccata*, *Paxillus involutus*, *Pluteus cervinus*, *Mycena galericulata*, *M. haematopoda*, *Phellinus punctatus*, *Ph. nigricans* var. *alni*, *Ph. laevigatus*, *Ph. lundellii*, *Peniophora erikssonii*	*Mycetentum galericulatae* Ricek (1967)
[14]. *Naucorio-Phellinetum* prov.	*Alnus incana*, *Sorbus aucuparia*, *Salix caprea*, *Padus avium*, *Sambucus racemosa* (*Alnetum incanae*; *Alno incanae-Padetum avii*)	*Naucoria alnetorum*, *Hebeloma crustuliniforme*, *Paxillus filamentosus*, *Xerocomus chrysenteron*, *Mycena haematopoda*, *Pleurotus dryinus*, *Phellinus punctatus*, *Ph. nigricans* var. *alni*, *Phellinus lundellii*, *Inonotus radiatus*, *Bisporella pallescens*, *Byssomerulius corium*, *Chondrostereum purpureum*, *Cylindrobasidium evolvens*, *Schizopora paradoxa*, *Hyphodontia crustosa*, *H. sambuci*	*Bisporetum antennatae* Jahn (1968)

[15]. *Thelephoro caryophillei-Meruliopsidetum* prov.	*Pinus sylvestris* (*Thymo serpylli-Pinetum*)	*Thelephora caryophillea* , *Suillus bovinus*, *Strobilurus tenacellus*, *Auriscalpium vulgare*, *Meruliopsis taxicola* , *Peniophora pini*	–
[16]. *Thelephoretum terrestridis* prov.	Gaps parcells (*Thymo serpylli-Pinetum*)	*Thelephora terrestris* var. *infundibuliformis*, *Th. caryophillea*, Amphinema byssoides	–
[17]. *Pilodermato byssini-Pluteetum cervinicis* prov.	*Picea abies* (*Maianthemo bifoliae-Piceetum abietis*)	*Piloderma byssinum*, *Boletus piceinus*, *Leccinum vulpinum*, *Lactarius* sp. div., *Cortinarius* sp. div., *Marasmius androsaceus*, *Phellinus chrysoloma*, *Gloeophyllum odoratum*, *Postia caesia*, *Phellinus ferrugineofuscus*, *Ph. viticola*, *Junghuhnia collabens*, *Pseudohydnum gelatinosum*, *Pluteus cervinus*	*Osmoporetum odorati* Ricek (1967); *Tyromycetum caesii* (ibid.)
[18]. *Pilodermato byssini-Phellinetum populicolis* prov.	*Populus tremula* (*Maianthemo bifoliae-Piceetum abietis*)	*Piloderma byssinum*, *Leccinum aurantiacum*, *Paxillus involutus*, *Lactarius controversus*, *Phellinus nigricans* var. *populicola*, *Ganoderma applanatum*, *Polyporus badius*, *Oxyporus obducens*, *Ceriporiopsis aneirina*, *C. resinascens*, *Postia alni*, *Pluteus cervinus*, *Peniophora nuda*, *Lentaria mucida*, *Clavicorona pyxidata*, *Xylaria hypoxylon*	*Xylarietum hypoxylonis* Ricek (1967)
[19]. *Laccario-Pluteetum umbrosatis* prov.	*Alnus* spp., *Acer platanoides*, *Ulmus glabra*, *Tilia cordata* (*Alno incanae-Fraxinetum*)	*Laccaria laccata*, *Lactarius obscuratus*, *Lepiota cristata*, *Marasmius epiphyllus*, *Collybia fusipes*, *Oxyporus populinus*, *Phellinus nigricans* var. *alni*, *Ganoderma adspersum*, *Polyporus squamosus*, *Lentinus schaefferi*, *Pluteus umbrosus*, *Chondrostereum purpureum*, *Cylindrobasidium evolvens*	–

* Underlined species resemble wood-inhabiting fungi.

conifers and being able to spread over throughout boreal, temperate as well as most tropical regions of the world are represented by fungi well adapted to environmental fluctuations. This group of fungi evolved by developing various survival strategies and is able to cause serious damage to forest and urban ecosystems (e.g., some of the most feared wood destroying fungi able to kill living trees, decompose their wood structure and remain in soil living on root fragments for several decades until new seedlings are planted: *Armillaria mellea*, *A. tabescens*, *Heterobasidion annosum*, *Ganoderma* spp., etc.), some other wood fungi spread over wood surfaces poor in moisture content [dry-rot fungi: e.g., *Serpula lacrymans*, *S. himantioides* (Fr.) P. Karst., *Coniophora puteana*, *Fibroporia vaillantii*, etc.] causing serious damage to material used in construction or wood made historical artifacts. In terms of disease control the most reliable strategy remains prevention by implementing detailed analyses and careful choice when planting new tree species or adopting "healthy" pruning habits, though when infection is observed correct fungus identification by observing both mycelia of fruitbody characters and tree disease symptoms are necessary followed by disease spreading control strategies. Making people aware of the ecological and economical impact correlated to such dangerous wood fungi species probably would ensure a higher degree of prevention, tree disease control and focus on adopting stronger regulations for dispersal of alien aggressive pathogenic species throughout borders. Another aspect refers to directing research towards finding practical solutions including eco-friendly biological control and not only by adopting the "chemical" approach as already experimented.

Regarding timber fungi, close attention is needed when selecting wood type for construction purposes. For example, *Cinnamomum osmophloeum* is one of the hardwood species known to show significant antifungal activity. However, on the first hand knowledge is a key factor in the selection of proper wood material. By taking into account wood preference of indoor dry-rot causing fungi or causes leading to their presence in an indoor environment combined with a careful selection of wood type used for construction seems to be a good prevention strategy.

Applied biotechnology uses wood-decaying fungi in many processes, mainly involving pharmacy, industry, environmental protection and cultivation. Therapeutically valuable by-products are increasingly demanded across the world, while enzymes resulted from wood-decaying fungi are widely required for diverse industrial applications. Medicinal wood-decaying fungi play a significant role in human health, demonstrating an increased scientific and public interest materialized through several thousands of publications worldwide; however, it is still unclear what type of extract is more potent to cure various diseases. This point is also supported by Wasser (2010) listing some other important unsolved issues of medicinal science including: the role of polysaccharide-protein or polysaccharide-peptide complexes in pharmacological activity of medicinal mushrooms; the development of new methods and processes in the study of medicinal mushrooms; high quality, long-term, double blinded, placebo-controlled studies with large trial populations; more attention must be paid to research on farm animals and medicinal mushrooms; and protection of intellectual properties of medicinal mushrooms' genetic resources for invention and innovation.

Some of the most important wood-decaying fungal species and their potential in biotechnology as a current research interest are oriented towards mycoremediation including bioconversion of agricultural wastes into eco-friendly valuable products, and the use of wood-inhabiting fungi in bioremediation of organo-pollutants, industrial contaminants or polyethylene degradation. Because such applications are strongly correlated to enzymatic activity of wood fungi, special attention is paid on how to increase enzyme production in various wood-inhabiting fungal species. We need to understand that the concept "everything is strain related" is strongly connected not only to quality and quantity of enzyme production but it is a powerful feature that is generally used in biotechnological companies and mushroom growing farms. Studies on nutrient requirements of some substrates for enhanced growth of some fungal species are still needed —such studies have value for industrial applications.

We will here briefly underline some important issues in cultivation of fungi.

1. Bioconversion of agricultural wastes should be a concept highly appreciated especially in poor and developing countries; however, there is a lack of knowledge on how this may be done. Therefore, we need to bring this knowledge and make it available to public, so that they can learn and change their habits of burning agricultural wastes or disposing unnecessary paper that represent a valuable source of protein and with a possible powerful local economic impact.
2. More attention must be paid to developing methods and encouraging fungi cultivation as an easy procedure without misleading the general public that this can be done only in highly equipped facilities. This is a significant step especially necessary in poor and developing countries.
3. A serious concern for public health is represented by the fact that the public is still unaware of the heavy metals accumulation potential and other undesirable pollutants in some fungal species. According to this, farmers should avoid usage of chemical substances in the cultivation process. Some books published 20–30 years ago still give instructions on how to cultivate mushrooms by using formaldehyde (currently known as a carcinogenic substance) in order to avoid contamination

during fungi cultivation and obtain higher yields and of course profits. Infact professional cultivators are encouraging the use of chemicals in mushroom cultivation. Therefore, making knowledge available for the public interested in mushroom cultivation is necessary.

4. Most publications on fungi cultivation are focused on commonly known cultivated species and little attention is paid to the harder to cultivate fungi (e.g., *Morchella* sp., or *Grifola frondosa*). New experiments should be employed in order to find easier ways to cultivate and obtain high biological efficiencies in the cultivation of such fungi.
5. Some cost effective mushroom growing methods are still missing for some commonly cultivated fungi: in the US widely vehiculated are the hydrated lime treatment method of substrate used for growing oyster mushrooms versus pasteurization of substrate. Which one is best over the other and for what type of fungi the hydrated lime method works?
6. Less information is available related to the phenomenon known as "fungal strain senescence". Some mushroom growers believe in this concept while others are rather passive when hearing about it. What are the factors involved producing strain senescence and how they can be observed at chemical and molecular level for most cultivated fungi and what strategies should be implemented to overcame this phenomenon? These are questions that interests mushroom growers and are still without a clear answer.
7. Not enough information is present on fungal strain preferences for cultivation; therefore mushroom growers are often puzzled when purchasing fungal strains from various fungal culture banks. Scientific studies on some noteworthy strains would help many growers in choosing what's best for them.

This chapter embodies the biodiversity aspects of wood-inhabiting fungi and is dedicated to the memory of Prof. E. Parmasto (1928–2012), a great mycologist.

References

Aisenstadt, M.A. and Bogolytzin, K.G. 2009. A peroxidase-dependent oxidation of lignin and its model substances. Chem. Pl. Mater., 2: 5–18 (in Russian).

Andersson, L., Alexeeva, N. and Kuznetsova, E. 2009. Revealing and investigation of biologically important forests in North-West Russia. T. 2. Guide to species identification. St Petersburg, Pobeda., pp. 139–217.

Arefiev, S.P. 2010. A system analysis of biota of xylotrophic fungi. Nauka, Novosibirsk., 260 pp. (in Russian).

Aro, N., Pakula, T. and Penttilä, M. 2005. Transcriptional regulation of plant cell degradation by filamentous fungi. FEMS Microbiol. Rev., 29: 719–739.

Bandoni, R.J. 1961. The genus *Naematelia*. Am. Midland Nat., 66: 319–328.

Bandoni, R.J. 1987. Taxonomic overwiev of the Tremellales. Stud. Mycol., 30: 87–110.
Binder, M. and Hibbett, D.S. 2006. Molecular systematics and biological diversification in Boletales. Mycologia, 98: 917–925.
Bon, M. 1981. *Lactarietum lacunarum*, nouvelle association fongique des lieux inondables. Docum. Mycol., 11: 19–28.
Bon, M. 1991. Les tricholomes et ressemblants (Tricholomoideae et Leucopaxilloideae). Genres: *Tricholoma, Tricholomopsis, Callistosporium, Porpoloma, Floccularia, Leucopaxillus* et *Melanoleuca*. Fl. Mycol. Eur., 2: 1–163.
Bondartsev, A.S. 1956. A guide to the house fungi. Academy of Sciences, Lenindrad., 80 pp. (in Russian).
Bondartsev, A.S. 1971. The Polyporaceae of the European USSR and Caucasia. Israel Program for scientific translations, Jerusalem, 896 pp.
Bondartseva, M.A. 1998. The handbook on fungi of Russia. Order Aphyllophorales. Ser. 2. 391 pp.
Bondartseva, M.A. 2001. Strategies of adaptation and functions of aphyllophoroid basidiomycetes in forest ecosystems. Kuprevich's lectures. III. Minsk., pp. 5–49.
Buchanan, P.K. and Ryvarden, L. 2000. An annotated checklist of polypore and polypore-like fungi recorded from New Zealand. N.Z. J. Bot., 38: 265–323.
Bujakiewicz, A. 1992. Macrofungi on soil in deciduous forests. Handbook of vegetation science founded by R. Tüxen. Vol. 19/1. Fungi in vegetation science. Dordrecht., pp. 49–78.
Burdsall, H.H. and Banik, M.T. 2001. The genus *Laetiporus* in North America. Harvard Papers in Botany, 6: 43–55.
Chadefaud, M. 1950. Les Psilotinées et l'évolution des Archégoniates. Bull. Soc. Bot. France., 97: 99–100.
Chamuris, G.P. 1988. The non-stipitate stereoid fungi in the Northern United States and adjacent Canada. Mycol. Mem., 14: 1–247.
Chen, D.M., Taylor, A.F.S., Burke, R.M. and Cairney, W.G. 2001. Identification of genes for lignin peroxidases and manganese peroxidases in ectomycorhizal fungi. New Phytol., 152: 151–158.
Church, A.H. 1919. Thalassiophyta and the subaerial transmigration. Bot. Mem., 3: 1–95.
Cooke, R.C. and Whipps, J.M. 1980. The evolution of modes of nutrition in fungi parasitic on terrestrial plants. Biol. Rev., 55: 341–362.
Cowling, E.G. 1961. Comparative biochemistry of the decay of sweetgum sapwood by white-rot and brown-rot fungi. US Dept. Agric. Tech. Bull., 258: 1–75.
D'Souza, T.M., Boominathan, K. and Reddy, C.A. 1996. Isolation of laccase gene-specific sequences from white-rot and brown-rot fungi by PCR. Appl. Environ. Microbiol., 62: 3739–3744.
Darimont, F. 1973. Recherches mycosociologiques dans les forêts de Haute Belgique. Inst. Roy. Sci. Nat. Belg. Mem., 170: 1–220.
David, A. and Rajchenberg, M. 1985. Pore fungi from French Antilles and Guiana. Mycotaxon, 22: 285–325.
Davydkina, T.A. 1980. Stereaceous fungi of Soviet Union. Nauka, Leningrad., 143 pp. (in Russian).
Dennis, R.W.G. 1978. British Ascomycetes. J. Cramer, Vaduz.
Dighton, J., White, J.F. and Oudemans, P. (eds). 2005. The fungal community. Its organization and role in the ecosystem. Third edition. Taylor & Francis, L.; N.Y.; Singapore., 936 pp.
Domański, S. 1988. Mała flora grzybów. Basidiomycetes (Podstawczaki). Aphyllophorales (Bezblaszkowce). 5. Corticiaceae: *Acanthobasidium-Irpicodon*. PWN, Warszawa-Krakow., 427 pp. (in Polish).
Domański, S. 1991. Mała flora grzybów. I. Basidiomycetes (Podstawczaki). Aphyllophorales (Bezblaszkowce). Stephanosporales (Stefanosporowce). 6. Corticiaceae: *Kavinia-Rogersella*, Stephanosporaceae: *Lindtneria*. PWN, Warszawa-Krakow., 272 pp. (in Polish).

Domański, S. 1992. Mała flora grzybów. I. Basidiomycetes (Podstawczaki). Aphyllophorales (Bezblaszkowce). 7. Corticiaceae: *Sarcodontia-Ypsilonidium*, *Christiansenia* and *Sygygospora*. W. Szafer Institute of Botany, Polish Academy of Sciences, Krakow., 258 pp. (in Polish).

Eriksson, J. and Ryvarden, L. 1973. The Corticiaceae of North Europe/With drawings by John Eriksson. Vol. 2: *Aleurodiscus-Confertobasidium*. Fungiflora, Oslo., pp. 60–261.

Eriksson, J. and Ryvarden, L. 1975. The Corticiaceae of North Europe/With drawings by John Eriksson. Vol. 3: *Coronicium-Hyphoderma*. Fungiflora, Oslo., pp. 287–546.

Eriksson, J. and Ryvarden, L. 1976. The Corticiaceae of North Europe/With drawings by John Eriksson. Vol. 4: *Hyphodermella-Mycoacia*. Fungiflora, Oslo., pp. 549–886.

Eriksson, J., Hjortstam, K. and Ryvarden, L. 1978. The Corticiaceae of North Europe/With drawings by John Eriksson. Vol. 5: *Mycoaciella-Phanerochaete*. Fungiflora, Oslo., pp. 889–1047.

Eriksson, K.E., Grunwald, A., Nilsson, T. and Vallander, L. 1980. A scanning electron microscopy study of the growth and attack on wood of three white-rot fungi and their cellulase-less mutants. Holzforschung, 34: 207–213.

Eriksson, J., Hjortstam, K. and Ryvarden, L. 1981. The Corticiaceae of North Europe/ With drawings by John Eriksson. Vol. 6: *Phlebia-Sarcodontia*. Oslo: Fungiflora., pp. 1051–1276.

Eriksson, J., Hjortstam, K. and Ryvarden, L. 1984. The Corticiaceae of North Europe/With drawings by John Eriksson. Vol. 7: *Schizopora-Suillosporium*. Fungiflora, Oslo., pp. 1281–1449.

Eriksson, K.E., Blanchette, R.A. and Ander, P. 1990. Microbial and enzymatic degradation of wood components. Springer, Berlin-Heidelberg-N.Y., 407 pp.

Erland, S. and Taylor, A.F.S. 1999. Resupinate ectomycorrhizal fungal genera. pp. 347–363. *In*: Ectomycorrhizal Fungi: Key Genera in Profile. Springer Verl., Heidelberg.

Ershov, R.V. and Ezhov, O.N. 2009. Aphyllophoroid fungi of aspen on Nort-West of Russian Plain. Arkhangelsk., 123 pp. (in Russian).

Ezhov, O.N., Ershov, R.V., Ruokolainen, A.V. and Zmitrovich, I.V. 2011. Aphyllophoraceous fungi of Pinega Reserve. Arkhangelsk., 147 pp. (in Russian).

Fidalgo, O. and Fidalgo, M.E.P.K. 1966. Polyporaceae from Trinidad and Tobago. Mycologia, 58: 862–904.

Fidalgo, O. and Fidalgo, M.E.P.K. 1968. Polyporaceae from Venezuele. 1. Mem. N.Y. Bot. Gdn., 17: 1–34.

Gamauf, C., Metz, B. and Seiboth, B. 2007. Degradation of plant cell wall polymers by fungi. pp. 325–340. *In*: Esser, K. (ed.). The Mycota. A comprehensive treatise of fungi as experimental systems for basic and applied research. Environmental and microbial relationships. 2nd ed. Springer, Heidelberg.

Ghobad-Nejhad, M. and Kotiranta, H. 2008. The genus *Inonotus* sensu lato in Iran, with keys to *Inocutis* and *Mensularia* worldwide. Ann. Bot. Fennici, 45: 465–476.

Ghobad-Nejhad, M. 2011. Updated checklist of corticioid and poroid basidiomycetes of the Caucasus region. Mycotaxon, 117: 1–70.

Gilbert, G.S. and Sousa, P. 2002. Host specialization among wood-decay polypore fungi in a Caribbean mangrove forest. Biotropica., 34: 396–404.

Gilbert, G.S., Gorospe, J. and Ryvarden, L. 2008. Host and habitat preferences of polypore fungi in Micronesian tropical flooded forests. Mycological Research, 112: 674–680.

Gilbertson, R.L. 1980. Wood-rotting fungi of North America. Mycologia, 72: 1–49.

Gilbertson, R.L. and Ryvarden, L. 1986. North American polypores. Vol. 1. Fungiflora, Oslo., pp. 1–436.

Gilbertson, R.L. and Ryvarden, L. 1987. North American polypores. Vol. 2. Fungiflora, Oslo., pp. 437–885.

Ginns, J. 1998. How many species are there? Folia cryptogamica Estonica, 33: 29–33.

González García, V., Portal Onco, M.A. and Rubio Susan, V. 2006. Review. Biology and systematics of the form genus *Rhizoctonia*. Spanish J. Agricultural Res., 4: 55–79.

Gottlieb, A.M., Wright, J.E. and Moncalvo, J.-M. 2002. *Inonotus* s. l. in Argentina—morphology, cultural characters and molecular analyses. Mycological Progress, 1: 299–313.

Hansen, L. and Knudsen, H. (eds.). 1997. Nordic macromycetes. Vol. 3: heterobasidioid, aphyllophoroid and gastromycetoid Basidiomycetes. Copenhagen: Nordsvamp, 445 pp.

Hansen, L. and Knudsen, H. (eds.). 2000. Nordic macromycetes. Vol. 1: Ascomycetes. Copenhagen: Nordsvamp, 309 pp.

Härkonen, M., Niemelä, T. and Mwasumbi, L. 2003. Tanzanian mushrooms. Edible, harmful and other fungi. Helsinki., 200 pp.

Heal, O.W. and Dighton, J. 1985. Resource quality and trophic structure of soil system. pp. 339–354. *In*: Fitter, A.H., Atkinson, D., Read, D.J. and Usher, M.B. (eds.). Ecological Interactions in Soil. Blackwell, Oxford.

Hibbett, D.S., Binder M. and Bischoff J.F. 2007. A higher-level phylogenetic classification of the Fungi. Mycological Research, 111: 509–547.

Hibbett, D.S. and Donoghue, M.J. 1995. Progress toward a phylogenetic classification of the Polyporaceae through parsimony analysis of mitochondrial ribosomal DNA-sequences // Can. J. Bot., 73: S853–S861.

Hood, I. 2003. An introduction to fungi on wood in Queensland. University of New England School of Environmental Sciences and Natural Resources Management, Armidale., 388 pp.

Imazeki, R., Otani, Y. and Hongo, T. 1988. Nihon no Kinoko [Fungi of Japan]. Tokyo, 623 pp.

Isikov, V.P. and Konoplya, N.I. 2004. Dendromycology. Alma Mater, Lugansk., 347 pp. (in Russian).

Jahn, H. 1966. Pilzgesellschaaften an *Populus tremula*. Z. Pilzk., 32: 26–42.

Jahn, H. 1968. Das *Bisporetum antennatae*, eine Pilzgesellschaft auf den Schnittflächen von Buchenholz. Westf. Plzbr., 7: 41–47.

Jahn, H. 1986. Der "Stanspilzhang" bei Glesse (Ottenstein), Süd-Niedersachsen Zur Pilzvegetation des Seggen-Hangbuchenwaldes (Carici-Fagetum) im Weserbergland und auβerhalb. Pilzbriefe, 10/11: 289–351.

James, T.Y., Kauf, F., Schoch, C.L. 2006. Reconstructing the early evolution of fungi using a six-gene phylogeny. Nature, 443: 818–822.

Jing, H.U.I., Wenjing, Z. and Zhiyan, Z. 2007. Changes in extracellular enzyme activities during submerged culture of *Tremella aurantialba*. Acta Edulis Fungi, 14: 33–36.

Jülich, W. 1972. Monographie der Athelieae (Corticiaceae, Basidiomycetes). Willdenowia Beih., 7: 1–283.

Jülich, W. and Stalpers, J.A. 1980. The resupinate non-poroid Aphyllophorales of the Northern Hemisphere. North-Holland Pub. Comp., Amst.; Oxf.; N.Y., 335 pp.

Jülich, W. 1984. Die Nichtblätterpilze, Gallertpilze und Bauchpilze. Aphyllophorales, Heterobasidiomycetes, Gastromycetes. Gustav Fischer, Jena: 626 S.

Karatygin, I.V. 1993. Coevolution of Fungi and Plants. St Petersburg, 118 pp. (in Russian).

Kirk, P.M., Cannon, P.F., David, J.C. and Stalpers, J.A. 2001. Ainsworth and Bisby's Dictionary of the Fungi. 9th edition. Oxford Univ. Press, N.Y. etc., 672 pp.

Kirk, T.K. 1983. Degradation and conversion of lignocelluloses. pp. 266–295. *In*: Smith, J.E. et al. (ed.). The Filamentous fungi. Vol. 4. Fungal technology. E. Arnold, London.

Knudsen, H. and Vestrholt, J. (eds.). 2008. Funga Nordica: New edition of "Nordic macromycetes, volume 2". Copenhagen: Nordsvamp, 968 pp.

Koenigs, J.W. 1972. Effects of Hydrogen peroxidase on cellulose and on its susceptibility to cellulose. Mater. Organismen, 7: 133–147.

Kotiranta, H. and Niemelä, T. 1993. Uhanalaiset käävät Suomessa (Threatened polypores in Finland)—Vesi-ja ympäristöhallinnon julkaisuja, sarja B 17: 1–116. Painatuskeskus, Helsinki.

Larsen, M. and Cobb-Poulle, L.A. 1990. *Phellinus* (Hymenochaetaceae). A survey of the world taxa. Synopsis Fungorum, 3: 1–206.

Larsson, K.-H., Parmasto, E., Fischer, M., Langer, E., Nakasone, K.K. and Redhead, S.A. 2006. Hymenochaetales: a molecular phylogeny for the hymenochaetoid clade. Mycologia, 98: 926–936.
Lindgren, M. 2001. Polypores (Basidiomycetes) species richness and community structure in natural boreal forest of NW Russian Karelia and adjacent areas in Finland. Acta Bot. Fennica, 170: 1–41.
Malysheva, V.F. and Malysheva, E.F. 2008. The higher basidiomycetes in forest and grassland communities of Zhiguli. St Petersburg, 242 pp. (in Russian).
Manskaya, S.M. and Kodina, L.A. 1975. Geochemistry of lignin. Nauka, Moscow, 229 pp. (in Russian).
Miettinen, O., Niemelä, T. and Spirin, W. 2006. Northern *Antrodiella* species, the identity of A. semisupina, and type studies of related taxa. Mycotaxon, 96: 211–239.
Miettinen, O. and Larsson, K.-H. 2011. *Sidera*, a new genus in Hymenochaetales with poroid and hydnoid species. Mycological Progress, 10: 131–141.
Nakasone, K.K. and Gilbertson, R.L. 1978. Cultural and other studies of fungi that decay Ocotillo in Arizona. Mycologia, 70(2): 266–299.
Nobles, M.K. 1958. Cultural characters as a guide to the taxonomy and phylogeny of the Polyporaceae. Can. J. Bot., 36: 883–926.
Núñez, M. and Ryvarden, L. 2001. East Asian polypores. Vol. 2. Fungiflora, Oslo., pp. 170–522.
Olive, L.S. 1946. New or rare Heterobasidiomycetes from Norh Carolina 2. J. Elisha Mitchell Sci. Soc., 62: 65–71.
Parmasto, E. 1986. Preliminary list of vietnamense Aphyllophorales and Polyporaceae s. str. Scripta Mycol., 14: 1–88.
Parmasto, E. 1998. *Athelia arachnoidea*, a lichenicolous basidiomycete in Estonia. Folia Cryptogamica Estonica., 32: 63–66.
Pirk, W. 1952. Die Pilzesellschaften der Baumweiden im mittleren Wesertal. Mitt. Flor.-Soz. Arbeist. N.F., 3: 93–96.
Pirozynski, K.A. 1983. Pacific Mycogeography: an appraisal. Aust. J. Bot. Suppl., 10: 137–159.
Polizeli, M.L., Rizzatti, A.C., Monti, R., Terenzai, H.F., Jorge, J.A. and Amorim, D.S. 2005. Xylanases from fungi: properties and industrial applications. Appl. Microbiol. Biotechnol., 67: 577–591.
Rabinovich, M.L., Bolobova, A.V. and Kondrashchenko, V.I. 2001. Theoretical bases for biotechnology of wood composites. Book 1: Wood and wood-destroying fungi. Nauka, Moscow, 264 pp. (in Russian).
Ragan, M.A. and Chapman, D.J. 1978. Biochemical phylogeny of the protists. N.Y., 127 pp.
Rajchenberg, M. 1989. Polyporaceae (Aphyllophorales, Basidiomycetes) from Southern South America: a mycogeographical view. Sydowia, 41: 277–291.
Renvall, P. 1995. Community structure and dynamics of wood-rotting Basidiomycetes on decomposing conifer trunks in northern Finland. Karstenia, 35: 1–51.
Ricek, E.W. 1967. Untersuchungen über die Vegetation auf Baumstümpfen 1. Jahrb. Obst Mus., 112: 185–252.
Richard, B. 2000. Les mycocoenoses des pelouses calcicoles du Barrois lorrain. Analyse inventoriale, patrimoniale et conservatoire. These pour obtenir le Diplôme d'Etat de Docteur en Pharmacie. Univ. H. Poincare, Nancy., 65 pp.
Roy, A. and De, A.B. 1996. Polyporaceae of India. R.P. Singh Gahlot, Dahra Dun., 287 pp.
Ryvarden, L. and Johansen, I. 1980. A preliminary polypore flora of East Africa. Fungiflora, Oslo, 225 pp.
Ryvarden, L. and Gilbertson, R.L. 1993. European polypores. Part 1. *Abortiporus-Lindtneria*. Fungiflora, Oslo., pp. 1–387.
Ryvarden, L. and Gilbertson, R.L. 1994. European polypores. Part 2. *Meripilus-Tyromyces*. Fungiflora: Oslo., pp. 388–743.

Schmidt, O. 2006. Wood and tree fungi. Biology, damage, protection, and use. Springer, Berlin-Heidelberg-N.Y., 334 pp.
Sinadsky, Yu.V. 1983. Pine: their diseases and destroyers. Moscow, 335 pp. (in Russian).
Šmarda, F. 1972. Pilzgesellschaften einiger Laubwälder Mährens. Acta Sc. Nat. Brno, 6: 1–53.
Šmarda, F. 1973. Die Pilzgesellschaften einiger Fichtenwälder Mährens. Acta. Nat. Acad.
Smirnova, O.V. (ed.). 2004. East European forests: Holocene history and modern state. Nauka, Moscow, 479 pp. (in Russian).
Spirin, W.A. 2002. Aphyllophoraceous fungi of Nizhegorod Region: species composition and ecological peculiarities. Thesis. St Petersburg, 242 pp. (in Russian).
Spirin, W.A., Zmitrovich, I.V. and Malysheva, V.F. 2006. On the systematics of *Inonotus* s.l. and *Phellinus* s.l. (Mucronoporaceae, Hymenochaetales). Nov. Syst. Pl. non Vasc., 40: 153–188 (in Russian).
Strid, A. 1975. Wood-inhabiting fungi of alder forests in North-Central Scandinavia 1. Aphyllophorales (Basidiomycetes). Taxonomy, ecology and distribution. University of Umea, Umea, 237 pp.
Stubblefield, S.P., Taylor, T.N. and Beck, C.B. 1985. Studies of Paleozoic fungi. V. Wood-decaying fungi in Callixylon newberryi from the Upper Devonian. Am. J. Bot., 72: 1165–1174.
Takhtajan, A.L. 1950. Phylogenetic bases of vascular plants system. Botanical Journal, 13: 135–139.
Teng, S.C. 1996. Fungi of China. Ithaka: Mycotaxon Ltd., pp. 1–586.
Torkelsen, A.-E. 1997. Tremellaceae Fr. pp. 86–90. *In*: Hansen, L. and Knudsen, H. (eds.). Nordic macromycetes 3. Heterobasidioid, aphyllophoroid and gastromycetoid Basidiomycetes. Nordsvamp, Copenhagen.
Ţura, D., Zmitrovich, I.V., Wasser, S.P., Spirin, W.A. and Nevo, E. 2011. Biodiversity of Heterobasidiomycetes and non-gilled Hymenomycetes (former Aphyllophorales) of Israel. A.R.A. Gantner Verlag K.-G., Ruggell, 566 pp.
Vasilyeva, L.N. and Stephenson, S.L. 2010. Biogeographical patterns in pyrenomycetous fungi and their taxonomy. 1. The Grayan disjunction. Mycotaxon, 114: P. 281–303.
Wright, J.E. and Deschamps, J.R. 1972. Basidiomycetes xilofilos de los Bosques Andinopatagonicos. Rev. Invest. Agrop. INTA, 9: 111–195.
Wright, J.E. and Deschamps, J.R. 1975. Basidiomycetes xilofilos de la region mesopotamica. II. Los generous *Daedalea, Fomitopsis, Heteroporus, Laetiporus, Nigroporus, Rigidoporus, Perenniporia* and *Vanderbylia*. Rev. Invest. Agrop. INTA, 12: 127–204.
Wright, J.E. 1985. Los hongos xilofagos: una revista. Anal. Acad. Nac. Cs. Ex. Fis. Nat., Buenos Aires, 37: 121–135.
Wu, S.H. 1990. The Corticiaceae (Basidiomycetes) subfamilies Phlebioideae, Phanerochaetoideae and Hyphodermoideae in Taiwan. Acta Bot. Fennica, 142: 1–123.
Yurchenko, E.O. 2001. Corticioid fungi on mosses in Belarus. Mycena, 1: 71–91.
Yurchenko, E.O. and Zmitrovich, I.V. 2001. Variability of *Hyphoderma setigerum* (Corticiaceae s. l., Basidiomycetes) in Belarus and northwest Russia. Mycotaxon, 78: 423–434.
Yurchenko, E.O. and Golubkov, V.V. 2003. The morphology, biology, and geography of a necrotrophic basidiomycete *Athelia arachnoidea* in Belarus. Mycological Progress, 2: 275–284.
Yurchenko, E.O. 2006. Natural substrata for corticioid fungi. Acta Mycol., 41: 113–124.
Zmitrovich, I.V., Malysheva, V.F. and Malysheva, E.F. 2003. Some concepts and terms of mycogeography: a critical review. Bulletin of Ecology, Forest management and Landscape management, 4: 173–188 (in Russian).
Zmitrovich, I.V., Psurtseva, N.V. and Belova, N.V. 2007. Evolutionary-taxonomical aspects of search and research of lignotrophic fungi as active producers of oxidative enzymes. Mycol. Phytopathol., 41: 57–78 (in Russian).
Zmitrovich, I.V. 2008. Definitorium fungorum Rossicum. Familia Atheliaceae et Amylocorticiaceae. KMK, Petropolis, 278 pp. (in Russian).

Zmitrovich, I.V. 2010. Epimorphology and tectomorphology of higher fungi. Folia Cryptogamica Petropolitana, 5: 1–279.

Zmitrovich, I.V. 2011. Middle taiga of Karelian Istmus: zonal, intrazonal and extrazonal phenomena. Bulletin of Ecology, Forest management and Landscape management, 12: 54–76 (in Russian).

Zmitrovich, I.V. 2012. Features of structure and dynamics of floodland gray alder forests in North-Wester European Russia. *In*: Human and North: Anthropology, Archaeology and ecology. Tjumen', 425 pp. (in Russian).

Zugmaier, W., Bauer, R. and Oberwinkler, F. 1994. Mycoparasitism of some *Tremella* species. Mycologia, 86: 49–56.

CHAPTER 3

Aquatic Hyphomycetes from Leaf Litter in Brazilian Urban Waters

Iracema Helena Schoenlein-Crusius,[1,*] *Carolina Gasch Moreira*[1] and *Elaine Malosso*[2]

ABSTRACT

Aquatic Hyphomycetes have been usually reported from clear, well-aerated, running waters, but since some years, their tolerance to unfavorable conditions was also noticed in temperate and tropical climates. The presence of aquatic Hyphomycetes in polluted environments reinforces their importance as effective decomposers of submerged leaf litter and the need to know more about their taxonomy, diversity and activity in several types of aquatic environments. The aquatic Hyphomycetes have been surveyed in streams, ponds, reservoirs and lakes in São Paulo city, Brazil, for the past 20 years. These studies were carried out in waters with different levels of eutrophication. The collections were done for both, mixed leaf litter and/or a specific leaf litter taking into account the season of collection. The richness and diversity of aquatic Hyphomycetes in urban waters were sometimes higher than initially expected as many species were recorded for the

[1] Instituto de Botânica, Núcleo de Pesquisas em Micologia, Av. Miguel Stéfano 3687, CEP: 04301-012 São Paulo, Brazil.

[2] Universidade Federal de Pernambuco, Departamento de Micologia, Av. Prof. Nelson Chaves, s/n., CEP: 50.670-420 Recife, Pernambuco, Brazil.
Email: elainemalosso@yahoo.com.br

* Corresponding author: iracema@crusius.com.br

first time in Brazil, while sometimes their number decreased and only a few ubiquitous species could be collected. Water eutrophication was found to affect the fungal composition regardless of the average number of species recovered. This chapter presents the state-of-the-art of the floristic surveys undertaken for aquatic Hyphomycetes in urban waters and perspectives of the new scientific approaches for the future.

Introduction

Conidial or anamorphic fungi reproduce characteristically by the formation of conidia or by the somatic hyphal differentiation (Alexopoulos et al., 1996; Kirk et al., 2008). Besides the formation of conidia, somatic structures such as chlamydospores or even hyphal fragments may also function as fungal multiplicants and dispersal structures (Kirk et al., 2008).

Once these fungi were classified as *Deuteromycota*, *Deuteromycotina* or *Deuteromycetes*, but since they are not included in any individual recognized phylum, genera and species are considered anamorphs or conidial phases of Ascomycota or Basidiomycota (Kirk et al., 2008). Among this large group of fungi, the species which produce conidia that can be totally free or contained in sporodochia were traditionally classified as Hyphomycetes (Alexopoulos et al., 1996).

The term, aquatic Hyphomycetes, is still interesting to be used as a conserved name in view of the great importance of the species that produce conidia exclusively in aquatic environments or in interstitial waters on terrestrial substrates. Regarding the dependence on water in the reproduction habitat, this fungal group parallels to truly aquatic organisms such as zoosporic fungi and other representatives of Ascomycota and Basidiomycota (Schoenlein-Crusius and Milanez, 1996; Schoenlein-Crusius and Malosso, 2007).

The first report of aquatic Hyphomycetes is given in the pioneering studies of Saccardo (1880) with the description of *Heliscus lugdunensis* Sacc. & Therry. During the 1940s, Ingold (Ingold, 1942, 1943, 1944, 1949) intensified the investigation of these fungi in Britain and afterwards in Africa, and described several species. To honor the pioneer mycologist-Ingold, aquatic Hyphomycetes are also referred to as "Ingoldean fungi".

In the literature, the term "tetraradiate fungi" has often been used because in several species the conidia are star-like formed by a central part from which three or four "arms" are projected in opposite directions. However, there are species with sigmoid conidia too. Because of the shape, their light weight these conidia easily disperse by wind or with the flow of water, enhancing the success of the species to reach long distances (Ingold, 1975).

The aquatic Hyphomycetes, in general, are found in clean, well aerated, moderately fast running waters, in natural foam or growing directly on submerged leaf litter (Ingold, 1975). However, their occurrences have also been reported from polluted waters (Au et al., 1992).

In temperate climates, the occurrence of aquatic Hyphomycetes may be more during autumn and the beginning of the winter, with the seasonal increase of the quantity of senescent leaves in the ecosystem (Nilsson, 1964; Ingold, 1975; Subramanian, 1983; Schoenlein-Crusius and Milanez, 1989, 1990a,b; Webster, 1980; Bärlocher, 2009). Furthermore, in the tropics, to assess the seasonal behavior of the aquatic Hyphomycetes, it should be taken into account at what time/season the increase of senescent leaves in the water bodies occur; in cold and dry seasons or in warm and rainy seasons depending on the type of ecosystem under observation (Schoenlein-Crusius and Milanez, 1996).

These fungi constitute an ecological group and have an important role in the decomposition of leaves and other organic allochtonous substrates, mainly in lotic waters (Tsui and Hyde, 2003). In general, the fungal biomass content of leaves recently fallen into the water increases proportionally with the decomposition, reaching 17% of the initial dry weight within a few weeks (Gessner, 1997).

The importance of the aquatic Hyphomycetes in the aquatic environments has also been related to the palatability of the leafy substrates to other components of the food chain. For example, some shredders have preference to consume leaves which are colonized by fungi, probably due to the nutritional enrichment of the decaying substrates (Kaushik and Hynes, 1971; Bärlocher and Kendrick, 1974; Cheng et al., 1997).

Among many important features of the aquatic Hyphomycetes it also needs to be mentioned that some species have the ability to decompose lignin; others may survive in anaerobic condition (Field and Webster, 1983); some reproduce in salty or marine waters (Sridhar and Kaveriappa, 1989); and some may tolerate high amounts of heavy metals, such as cadmium (Abel and Bärlocher, 1984, 1988).

Furthermore, the type of substrate available for decomposition may also be relevant to determine the richness, distribution pattern and diversity of fungi in aquatic ecosystems (Canhoto and Graça, 1996; Gulis and Suberkropp, 2003a,b), probably due to lignin, cellulose, hemicelluloses, tannins, terpenoids, waxes and oils which can hamper the growth of the microorganisms. Thus, the type of plant species that surrounds aquatic environments is a relevant factor to the native mycota, connecting fungal richness to the community structure of the riparian forest (Laitung and Chauvet, 2005). Unfortunately, in urban areas, landscaping projects have sometimes taken the aesthetics of architecture more into account than the need to provide different substrates to the local aquatic community.

In tropical ecosystems, where the vegetation is usually highly diverse and provides a range of compounds and nutrients to the lower levels of the trophic chain, the replacement of original forests by species-poor gardens may be a problem. There is a tendency that water bodies which are surrounded by vegetation that is similar to the original forest reveal a higher fungal diversity than those that are unnatural (Schoenlein-Crusius et al., 2009a,b).

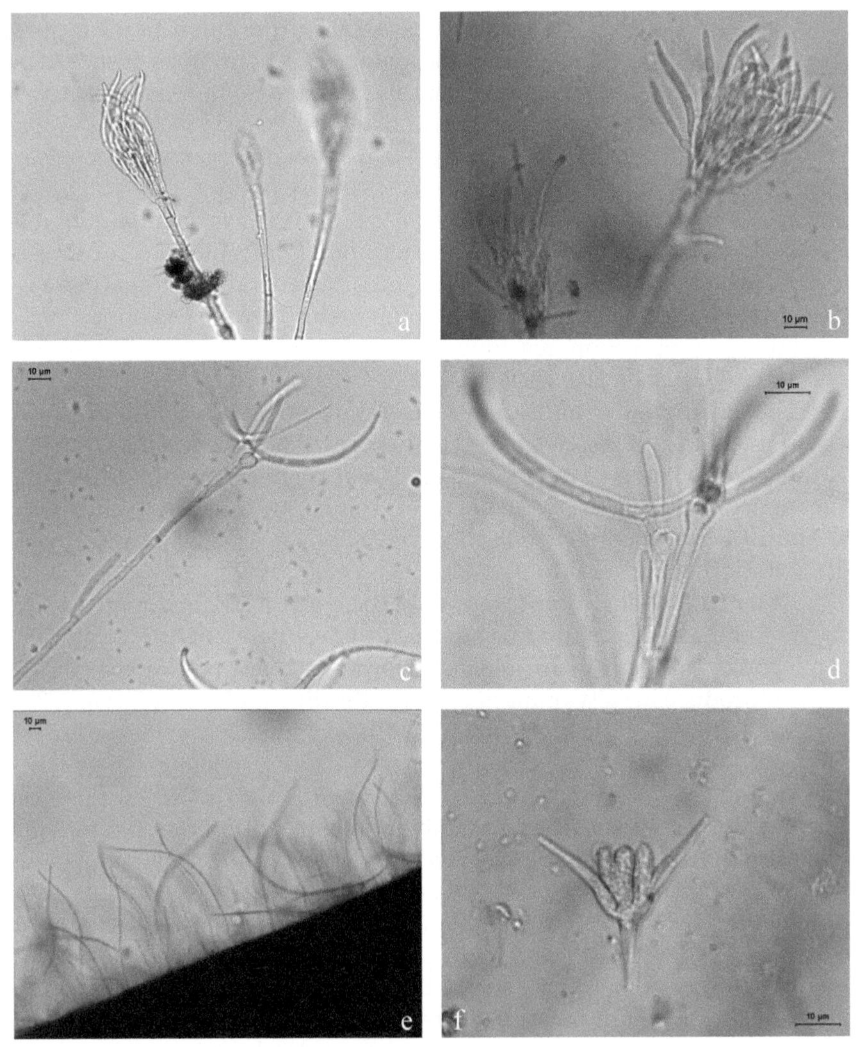

Figure 3.1. Overview of genera of aquatic Hyphomycetes observed in Brazilian urban waters: (a, b) *Flagellospora*; (c, d) *Lunulospora*; (e) *Anguillospora*; (f) *Tricladium*. Images by Carolina Gasch Moreira.

According to Kirk et al. (2001) and Gulis et al. (2005), approximately 100 anamophic genera and 300 species of aquatic Hyphomycetes are described, mainly from temperate climates. A substantial increase of these numbers is expected if more fungal surveys are undertaken for these fungi in the tropics. Several of these species may be ubiquitous, some more frequent for the temperate regions while others may be found in the tropics (Marvanová, 1997).

According to Schoenlein-Crusius and Grandi (2003), in the South American continent (Argentina, Chile, Ecuador, Peru, Venezuela and Brazil), approximately 90 taxa of Ingoldean fungi are recorded that are relatively fewer as compared with other regions like Europe.

With the above background in mind, investigations to study aquatic Hyphomycetes were undertaken for the last few years and this chapter embodies the findings of the surveys conducted for these fungi in urban waters in Brazil and throws some light on the approaches need to be adapted for the future.

Methodology

To isolate the fungi from the leaves collected from aquatic ecosystems, the leaf washing method, as described by Pugh et al. (1972) for terrestrial fungi, was adapted in combination with multiple baiting methods used for zoosporic organisms. The growth of tetraradiate conidia was observed on the margins of washed leaf disks which were incubated in sterile distilled water together with cellulosic, chitinous and keratinous baits (Schoenlein-Crusius and Milanez, 1989). However, to obtain the Ingoldean fungi, methods as described by Ingold were also followed (Ingold, 1975). Isolation of aquatic Hyphomycetes is relatively simple. Leaves or leaf fragments of approximately 1 cm^2 are incubated in Petri dishes containing pond or sterile water for 5–7 days at temperatures ranging 15 to 20°C. After which time, the fungal structures were observed on slides in distilled water or 2% lacto-phenol cotton blue. The growth of conidiophores, tetraradiate or sigmoid fungi was detected on veins and margins of the leaf fragments. Drawings were made with a *camera lucida* and photographs were taken. Taxonomical keys (Nilsson, 1964; Ingold, 1975; Subramanian, 1983; Bärlocher, 1992; Marvanová, 1997) were used to identify the species by their morphological characteristics.

To obtain cultures of these fungi, their spores were picked up using Pasteur pipettes and transferred to a culture medium (a) such as the malt extract agar. We used this method most frequently to recover fungi during most surveys of aquatic Hyphomycetes in Brazilian urban waters. In more recent studies, the collected leaves were placed in aquaria containing sterile water that was aerated by an aerator for two to three days at 20°C to

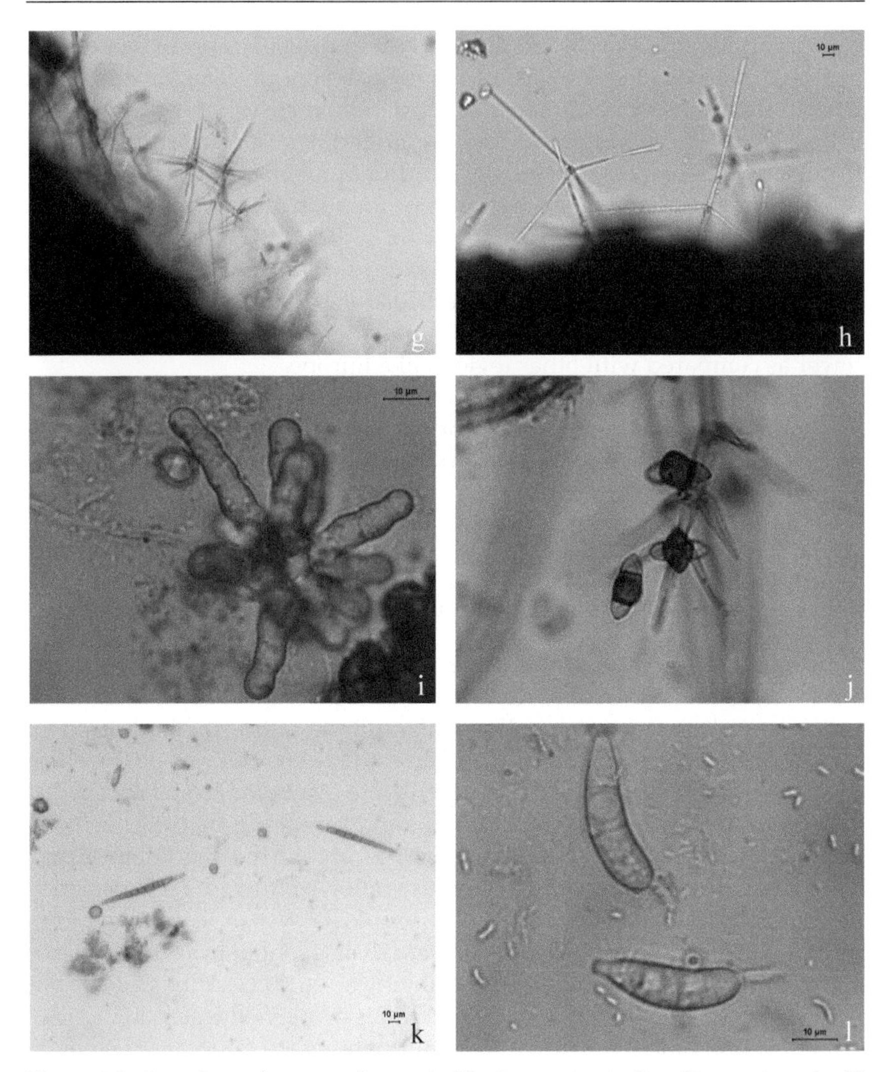

Figure 3.2. Overview of genera of aquatic Hyphomycetes in Brazilian waters: (g, h) *Triscelophorus*; (i) *Pyramidospora*; (j) *Uberispora*; (k) *Camposporium*; (l) *Margaritispora*. Images by Carolina Gasch Moreira.

Color image of this figure appears in the color plate section at the end of the book.

stimulate the production of viable conidia. The conidia were picked up and transferred to culture media for growth. The "leader" method of Descals (2005) has also been successfully used in order to obtain pure cultures but the contaminants of the tropical water make the purification of cultures, a difficult and a very exhaustive task.

In some ecological studies, senescent leaves from a given tree species or mixed leaf litter were collected, air dried, placed in several nylon net litter bags (in general with 1 mm diam. mesh) and then submerged completely in the water (Bärlocher, 2005). Samples of these litter bags were collected monthly during the entire period of study. The leaves so collected were washed with clean water to get rid of animal and/or plant debris and then subjected to basic methods of isolation of aquatic Hyphomycetes (Moreira, 2011). The advantage of this method is that it enables the investigator to observe variations of the fungal diversity during the time of decomposition of the leaves under standard conditions. However, the disadvantage is that the leaves are very grouped/packed and that might hinder the development of more delicate fungal structures. More so, instead of the leaf litter of a given plant species, sometimes free mixed leaf litter samples were collected as substrates for studies concerned to the diversity of Ingoldean fungi in urban waters. For example, in the survey of Ingoldean fungi in the Botanical Garden of São Paulo, samples of submerged leaf litter samples were directly collected from the margins of a water body, taken to the laboratory and submitted to the same procedure described above to obtain aquatic Hyphomycetes. These samples yielded 24 fungal taxa (Schoenlein-Crusius et al., 2009).

The total time period during which surveys of Ingoldean fungi were done varied from six months (Schoenlein-Crusius et al., 1990), seven to eight months (Schoenlein-Crusius and Milanez, 1989; Schoenlein-Crusius et al., 1992, 1998), and one to two years (Schoenlein-Crusius and Milanez, 1998; Schoenlein-Crusius et al., 2009). However, in general, the collections were made monthly (Schoenlein-Crusius and Milanez, 1989; Schoenlein-Crusius et al., 2009; Moreira, 2011).

Observations and Discussion

Floristic Studies

Although recent, the contribution of the Brazilian studies for the aquatic fungal flora of the South American region has been encouraging. Until 2003, around 50 taxa were reported, of which more than 60% of the reports of aquatic Hyphomycetes were made from Brazil (Schoenlein-Crusius and Grandi, 2003).

Search for aquatic Hyphomycetes in Brazil was initiated in urban systems. The fungal genera, *Lemonniera* and *Triscelophorus*, were first observed on decomposing submerged leaves of *Ficus microcarpa* L. f. in an artificial and highly eutrophicated lake (Schoenlein-Crusius and Milanez, 1989). Later on, *Quercus robur* L. leaves were submerged in a less eutrophicated artificial lake in the region of Itapecerica da Serra, in São Paulo state, to verify whether the better quality of water and the use of a specific leaf type would provide a higher fungal diversity (Schoenlein-Crusius et al., 1990). As the reasons to explain the pattern of aquatic Hyphomycetes occurrence were not clear, several small collections of submerged leaf litter were randomly made in São Paulo state. This exercise resulted in the publication of 11 species: *Anguillospora crassa* Ingold, *Anguillospora longissima* (Sacc. & P. Syd.) Ingold. *Camposporium pellucidum* (Grove) Hughes, *Campylospora chaetocladia* Ranzoni, *Flabellospora crassa* Alasoadura, *Clavariopsis aquatica* De Wild., *Heliscella stellata* (Ingold & Cox) Marv., *Lemonniera aquatica* de Wild., *Lunulospora curvula* Ingold, *Triscelophorus monosporus* Ingold and *Tetrachaetum elegans* Ingold (Schoenlein-Crusius and Milanez, 1990a,b).

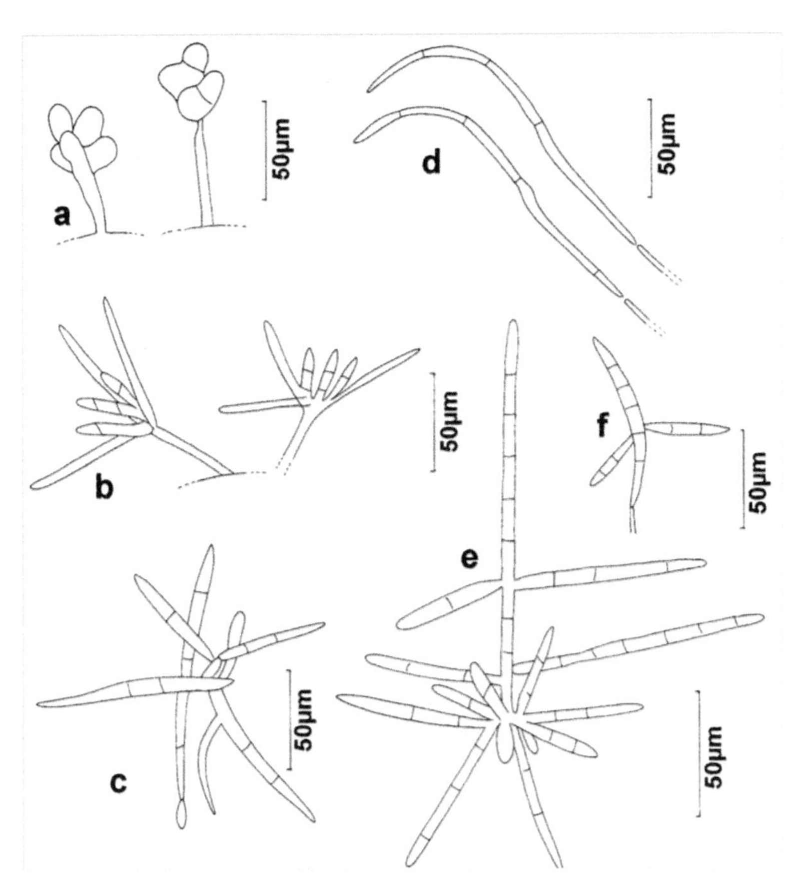

Figure 3.3. Conidia from some aquatic Hyphomycetes in Brazilian waters. (a) *Pyramidospora casuarinae* Nilsson; (b) *Tetracladium setigerum* (Grove) Ingold; (c) *Varicosporum elodae* Kegel; (d) *Anguillospora filiformis* Greathead; (e) *Dendrospora erecta* Ingold; (f) *Tricladium splendens* Ingold.

In the "Cerrado" biome, a type of Brazilian Savannah, tax-ecological studies were carried out in the municipality of São Carlos (SP). The results of the study were published by Malosso (Malosso, 1995, 1999). Later on a detailed taxonomical descriptions of the species found in the surveys done at several locations in São Paulo state (Mogi-Guaçú, Luís Antonio, Jataí and Itirapina) were published (Schoenlein-Crusius, 2002).

One of the authors, while working in the Atlantic rainforest of Paranapiacaba of São Paulo state, that comprises of diverse plant community, has heavy rainfall round the year and different running water bodies, collected many species of aquatic Hyphomycetes. Of which some were taxonomically described. Some of the described species were found in a fungal succession study during decomposition of submerged *Alchornea triplinervia* Spreng. (M.) Arg. leaves in streams, along with zoosporic and geofungi, showing their excellent competitive saprophytic abilities (Schoenlein-Crusius, 1998). In another experiment when additional leaves of *Quercus robur* L. and *Ficus microcarpa* L. f. were placed together with *A. triplinervia* leaves in the same stream, to verify a possible selective effect of the type of leaf species on the fungal community, no conclusive differences were observed in terms of richness (Schoenlein-Crusius et al., 1992), however, the quantity of aquatic Hyphomycetes propagules was higher (Pires-Zottarelli et al., 1993).

Using ergosterol to estimate the quantity of fungal biomass during the decomposition process and fungal succession on mixed leaf litter, submerged in a stream in the same Atlantic Rainforest, a great variation in the fungal biomass was observed when data for rainy and dry seasons were compared. In the leaf litter samples, ergosterol content varied from 2907 to 45809 µg/g. However, the seasonal variations of ergosterol content and the biodiversity of the aquatic Hyphomycetes were not concomitant.

The set of results from the aforesaid studies in the Atlantic forest allows one to conclude that regardless to the type of leaf substrate, the diversity of aquatic Hyphomycetes tends to be high. The higher production of fungal biomass and the consequent decomposition of allochtonous organic matter make the Atlantic rainforest a very suitable biome for the study of fungi (Schoenlein-Crusius et al., 2009; Capelari et al., 2009).

A comparative study for non-polluted with polluted (with domestic sewage) streams of the Atlantic rainforest on the diversity and activity of macro invertebrates, algae and fungi in the streams of the "Reserva Pedra Branca" in the city of Rio de Janeiro indicated that at the highly polluted sites, incidence of *Flabellospora crassa* on the submerged leaves *Myrcia rostrata* was higher, besides many zoosporic organisms and geofungi (Sandra Padilha Magalhães and Iracema H. Schoenlein-Crusius, pers. comm.).

Surveys conducted for fungi by Wellbaum et al. (1999) and Wellbaum (2000, 2007) for the aquatic and terrestrial environment of a small island called, "Ilha dos Eucaliptos" in the "Guarapiranga" reservoir, verifying the influence of the type of vegetation (*Eucalyptus*, native and transitory forest) on the diversity of fungi, using the poured plate method to isolate the fungi concluded that the number of taxa and fungal occurrences were higher in the submerged leaves during summer than in the terrestrial leaves during the winter. In these surveys, the aquatic Hyphomycetes could not be detected because only culture media were used for isolation, in which the ingoldian fungi grow only in the mycelia form, but no conidia are produced that help in the identification.

However, Malosso (1999) compared the diversity of the aquatic Hyphomycetes and the ergosterol content of leaves submerged in the "Guarapiranga" reservoir with that of a river continuous system "Rios Jacaré-Monjolinho" in the municipality of São Carlos, SP and recorded higher diversity and number of occurrences of fungi in the lotic system.

In another study in the "Parque Estadual das Fontes do Ipiranga", which is one of the last reserves of Atlantic rainforest in the state of São Paulo, falling in the metropolitan area of São Paulo city, senescent leaves of *Tibouchina pulchra* Cogn. were placed in the nylon net litter bags and submerged in the hypereutrophic "Lago das Garças" and in the "Lago das Ninféias" (meso-oligotrophic). This study resulted into 63 fungal taxa from the leaves submerged in the hypereutrophic lake and 79 taxa from the meso-oligotrophic lake. Despite higher species richness, the leaves collected from the meso-oligotrophic lake had lower decomposition rate and ergosterol content. Thus, it was concluded that the trophic condition of lakes did not only affect quantitatively, but also qualitatively the distribution pattern of species during the fungal succession, thereby influencing ergosterol concentration and decomposition rate of the submerged leaves (Moreira, 2006; Moreira and Schoenlein-Crusius, 2007).

At the same reserve, the diversity of aquatic Hyphomycetes associated with mixed decomposing leaf litter was compared with different trophic levels of the aquatic systems. The results were that the number of fungal taxa was higher in the mesotrophic than in the oligotrophic lake and lower in the areas contaminated with organic pollutants (Schoenlein-Crusius et al., 2009b), and a total of 24 fungal taxa were recorded with predominance of *Anguillospora crassa* Ingold, *Lunulospora curvula* Ingold, *Tetrachaetum elegans* Ingold and *Camposporium pellucidum* (Grove) Hughes.

Gruppi (2008) evaluated the diversity of microscopic fungi in mixed submerged leaves in some lakes in the Central Park in the municipality of Santo André, SP, with a view to verify differences among the sampling points, isolation methods and interactions between biotic and abiotic parameters of water. For isolating the fungi, the collected leaf litter fragments, both unwashed and washed were placed in sterile distilled water and also onto the potato-dextrose-agar medium. Entire leaves were also collected both from the areas with dense and thinner vegetation, and incubation in a fish aquarium containing pond water. A total of 38 fungal taxa were recovered. Of which 26 were Hyphomycetes, eight Ingoldean fungi, one Coelomycete, two belonged to Zygomycota and one to a zoosporic organism. However, from culture media, 28 taxa were isolated. Leaf fragments placed in water yielded seven fungi while from the leaves incubated in the aquarium five taxa were obtained. The fact that the presence of riparian vegetation at the sampling sites might have influenced the fungal diversity associated with the decomposing leaves cannot be ignored. This reinforces the importance of the landscape planning around the aquatic environment of the park.

In the "Parque Municipal Alfredo Volpi", two studies regarding the decomposition rate and geofungi associated with submerged leaves of *Caesalpinia echinata* Lam. and the diversity of aquatic Hyphomycetes associated with submerged leaves of *Campomanesia phaea* (O. Berg.) Lam. were carried out by Fagundes (2008) and Dias (2008), respectively. The leaves of *Caesalpinia echinata* decomposed faster initially, but with slower weight loss, with a quicker replacement of geofungi during fungal succession (Fagundes, 2008). Similarly, *Campomanesia phaea* also decomposed slowly and 21 fungal taxa (68 occurrences) were recorded during the decay period (Dias, 2008). These results are indicative of the fact that even if the leaves were from native trees, the aquatic environment tropical, and the associated mycota fairly diversified, the decomposition rates were not always high.

Some commonly found Ingoldian fungi along with some interesting species, from a taxonomic point of view, found in urban areas are tabulated below (Table 3.1), including *Naiadella fluitans* Marvanová & Bandoni, *Triscelosporium verrucosum* Nawawi & Kuthubuteen, *Blodgettia indica* Subramanian and *Tetracladium marchalianum* De Wild which were only found in urban waters until now. Besides these, many others with striking characteristics have also been studied to confirm the taxonomical identity. If the significant morphological modifications/variations of these unknown taxa are a result of environmental pressure, or they are the urban ecotypes or simply new species is a subject for further studies. It is also important to know the actual relevance of having a rich Ingoldian mycota in urban waters and the factors that must be taken into account to conserve the original diversity of these fungi. Urban areas have to be considered not only as polluted, disturbed and diversity poor, but also as systems which may offer a new kind of environment. Pools can also have unusual species of fungi.

Table 3.1. Aquatic Hyphomycetes commonly observed with different submerged leaf litter collected from aquatic ecosystems in the State of São Paulo, Brazil.

Fungal species	**Type of substrate**
Alatospora acuminata Ingold	mixed leaf litter
Anguillospora crassa Ingold	leaf litter of *Caesalpinia echinata* Lam., *Tibouchina pulchra* Cogn. and mixed leaf litter
A. longissima (Sacc. and Syd.) Ingold	leaf litter of *Ficus microcarpa* L. f., *Caesalpinia echinata, Campomanesia phaea* (O. Berg.) and mixed leaf litter
Blodgettia indica Subramanian	Mixed leaf litter
Campylospora chaetocladia Ranzoni	mixed leaf litter
Camposporium pellucidum (Grove) Hughes	mixed leaf litter
Dactylella aquatica (Ingold) Ranzoni	mixed leaf litter
Dactylella submersa (Ingold) Nilsson	mixed leaf litter
Dendrospora erecta Ingold	mixed leaf litter
Flabellopsora crassa Alasoadura	mixed leaf litter
Flagellospora curvula Ingold	mixed leaf litter
Flagellospora penicillioides Ingold	mixed leaf litter
Lemmoniera aquatica De Wild.	mixed leaf litter
Lunulospora curvula Ingold	mixed leaf litter, leaf litter of *Campomanesia phaea*
Lunulospora cymbiformis Miura	mixed leaf litter
Margaritispora aquática Ingold	mixed leaf litter
Monotosporella microaquatica (Tubaki) S. Nilsson	mixed leaf litter
Naiadella fluitans Marvanová & Bandoni	mixed leaf litter
Pyramidospora casuarine Nilsson	leaf litter of *Caesalpinia echinata, Campomanesia phaea* and mixed leaf litter
Tetrachaetum elegans Ingold	leaves of *F. microcarpa* and mixed leaf litter
Tetracladium marchalianum De Wild.	mixed leaf litter
Tetracladium setigerum (Grove) Ingold	mixed leaf litter
Tetracladium maxilliforme (Rostr.) Ingold	mixed leaf litter
Tripospermum camelopardus Ingold, Dann and Mac Dougall	mixed leaf litter
Tripospermum myrtii (Lind.) S.J. Hughes	mixed leaf litter
Triscelophorus acuminatus Nawawi	mixed leaf litter, leaf litter of *Caesalpinia echinata* and *Campomanesia phaea*
Triscelophorus magnificus Petersen	mixed leaf litter
Triscelophorus monosporus Ingold	mixed leaf litter, leaf litter of *Caesalpinia echinata* and *Campomanesia phaea*
Triscelosporium verrucosum Nawawi & Kuthubuteen	mixed leaf litter
Varicosporium elodeae Kegel	mixed leaf litter and leaf litter of *Caesalpinia echinata*

Ecological Aspects

Fungal diversity is affected by many abiotic, climatic and edaphic factors. Aquatic Hyphomycetes were observed for the first time in small, shallow, moderately turbulent and well-aerated cold streams (Ingold, 1942). Thereafter, several surveys and ecological studies were conducted, mainly in temperate climates, to evaluate the influence of the environmental and climatic condition on the fungal diversity. One of the classic statements about the influence of abiotic factors was given by Suberkropp (1984). According to him, the diversity of the fungi in/on submerged leaves may have different assemblages according to the temperature of the water. Temperature may affect the metabolisms of a single species or define the structure of a whole fungal community (Fernandes et al., 2009).

Medeiros et al. (2009) verified that the decrease in oxygen content of streams may affect the diversity and activity of some aquatic Hyphomycetes, and consequently, the decomposition rate of *Alnus glutinosa* leaves. The sporulation rate was strongly inhibited by the decrease of dissolved oxygen content in water. Species such as *Flagellospora curvula* and *Anguillospora filiformis* were predominant in waters with higher oxygen content, whereas *Articulospora tetracladia*, *Cylindrocarpon* sp. and *Flagellospora curvula* were more frequent in waters with depleted oxygen content.

Nutrients present in the aquatic environment, temperature, pH and oxygen level may limit or may stimulate the occurrence of some fungi (Suberkropp and Chauvet, 1995). However, it is difficult to define precisely the range of the influence of these factors on the fungal diversity, but the increase of water temperature may stimulate general microbial metabolism, enhancing competition among several types of organisms that may hamper growth or germination of the Ingoldean fungal spores.

In temperate climates, the aquatic Hyphomycetes present a well defined seasonal behavior, occurring mainly at the end of the summer and autumn, when leaves are mostly senescent (Bärlocher, 1992), whereas, the studies conducted in the tropics (Betancourt et al., 1987; Iqbal et al., 1980; Sridhar and Kaveriappa, 1989), including the Brazilian studies (Schoenlein-Crusius et al., 1990, 1992, 1999), concluded that the occurrence of aquatic Hyphomycetes is greatly influenced by the status of decomposition of the leaves, especially with regards to the nutrient content or trophic level of aquatic environment, rather than seasonal changes (dry or rainy seasons).

It is assumed that in the lentic systems, aquatic Hyphomycetes are less diverse and abundant due to their replacement by other fungal groups (Webster and Descals, 1981). However, many species have been found as integrants of the mycota associated to several types of leaves submerged in many artificial lakes (Schoenlein-Crusius and Milanez, 1989; Schoenlein-Crusius, 2002; Schoenlein-Crusius et al., 2004).

Considering that some aquatic Hyphomycetes may have substrate specificity, many studies have aimed to evaluate if the leaf species had any influence on the diversity of leaf mycota. It has been found that different leaf species may have varying decomposition rates and fungi associated with them, even if they are submerged in the same aquatic environment (Canhoto and Graça, 1996; Gulis and Suberkropp, 2003a,b).

Intrinsic features of the leaves, such as nutrient, lignin, cellulose, hemicelulose, tannins, terpenoids, waxes and oils, influence the rate of leaf decay and the diversity of aquatic Hyphomycetes, justifying the importance of the plant species composition of riparian forests to aquatic environments (Bärlocher and Graça, 2002; Laitung and Chauvet, 2005). If the decomposition of leaves takes a longer period, rare species or taxa with reduced saprophytic ability may have a greater opportunity to colonize the substrate, contributing to increase the fungal richness (Gessner and Chauvet, 1994; Canhoto and Graça, 1999).

In general, the diversity and taxonomical composition of the fungal community changes with time when persistent species which are present in the substrates since the beginning of the decomposition process finally "allow" the colonization by the outsiders. Rossett and Bärlocher (1985) observed that when inoculated pure cultures of aquatic Hyphomycetes in substrates at the beginning of the decomposition, a significant decrease of all other species was noticed.

The water's trophic state may directly influence taxonomic composition of the aquatic mycota. For instance, when Suberkropp (1984) incubated leaves in waters with low nitrogen and phosphorus content, and afterwards transferred them to a nutrient-rich stream, species from the environment rapidly joined the fungal community of the decomposing leaves. However, the contrary did not happen when leaves were transferred from nutrient-rich to nutrient-poor waters. In a similar way, when leaves were submerged *"in vitro"* in waters with different nutrient contents, the increase of nitrogen and phosphorus stimulated biomass production and reproduction of the fungi, but not species richness (Sridhar and Bärlocher, 2000).

The aquatic Hyphomycetes have the ability to consume particulate and dissolved organic matter, as well as mineralized elements in the water. The lack or the excess of nutrients in the water may compromise the fungal diversity (Gulis and Suberkropp, 2003a). Thus, water sanitation treatments that intend to reduce the particulate organic matter may negatively affect the diversity of aquatic Hyphomycetes if not adequately monitored. Considering the great importance of aquatic Hyphomycetes for the maintenance of aquatic environment, it is important to evaluate the intensity of the impact of water treatment on their diversity and taxonomic richness.

Why is it important to study Ingoldean fungi in polluted waters? The answer is, if the comprehension of the influence of abiotic factors on the fungal diversity is difficult, the recognition of the impact of anthropic impacts is far more complex. Human interference, expressed as eutrophication may affect leaves decomposition and, consequently, the activity and diversity of aquatic Hyphomycetes. The increase or suppression of fungal species may occur according to nutrient status (Gulis and Suberkropp, 2003b) or the presence of toxic compounds (Shridhar et al., 2001). For instance, Greathead (1961) and Conway (1970) verified a drastic decrease of the number of aquatic Hyphomycetes in streams contaminated with domestic sewage. Burgos and Castillo (1986) made comparative studies before and after the release of organic pollution in the streams. The leaves before the introduction of pollutants were colonized by 17 fungal species, whereas after the addition of the pollutants the leaves were colonized only by four species. Au et al. (1992) also noted significant decrease of the fungal diversity in streams in Hong Kong. On the other hand, Sladechova (1963) reported that the growth of some aquatic Hyphomycetes was stimulated by the presence of starch in the effluent released in a stream. Schoenlein-Crusius and Milanez (1989) followed the fungal succession on leaves of *Ficus microcarpa* submerged in a eutrophic lake, observing many species of aquatic Hyphomycetes that were present in oligotrophic or mesotrophic lakes in further studies (Schoenlein-Crusius et al., 1992; Schoenlein-Crusius et al., 2009b).

Raviraja et al. (1998) considered that the decrease of 80% of the diversity of aquatic Hyphomycetes in an Indian stream was due to organic pollution. The typical situation for the fungi in impacted environments is that a few species succeed to grow intensively for a short period, keeping the fungal diversity very low.

Furthermore, in general, water bodies situated in urban or agricultural regions are under constant risk of being affected by the use of pesticides, fertilizers, sewage and other compounds that usually cause an increase in phosphorus and nitrogen contents of the environment (Carpenter et al., 1998). Added nutrients may stimulate the growth of many species that were earlier limited by the absence or scarcity of some nutritional elements and thereby increase their diversity and activity (Gulis and Suberkropp, 2003b). Not only this, but the intense proliferation of some species may enhance competition among the species for space and nutrients resulting in the establishment of a very competitive, but poorly diversified mycota (Webster and Benfield, 1986).

That each taxonomical identity in the ecosystem has a certain function in the food chain and, therefore, in the ecological balance, the decrease of diversity has dramatic effects on many populations, with many irreversible consequences (Webster and Benfield, 1986; Maltby, 1992). In polluted water bodies, it is common to have oxygen contents that are limiting to the mycota. Especially, aquatic Hyphomycetes have been observed as fairly tolerant to different types of pollutants, to temperature and pH variations, and several trophic levels, but are sensitive to low oxygen contents in the water (Carpenter et al., 1998).

Considering that aquatic Hyphomycetes have a significant role in the decomposition process of submerged leaf litter, and consequently the trophic chain, the evaluation of species richness and quantity of biomass may contribute to indication of environmental stress (Gessner and Chauvet, 1997). Also, aquatic Hyphomycetes promote nutritional enrichment of the decomposing leaves, increasing palatability of the substrates to detritivorous organisms (Bärlocher and Kendrick, 1974; Graça, 2001).

In urban systems, the presence of heavy metals in water or accumulated in organic substrates such as leaves and roots is very common. The presence of heavy metals in water may affect negatively the decomposition rate of the leafy substrates (Abel and Bärlocher, 1984, 1988), influencing the decomposition rate and microbial respiration (Niyogi et al., 2001), sporulation rate (Shridhar et al., 2001) and the taxonomic diversity of aquatic Hyphomycetes (Bermingham et al., 1996; Krauss et al., 1998, 2005a,b).

In spite of the adverse condition that urban polluted waters offer to aquatic Hyphomycetes, many species may survive, resisting organic pollution (Au et al., 1992; Raviraja et al., 1998; Sridhar and Raviraja, 2001) and heavy metals (Krauss et al., 1998; Sridhar et al., 2000, 2001), in shallow (Sridhar et al., 2000, 2001) and in groundwater (Krauss et al., 2003a,b, 2005a,b).

Some species of aquatic Hyphomycetes have the ability to tolerate and survive in environments with high heavy metal contents, such as *Heliscus lugdunensis*, which was frequently observed in leaves submerged in zinc and cadmium rich waters (Krauss et al., 2001). The resistance and tolerance of this species depend on the kind of heavy metals present in the environment. The fact that a species is tolerant to a type of metal does not necessarily mean that it will react so to every kind of metallic element (Jaeckel et al., 2005).

Bärlocher (1992) considered that the addition of organic pollution to the water may affect the abundance of individual fungal species, hampering the leaf decomposition rate, not necessarily decreasing the activity of the Ingoldian fungi community. However, Ingoldian fungi may be affected if the dissolved oxygen level is critical. The ideal situation is to know the fungal diversity before the ecosystem is subjected to a more intense anthropogenic influence. This facilitates the analysis of the dimensions of the impact, and verification if environmental remediation measures are indeed returning the diversity of the mycota, approximately as much as possible, to what it was previously.

Due to their sensitivity to lower levels of oxygen in water, aquatic Hyphomycetes may seem to be efficient bioindicators of water quality. Another aspect to be taken into account to justify the study of aquatic Hyphomycetes in adverse conditions is that the decrease or absence of the species may indicate a situation of environmental impact, while the return of some species in sites that are being dispolluted may indicate the recovering level of impacted environments.

Concluding Remark

The finding presented in the aforesaid pages clearly indicate that the assessment of the diversity of the mycota of urban aquatic environments that undergo intense anthropogenic influence, pollution, etc. can still bring surprises and many novelties. Therefore, it is worth studying the aquatic fungi in such environments, comparing several sites with different types of water quality in order to discover which species are resistant or need preservation. This information will help educate people that there is an active and important mycota in urban waters—water bodies of parks, public forests, reserves and reservoirs, which may be seen as sources of microorganisms, beyond considering them only as aesthetic elements of the landscape.

Acknowledgements

The authors are grateful to the "Conselho Nacional de Desenvolvimento Científico e Tecnológico"—CNPq for grants and financial support **(processo 304526/2009-6)**.

References

Abel, T.H. and Bärlocher, F. 1984. Effects of cadmium on aquatic hyphomycetes. Applied Environmental Microbiology, 48: 245–251.

Abel, T.H. and Bärlocher, F. 1988. Uptake of cadmium by *Gammarus fossarum* (Amphipoda) from food and water. Journal of Ecology, 25: 223–231.

Alexopoulos, C.J., Mims, C.W. and Blackwell, M. 1996. Introductory Mycology. John Wiley & Sons, Inc., 4° ed., New York, USA.

Au, D.W., Hodkiss, I.J. and Vrijmoed, L.L.P. 1992. Fungi and cellulolytic activity with decomposition of *Bauhinia purpurea* leaf litter in a polluted and unpolluted Hong Kong waterway. Canadian Journal of Botany, 70: 1071–1079.

Bärlocher, F. 1992. Research on aquatic hyphomycetes: historical background and overview. pp. 1–15. *In*: Bärlocher, F. (ed.). The ecology of aquatic Hyphomycetes. Springer Verlag, Berlin, Germany.

Bärlocher, F. 2005. Leaf mass loss estimated by litter bag technique. pp. 37–42. *In*: Graça, M.A.S., Bärlocher, F. and Gessner, M.O. (eds.). Methods to Study Litter Decomposition. Springer Verlag Dordrecht, The Netherlands.

Bärlocher, F. 2009. Reproduction and dispersal in aquatic hyphomycetes. Mycoscience, 50: 3–8.
Bärlocher, F. and Graça, M.S. 2002. Exotic riparian vegetation lowers fungal diversity but not leaf decomposition in Portuguese streams. Freshwater Biology, 47: 1123–1135.
Bärlocher, F. and Kendrick, B. 1974. Dynamics of the fungal population on leaves in the stream. Journal of Ecology, 62: 761–791.
Berminghan, S., Maltby, L. and Coole, R.C. 1996. Effects of a coal mine effluent on aquatic hyphomycetes. I. Field study. Journal of Applied Ecology, 33: 1311–1321.
Betancourt, C., Cruz, J. and Garcia, J. 1987. Los hifomicetos acuaticos de la quebrada Doana Juana em el bosque estatal de Toro negro, Villalba, Puerto Rio. Caribbean Journal of Science, 23: 278–284.
Burgos, E.J. and Castillo, P.H. 1986. Hyphomycetes acuaticos como indicadores de contaminacion. Biota, 2: 1–10.
Canhoto, C. and Graça, M.A.S. 1996. Decomposition of *Eucalyptus globulus* leaves and three native leaf species (*Alnus glutinosa, Castanea sativa* and *Quercus faginea*) in a Portuguese low order stream. Hydrobiologia, 333: 79–85.
Canhoto, C. and Graça, M.A.S. 1999. Leaf barriers to fungal colonization and shreeders (*Tipula lateralis*) consumption of decomposing *Eucalyptus globulus*. Microbial Ecology, 37: 163–172.
Capelari, M., Grandi, R.A.P., Gugliotta, A.M., Pires-Zottarelli, C.L.A. and Schoenlein-Crusius, I.H. 2009. Diversidade de Fungos, Chapter 12. pp. 216–228. *In*: Lopes, M.I.S., Kirizawa, M. and Fiúza, M.M. (eds.). Patrimônio da Reserva Biológica do Alto da Serra de Paranapiacaba, a antiga Estação Biológica do Alto da Serra. Governo do Estado de São Paulo, Secretaria do Meio Ambiente, Instituto de Botânica, São Paulo, Brasil.
Carpenter, S.R., Caraco, N.F., Correll, D.L., Howarth, R.W., Sharpley, A.N. and Smith, V.H. 1998. Nonpoint pollution of surface waters with phosphorus and nitrogen. Ecological Applications, 8: 559–568.
Cheng, Z.L., Andre, P. and Chiang, C. 1997. Hyphomycetes and macroinvertebrates colonizing leaf litter in two Belgian streams with contrasting water quality. Limnetica, 13(2): 57–63.
Conway, K.E. 1970. The aquatic hyphomycetes of central New York. Mycologia, 62: 16–530.
Descals, E. 2005. Techniques for handling Ingoldian Fungi. pp. 129–141. *In*: Graça, M.A.S., Bärlocher, F. and Gessner, M.O. (eds.). Methods to Study Litter Decomposition. Springer Verlag Dordrecht, The Netherlands.
Dias, A.I. 2008. Avaliação da velocidade de decomposição e da riqueza de Hyphomycetes aquáticos associados às folhas de *Campomanesia phaea* (O. Berg.) Landrum (Cambuci) submersas em um lago artificial no Parque Municipal Alfredo Volpi, SP. Graduation Monograph, Universidade de Santo Amaro, São Paulo, Brasil.
Fagundes, J.C. 2008. Avaliação da velocidade de decomposição e da riqueza de geofungos associados às folhas de *Caesalpinia echinata* Lam. (pau-brasil) submersas em um lago artificial no Parque Municipal Alfredo Volpi, SP. Graduation Monograph, Universidade de Santo Amaro, São Paulo, Brasil.
Fernandes, I., Uzun, B., Pascoal, C. and Cássio, F. 2009. Responses of aquatic fungal communities on leaf litter to temperature-change events. International Review of Hydrobiology, 94: 410–418.
Field, J.I. and Webster, J. 1983. Anaerobic survival of aquatic fungi. Transaction of the British Mycological Society, 81: 365–369.
Gessner, M.O. 1997. Fungal biomass, production and sporulation associated with particulate organic matter in streams. Limnetica, 13(2): 33–44.
Gessner, M.O. and Chauvet, E. 1994. Importance of stream microfungi in controlling breakdown rates of leaf litter. Ecology, 75: 1807–1817.
Gessner, M.O. and Chauvet, E. 1997. Growth and production of aquatic hyphomycetes in decomposing leaf litter. Limnology and Oceanography, 42: 496–505.

Gessner, M.O., Bauchrowitz, M.A. and Escautier, M. 1991. Extraction and quantification of ergosterol as a measure of fungal biomass in leaf litter. Microbial Ecology, 22: 285–291.

Goh, T.K. 1997. Tropical freshwater hyphomycetes. pp. 189–227. *In*: Hyde, K.D. (ed.). Biodiversity of Tropical Microfungi. Hong Kong University Press, Hong Kong, China.

Graça, M.A.S., Ferreira, R.C. and Coimbra, C.N. 2001. Litter processing along a stream gradient: the role of invertebrates and decomposers. Journal of the North American Benthological Society, 20: 408–420.

Greathead, S.K. 1961. Some aquatic hyphomycetes in South Africa. Journal of Southafrican Botany, 27: 195–228.

Gruppi, V. 2008. Diversidade de fungos microscópicos em folhedo submerso de lagos do Parque Central no município de Santo André, SP, Brasil. Graduation Monograph, Universidade Metodista, São Bernardo do Campo, São Paulo, Brasil.

Gulis, V. and Suberkropp, K. 2003a. Interactions between stream fungi and bacteria associated with decomposing leaf litter at different levels of nutrient availability. Aquatic Microbial Ecology, 30: 149–157.

Gulis, V. and Suberkropp, K. 2003b. Leaf litter decomposition and microbial activity in nutrient-enriched and unaltered reaches of a headwater stream. Freshwater Biology, 48: 123–134.

Gulis, V., Marvanová, L. and Descals, E. 2005. An illustrate key to the common temperate species of aquatic hyphomycetes. pp. 153–168. *In*: Graça, M.A.S., Bärlocher, F. and Gessner, M.O. (eds.). Methods to Study Litter Decomposition: A Practical Guide. Springer Verlag Dordrecht, The Netherlands.

Ingold, C.T. 1942. Aquatic hyphomycetes of decaying alder leaves. Transactions of the British Mycological Society, 25: 339–417.

Ingold, C.T. 1943. Further observations on aquatic hyphomycetes. Transactions of the British Mycological Society, 26: 104–115.

Ingold, C.T. 1944. Some new aquatic hyphomycetes. Transactions of the British Mycological Society, 28: 35–43.

Ingold, C.T. 1949. Aquatic hyphomycetes from Switzerland. Transactions of the British Mycological Society, 32: 341–345.

Ingold, C.T. 1975. An Illustrated Guide to Aquatic and Water-borne Hyphomycetes (Fungi Imperfecti) with notes on their Biology. Freshwater Biological Association n. 30, 96 p. Windermere, United Kingdom.

Iqbal, S.H., Bhatty, S.F. and Malik, K.S. 1980. Freshwater hyphomycetes on submerged decaying pine needles in Pakistan. Transactions of the Mycological Society of Japan, 21: 321–327.

Jaeckel, P., Krauss, G.J. and Krauss, G. 2005. Cadmium and zinc response of the fungi *Heliscus lugdunensis* and *Verticillium* cf. *Alboatrum* isolated from highly polluted water. Science of the Total Environment, 346(1-3): 274–279.

Kaushik, N.K. and Hynes, H.B. 1971. The fate of dead leaves that fall into streams. Archiv für Hydrobiologie, 68: 465–515.

Kirk, P.M., Cannon, P.F., David, J.C. and Stalpers, J.A. 2001. Ainsworth & Bisby's Dictionary of the Fungi. 9th. ed. CAB International, 655 p., United Kingdom.

Kirk, P.M., Cannon, P.F., Minter, D.W. and Stalpers. J.A. 2008. Dictionary of the Fungi. 10th ed. CAB International, Wallingford, United Kingdom.

Krauss, G., Bärlocher, F. and Krauss, G.-J. 2003a. Effects of pollution on aquatic hyphomycetes. pp. 211–230. *In*: Tsui, C.K.M. and Hyde, K.D. (eds.). Freshwater Mycology: A Practical Approach. Fungal Diversity Research Series 10, Fungal Diversity Press, Hong Kong, China.

Krauss, G., Sridhar, K.R., Jung, K., Wennrich, R., Ehrman, J. and Bärlocher, F. 2003b. Aquatic Hyphomycetes in polluted groundwater habitats of central Germany. Microbial Ecology, 45: 329–339.

Krauss, G., Schlosser, D. and Krauss, G.-J. 2005a. Aquatic fungi in heavy metal and organically polluted habitats. pp. 221–246. *In*: Deshmukh, S.K. and Rai, M.K. (eds.). Biodiversity of Fungi: Their Role in Human Life. Oxford and IBA Publishing Co. New Delhi, India.

Krauss, G., Sridhar, K.R. and Bärlocher, F. 2005b. Aquatic hyphomycetes and leaf decomposition in contaminated groundwater wells in central Germany. Archives of Hydrobiology, 162(3): 417–429.

Laitung, B. and Chauvet, E. 2005. Vegetation diversity increases species richness of leaf-decaying fungal communities in woodland streams. Archiv für Hydrobiologie, 164: 217–235.

Malosso, E. 1995. Ocorrência de Hyphomycetes (Fungi Imperfecti) e Fungos Zoospóricos em Ambientes Aquáticos (Rio do Monjolinho, São Carlos, SP). Graduation Monograph, Universidade Federal de São Carlos, São Carlos, Brasil.

Malosso, E. 1999. Hyphomycetes em ambientes aquáticos lótico e lêntico—ocorrência e biomassa. M.S. Thesis. Universidade Federal de São Carlos, São Carlos, Brasil.

Maltby, L. 1992. Heterotrophic microbes. pp. 165–194. *In*: Calow, P. and Pettes, G.E. (eds.). The Rivers Handbook: Hydrological and Ecological Principles. Vol. 1. Blackwell Scientific, London, United Kingdom.

Medeiros, A.O., Pascoal, C. and Graça, M.A. 2009. Diversity and activity of aquatic fungi under low oxygen conditions. Freshwater Biology, 54: 142–149.

Moreira, C.G. 2006. Avaliação da diversidade e biomassa de fungos associados a folhas em decomposição de *Tibouchina pulchra* Cogn. submersas em reservatórios do Parque Estadual das Fontes do Ipiranga (PEFI), São Paulo, SP. MS. Thesis, Instituto de Botânica, São Paulo, Brasil.

Moreira, C.G. 2011. Sucessão de hifomicetos e avaliação da biomassa fúngica durante a decomposição de folhedo de *Caesalpinia echinata* Lam. e *Campomanesia phaea* (O. Berg.) Landrum submersos em lagos artificiais na cidade de São Paulo, SP. Doctor´s Thesis. Instituto de Botânica, São Paulo, Brasil.

Moreira, C.G. and Schoenlein-Crusius, I.H. 2007. Fungos decompositores de substratos foliares submersos em ambientes aquáticos continentais: estado da arte e novos dados obtidos para o Brasil. pp. 172–176. *In*: Barbosa, L.M. and Santos Junior, N.A. (org.). A Botânica no Brasil: pesquisa, ensino e políticas públicas ambientais, Sociedade Botânica do Brasil, Vol. 1, São Paulo, Brasil.

Nilsson, S. 1964. Freshwater Hyphomycetes. Symbolae Botanicae Upsalienses, 18: 1–130.

Niyogi, D.K., Lewisand, W.M. and McKnight, D.M. 2001. Litter breakdown in mountain streams affected by mine drainage: biotic mediation of abiotic controls. Journal of Applied Ecology, 11: 506–516.

Pires-Zottarelli, C.L., Schoenlein-Crusius, I.H. and Milanez, A.I. 1993. Quantitative estimation of zoosporic fungi and aquatic hyphomycetes on leaves submerged in a stream in the Atlantic rainforest in the state of São Paulo, Brazil. Revista de Microbiologia, 24(3): 192–197.

Pugh, G.J.F., Buckey, N.G. and Mulder, J. 1972. The role of phylloplane fungi in the early colonization of leaves. Symposia Biologica Hungarica, 11: 329–333.

Raviraja, N.S., Sridhar, K.R. and Bärlocher, F. 1998. Breakdown of *Fícus* and *Eucalyptus* leaves in an organically polluted river in India: fungal diversity and ecological functions. Freshwater Biology, 39: 537–545.

Rossett, J. and Bärlocher, F. 1985. F. Transplant experiments with aquatic hyphomycetes. Verhandlungen von der Internationallen Vereinigung für Limnologie, 22: 2786–2790.

Saccardo, P.A. 1880. Sylloge Fungorum, Padua.

Schoenlein-Crusius, I.H. 2002. Aquatic hyphomycetes from cerrado regions in the state of São Paulo, Brazil. Mycotaxon, 81: 457–462.

Schoenlein-Crusius, I.H. and Grandi, R.A.P. 2003. The diversity of aquatic hyphomycetes in South America. Brazilian Journal of Microbiology, 34: 183–193.

Schoenlein-Crusius, I.H. and Malosso, E. 2007. Diversity of Aquatic hyphomycetes in the tropics. pp. 61–81. *In*: Ganguli, B.N. and Deshmukh, S.K. (eds.). Fungi: multifaceted Microbes. Anamaya Publishers, New Delhi, India.

Schoenlein-Crusius, I.H. and Milanez, A.I. 1989. Sucessão fúngica em folhas de *Ficus microcarpa* L.f. submersas no lago frontal situado no Parque Estadual das Fontes do Ipiranga, São Paulo, SP. Revista de Microbiologia, 20: 95–101.

Schoenlein-Crusius, I.H. and Milanez, A.I. 1990a. Aquatic hyphomycetes in São Paulo State, Brazil. I. First observations. Hoehnea, 17(2): 111–115.

Schoenlein-Crusius, I.H. and Milanez, A.I. 1990b. Hyphomycetes aquáticos no Estado de São Paulo, Brasil. Revista Brasileira de Botânica, 13: 61–68.

Schoenlein-Crusius, I.H. and Milanez, A.I. 1996. Diversity of aquatic fungi in Brazilian Ecosystems. Chapter 4. pp. 31–48, *In*: Bicudo, C.E.M. and Menezes, N.A. (eds.). Biodiversity in Brazil, A First Approach, CNPq, São Paulo, Brasil.

Schoenlein-Crusius, I.H. and Milanez, A.I. 1998. Fungal succession on leaves of *Alchornea triplinervia* (Spreng.) M. Arg. Submerged in a stream of an Atlantic Rainforest in the State of São Paulo, Brazil. Revista Brasileira de Botânica, 21(3): 253–259.

Schoenlein-Crusius, I.H., Pires-Zottarelli, C.L.A. and Milanez, A.I. 1990. Sucessão fúngica em folhas de *Quercus robur* L. (carvalho) submersas em um lago situado no município de Itapecerica da Serra, SP. Revista de Microbiologia, 21(1): 61–67.

Schoenlein-Crusius, I.H., Pires-Zottarelli, C.L.A. and Milanez, A.I. 1992. Aquatic fungi in leaves submerged in a stream in the Atlantic rainforest. Revista de Microbiologia, 23(3): 167–171.

Schoenlein-Crusius, I.H., Pires-Zottarelli, C.L.A. and Milanez, A.I. 2004. Amostragem em Limnologia: os Fungos Aquáticos. pp. 179–191. *In*: Bicudo, C.E.M. and Bicudo, D.C. (org.). RiMa Editora, São Carlos, São Paulo, Brasil.

Schoenlein-Crusius, I.H., Pires-Zottarelli, C.L.A., Milanez, A.I. and Humphreys, R.D. 1999. Interaction between the mineral content and the occurrence number of aquatic fungi in leaves submerged in a stream in the Atlantic rainforest, São Paulo, Brazil. Revista Brasileira de Botânica, 22(2): 133–139.

Schoenlein-Crusius, I.H., Milanez, A.I., Trufem, S.F.B. and Pires-Zottarelli, C.L.A. 2009a. Fungos: estudos ecológicos. pp. 233–241. *In*: Lopes, M.I.S., Kirizawa, M. and Fiúza, M.M. (eds.). Patrimônio da Reserva Biológica do Alto da Serra de Paranapiacaba, a antiga Estação Biológica do Alto da Serra, Governo do Estado de São Paulo, Secretaria do Meio Ambiente, Instituto de Botânica, São Paulo, Brasil.

Schoenlein-Crusius, I.H., Moreira, C.G. and Bicudo, D.C. 2009b. Aquatic hyphomycetes in the "Parque Estadual das Fontes do Ipiranga"—PEFI, São Paulo, Brazil. Revista Brasileira de Botânica, 32(3): 411–426.

Schoenlein-Crusius, I.H., Moreira, C.G. and Pires-Zottarelli, C.L.A. 2007. O papel dos fungos nos ecossistemas aquáticos. Boletim da Sociedade Brasileira de Limnologia, 36(1): 26–30.

Sládecková, A. 1963. Aquatic deuteromycetes as indicators of starch campaign pollution. Internationale Revue der Gesamten Hydrobiologie, 48: 35–42.

Sridhar, K.R. and Bärlocher, F. 2000. Initial colonization, nutrient supply, and fungal activity on leaves decaying in streams. Applied and Environmental Microbiology, 66(3): 1114–1119.

Sridhar, K.R. and Kaveriappa, K.M. 1989. Observations on aquatic hyphomycetes of the western Ghat streams, India. Nova Hedwigia, 49(3-4): 455–467.

Sridhar, K.R. and Raviraja, N.S. 2001. Aquatic hyphomycetes and leaf litter processing in polluted and unpolluted habitats. pp. 293–314. *In*: Misra, J.K. and Horn, B.W. (eds.). Trichomycetes and Other Fungal Groups. Science Publishers, New York & London.

Sridhar, K.R., Krauss, G., Bärlocher, F., Wennrich, R. and Krauss, G.-J. 2000. Fungal diversity in heavy metal polluted water in central Germany. Fungal Diversity, 5: 119–129.

Sridhar, K.R., Krauss, G., Bärlocher, F., Raviraja, N.S., Wennrich, R., Baumbach, R. and Krauss, G.-J. 2001. Decomposition of alder leaves in two heavy metal polluted streams in central Germany. Aquatic Microbial Ecology, 26: 73–80.

Suberkropp, K. 1984. The effect of temperature on the seasonal occurrence of aquatic hyphomycetes. Transactions of the British Mycological Society, 82: 53–62.

Suberkropp, K. and Chauvet, E. 1995. Regulation of leaf breakdown by fungi in streams: influences of water chemistry. Ecology, 76: 1433–1445.
Subramanian, C.V. 1983. Hyphomycetes. Taxonomy and biology. London: Academic Press, 502 p.
Tsui, C.K.M. and Hyde, K.D. 2003. Freshwater Mycology. Fungal Diversity Research Series 10, Hong Kong.
Webster, J. 1980. Introduction to fungi. Cambridge, Cambridge University Press, 669 p.
Webster, J. and Benfield, E.F. 1986. Vascular plant breakdown in freshwater ecosystems. Annual Review of Ecology and Systematics, 17: 567–594.
Webster, J. and Descals, E. 1981. Morphology, distribution and ecology of conidial fungi in freshwater habitats. pp. 295–355. *In*: Cole, G.C. and Kendrick, B. (eds.). Biology of the Conidial Fungi, Vol. 1. Academic Press, London, United Kingdom.
Wellbaum, C., Schoenlein-Crusius, I.H. and Santos, V.B. 1999. Fungos filamentosos em folhas do ambiente terrestre e aquático da Ilha dos Eucaliptos, Represa do Guarapiranga, SP. Revista Brasileira de Botânica, 22: 69–74.
Wellbaum, C., Schoenlein-Crusius, I.H., Malosso, E. and Tauk-Tornisiello, S.M. 2007. Fungos filamentosos de folhas em decomposição na Represa de Guarapiranga, São Paulo, SP. Holos Environment, 7(2): 171–190.

CHAPTER 4

Fungi from Substrates in Marine Environment

Zhu-Hua Luo[1] and *Ka-Lai Pang*[2,*]

ABSTRACT

Fungi are one of the main decomposers in marine environment, transforming complex organic matter into simpler molecules for their own nutrition and/or for other organisms. Major growth substrates of marine fungi include wood (drift timber, mangrove wood), marine algae, saltmarsh grasses (*Spartina* spp., *Juncus roemerianus, Phragmites australis*) and *Nypa fruticans*. Marine fungi also occur on other substrata such as animal exoskeletons and keratinaceous materials, but recent research has been focused on fungi in deep-sea floor and seawater, sediments, minerals, and animals around hydrothermal vents. In this chapter, fungi associated with these substrates have been briefly reviewed, with an emphasis on their diversity.

Introduction

Natural products of marine origin have been the main targets in recent years for new compounds for industrial and pharmaceutical purposes. Microbes associated with marine animals including sponges and corals are extensively explored for secondary metabolites (Blunt et al., 2012). Marine fungi occur on a range of substrata/substrates in marine environments including woody

[1] Key Laboratory of Marine Biogenetic Resources, Third Institute of Oceanography, State Oceanic Administration, 178 Daxue Road, Xiamen 361005, PR China.
[2] Institute of Marine Biology and Centre of Excellence for the Oceans, 2 Pei-Ning Road, Keelung 202-24, Taiwan (R.O.C.).
* Corresponding author

tissues, leaves and fruits of marine plants, sea grasses, algae, seaweeds, animal exoskeletons and keratinaceous materials, sediments, sea foam and seawater (Vrijmoed, 2000).

Early research on marine fungi was mainly on those growing on seaweeds and wood (Sutherland, 1916b; Barghoorn and Linder, 1944). In recent years, studies of marine fungi associated with marsh grasses such as *Spartina* spp., *Juncus roemerianus* and *Phragmites australis* are also underway in different parts of the world (Calado and Barata, 2012). During the last three decades, many new marine fungi have been described on mangrove flora, which also include the mangrove palm *Nypa fruticans* (Hyde et al., 1999). The total number of marine fungi described from these substrates currently stands at 530, of which 424 species belong to the Ascomycota, 12 to the Basidiomycota and 94 to the asexual fungi (Jones et al., 2009). However, this number does not include the traditionally known phycomycetes (or "lower" marine fungi) and "facultative" marine fungi. "Lower" marine fungi are predominantly zoosporic true fungi (Chytridiomycota, Neocallimastigomycota, Olpidium group) and fungus-like organisms producing motile spores, including the opisthokonta (Mesomycetozoea, Cryptomycota), the chromalveolata (Oomycetes, Labyrinthulomycota, Hyphochytridiomycota) and the Rhizaria (Phytomyxea) (Adl et al., 2005; Neuhauser et al., 2012). "Facultative" marine fungi are those that originate from terrestrial or freshwater environments, exist as propagules in the marine habitats, but do not complete their life cycles in the marine environments. Typical examples of this group of fungi are terrestrial taxa *Aspergillus* spp., *Cladosporium* spp. and *Penicillium* spp. but *Aspergillus sydowii*, a fungus causing death of sea fans, may be a true marine fungus (Geiser et al., 1998). Many of the "facultative" marine fungi have so far been isolated from seaweeds and sediments and may be actively involved in the degradation of complex organic matters into simpler molecules for use as their own nutrients and for other organisms in the food web.

In this chapter, marine fungi associated with various major substrates in the sea, i.e., decaying wood, saltmarsh grasses, marine algae and the mangrove palm *Nypa fruticans* are reviewed (Figs. 4.1–4.4). In addition, fungi reported from the deep-sea floor and hydrothermal vents are also overviewed with the current surge of research activity in this field.

Decaying Wood

Fungi associated with decaying wood are the best studied ecological group of the sea. Wood-decaying enzymes such as cellulases, hemicellulases and laccases are produced by marine fungi to mineralize wood for nutrition, causing a soft-rot type of decay (Bucher et al., 2004; Luo et al., 2005). Wood is available as trapped wood in between rocks and driftwood for marine

Figures. 4.1–4.4. Major habitats and substrates of marine fungi. **Figure 4.1.** A mangrove forests in the Philippines with intertidal driftwood and dead attached wood for colonization of marine fungi. **Figure 4.2.** An estuarine palm forest (*Nypa fruticans*) in Thailand with intertidal dead fronds providing substrates for growth of marine fungi. **Figure 4.3.** A marsh of *Spartina* and *Salicornia* in England with dead plants as the substrates for growth. **Figure 4.4.** A rocky shore in Taiwan with washed up seaweeds.

fungi to colonize; in the tropics/subtropics, decaying wood in mangroves is abundant for colonization of marine fungi (Jones and Pang, 2012). Earlier studies were mainly focused on trapped wood, wood panels and driftwood in temperate areas (Barghoorn and Linder, 1944) but during the last few decades, extensive studies have been done on mangrove fungi (Schmit and Shearer, 2003, 2004). Marine fungi have also been reported from very cold locations: Grytviken, South Georgia (Antarctic) (Pugh and Jones, 1986), and Tromsøe and Longyearbyen, Norway (Arctic) (Pang et al., 2011). However, there is little overlap in fungal species from tropical and temperate regions (Hyde et al., 2005). But there are a number of cosmopolitan species such as *Ceriosporopsis halima* (Fig. 4.5), *Lignincola laevis* and *Torpedospora radiata* (Fig. 4.6) (Jones and Pang, 2012). Temperature is a key factor in controlling the distribution of marine fungi (Jones, 2000). Mouzouras (1989) and Mouzouras et al. (1988) showed that temperature affected the degree of weight loss of balsa (*Ochroma lagopus*) test blocks after 24 weeks exposure to selected marine fungi: *Digitatispora marina* (a cold water basidiomycete) causing 14 and 5% weight losses at 10°C and 22°C, respectively; *Halocyphina villosa* (a tropical basidiomycete) causing 0 and 23% losses at 10°C and 22°C, and for *Nia vibrissa* (a cosmopolitan species), the losses were 13 and 28% at 10°C and 22°C, respectively. Fruiting bodies of *D. marina* were only visible when the temperature dropped to below 10°C (Byrne and Jones, 1974).

Drift Timber in Temperate Areas

The Ascomycota is the dominant group on wood with many asexual fungi, while the Basidiomycota is not common (Jones et al., 1998). In San Juan Island, USA, the most frequently collected marine fungi were *Halosphaeria appendiculata* (Fig. 4.7), *Monodictys pelagica, Halosarpheia trullifera* and *Ceriosporopsis halima* (Jones et al., 1998). In Denmark, 19 species of marine fungi were regarded as being frequent, e.g., *Arenariomyces trifurcatus* (Fig. 4.8), *Nereiospora comata, Corollospora maritima* (Fig. 4.9), *Remispora stellata, R. maritima* (Koch and Petersen, 1996). Early colonizers of wood in temperate areas include *C. halima, Lulworthia* spp., while others are late colonizers such as *C. maritima, H. appendiculata* and *R. maritima* (Jones, 1976). Panebianco et al. (2002) pre-inoculated test blocks with *C. halima, C. maritima, Halosphaeriopsis mediosetigera* (Fig. 4.10) and *Marinospora calyptrata* and exposed the test blocks at Langstone Harbour, Portsmouth, UK. The pre-inoculated fungi were the dominant form throughout the 15-month incubation period, suggesting possible interference competition.

Mangrove Wood

Since 1980, much work has been done on fungi growing on mangrove wood in the tropical/sub-tropical areas (Brunei, Hong Kong, Malaysia, Singapore, Taiwan, and Thailand) and many new taxa have been described (Jones et al., 2009). Schmit and Shearer (2003) listed 625 taxa, but this figure included those growing on terrestrial parts of mangrove trees while Alias et al. (2010) provided a recent figure of 287 fungi growing on submerged mangrove substrata. Mangrove fungi constitute the second largest ecological group of marine fungi (Hyde and Jones, 1988) and again the Ascomycota is dominant with a few basidiomycetes (Jones et al., 2009). Arfi et al. (2012) used tag-encoded 454 pyrosequencing of the nuclear ribosomal internal transcribed spacer-1 to study fungi associated with the exposed and submerged parts of *Avicennia marina* and *Rhizophora stylosa* in New Caledonian mangroves. They confirmed that the Ascomycota is the main group associated with mangrove trees with Dothideomycetes and Sordariomycetes being dominant in the submerged samples. Unknown sequence types were found and did not match with those in the GenBank, suggesting either imbalanced representation of sequences in the GenBank or novel fungal lineages. Jones and Pang (2012) listed 16 core mangrove fungi including *Marinosphaera mangrovei*, *Lignincola laevis* (Fig. 4.11), *Kallichroma tethys*, *Verruculina enalia* (Fig. 4.12). The genera *Acrocordiopsis*, *Aigialus*, *Ascocratera*, *Belizeana*, *Halomassarina*, *Helicascus*, *Julella*, *Morosphaeria*, *Paraliomyces*, *Pyrenographa*, and *Salsuginea* are unique to mangrove habitats and have never been reported from temperate regions (Jones and Pang, 2012). On *Avicennia alba* and *A. lanata*, wood blocks exposed at the Mandai mangrove in Singapore, *L. laevis* was the early colonizer, *V. enalia* and *Lulworthia* sp. were the intermediate colonizers and *Aigialus parvus* was the late colonizer (Tan et al., 1989).

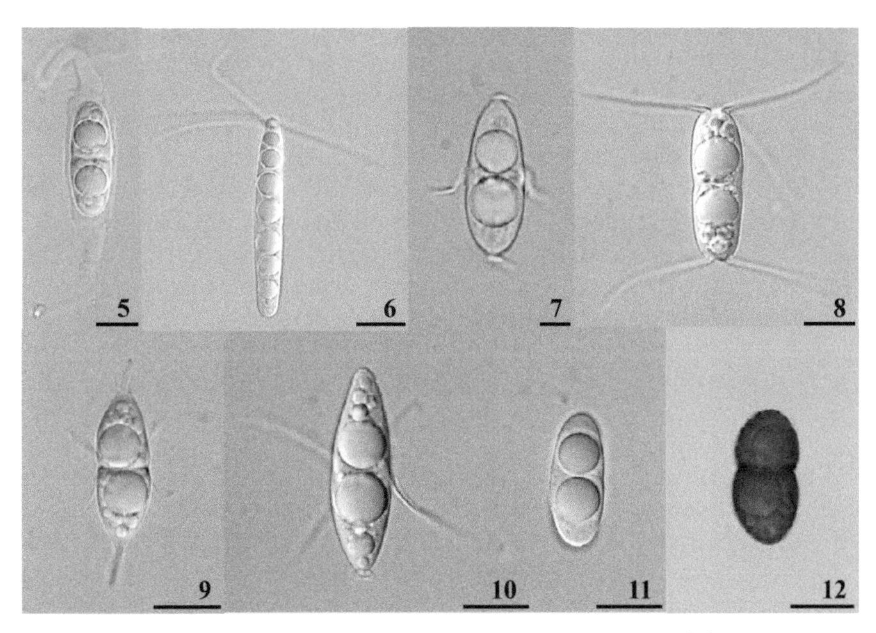

Figures 4.5–4.12. Lignicolous marine fungi. **Figure 4.5.** *Ceriosporopsis halima*. **Figure 4.6.** *Torpedospora radiata*. **Figure 4.7.** *Halosphaeria appendiculata*. **Figure 4.8.** *Arenariomyces trifurcatus*. **Figure 4.9.** *Corollospora maritima*. **Figure 4.10.** *Halosphaeriopsis mediosetigera*. **Figure 4.11.** *Lignincola laevis*. **Figure 4.12.** *Verruculina enalia*. Scale bars = 10 μm.

Marine Algae

Early marine fungi were described from marine algae (Rostrup, 1889; Patouillrad, 1897) and Sutherland (1915, 1916a,b) described numerous algicolous fungi from collections made in Scotland and south coast of England. Kohlmeyer and Volkmann-Kohlmeyer (2003) tallied 79 species of filamentous fungi associated with marine algae, which either were parasitic or had symbiotic relationships occurring as endophytes, parasites, primitive marine lichens or mycophycobioses (Kohlmeyer and Kohlmeyer, 1979). Jones et al. (2012) recently reviewed fungi associated with algae and listed 112 taxa that are either parasitic, saprophytic or forming lichenized relationships with algal hosts.

Fungi associated with algae are taxonomically very different from those on wood, although they are mostly ascomycetes and asexual fungi (Kohlmeyer, 1973). Marine fungi on wood are primarily members of the Halosphaeriaceae, Microascales while *Lulworthia*, *Haloguignardia* and *Spathulospora* are typical marine fungal genera from algae, belonging to the Lulworthiales (Jones et al., 2009). Many taxa await a molecular study to confirm their taxonomic placement, e.g., *Chadefaudia* spp., *Didymella magnei*, *Pontogenia* spp. and *Turgidosculum ulvae* (Kohlmeyer and Volkmann-Kohlmeyer, 2003; Jones et al., 2009). Many zoosporic organisms including the Chytridiomycota and the Chromista can colonize seaweeds and cause significant mortality on commercially important seaweeds such as *Porphyra* (Gachon et al., 2010).

Zuccaro et al. (2003) initiated a cultural and molecular study to investigate fungi associated with healthy and dead fronds of the brown alga *Fucus serratus*. Isolated fungi were common hyphomycetes including *Fusarium* sp. and *Penicillium* sp. although *Corollospora angusta*, a typical marine fungus, was also cultured. Denaturing gradient gel electrophoresis in conjunction with sequencing of the partial 28S rDNA revealed four main groups of the Ascomycota: the Halosphaeriaceae, the Hypocreales, the Lulworthiales and the Pleosporales. Zuccaro et al. (2008) found similar results and concluded that there was a change in fungal community from healthy to dead thalli. Recently, Loque et al. (2010) surveyed fungi associated with healthy *Adenocystis utricularis*, *Desmarestia anceps*, and *Palmaria decipiens* from Antarctica using cultural and subsequent sequencing of the intergenic spacer regions and partial 28S rDNA. Fungi recovered are common soil taxa with psychrophilic species such as *Geomyces pannorum*.

The number of algicolous fungi accounts for around 15% of the total obligate marine fungi described (Jones et al., 2009) but in view of the number of described seaweeds (9,200 to 12,500), this ecological group of marine fungi is understudied. Many species isolated from marine algae are saprobes which degrade the complex polymeric cell wall of the algae into simple sugars which may have implications in the biofuel industry.

Saltmarsh Grasses

Salt marshes are intertidal areas between terrestrial and marine environments, which are common in temperate regions (Allen and Pye,

1992). In these areas, the herbaceous plants are dominant with a few shrubs and grasses like *Spartina* spp., *Phragmites australis* and *Juncus roemerianus* that can form dense vegetation. Due to their high primary production rate, the salt marshes play a major role in the aquatic food web and in exporting of nutrients to adjacent ecosystems (Teal, 1962).

Fungi play an important role in the cycling of nutrients as saprobes in salt marshes by producing degradative enzymes (e.g., laccases) to mineralize complex plant polymers such as lignin and cellulose into simpler molecules for their own nutrition or for other organisms in the food web in this ecosystem (Lyons et al., 2003). Fungi also form mutualistic relationships with other organisms, indicating their ecological importance in the salt marshes. In a study to investigate the relationship between the snail *Littoraria* and intertidal fungi in the *Spartina* zone of many US salt marshes, it was observed that the snail grazes live grass for fungi to infect and the fungi themselves are food for the snail (Silliman and Newell, 2003).

Bacteria and fungi are the two major decomposers in saltmarsh environments (Buchan et al., 2003). Degradation of grasses starts on the standing dead tissues and fungi, especially ascomycetes, are active in the early stage of decay (Lyons et al., 2010). Ergosterol concentration on dead aerial blades of *S. maritima* was always higher than those on the sediment surface, supporting fungi as early decomposers of *S. maritima* in Portugal (Castro and Freitas, 2000). A total of 132 marine fungi have been reported from various *Spartina* spp. (Calado and Barata, 2012). A high lignocellulose content and the non-lignin cinnamyl phenols of *Spartina* spp. support higher numbers of marine fungi when compared with other grass species (Newell et al., 1996). *Phaeosphaeria spartinicola, P. halima* and a *Mycosphaerella* sp. are common ascomycetes on *Spartina* spp. in the US east coast (Buchan et al., 2002; Lyons et al., 2010). Additional species may include *Buergenerula spartinae* and an unidentified species "4clt" (Lyons et al., 2003; Walker and Campbell, 2010; Calado and Barata, 2012). In the west coast of Portugal, *Natantispora retorquens, Phialophorophoma litoralis, Sphaerulina oraemaris* and *Phoma* sp. had higher percentages of occurrences (Barata, 2002). Walker and Campbell (2010) found similar fungal assemblages on *S. alterniflora* between natural and created salt marshes in the southeastern salt marshes of US, possibly suggesting that these fungi have adapted to this environment. These fungi occurred throughout the decay process of around 10 weeks during summer in southeastern salt marshes, but some fungi appeared late in the decay process (Buchan et al., 2002).

Juncus roemerianus is another common marsh plant species that has been extensively explored for fungi. In the eastern US coast, *J. roemerianus* covers about 50% of the salt marshes and a series of papers have been published by Jan Kohlmeyer and his co-workers documenting fungi growing on the exposed, intertidal and submerged parts of this plant (Kohlmeyer and

Volkmann-Kohlmeyer, 2001). One hundred and thirty-six fungi have been documented on *J. roemerianus* and Jan Kohlmeyer's research team alone has reported 48 new species, 14 new genera, and one family new to science (Kohlmeyer and Volkmann-Kohlmeyer, 2001; Calado and Barata, 2012). It should be noted that many of the fungi reported were not obligately marine, since many of these were isolated from the exposed parts of the plant. Ascomycetes and asexual fungi were the main groups and coelomycetes were the dominant asexual fungi (Kohlmeyer and Volkmann-Kohlmeyer, 2001). Common ascomycetes included *Loratospora aestuarii, Papulosa amerospora, Aropsiclus junci, Anthostomella poecila, Physalospora citogerminans, Scirrhia annulata, Massarina ricifera,* and *Tremateia halophila* (Newell and Porter, 2002). Although *J. roemerianus* and *S. alterniflora* dominated in the US east coast, fungal communities involved in the decomposition of these two plants were significantly different based on molecular analysis (i.e., ARISA, T-RFLP) (Blum et al., 2004; Torzilli et al., 2006).

Phragmites australis is a grass with cosmopolitan distribution and can grow in terrestrial, freshwater and intertidal environments. There are over 300 fungi reported from *P. australis* but many have been isolated from terrestrial or freshwater habitats (Wong and Hyde, 2002), among which 109 species are from intertidal *P. australis* (Calado and Barata, 2012). As in *Spartina* marshes, fungi are essential in the early decay of the plant above ground (van Ryckege et al., 2007). Common fungi found on intertidal *P. australis* are markedly different in different geographical locations as *Septoriella* spp., *Phoma* sp., *Cladosporium* spp., *Phomatospora berkeleyi, Phaeosphaeria* sp. and *Didymella glacialis* are found in the Netherlands (van Ryckege et al., 2007) while *Ligníncola laevis, Collectotrichum* sp., *Halosarpheia phragmiticola, Phomopsis* sp., *Cladosporium* sp. and *Trichoderma* sp. are found in Hong Kong (Poon and Hyde, 1998).

Mangrove Palms

In recent years, there have been many studies on marine fungi on the intertidal mangrove palm, *Nypa fruticans*, which is one of the few brackish palms that grow at the upper part of mangroves. However, this palm can be found in freshwater habitats as well. Marine fungi occur on the intertidal leaf veins, rachides, petiole bases and inflorescences, but mostly are on rachides (Hyde and Alias, 2000). Hyde (1992) was the first to look at marine fungi on *N. fruticans* collected from beaches and intertidal zone of mangroves in Brunei and found 43 species of fungi mostly ascomycetes and asexual fungi; among them, four were new species. Hyde and Sarma (2006) later reported a similar number of fungi (46 species) from Tutong River, Brunei. Subsequent papers on the diversity of marine fungi on *N. fruticans* mainly described the new taxa *Arecophila nypae* (Hyde, 1996), *Aniptodera intermedia*, and *Anthostomella* spp., etc. (Hyde et al., 1999). Pilantanapak et al. (2005) identified 81 species of marine fungi on this palm from the samples collected in Thailand. Substrate preferences have been noticed among the marine fungi particularly for *N. fruticans*. Studies of marine fungi on intertidal substrates such as *Rhizophora apiculata*, *Xylocarpus granatum* and *N. fruticans* indicated highly significant variations in the fungal communities, for example, *Linocarpon appendiculatum*, *Microthyrium* sp. and *Oxydothis nypicola* only occurred on *N. fruticans* (Besitulo et al., 2010). According to Hyde and Alias (2000), more than 40 species are unique to *N. fruticans*. The Ascomycota and especially asexual fungi are dominant on this substrate and the dominant fungi include *L. appendiculatum*, *L. nypae*, *Oxydothis nypae* and *Astrosphaeriella striatispora* (Loilong et al., 2012). A few basidiomycetes have also been isolated from the palm, e.g., *Calathella marina* and *Halocyphina villosa* (Nor et al., unpubl.).

Considering the high diversity of marine fungi on *N. fruticans*, this is an area requiring further studies. Although Loilong et al. (2012) have recorded 135 taxa (90 Ascomycota, three Basidiomycota, 42 asexual taxa) from *Nypa* fronds in Asia, 38 of them have not been formally described. Furthermore, many mangrove palms have not been examined for marine fungi, e.g., *Calamus erinaceus*, *Oncosperma tigillarium* (Tomlinson, 1986). *Lignincola conchicola* has recently been described from *Phoenix paludosa* forming ascomata on the adhesive pad of a marine invertebrate and it is the first report of a marine fungus on this palm (Liu et al., 2011).

Deep-sea Floor

The deep sea usually refers to the oceans with depths greater than 1000 m, which covers approximately 60% of the Earth's surface (Danovaro et al., 2008). It is recognized as an extreme environment characterized by an absence of sunlight, low nutrient input, high hydrostatic pressures (up to 1,100 bars) and low temperatures (less than 4°C), except in the vicinity of hydrothermal vents (temperatures of up to 400°C) (Rothschild and Mancinelli, 2001; Jørgensen and Boetius, 2007). The deep-sea ecosystems, once thought to be uninhabitable to life due to its extreme environment, are now known to harbour high species diversity. Some studies suggest that there could be over one million species globally forming one of the largest biological ecosystems on Earth (Kelly et al., 2010). However, the deep sea still remains largely under-explored and many fundamental ecological questions have not been addressed. Microorganisms play an important role in the deep-sea ecosystem. With the advancement of sampling instruments and culture-independent molecular techniques, the presence and ecological significance of bacteria and archaea in deep-sea environments have been well recognized. In contrast, the eukaryotic microorganisms, such as fungi, have been rarely reported from the same environment.

Based on culture-dependent and -independent methods, the presence of fungi has been sporadically reported in deep-sea floor habitats (water depths ranging from a few hundreds to 10000 m). The barotolerance of cultivable fungi isolated from deep-sea sediments of the Central Indian Basin (~5000 m depth) has been reported recently (Damare et al., 2006; Singh et al., 2010). Damare et al. (2006) recovered 181 cultivable fungi from sediments of the Central Indian Basin and found that most of them are terrestrial sporulating species. *Aspergillus* species were the most commonly isolated fungi followed by non-sporulating and unidentified sporulating fungi. Singh et al. (2010) isolated 16 filamentous fungi and 12 yeasts from the same geographical location. Filamentous ascomycetes were dominant while most of the isolated yeasts were basidiomycetes. Although most of fungi isolated in both studies were terrestrial species, they grew well under elevated hydrostatic pressure (200 bar) and low temperature (5°C), indicating that these fungi have evolved into distinct physiological types to adapt to the deep-sea conditions. Nagahama et al. (2001) investigated the diversity and distribution of yeasts in deep-sea environments around the northwest Pacific Ocean (depths ranging from 1050 to 10897 m). A total of 99 yeast strains, including 44 ascomycetous yeasts and 45 basidomycetous yeasts, were recovered from benthic animals and sediments collected from the deep-sea floor. Ascomycetous yeasts belonged to the genera *Candida, Debaryomyces, Kluyveromyces, Saccharomyces* and *Williopsis*, while basidomycetous yeasts included *Rhodotorula, Sporobolomyces, Cryptococcus* and *Pseudozyma*. Ascomycetous yeasts were more frequently recovered from sediments located at shallower regions (89.4% of the total yeast isolates) than from sediments collected at depths greater than 2000 m (36% of the total yeast isolates). Basidomycetous yeasts accounted for 88.5% of the total yeast strains isolated from benthic animals, much higher than that in the case of yeasts from sediments (29.2%). These observations suggested that diversity and distribution of yeasts in deep-sea environments were strongly influenced by the nutrient conditions and hydrostatic pressure of the habitats.

So far, few studies have investigated fungal diversity in deep-sea floor sediments using culture-independent molecular methods (Lai et al., 2007; Takishita et al., 2007; Nagano et al., 2010; Nagahama et al., 2011). Lai et al. (2007) studied the fungi in methane hydrate-bearing deep sea marine sediments in the South China Sea using Internal Transcribed Spacer (ITS) regions of rRNA gene clone analysis and found that most of the sequences recovered were affiliated with orders of the Ascomycota and Basidomycota, including *Phoma, Lodderomyces, Malassezia, Cryptococcus, Cylindrocarpon, Hortaea, Pichia, Aspergillus,* and *Candida*. *Phoma glomerata* was the dominant species recovered, accounting for 80% of the total clones. Nagano et al. (2010) studied fungal diversity in deep-sea sediment samples collected at several locations off Japanese islands (depths ranging from 1200 to 10000 m, mainly below 7000 m) based on ITS regions of rRNA gene clone analysis. Although some of the amplified sequences were identified as common terrestrial fungal species, such as *Penicillium*, *Aspergillus*, *Trichosporon*, and *Candida*, the majority (34 out of 43) were novel sequences showing a very low similarity with previously identified fungal sequences in public databases. A highly novel group named DSF-group 1 belonging to the phylum Ascomycota was frequently observed in almost all sediment samples examined, and phylogenetic analysis suggested the affiliation of this group to genera *Candida* and *Metschnikowia* but with low sequence similarity (about 70%). Based on culture-independent methods, this fungal group has been widely detected in oxygen-depleted deep-sea environments, such as methane cold-seeps (Takishita et al., 2007; Nagahama et al., 2011), anaerobic bacterial mats (Bass et al., 2007), and methane hydrate-bearing deep-sea sediments (Lai et al., 2007), suggesting that they may be anaerobic or facultatively anaerobic fungi (Nagano and Nagahama, 2012). Another novel fungal phylotype was affiliated to the phylum Chytridiomycota with *Rozella* spp. as the closest related fungal organisms. The presence of Chytridiomycota has been detected in several deep-sea environments such as anaerobic bacterial mats and deep-sea hydrothermal vents using culture-independent methods (Bass et al., 2007; Le Clavez et al., 2009), but no cultures were isolated from deep-sea environments so far. Fungal diversity in deep-sea sediments at methane cold-seeps of Sagami Bay, Japan, was surveyed by Takishita et al. (2007) and Nagahama et al. (2011) based on small-subunit ribosomal DNA (SSU rDNA) clone analysis. Takishita et al. (2007) reported that the basidiomycetous yeast *Cryptococcus curvatus* was the most dominant species recovered from the cold-seep sediments. Nagahama et al. (2011) revealed the presence of novel deep-branching clades as major fungal components in methane cold-seep sediments.

Deep-sea Hydrothermal Vents

Deep-sea hydrothermal vents are underwater geysers on the seafloor which are characterized by high temperatures of 300–400°C, a lack of dissolved oxygen, strong acidity (pH 2 to 3), a high (but highly variable) concentration of electron donors (sulphide, methane, and hydrogen), the presence of heavy metals, and an absence of sulphate (Jørgensen and Boetius, 2007; Le Calvez et al., 2009). The discovery of deep-sea hydrothermal vents has resulted in a completely new concept of energy sources available for life in the deep-sea environments (Jørgensen and Boetius, 2007). An amazingly rich diversity of bacteria and archaea has been described from deep-sea hydrothermal vent ecosystems, while investigation on fungi has been uncommon.

Fungi can be associated with a large variety of substrata in deep-sea hydrothermal vent systems, including seawater, sediments, minerals, and animals such as shrimps, mussels, tubeworms, corals, sponges and gastropods. Gadanho and Sampaio (2005) investigated diversity of culturable yeasts in seawater samples of deep-sea hydrothermal vents and found that non-pigmented yeasts were much more abundant than pink-pigmented yeasts. The non-pigmented yeasts belonged to the genera *Candida, Exophiala, Pichia,* and *Trichosporon,* whereas pigmented yeasts only included two genera, *Rhodosporidium* and *Rhodotorula.* A large number of undescribed species were recovered, accounting for 33% of the total number of yeast taxa isolated. Burgaud et al. (2010) recovered 32 yeast isolates from various substrata of deep-sea hydrothermal vents; the genera represented were *Rhodotorula, Rhodosporidium, Candida, Debaryomyces* and *Cryptococcus.* Twenty-seven isolates were recovered from animal substrates, while only four were isolated from mineral substrates, suggesting that yeasts are mostly associated with animals near deep-sea vents. Basdiomycetous yeasts accounted for about 66% of the total yeast isolates, whereas species richness of ascomycetous yeasts was slightly higher than that of basdiomycetous yeasts. Bugrad et al. (2009) focused on the occurrence and diversity of culturable filamentous fungi in the deep-sea hydrothermal vents. A total of 62 filamentous fungi were isolated from various substrata, most of which were from mussels (36 isolates) and shrimps (14 isolates), suggesting that fungi in deep-sea hydrothermal vents were mostly associated with animals. Most of the filamentous fungi isolated were ascomycetes (61 isolates); only one isolate was affiliated to Basidiomycota and was identified as *Tilletiopsis pallescens.* Filamentous fungi of the Ascomycota were grouped into 13 different taxonomic groups, in which Helotiales (14 strains) was the most dominant order, followed by Chaetothyriales (11 strains), Hypocreales (nine strains), Coniochaetales (eight strains), Eurotiales (six strains), Onygenales (two strains), Dothideomycetes (two strains), Xylariales (one strain), Orbiliales (one strain), and Capnodiales (one strain). Some isolates were novel species/genera without similar sequences in the public databases. Le Calvez et al. (2009) assessed the fungal diversity in animals and rocks from deep-sea hydrothermal vents using both culture-dependent and -independent methods. The work revealed an unexpected diversity of species belonging to three fungal phyla, Ascomycota, Basidiomycota, and Chytridiomycota. A total of 20 distinct fungal phylotypes were obtained, nine of which were new at the genus level or a higher taxonomic level, including two in the Chytridiomycota, three in the Basidiomycota, and four in the Ascomycota. The phylotypes affiliated with the Chytridiomycota formed an ancient evolutionary lineage. All the culturable fungi were affiliated with the Ascomycota, whereas only Chytridiomycota and Basidiomycota were recovered by culture-independent molecular method in that study, suggesting that the abundance of ascomycetous fungi was extreme low in deep-sea hydrothermal vent ecosystems and cultivation led to bias amplification of this phylum from the environments. Cornell et al. (2009) investigated fungal diversity in actively growing Fe-oxide mats and basalt rock surfaces from deep-sea hydrothermal vents. A diverse fungal community was recovered, belonging to the genera *Pichia*, *Cryptococcus*, *Rhodospordium*, *Rhodotorula*, *Dioszegia*, *Clavispora*, *Sporidiobolus*, *Aremonium*, *Geomyces*, *Aspergillus* and *Fomitopsis*. All yeast species obtained were able to produce siderophores for Fe acquisition and utilization, and one isolate *Rhodotorula graminis* was able to oxidize Mn (II), suggesting that fungi in deep-sea hydrothermal vent ecosystems may play an important role in the process of biomineralization.

Concluding Remarks

Studies of marine fungi have revealed significant differences in diversity on various substrates (Table 4.1). Recent research activity in deep-sea habitats has suggested exciting fungal occurrence on substrates in such a once-thought inhospitable environment. Although a lack of interest for the studies of biodiversity of marine fungi has been witnessed in recent years, an estimate that there can be as high as 10,000 species of marine fungi, if wider geographical locations are explored (Jones, 2011), is enough to inspire many mycologists to study these organisms from such an exciting and varied environment. The possibility, that many fungi new to science can be discovered from marine ecosystem, cannot be overlooked. Also ecologically, marine fungi by their ability to produce degradative enzymes are quite significant for these extreme environments. Further, the studies of fungi associated with the substrates of deep-sea habitats would also be a challenging venture.

Acknowledgements

K.L. Pang would like to thank Ministry of Science, Taiwan for financial support (NSC101-2621-B-019-001-MY3). Z.H. Luo would like to thank China Ocean Mineral Resources R&D Association (COMRA) Program (DY125-15-R-01) and National Natural Science Foundation of China (41376171) for financial support.

Table 4.1. Common fungal taxa found on various substrates in the sea.

Substrata	Key fungal taxa	References
Temperate timbers	*Ceriosporopsis halima, Corollospora maritima, Halosphaeria appendiculata, Monodictys pelagica, Remispora* spp.	Koch and Petersen (1996); Jones et al. (1998)
Mangrove wood	*Halocyphina villosa, Kallichroma tethys, Marinosphaera mangrovei, Lignincola laevis, Saagaromyces* spp., *Verruculina enalia*	Jones and Pang (2012)
Marine algae	*Chadefaudia* spp., *Halosigmoidea* spp., *Haloguignardia* spp., *Lulworthia* spp.	Kohlmeyer and Volkmann-Kohlmeyer (2003); Jones et al. (2009)
Spartina spp.	*Buergenerula spartinae, Mycosphaerella* sp., *Natantispora retorquens, Phaeosphaeria spartinicola, P. halima, Phialophorophoma litoralis, Sphaerulina oraemaris*	Calado and Barata (2012)
Juncus roemerianus	*Anthostomella poecila, Aropsiclus junci, Loratospora aestuarii, Massarina ricifera, Papulosa amerospora, Physalospora citogerminans, Scirrhia annulata, Tremateia halophila*	Kohlmeyer and Volkmann-Kohlmeyer (2001); Calado and Barata (2012)
Phragmites australis	*Cladosporium* spp., *Collectotrichum* sp., *Didymella glacialis, Halosarpheia phragmiticola, Lignincola laevis, Phaeosphaeria* sp., *Phoma* sp., *Phomatospora berkeleyi, Phomopsis* sp., *Septoriella* spp., *Trichoderma* sp.	Poon and Hyde (1998); van Ryckege et al. (2007)
Nypa fruticans	*Astrosphaeriella striatispora, Linocarpon appendiculatum, L. nypae, Oxydothis nypae*	Loilong et al. (2012)
Deep-sea sediment	*Aspergillus, Candida, Penicillium, Trichosporon*	Nagahama et al. (2001); Damare et al. (2006); Nagano et al. (2010)
Deep-sea hydrothermal vents	*Candida, Cryptococcus, Debaryomyces, Exophiala, Pichia, Rhodosporidium, Rhodotorula, Trichosporon*	Gadanho and Sampaio (2005); Burgaud et al. (2010)

References

Adl, S.M., Simpson, A.G.B., Farmer, M.A., Andersen, R.A., Anderson, O.R., Barta, J.R., Bowser, S.S., Brugerolle, G., Fensome, R.A., Fredericq, S., James, T.Y., Karpov, S., Kugrens, P., Krug, J., Lane, C.E., Lewis, L.A., Lodge, J., Lynn, D.H., Mann, D.G., McCourt, R.M., Mendoza, L., Moestrup, O., Mozley-Standridge, S.E., Nerad, T.A., Shearer, C.A. and Smirnov, A.V. 2005. The new higher level classification of Eukaryotes with emphasis on the taxonomy of protists. J. Eukaryot. Microbiol., 52: 399–451.

Alias, S.A., Zainuddin, N. and Jones, E.B.G. 2010. Biodiversity of marine fungi in Malaysian mangroves. Bot. Mar., 53: 545–554.

Allen, J.R.L. and Pye, K. 1992. Coastal saltmarshes: their nature and importance. pp. 1–18. *In*: Allen, J.R.L. and Pye, K. (eds.). Saltmarshes: Morphodynamics, Conservation, and Engineering Significance. Cambridge University Press, Cambridge, UK.

Arfi, Y., Buée, M., Marchand, C., Levasseur, A. and Record, E. 2012. Multiple markers pyrosequencing reveals highly diverse and host-specific fungal communities on the mangrove trees *Avicennia marina* and *Rhizophora stylosa*. FEMS Microbiol. Ecol., 79: 433–444.

Barata, M. 2002. Fungi on the halophyte *Spartina maritima* in salt marshes. pp. 179–193. *In*: Hyde, K.D. (ed.). Fungi in Marine Environments. Fungal Diversity Press, Hong Kong.

Barghoorn, E.S. and Linder, D.H. 1944. Marine fungi: their taxonomy and biology. Farlowia, 1: 395–467.

Bass, D., Howe, A., Brown, N., Barton, H., Demidova, M., Michelle, H., Li, L., Sanders, H., Watkinson, S.C., Willcock, S. and Richards, T.A. 2007. Yeast forms dominate fungal diversity in the deep oceans. Proc. Biol. Sci., 274: 3069–3077.

Besitulo, A., Moslem, M.A. and Hyde, K.D. 2010. Occurrence and distribution of fungi in a mangrove forest on Siargao Island, Philippines. Bot. Mar., 54: 535–544.

Blum, L.K., Roberts, M.S., Garland, J.L. and Mills, A.L. 2004. Distribution of microbial communities associated with the dominant high marsh plants and sediments of the United States East Coast. Microb. Ecol., 48: 375–388.

Blunt, J.W., Copp, B.R., Keyzers, R.A., Munro, M.H.G. and Prinsep, M.R. 2012. Marine natural products. Nat. Prod. Rep., 29: 144–222.

Buchan, A., Newell, S.Y., Moreta, J.I.L. and Moran, M.A. 2002. Analysis of internal transcribed spacer (ITS) regions of rRNA genes in fungal communities in southeastern US salt marsh. Microb. Ecol., 43: 329–340.

Buchan, A., Newell, S.Y., Butler, M., Biers, E.J., Hollibaugh, J.T. and Moran, M.A. 2003. Dynamics of bacterial and fungal communities on decaying salt marsh grass. Appl. Environ. Microbiol., 69: 6676–6687.

Bucher, V.V.C., Hyde, K.D., Pointing, S.B. and Reddy, C.A. 2004. Production of wood decay enzymes, mass loss and lignin solubilization in wood by marine ascomycetes and their anamorphs. Fungal Divers., 15: 1–14.

Burgaud, G., Arzur, D., Durand, L., Cambon-Bonavita, M.A. and Barbier, G. 2010. Marine culturable yeasts in deep-sea hydrothermal vents: species richness and association with fauna. FEMS Microbiol Ecol., 73: 121–133.

Burgaud, G., Le Calvez, T., Arzur, D., Vandenkoornhuyse, P. and Barbier, G. 2009. Diversity of culturable marine filamentous fungi from deep-sea hydrothermal vents. Environ. Microbiol., 11: 1588–1600.

Byrne, P.J. and Jones, E.B.G. 1974. Lignicolous marine fungi. Veröff. Inst. Meeresforsch. Bremerhaven, Suppl., 5: 301–320.

Calado, M.D.L. and Barata, M. 2012. Salt marsh fungi. pp. 345–381. *In*: Jones, E.B.G. and Pang, K.L. (eds.). Marine Fungi and Fungal-like Organisms. De Gruyter, Berlin, Germany.

Castro, P. and Freitas, H. 2000. Fungal biomass and decomposition in *Spartina maritima* leaves in the Mondego salt marsh (Portugal). Hydrobiol., 428: 171–177.

Connell, L., Barrett, A., Templeton, A. and Staudigel, H. 2009. Fungal diversity associated with an active deep sea volcano: Vailulu'u Seamount, Samoa. Geomicrobiol. J., 26: 597–605.

Damare, S., Raghukumar, C. and Raghukumar, S. 2006. Fungi in deep-sea sediments of the Central Indian Basin. Deep-Sea Res., I 53: 14–27.

Danovaro, R., Gambi, C., Lampadariou, N. and Tselepides, A. 2008. Deep-sea nematode biodiversity in the Mediterranean basin: testing for longitudinal, bathymetric and energetic gradients. Ecography, 31: 231–244.

Gachon, M.M., Sime-Ngando, T., Strittmatter, M., Chambouvet, A. and Kim, G.H. 2010. Algal diseases: spotlight on a black box. Trends Plant Sci., 15: 633–640.

Gadanho, M. and Sampaio, J.P. 2005. Occurrence and diversity of yeasts in the mid-atlantic ridge hydrothermal fields near the Azores Archipelago. Microb. Ecol., 50: 408–417.

Geiser, D.M., Taylor, J.W., Ritchie, K.B. and Smith, G.W. 1998. Cause of sea fan death in the West Indies. Nature, 394: 137–138.

Hyde, K.D. 1992. Fungi from decaying intertidal fronds of *Nypa fruticans*, including three new genera and four new species. Bot. J. Linn. Soc., 110: 95–110.

Hyde, K.D. 1996. Fungi from palms. XXIX. *Arecophila* gen. nov. (Amphisphaeriaceae, Ascomycota), with five new species and two new combinations. Nova Hedw., 63: 81–100.

Hyde, K.D. and Jones, E.B.G. 1988. Marine mangrove fungi. Mar. Ecol., 9: 15–33.

Hyde, K.D. and Alias, S.A. 2000. Biodiversity and distribution of fungi associated with decomposing *Nypa* palm. Biodivers. Conserv., 9: 393–402.

Hyde, K.D. and Sarma, V.V. 2006. Biodiversity and ecological observations on filamentous fungi of mangrove palm *Nypa fruiticans* Wurumb (Liliosida-Arecales) along the Tutong River, Brunei. Ind. J. Mar. Sci., 35: 297–307.

Hyde, K.D., Cai, L. and Jeewon, R. 2005. Tropical fungi. pp. 93–109. *In*: Dighton, J., White, J.F. and Oudemans, P. (eds.). The Fungal Community: its Organization and Role in the Ecosystem. CRC Press, New York, USA.

Hyde, K.D., Goh, T.K., Lu, B.S. and Alias, S.A. 1999. Eleven new intertidal fungi from *Nypa fruticans*. Mycol. Res., 103: 1409–1422.

Jones, E.B.G. (ed.). 1976. Recent Advances in Aquatic Mycology. Elek, London, UK.

Jones, E.B.G. 2000. Marine fungi: some factors influencing biodiversity. Fungal Divers., 4: 53–73.

Jones, E.B.G. 2011. Are there more marine fungi to be described? Bot. Mar., 54: 343–354.

Jones, E.B.G. and Pang, K.L. 2010. 11th International Marine and Freshwater Mycology Symposium, Taichung, Taiwan ROC, November 2009. Bot. Mar., 53: 475–478.

Jones, E.B.G. and Pang, K.L. 2012. Tropical aquatic fungi. Biodivers. Conserv. 21: 2403–2423.

Jones, E.B.G. 1976. Lignicolous and algicolous fungi. pp. 1–51. *In*: Jones, E.B.G. (ed.). Recent Advances in Aquatic Mycology. Elek, London, UK.

Jones, E.B.G., Sakayaroj, J., Suetrong, S., Somrithipol, S. and Pang, K.L. 2009. Classification of marine Ascomycota, anamorphic taxa and Basidiomycota. Fungal Divers., 35: 1–187.

Jones, E.B.G., Pang, K.L. and Stanley, S. 2012. Fungi from marine algae. pp. 329–344. *In*: Jones, E.B.G. and Pang, K.L. (eds.). Marine Fungi and Fungal-like Organisms. De Gruyter, Berlin, Germany.

Jones, E.B.G., Vrijmoed, L.L.P. and Alias, S.A. 1998. Intertidal marine fungi from San Juan Island and comments on temperate water species. Bot. J. Scotl., 50: 177–184.

Jørgensen, B.B. and Boetius, A. 2007. Feast and famine-microbial life in the deep-sea bed. Nat. Rev. Microbiol., 5: 770–781.

Kelly, N.E., Shea, E.K., Metaxas, A., Haedrich, R.L. and Auster, P.J. 2010. Biodiversity of the deep-sea continental margin bordering the Gulf of Maine (NW Atlantic): relationships among sub-regions and to shelf systems. PLoS One, 5: e13832.

Koch, J. and Petersen, K.R.L. 1996. A check list of higher marine fungi on wood from Danish coasts. Mycotaxon, 60: 397–414.

Kohlmeyer, J. and Kohlmeyer, E. 1979. Marine Mycology—the Higher Fungi. Academic Press, New York, USA.

Kohlmeyer, J. 1973. Fungi from marine algae. Bot. Mar., 16: 201–215.

Kohlmeyer, J. and Volkmann-Kohlmeyer, B. 2001. The biodiversity of fungi on *Juncus roemerianus*. Mycol. Res., 105: 1411–1412.

Kohlmeyer, J. and Volkmann-Kohlmeyer, B. 2003. Marine ascomycetes from algae and animal hosts. Bot. Mar., 46: 285–306.

Lai, X., Cao, L., Tan, H., Fang, S., Huang, Y. and Zhou, S. 2007. Fungal communities from methane hydrate-bearing deep-sea marine sediments in South China Sea. ISME J., 1: 756–762.

Le Calvez, T., Burgaud, G., Mahé, S., Barbier, G. and Vandenkoornhuyse, P. 2009. Fungal diversity in deep-sea hydrothermal ecosystems. Appl. Environ. Microbiol., 75: 6415–6421.

Liu, J.K., Jones, E.B.G., Chukeatirote, E., Bahkali, A.H. and Hyde, K.D. 2011. *Lignincola conchicola* from palms with a key to the species of *Lignincola*. Mycotaxon, 117: 343–349.
Loilong, A., Sakayaroj, J., Rungjindamai, N., Choeyklin, R. and Jones, E.B.G. 2012. Biodiversity of fungi on the palm *Nypa fruticans*. pp. 273–290. *In*: Jones, E.B.G. and Pang, K.L. (eds.). Marine Fungi and Fungal-like Organisms. De Gruyter, Berlin, Germany.
Loque, C.P., Medeiros, A.O., Pellizzari, F.M., Oliveira, E.C., Rosa, C.A. and Rosa, L.H. 2010. Fungal community associated with marine macroalgae from Antarctica. Polar Biol., 33: 641–648.
Luo, W., Vrijmoed, L.L.P. and Jones, E.B.G. 2005. Screening of marine fungi for lignocellulose-degrading enzyme activities. Bot. Mar., 48: 379–386.
Lyons, J.I., Alber, M. and Hollibaugh, J.T. 2010. Ascomycete fungal communities associated with early decaying leaves of *Spartina* spp. from central California estuaries. Oecologia, 162: 435–442.
Lyons, J.I., Newell, S.Y., Buchan, A. and Moran, M.A. 2003. Diversity of ascomycete laccase gene sequences in a southeastern US salt marsh. Microb. Ecol., 45: 270–281.
Mouzouras, R. 1989. Soft rot decay of wood by marine microfungi. J. Inst. Wood Sci., 11: 193–201.
Mouzouras, R., Jones, E.B.G., Venkatasamy, R. and Holt, D.M. 1998. Microbial decay of lignocellulose in the marine environment. pp. 329–335. *In*: Thompson, M.F., Sarojini, R. and Nagabhushanaim, R. (eds.). Marine Biodeterioration. Oxford and JBH Publishing, New Delhi, India.
Nagahama, T., Takahashi, E., Nagano, Y., Abdel-Wahab, M.A. and Miyazaki, M. 2011. Molecular evidence that deep-branching fungi are major fungal components in deep-sea methane cold-seep sediments. Environ. Microbiol., 13: 2359–2370.
Nagahama, T., Hamamoto, M., Nakase, T., Takami, H. and Horikoshi, K. 2001. Distribution and identification of red yeasts in deep-sea environments around the northwest Pacific Ocean. Antonie Van Leeuwenhoek, 80: 101–110.
Nagano, Y. and Nagahama, T. 2012. Fungal diversity in deep-sea extreme environments. Fungal Ecol., 5: 463–471.
Nagano, Y., Nagahama, T., Hatada, Y., Nunoura, T., Takami, H., Miyazaki, J., Takai, K. and Horikoshi, K. 2010. Fungal diversity in deep-sea sediments—the presence of novel fungal groups. Fungal Ecol., 3: 316–325.
Neuhauser, S., Glockling, S.L., Leaño, E.M., Lilje, O., Marano, A.V. and Gleason, F.H. 2012. An introduction to fungus-like microorganisms. pp. 137–151. *In*: Jones, E.B.G. and Pang, K.L. (eds.). Marine Fungi and Fungal-like Organisms. De Gruyter, Berlin, Germany.
Newell, S.Y. and Porter, D. 2002. Microbial secondary production from saltmarsh-grass shoots, and its known and potential fates. pp. 159–185. *In*: Weinstein, M.P. and Kreeger, D.A. (eds.). Concepts and Controversies in Tidal Marsh Ecology. Kluwer, Amsterdam, Netherland.
Newell, S.Y., Porter, D. and Lingle, W.L. 1996. Lignocellulolysis by ascomycetes (Fungi) of a saltmarsh grass (smooth cordgrass). Microsc. Res. Techniq., 33: 32–46.
Panebianco, C., Tam, W.Y. and Jones, E.B.G. 2002. The effect of pre-inoculation of balsa wood by selected marine fungi and their effect on subsequent colonisation in the sea. Fungal Divers., 10: 77–88.
Pang, K.L., Chow, R.K.K., Chan, C.W. and Vrijmoed, L.L.P. 2011. Diversity and physiology of marine lignicolous fungi in Arctic waters: a preliminary account. Polar Res., 30: 1–5.
Patouillrad, N. 1897. J. Bot. Paris, 11: 242.
Pilantanapak, A., Jones, E.B.G. and Eaton, R.A. 2005. Marine fungi on *Nypa fruiticans* in Thailand. Bot. Mar., 48: 365–373.
Poon, M.O.K. and Hyde, K.D. 1998. Biodiversity of intertidal estuarine fungi on *Phragmites* at Mai Po Marshes, Hong Kong. Bot. Mar., 41: 141–155.
Pugh, G.J.F. and Jones, E.B.G. 1986. Antarctic marine fungi: a preliminary account. pp. 323–330. *In*: Moss, S.T. (ed.). The Biology of Marine Fungi. Cambridge University Press, Cambridge, UK.
Rostrup, E. 1889. *Leptosphaeria marina*. Bot. Tidsskr., 17: 234.

Rothschild, L.J. and Mancinelli, R.L. 2001. Life in extreme environments. Nature, 409: 1092–1101.
Schmit, J.P. and Shearer, C.A. 2003. A checklist of mangrove-associated fungi, their geographical distribution and known host plants. Mycotaxon, 85: 423–477.
Schmit, J.P. and Shearer, C.A. 2004. Geographic and host distribution of lignicolous mangrove microfungi. Bot. Mar., 47: 496–500.
Silliman, B.R. and Newell, S.Y. 2003. Fungal farming in a snail. PNAS, 100: 15643–15648.
Singh, P., Raghukumar, C., Verma, P. and Shouche, Y. 2010. Phylogenetic diversity of culturable fungi from the deep-sea sediments of the Central Indian Basin and their growth characteristics. Fungal Divers., 40: 89–102.
Sutherland, G.K. 1915. New marine fungi on *Pelvetia*. New Phytol., 14: 33–42.
Sutherland, G.K. 1916a. Marine Fungi Imperfecti. New Phytol., 15: 35–48.
Sutherland, G.K. 1916b. Additional notes on marine pyrenomycetes. Trans. Br. Mycol. Soc., 5: 257–263.
Takishita, K., Yubuki, N., Kakizoe, N., Inagaki, Y. and Maruyama, T. 2007. Diversity of microbial eukaryotes in sediment at a deep-sea methane cold seep: surveys of ribosomal DNA libraries from raw sediment samples and two enrichment cultures. Extremophiles, 11: 563–576.
Tan, T.K., Leong, W.F. and Jones, E.B.G. 1989. Succession of fungi on wood of *Avicennia alba* and *A. lanata* in Singapore. Can. J. Bot., 67: 2686–2691.
Teal, J.M. 1962. Energy flow in the salt marsh ecosystem of Georgia. Ecology, 43: 614–624.
Tomlinson, P.B. 1986. The Botany of Mangroves. Cambridge University Press, London, UK.
Torzilli, A.P., Sikaroodi, M., Chalkley, D. and Gillevet, P.M. 2006. A comparison of fungal communities from four salt marsh plants using automated ribosomal intergenic spacer analysis (ARISA). Mycologia, 98: 690–698.
Van Ryckegem, G., Gessner, M.O. and Verbeken, A. 2007. Fungi on leaf blades of *Phragmites australis* in a brackish tidal marsh: diversity, succession, and leaf decomposition. Microb. Ecol., 53: 600–611.
Vrijmoed, L.L.P. 2000. Isolation and culture of higher filamentous fungi. pp. 1–20. *In*: Hyde, K.D. and Pointing, S.B. (eds.). Marine Mycology—A Practical Approach. Fungal Diversity Press, Hong Kong.
Walker, A.K. and Campbell, J. 2010. Marine fungal diversity: a comparison of natural and created salt marshes of the north-central Gulf of Mexico. Mycologia, 102: 513–521.
Wong, M.K.M. and Hyde, K.D. 2002. Fungal saprobes on standing grasses and sedges in a subtropical aquatic habitat. pp. 195–212. *In*: Hyde, K.D. (ed.). Fungi in Marine Environments. Fungal Diversity, Hong Kong.
Zuccaro, A., Schulz, B. and Mitchell, J.I. 2003. Molecular detection of ascomycetes associated with *Fucus serratus*. Mycol. Res., 107: 1451–1466.
Zuccaro, A., Schoch, C.L., Spatafora, J.W., Kohlmeyer, J., Dreager, S. and Mitchell, J.I. 2008. Detection and identification of fungi intimately associated with the brown seaweed *Fucus serratus*. Appl. Environ. Microbiol., 74: 931–941.

CHAPTER 5

Diversity of Micro-Fungi in Streams Receiving Coal Mine Drainage from Jaintia Hills, Meghalaya, a North Eastern State of India

Ibandarisuk Lyngdoh and *Highland Kayang**

ABSTRACT

The diversity of fungi in the acidic streams [Um-Mynkseh (UMR) and Um-Rimet (RR)] of the Jaintia hills district of Meghalaya, receiving run-off water of coal mining areas were studied. Um-Myntdu (MR), a stream which is located away from the coal mining areas was taken as a reference site (unpolluted). Highest fungal diversity was observed in Um-Myntdu (reference site), lowest in Um-Mynkseh and intermediary in Um-Rimet sites. Among the 83 species of fungi isolated, 12 species were found to be common to both the polluted and reference sites, while three species were found to be present only at the polluted sites. The aquatic hyphomycetes were confined only to the reference site. Negative correlation between fungal diversity and some physicochemical parameters of the acidic streams of (Um-Mynkseh and Um-Rimet) were found.

Microbial Ecology Laboratory, Department of Botany, North Eastern Hill University, Shillong, India.

* Corresponding author: hkayang@yahoo.com

Introduction

Mine drainage is one of the main problems associated with coal mining activities that results in the accumulation of toxic elements in the environment. Consequently, these sites become inhospitable for most of the life forms, except for those microorganisms that are able to tolerate the acidity and the high concentration of heavy metals (Castro-Silva et al., 2003). However, archaea, bacteria, fungi, algae and protozoa have been isolated from mining areas, and have been shown to be active at these sites (Johnson, 1991).

Although the mechanisms with which hyper-acidophiles such as bacteria and fungi cope with acidic environment need further investigation, these organisms are known to maintain a relatively neutral pH within themselves by pumping protons out of the cell and by establishing low proton membrane permeability (Nicolay et al., 1987). Raven (1990) provided detailed information on internal and external pH regulations. Microorganisms are probably common in the acidic environment because of this internal pH regulation.

Several studies have described microbial diversity in acid environments (Johnson, 1998; Baker and Banfield, 2003), and microbial community composition in acid-sulphate systems has been characterized using traditional culturing techniques (Webb et al., 1998; Wielinga et al., 1999). For the Indian environment, we have data for fungal diversity from other extreme environments like "Usar" soils and alkaline waters (Rai et al., 1971; Rai and Chowdhery, 1979; Misra, 1983a,b, 1985a,b), but we lack data on the occurrence and diversity of fungi from streams receiving coal mine drainage. Hence a stream of Jaintia hills, Meghalaya, a Northeastern state of India has been studied. This chapter provides information on fungal communities associated with coal mine drainage in the affected streams.

Methodology

Study Site

For the present study two streams namely, Um-Mynkseh (25° 22.436′ N to 090° 20.360′ E) and Um- Rimet (25° 21.702′ N to 092° 20.495′ E), in the coal mining areas were selected to study the diversity of micro-fungi in streams affected by coal mine drainage of the Jaintia hills, Meghalaya. Um-Myntdu (25° 27.128′N to 90° 11.475′E) which is located away from the mining area was taken as a control (reference site). The study was conducted for a period of two years starting from January 2006 to December 2007 at monthly intervals in the selected study sites.

Enumeration of Fungi

Isolation of fungal and bacterial population was done by dilution plate methods (APHA, 1998). For fungi, 1 ml of the water sample was diluted (1/1000) and inoculated in suitable culture media such as Rose Bengal agar (Peptone 15 g, Magnesium sulphate 0.5 g, Pot. Hydrogen phosphate 1 g, water 1000 ml, Rose Bengal 1: 30,000) or Malt extract yeast agar (Malt extract 3.0 g, Yeast extract 3.0 g, Peptone 5.0 g, Glucose 10.0 g, Dw 1000 ml, Agar 15 g) medium. Three replicates were maintained for each sample and Petri plates were incubated at 24 ± 2°C. The Colony Forming Units (CFU) of fungi were estimated by counting the number of fungal colonies. CFU was then expressed as CFU/ml. The fungal colonies were identified based on their morphology and reproductive structures consulting the monographs of Subramaniam (1971), Barnet and Hunter (1972), and Domsch et al. (1980). Determination of the relative abundance of fungal species was done by using the formula below:

$$\textit{Relative abundance (\%)} = \frac{\textit{Total number of the colonies of individual species}}{\textit{Total number of colonies of all species}} x\ 100$$

Statistical Analysis

Using the data obtained, the following indices of fungal and bacterial species were evaluated:

1. Index of general diversity (H'); Shannon and Weaver (1949) cited in Odum (1971).
   $H' = -\Sigma\ (Pi \ln Pi)$
   (Where $Pi = n_i/N$; and n_i is the importance value of each species and *N* is the total importance value)
2. Index of dominance (C); (Simpson, 1949)
   $C = \Sigma\ (n_i/N)^2$
   (Where $Pi = n_i/N$ is the number of individuals of each species and *N* is the total number of individuals in that location).
3. Index of similarity: Sorensen (S) and Jaccard (J) (Krebs, 1989)
   $$S = \frac{2C}{S_1 + S_2}$$
   Where, S_1 = number of species in site 1
   S_2 = number of species in site 2
   C= number of species that are common to both site 1 and site 2
   The Jaccard coefficient is calculated using the equation:
   $$Cj = \frac{a}{(a + b + c)}$$

Where, a= number of species common to both sites
b= number of species in site B, but not in A
c= number of species in site A, but not in B

Analysis of variance (ANOVA) was carried out for both fungi and bacteria and the means were separated by the Tukey test. Pearson correlation coefficients were computed between physico-chemical properties and H′ of study sites.

Results

Fungal Population

The colony forming units of fungi increased in the second year of study in all the three sites studied. In the first year of study at UMR and RR in coal mining sites the fungal population ranged from 7.0 to 33.0 x 10^3 and 7.6 to 51.0 x 10^3 CFU/ml, respectively. In MR, fungal population ranged from 4.0 to 16.0 x 10^3 CFU/ml. During the second year of study, UMR and RR fungal population ranged from 8.3 to 28.3 x 10^3 and 7.0 to 31.0 x 10^3 CFU/ml, respectively whereas fungal population in MR ranged from 5.3 to 24.3 x 10^3 CFU/ml (Fig. 5.1).

A total of 81 species of fungi were isolated of which 12 species were found to be common to UMR, RR and MR sites, while three species were found to be present only in UMR and RR. The fungi isolated and their relative abundance are given in Table 5.1. Shannon's diversity indices of fungi during the first year of study ranged from 1.1 to 2.5 in MR, 0.8 to 1.7 in UMR and 0.3 to 2.0 in RR. During the second year of study, Shannon's indices ranged from 1.5 to 2.5 in MR, 0.03 to 1.8 in UMR and 0.9 to 1.7 in RR (Fig. 5.2). Simpson's dominance indices of fungal species during the first year of study ranged from 0.08 to 0.34 in MR, 0.18 to 0.54 in UMR and 0.14

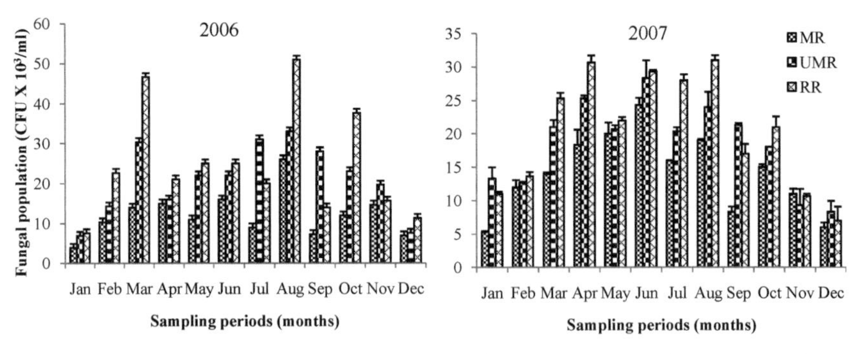

Figure 5.1. Fungal population of water samples in Um-Myntdu (MR), Um-Mynkseh (UMR) and Um-Rimet (RR).

Table 5.1. List of fungi isolated from the three sites (Jan., 2006 to Dec., 2007).

Sl no.	Fungal species	MR	UMR	RR
1.	*Absidia spinosa* Lendn.	+	–	–
2.	*Achlya* sp. (fungi-like organism)	+	–	–
3.	*Acremonium butyri* (van Beyma) W. Gams	–	+	+
4.	*A. cerealis* (Karst.) W. Gams	+	–	–
5.	*A. fuscidioides* (Nicot) W. Gams	+	–	–
6.	*A. kiliense* Grütz	+	–	–
7.	*A. strictum* W. Gams	+	–	–
8.	*Alternaria alternata* (Fr.) Keissler	–	+	+
9.	*A. longipes* (Ellis & Everhart) Mason	+	–	+
10.	*A. macrospora* Zimm.	+	–	–
11.	*Anguillospora longissima*	+	–	–
12.	*Aspergillus flavus* Link ex Gray	+	–	+
13.	*A. fumigatus* Fres.	+	+	+
14.	*A. japonicus* Saito	+	–	–
15.	*A. niger* van Tieghem	+	+	+
16.	*A. sydowii* (Bainier & Sartory) Thom & Church	+	+	+
17.	*A. terreus* Thom	+	–	–
18.	*A. ustus* (Bainier) Thom & Church	+	–	+
19.	*A. versicolar* (Vuill.) Tiraboschi	+	–	–
20.	*Aureobasidium pullullans* (de Bary) G. Arnaud	+	+	–
21.	*Broomella acuta* Shoemaker & E. Müll.	+	–	–
22.	*Candida* sp.	+	+	+
23.	*Chaetomium nozdrenkoae* Sergeeva	+	–	+
24.	*Chloridium* sp.	–	–	+
25.	*Cladosporium herbarum* (Pers.) Link ex Gray	+	+	+
26.	*C. sphaerospermum* Penz. 1882	–	+	–
27.	*Cryptococcus humicola* (Dasz.) Gobulev	–	+	+
28.	*C. laurentii* (Kuff.) C.E. Skinner	–	+	–
29.	*Exophalia jeanselmei*	–	–	+
30.	*Helicosporium* sp.	–	+	–
31.	*Humicola fuscoatra* Traaen	+	–	–
32.	*H. grisea* Traaen	+	+	+
33.	*Ingoldia* sp.	+	–	–
34.	*Monilia* sp.	–	–	+
35.	*Mortierella vinaceae* Dixon-Stewart	–	–	+
36.	*Mucor hiemalis f. corticola* (Hagem) Schipper	+	+	–
37.	*M. hiemalis f. silvaticus* Wehmer	–	–	+
38.	*Myrothecium verrucaria* (Alb. & Schwein.) Ditmar	+	–	+
39.	*Nectria ventricosa* C. Booth	+	–	–
40.	*Oidiodendron griseum* Robak	+	+	+
41.	*O. truncatum* G.L. Barron	+	+	+
42.	*Paecilomyces* sp.	–	–	+

Table 5.1. contd....

Table 5.1. contd.

Sl no.	Fungal species	MR	UMR	RR
43.	*Penicillium chrysogenum* Thom	+	+	–
44.	*P. digitatum* (Pers.) Sacc.	+	–	–
45.	*P. expansum* Link ex Gray	+	+	+
46.	*P. frequentans* Westling	–	+	+
47.	*P. italicum* Wehmer	+	–	–
48.	*P. janthinellum* Biourge	+	+	–
49.	*P. jensenii* Zaleski	+	–	–
50.	*P. lanosum* Westling	+	–	–
51.	*P. nigricans* Bain. Ex Thom	+	–	–
52.	*P. purpurogenum* Stoll	+	+	–
53.	*P. restrictum* Gilman & Abbott	–	–	+
54.	*P. rubrum* Stoll	+	–	+
55.	*P. simplicissimum* (Oudem.) Thom	+	+	+
56.	*P. stoloniferum* Thom	–	–	+
57.	*P. variable* Sopp.	+	–	–
58.	*Phialophora fastigiata* (Lagerb. & Melin) Conant	+	+	+
59.	*Phoma eupyrena* Sacc.	+	–	–
60.	*Phoma* sp.	+	–	–
61.	*Pythium aphanidermatum* (Edson) Fitzp. (fungi-like)	+	–	+
62.	*P. intermedium* de Bary	+	–	–
63.	*P. irregular* Buisman	+	+	–
64.	*Pythium* sp.	–	–	–
65.	*Rhizopus stolonifer*(Ehrenb. Ex Link) Lind	+	+	+
66.	*Rhodotorula aurantiaca* (Saito) Lodder	–	+	+
67.	*R. mucilaginosa* (A. Jorg)	–	–	+
68.	*Saccharomyces cerevisiae* Meyen ex E.C. Hansen	+	+	–
69.	*Staphylotrichum coccosporum* J.A. Mey. & Nicot	+	–	+
70.	*Talaromyces emersoni* Stolk	+	–	–
71.	*T. helicus* (Raper & Fennell) C.R. Benj.	+	–	–
72.	*T. trachyspermus* (Shear) Stolk & Samson	–	–	+
73.	*Tetrachaetum elegans*	+	–	–
74.	*T. wortmanii* (Klocker) C.R. Banjamin	+	–	–
75.	*Trichoderma harzianum* Rifai	+	–	–
76.	*T. koningii* Oudem.	+	+	+
77.	*T. polysporum* (Link ex Pers.) Rifai	–	–	+
78.	*T. pseudokoningii* Rifai	+	–	–
79.	*T. viride* Pers. ex Gray	+	–	+
80.	*Trichosporon dulcitum* (Berkhout) Weijman	–	–	+
81.	*Verticillium alboatrum* Reinke & Berthold	+	–	–
82.	*V. nigrescens* Pethybr.	–	–	+
83.	White sterile mycelium	+	–	–

Note: Species present (+) and absence (–)

to 0.81 in RR. During the second year of study, it ranged from 0.11 to 0.23 in MR, 0.19 to 0.89 in UMR and 0.23 to 0.55 in RR (Fig. 5.3).

Sorenson's and Jaccard's similarity indices were calculated for fungi to find out the similarity between the study sites. Sorenson's similarity index during the first year of study showed a similarity of 0.42 (MR x UMR), 0.38 (MR x RR), 0.52 (UMR x RR) and Jaccard's similarity index showed a similarity of 0.17 (MR x UMR), 0.16 (MR x RR) and 0.21 (UMR x RR). During the second year, Sorenson's similarity index showed a similarity of 0.36 (MR x UMR), 0.41 (MR x RR), 0.44 (UMR x RR) and Jaccard's similarity index showed a similarity of 0.15 (MR x UMR), 0.17 (MR x RR) and 0.18 (UMR x RR) (Fig. 5.4).

Pearson's correlation coefficient values (r) of fungal population with various physico-chemical parameters are given in Table 5.2. One-way analysis of variance (ANOVA) showed no significant variation ($p<0.05$)

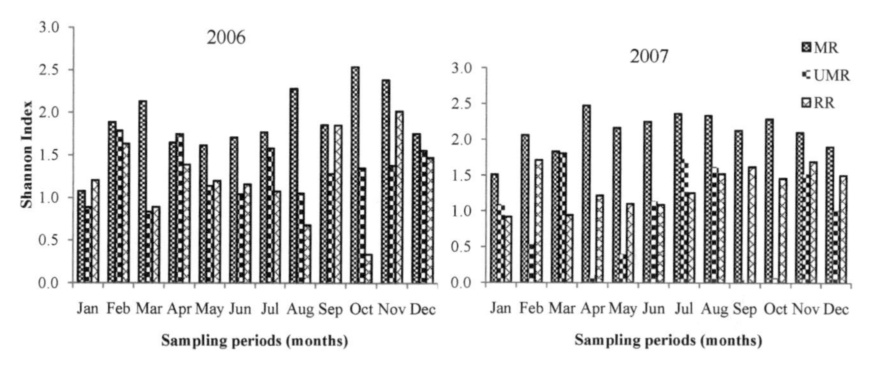

Figure 5.2. Shannon's indices of general diversity in Um-Myntdu (MR), Um-Mynkseh (UMR) and Um-Rimet (RR).

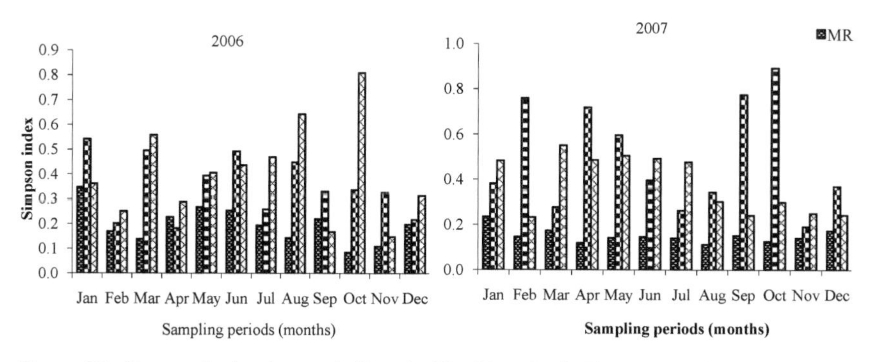

Figure 5.3. Simpson's dominance indices in Um-Myntdu (MR), Um-Mynkseh (UMR) and Um-Rimet (RR).

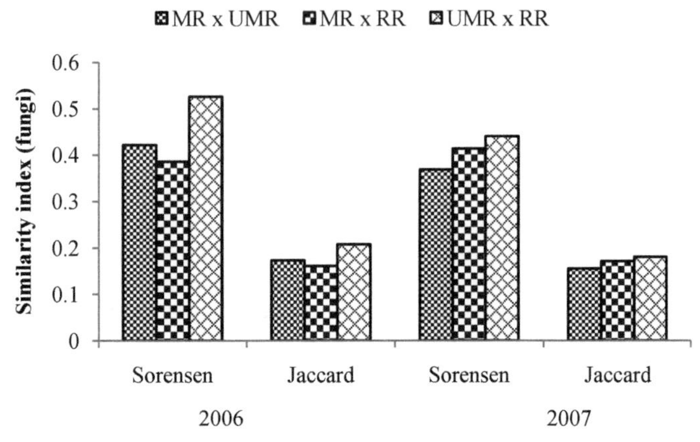

Figure 5.4. Sorenson's and Jacarrd's similarity indices of fungi in Um-Myntdu (MR), Um-Mynkseh (UMR) and Um-Rimet (RR).

between MR and UMR. MR also showed no significant variation ($p<0.05$) with RR. No significant variation was also observed between the three study sites.

Discussion

In natural environments, fungi rarely encounter conditions that allow optimal growth, since they are conditioned for abiotic factors, nutrient availability and pollutants (Gadd et al., 2001). Pollutants such as discharges from mining activities, all have deleterious effects on the fungal communities (Barlocher, 1992; Niyogi et al., 2002). The effect of mine drainage including acidic and heavy metal effects reduced biodiversity of fungi (Maltby and Booth, 1991). Nevertheless, fungi are ubiquitous and can occur in metal-polluted habitats (Sridhar and Barlocher, 2000). In the present work, the impact of coal mine drainage was assessed on fungal communities and diversity to understand their ability to survive in mine drainage polluted water bodies.

The zygomycetous fungi such as *Rhizopus stolinifer* and *Mucor hiemalis* were present at UMR, RR and MR, whereas, aquatic fungi such as *Anguillospora longissima* and *Tetrachaetum elegans* were restricted only at MR. The non-aquatic hyphomycetous genera such as *Aspergillus, Alternaria, Aureobasidium, Cladosporium, Oidiodendron, Penicillium, Phialophora* and *Trichoderma* occurred in all the sites irrespective of the water quality. It could be assumed that they are tolerant to pollution (Ravikumar et al., 2000). Traditionally yeast is defined as fungi belonging to the Ascomycota and Basidiomycota. Two yeast species, i.e., *Rhodotorula* and *Cryptococcus*

Table 5.2. Correlation coefficient values of microbiological (bacteria and fungi) with various physico-chemical parameters ($p<0.05$).

Sites	Water properties	BP	WT	pH	Cl	Am	NO_3	Phos	Mg	Ca	Cu	K	TH	Zn
MS	FP	0.41^{a}	0.54^{b}	-	-	-	-	-	0.62^{b}	0.59^{b}	-	-	0.56^{b}	-
UMS	FP	-	0.56^{b}	0.46^{a}	-	-	0.60^{b}	-	0.63^{b}	0.73^{c}	0.42^{a}	0.61^{b}	0.52^{b}	-
RS	FP	-	-	-	0.48^{a}	-	0.42^{a}	-	0.41^{a}	0.61^{b}	-	-		-

(Note: BP= Bacterial population, FP= Fungal population, WT= Water temperature, Con= Conductivity, DO= Dissolved oxygen, Phos= Phosphate: SO_4= Sulphate, Amm = Ammonia, NO_3 = Nitrate, Mg = Magnesium, Ca= Calcium, Cl = Chloride, Cu = Copper, Fe = Iron, K = Potassium, Cd = Cadmium, Zn = Zinc). Values marked with a, b, and c are significant at $P \leq 0.05$, $P \leq 0.01$ and $P \leq 0.001$, respectively.

isolated were confined to the sites from coal mining areas and maybe considered as indicators of mine drainage. The role of fungi as pollution indicators has been discussed by earlier workers (Ravikumar et al., 2000; Manoharachary, 1985). This study was in accordance with Lopez-Archilla et al. (2004) who reported that both yeast and filamentous fungi have been found at the acidic mine polluted aquatic environment.

Shannon's diversity indices (*H*) and species richness of fungi were significantly higher ($p< 0.05$) at site MR followed by RR and UMR. Simpson's Indices (D), which measures the probability that two individuals randomly selected from a sample will belong to the same species accounts for both abundance and evenness of the species present, were relatively higher at UMR and RR compared to MR. All metrics reflect the same in that sampling site MR was the best in both abundance and evenness, and the sampling sites UMR and RR which were affected by coal mine drainage were dominated by related species and were poor in diversity. Niyogi et al. (2002) assessed fungal communities in 20 mountain streams (Colorado, USA) influenced by mine drainage and found that, similar to other stream communities, the diversity of fungi was sensitive to low pH (<6) or high dissolved zinc (>1mg/l). A more detailed study on microbial taxa may help in developing scientific criteria for pollution monitoring and in developing remediation strategies in the coal mine affected sites.

References

APHA. 1998. Standard methods for examination of water and waste water. 20th ed. American Public Health Association, Washington, DC.

Baker, B.J. and Banfield, J.F. 2003. Microbial communities in acid mine drainage. FEMS Microbial Ecology, 44: 139–152.

Barlocher, F. 1992. Human interference. pp. 173–181. *In*: Barlocher, F. (ed.). Ecology of Aquatic Hyphomycetes. Springer verlag, New York.

Barnet, K.L. and Hunter, B.B. 1972. Illustrated Genera of Imperfect Fungi. Burgess Publishing Company, Minneapolis.

Castro-Silva, M.A., Lima, A.O.S., Gerchenski, A.V., Jaques, D.B., Rodriques, A.L., Lima, P. and Rorig, L.R. 2003. Heavy metal resistance of microorganisms isolated from coal mining environments of Santa Catarina. Brazilian Journal of Microbiology, 34(1): 45–47.

Domsch, K.H., Gams, W. and Anderson, T.H. 1980. Compendium of soil fungi. Academic Press, London.

Gadd, G.M., Ramsay, L., Crawford, J.W. and Ritz, K. 2001. Nutritional influence on fungal colony growth and biomass distribution in response to toxic metals. FEMS Microbiology Letters, 204: 311–316.

Hammer, O., Harper, D.A.T. and Ryan, P.D. 2001. PAST: Paleontological statistics software package for education and data analysis. Palaeontologia Electronica, 4: 1–9.

Johnson, D.B. 1991. Diversity of microbial life in highly acidic, mesophilic environments. *In*: Bertha, J. (ed.). Diversity of Environmental Biogeochemistry.

Johnson, D.B. 1998. Biodiversity and ecology of acidophilic microorganisms. FEMS Microbial Ecology, 27: 307–317.

Krebs, C.J. 1989. Ecological Methodology. Harper and Row, New York.

Lopez-Archilla, A.I., Gonzalez, A.E., Terron, M.C. and Amils, R. 2004. Ecological study of fungal population of the acidic Tinto River in Southwestern Spain. Canadian Journal of Microbiology, 50: 923–934.
Maltby, L. and Booth, R. 1991. The effects of coal mine effluent on fungal assemblages and leaf breakdown. Water Research, 25: 247–250.
Manoharachary, C. 1985. Aspects and prospects of water-borne conidial fungi from India. Advanced Biological Research, 4: 160–183.
Misra, J.K. 1983a. Aquatic mycoflora of alkaline ponds and soils. Bibliotheca Mycologica, J.
Misra, J.K. 1983b. Geofungi inhabiting alkaline ponds. Geophytology, 13: 98–110.
Misra, J.K. 1985a. Active mycoflora from alkaline water and its significance. Biol. Mem., 11: 171–176.
Misra, J.K. 1985b. Fungi from the plant detritus under alkaline water and their ecology. Indian J. Pl. Pathol., 3: 25–32.
Nicolay, K., Veenhuis, M., Douma, A.C. and Harder, W. 1987. A 31P NMR study of the internal pH of yeast peroxisomes. Archives of MicrobiologyI, 147: 37–41.
Niyogi, D.K., McKnight, D.M. and Lewis, W.M., Jr. 2002. Fungal communities and biomass in mountain streams affected by mine drainage. Archives of Hydrobiology, 155(2): 255–271.
Odum, H.T. 1971. Environment, Power and Society. Wiley/Interscience, New York.
Rai, J.N., Agarwal, S.C. and Tewari, J.P. 1971. Fungal microflora of 'Usar' soils of India. Jour. Indian Bot. Soc., 50: 59–69.
Rai, J.N. and Chowdhery, H.J. 1979. Fungi occurring in active state in Indian 'Usar' soils. Nova Hedwigia, 63: 59–69.
Raven, J.A. 1990. Predictions of Mn and Fe use efficiencies of phototrophic growth as a function of light availability for growth and C assimilation pathway. New Phytologist, 116: 1–18.
Ravikumar, M., George, V.K., Selvaraj, R. and Senthilkumar, S. 2000. Distribution of fungi in river Kaveri, Kollidam and Uyyakkondan canal in relation to water pollution. Asian Journal of Microbial Biotechnology and Environmental Research, 2 (3-4): 209–214.
Simpson, E.H. 1949. Measurement of diversity. Nature, 163: 688.
Sridhar, K.R. and Barlocher, F. 2000. Initial colonization, nutrient supply and fungal activity on leaves decaying in streams. Applied and Environmental Microbiology, 66: 1114–1119.
Subramaniam, C.V. 1971. Hyphomycetes: An account of Indian species except Cercosporiae, ICAR, New Delhi. 930.
Webb, J.S., McGiness, S. and Lappin-Scott, H.M. 1998. Metal removal by sulphate reducing bacteria from natural and constructed wetlands. Journal of Applied Microbiology, 84: 240–248.
Wielinga, B., Lucy, J.K., Moore, J.N., Seastone, O.F. and Gannon, J.E. 1999. Microbiological and geochemical characterization of fluvially deposited sulfidic mine tailings. Applied and Environmental Microbiology, 65: 1548–1555.

CHAPTER 6

Distribution of Myxomycetes Among the Microhabitats Available for these Organisms in Tropical Forests

Carlos Rojas,[1,*] *Adam W. Rollins*[2] and *Steven L. Stephenson*[3]

ABSTRACT

Myxomycetes (plasmodial slime molds or myxogastrids) are a group of fungus-like organisms common and often abundant in all types of terrestrial ecosystems. However, only recently have ecological studies of the group been carried out in tropical forests. The results of these studies indicate that myxomycetes are associated with a number of different microhabitats in tropical forests. Among these are coarse woody debris, ground litter and the bark surface of living trees, all of which can be found in temperate and boreal forests. However, myxomycetes also are associated with several other microhabitats (e.g., aerial litter, lianas and canopy soil) that are essentially limited to tropical forests. The objective of this chapter is first to describe all of the microhabitats potentially available for myxomycetes in tropical forests and then to discuss various aspects of the ecology of the assemblage of species associated with each of these.

[1] Engineering Research Institute, University of Costa Rica, 11501 San Pedro de Montes de Oca, Costa Rica.
[2] Department of Biology, Lincoln Memorial University, Harrogate, TN 37752.
[3] Department of Biological Sciences, University of Arkansas, Fayetteville, Arkansas 72701.
* Corresponding author

Introduction

The myxomycetes (plasmodial slime molds or myxogastrids) are a group of eukaryotic microorganisms that belong to the supergroup Amoebozoa (Adl et al., 2012). Their phylogenetic position is currently supported by molecular studies that clearly show their monophyletic character within the Amoebozoa (see Pawlowski and Burki, 2009). In the past, myxomycetes usually have been considered as members of a group known as the Eumycetozoa (Olive, 1975). This hypothetical group included, along with the myxomycetes, two other groups of amoebae known as dictyostelids and protostelids (also called protosteloid amoebae). However, the integrity of the Eumycetozoa as a natural group has been questioned recently on the basis of molecular evidence indicating that the protostelids are probably not monophyletic (Shadwick et al., 2009). This fact does not affect the currently recognized position of myxomycetes within the Amebozoa, but it changes the concepts and nomenclatural treatments of the particular subgroups of organisms to which they are related. Given this situation and for the purpose of this chapter, myxomycetes simply will be treated herein as a group of amoebozoans taxonomically distinct from both dictyostelids and protostelids.

Life Cycle

The generalized ("textbook") life cycle of myxomycetes (Fig. 6.1) has been described in detail by a number of authors (e.g., Martin and Alexopoulos, 1969; Stephenson and Stempen, 1994; Everhart and Keller, 2008). In brief, the life cycle involves two trophic stages—one consisting of microscopic uninucleate amoeboflagellate cells and the other of a multinucleate plasmodium—along with a reproductive stage (fruiting body or sporocarp) somewhat similar to the spore-producing fruiting bodies of certain macrofungi, albeit much smaller. Although myxomycetes are fundamentally amoeboid organisms, their capability to produce a fruiting body that is large enough to be observed directly in nature (the exceptionally large fruiting bodies produced by some species can be the size of a dinner plate!) is truly remarkable (Martin and Alexopoulos, 1969; Stephenson and Stempen, 1994). This capacity is shared with dictyostelids, protostelids and members of the genera *Ceratiomyxa* and *Copromyxa*, the only other "fruiting-capable" groups within the Amebozoa (Cavalier-Smith, 2003; Adl et al., 2012).

Fruiting bodies exhibit considerable diversity with respect to their morphology, but the most common type is the sporangium (Fig. 6.2), which can be stalked or sessile. The other types of fruiting bodies usually recognized are plasmodiocarps, aethalia and pseudoaethalia (Stephenson and Stempen, 1994). As a fruiting body develops from a plasmodium, the

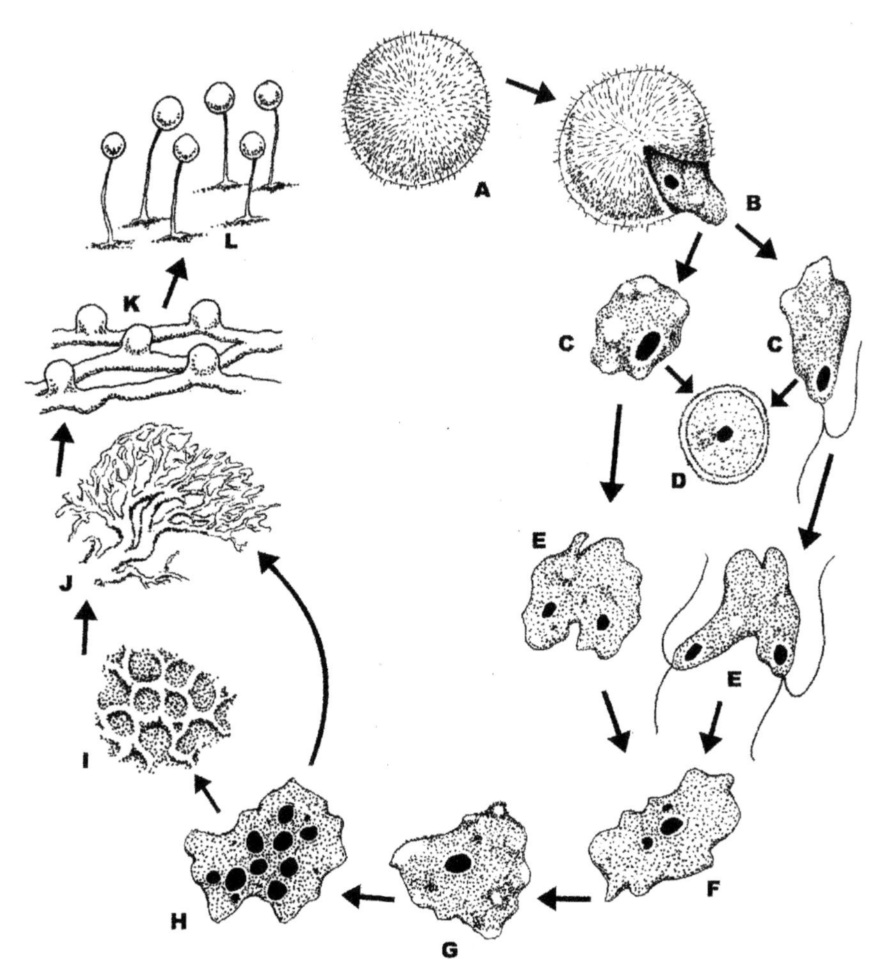

Figure 6.1. Generalized life cycle in the myxomycetes. A-B. A protoplast emerges from the spore. C. The protoplast can take the form of an amoeba or a flagellated cell during the first trophic stage. D. Under dry conditions or in the absence of food, an amoeba can form a microcyst, or resting stage. E-F. Compatible amoeboflagellates fuse to form a zygote (G). H. The nucleus of the zygote divides by mitosis and each subsequent nucleus also divides without being followed by cytokinesis, thus producing a single large cell (J), the plasmodium, that represents the second trophic stage. Under adverse conditions, the plasmodium can form the second type of resting stage found in myxomycetes, the sclerotium (I). K-L. Fruiting bodies are formed from the plasmodium. During formation of the fruiting bodies, spores are produced. Adapted from Stephenson (2003a).

process of spore formation takes place. The newly formed spores, which are usually produced as a result of meiotic divisions, are released into the environment once the fruiting body is fully mature. The overall similarity (at least in many of the more common myxomycetes) of the mature fruiting

Figure 6.2. Fruiting bodies of *Hemitrichia calyculata*, one of the more common myxomycetes associated with coarse woody debris in tropical forests (photo by Kim Fleming). Scale bar = 1.0 mm.

body to those of certain fungi accounts for the fact myxomycetes were once considered part of the kingdom Fungi (Alexopoulos et al., 1996).

The spores of myxomycetes are typically more or less round, usually possess some type of ornamentation on the spore wall, and generally range in size from 5 to 15 µm (Tesmer and Schnittler, 2007). Presumably, most spores are wind-dispersed and continue the life cycle by germinating to produce uninucleate amoeboflagellate cells (Stephenson et al., 2008a). However, it has been observed that various animals can serve as vectors for spores, including both invertebrates (e.g., Blackwell, 1984) and vertebrates (e.g., Townsend et al., 2005). Amoeboflagellate cells feed and divide by binary fission to build up large populations in the various microhabitats in which myxomycetes occur. Ultimately, this initial trophic stage in the life cycle gives rise to the second tropic stage, the plasmodium. This process can result from gametic fusion between compatible amoeboflagellates or it can be apomictic (Collins, 1980, 1981). Bacteria apparently represent the main food resource for both trophic stages, but plasmodia are also known to feed upon yeasts, algae (including cyanobacteria), fungal spores and hyphae (Stephenson and Stempen, 1994; Smith and Stephenson, 2007).

Under adverse conditions, such as drying out of the immediate environment or low temperatures, a plasmodium may convert into a hardened, resistant structure called a sclerotium, which is capable of reforming the plasmodium upon the return of favorable conditions. Moreover, amoeboflagellate cells can undergo a reversible transformation to dormant structures called microcysts. Both sclerotia and microcysts can remain viable for long periods of time and are probably very important in the continued survival of myxomycetes in some ecological situations and/or habitats.

Distribution and Ecology

Myxomycetes have been recorded from every major type of terrestrial ecosystem examined to date, and at least a few species have been recovered from aquatic habitats (Lindley et al., 2007). Temperature and moisture are thought to be the main factors limiting the occurrence of myxomycetes in nature (Alexopoulos, 1963), and species richness tends to increase with increasing diversity and biomass of the vascular plants providing the resources (various types of detritus) that support the bacteria and other microorganisms upon which the two trophic stages in the myxomycete life cycle feed (Madelin, 1984; Stephenson, 1989). The pH of the substrates potentially available to myxomycetes in a particular habitat also represents an important factor influencing the distribution of these organisms (Härkönen, 1977; Stephenson, 1989; Wrigley de Basanta, 2000; Mosquera et al., 2000). Although many myxomycetes appear to have a relatively wide pH tolerance, this is not the case for all species. Härkönen (1977), who measured the pH of substrates upon which fruitings occurred in a study of the distribution patterns of myxomycetes associated with the bark of living trees in southern Finland, concluded that species of myxomycetes have different pH optima and amplitudes. In her study, some species seemed to prefer an acidic substrate, whereas others never developed under low pH conditions. Stephenson (1989) found the same to be true for both bark and forest floor litter in a study carried out in the eastern United States.

Previous Ecological Studies

Historically, myxomycetes have been studied much more often by trained taxonomists than by ecologists. Although early publications (e.g., Lister, 1894) frequently contained basic ecological observations relating to particular species of myxomycetes, virtually all of the truly ecological investigations of the group have been carried out during the past few decades (e.g., Maimoni-Rodella and Gottsberger, 1980; Eliasson, 1981; Stephenson, 1988, 1989; Wrigley de Basanta et al., 2010). The majority of such studies have taken place in temperate areas of the Northern Hemisphere (Stephenson et al., 2004), most notably in Europe and North America. Although a number of other localities throughout the world have been studied more intensively in the last 50 years or so, there are still areas that are highly understudied. Unfortunately, this is a statement that can be made for most groups of organisms, especially microorganisms. Based on the information currently available, biodiversity of myxomycetes appears to be higher in temperate than in tropical or boreal regions of the world (Schnittler, 2001b). However, whether or not this pattern is real has yet to be determined. For example, it is possible that studies carried out thus far in tropical regions have captured only a portion of the total biodiversity of the assemblage of species present. This would be the case if many of the species involved in this "hidden biodiversity" persist largely as amoeboflagellates and only rarely form plasmodia and fruiting bodies.

Microhabitats

Myxomycetes have long been known to occur on decaying wood, litter, soil, herbivore dung and the bark surface of living trees (Stephenson and Stempen, 1994). Each of these microhabitats is known to support a different assemblage of species (Stephenson, 2003a). More recently, based largely on studies carried out in tropical regions of the world, myxomycetes have been reported from a number of other microhabitats (Fig. 6.3), including aerial

Figure 6.3. Some of the microhabitats available to myxomycetes in tropical forest. (A) Dead but still attached leaves at the base of this giant bromeliad represent aerial litter; (B) Dead leaves on the forest floor are an example of ground litter; (C) *Heliconia* inflorescence; (D) Woody vine (liana); (E) Bark surface of a living trees (note how thin the bark of a tropical tree can be); (F) Coarse woody debris on the forest floor.

litter (Stephenson, 2003b), inflorescences of large herbaceous plants (e.g., Schnittler and Stephenson, 2002), woody twigs (Stephenson et al., 2008b), bryophytes (e.g., Schnittler, 2001a) and lianas (e.g., Wrigley de Basanta et al., 2008). Several of these microhabitats are common in tropical forests but have no ecological counterparts in temperate forests. The objective of this chapter is first to describe each of the microhabitats potentially available for myxomycetes in tropical forests and then to discuss various aspects of the ecology of the assemblage of species associated with the microhabitat in question.

Coarse Woody Debris

Lignicolous myxomycetes associated with coarse woody debris are the best known of all myxomycetes, since the species typically associated with this microhabitat tend to be among those characteristically producing fruiting bodies of sufficient size to be detected easily in the field (Martin and Alexopoulos, 1969). Many of the more common and widely known myxomycete taxa, including various species of *Arcyria*, *Lycogala*, *Stemonitis* and *Trichia*, are predominantly lignicolous. In one study carried out in temperate deciduous forests in the eastern United States, Stephenson (1988) reported that more than 90% of all specimens of myxomycetes collected in the field over a five year period were associated with coarse woody debris. In a tropical forest, the percentage of specimens associated with this microhabitat is usually much lower. This may be due to the greater diversity of other microhabitats available for myxomycetes in tropical forests and the relatively more rapid decomposition rate of coarse woody debris. A number of studies have indicated that particular species of myxomycetes tend to be associated with a certain stage in the decomposition of logs and large branches of trees. Some species tend to occur on such substrates at a very early stage, when the bark is still present, while other species are invariably present only after the wood is completely decorticated. Although this fact has not always been recognized, some species of myxomycetes (e.g., members of the genus *Badhamia*) are almost always found on the bark, while others are usually collected from wood. This is not surprising, since bark and wood are actually quite different with respect to their composition. The pH of wood varies considerably, and this factor alone is important in determining the distribution of myxomycetes. For example, *Ceratiomyxa morchella* seems to prefer decaying wood that is fairly acidic, in contrast to the majority of tropical myxomycetes. Some woody or semi-woody substrates found in tropical forests are exceedingly poor substrates for myxomycetes, for reasons that are not clearly apparent. This is the case for the trunks of tree ferns and palms. In contrast, old decaying palm fronds can be quite productive at times (Stephenson, 2003b). One of the more commonly encountered species in the latter microhabitat is *Perichaena depressa* (Fig. 6.4), which also occurs on twigs, ground litter and the bark surface of living trees.

Among the species most likely to be associated with coarse woody debris in tropical forests are *Arcyria denudata*, *Cribraria cancellata*, *Hemitrichia calyculata*, *Lycogala epidendrum*, *Physarella oblonga*, *Physarum stellatum* and *Stemonitis fusca*. The majority of these are also common in temperate forests, but *H. calyculata* is noticeably more abundant in tropical forests than temperate forests. *P. oblonga*, which sometimes occurs in extensive fruitings on large, moist decaying logs, is only rarely collected outside of the tropics.

Although the authors are not aware of any studies that have considered the evidence for a possible correlation between the presence of termites or ants and the absence of myxomycetes on or within a particular log, anyone who has done extensive collecting in tropical forests is likely to have noticed that logs colonized by either group of insects tend to be devoid of myxomycetes. It is well-known that various beetles feed upon plasmodia, so insect predation could be a factor of some consequence.

Figure 6.4. Fruiting bodies of *Perichaena depressa*, a myxomycete typically associated with several different types of microhabitats in tropical forests (photo by Clive Shirley). Scale bar = 3.0 mm.

Bark Surface of Living Trees

A distinctive ecological group of myxomycetes is associated with the microhabitat represented by the bark surface of living trees, but the species involved are not as well known as the myxomycetes associated with such microhabitats as coarse woody debris or litter. The primary reason for this is that many of the species involved produce small and thus relatively inconspicuous fruiting bodies and are rather sporadic in their occurrence, thus making them especially difficult to detect in the field. However, the moist chamber culture technique as it applies to myxomycetes (Gilbert and Martin, 1933) provides a convenient and often very productive method of supplementing field collections when studying bark as well as a number of the other microhabitats known to support myxomycetes. Since its introduction, the technique has been used with considerable success by many researchers (e.g., Keller and Brooks, 1976; Blackwell and Gilbertson, 1980; Härkönen, 1981; Stephenson, 1989). More than 100 species of "corticolous" (bark-associated) myxomycetes have been reported from the bark microhabitat as field and/or moist chamber collections (Mitchell, 1980). Many of these are also known to occur in other microhabitats, but at least some species seem restricted to bark of living trees. Prominent examples include various species of *Echinostelium*, *Licea* and *Macbrideola* (Alexopoulos, 1964; Mitchell, 1980).

Based on the data available from studies of corticolous myxomycetes in tropical forests (e.g., Schnittler and Stephenson, 2000), both the number of species and their relative abundance appear to be lower for the bark of tropical trees than the bark of trees that occur in temperate forests of the Northern Hemisphere (Stephenson, 1989). Although several factors are likely to be involved, the generally smooth surface of many tropical trees may be a factor. Presumably, a smooth bark surface would represent a less effective "spore trap" for air-borne spores of myxomycetes if indeed most myxomycetes are introduced to a particular microhabitat as the result of being dispersed by air currents.

Lianas

Woody vines (usually called lianas) are common and often conspicuous features of moist tropical forests throughout the world. These long-stemmed plants are rooted in the soil at ground level and climb up to well-lit areas of the forest canopy by using the surrounding trees for support. Lianas do not represent a single taxonomic grouping (examples are found in a number of different families of flowering plants), and the term "liana" simply denotes a general growth habitat. Although lianas can be described as having bark, the latter is sometimes quite different from that of most tropical trees. The bark surfaces of different types of lianas can vary considerably with respect to pH, texture and water-holding capacity, and these differences can give rise to compositional differences in the assemblages of myxomycetes present (Wrigley de Basanta et al., 2008). Moreover, the occurrence of fruiting bodies that have developed under natural conditions in the field appears to be relatively more common for the bark of lianas than is the case for the bark of trees. However, only recently (Lado et al., 2003; Wrigley de Basanta et al., 2008; Ko Ko et al., 2010) have the myxomycetes associated with this microhabitat been investigated. Based on the results obtained in the relatively few studies that have been carried out, lianas appear to support a diverse assemblage of myxomycetes. In all of these studies, data were obtained largely with the use of moist chamber cultures. For example, Ko Ko et al. (2010) recorded 30 different species representing 15 genera from a single small area of northern Thailand. These authors collected samples from both dead lianas and living lianas, and these samples were collected at two different heights above the ground. Interestingly, dead lianas yielded more species at both of the heights sampled, with the higher total recorded for samples taken from a height of at least 2 m.

Ceratiomyxa fruticulosa, a species usually confined to coarse woody debris, occasionally occurs on samples of lianas placed in moist chamber cultures. *Physarum pusillum*, which also can be found on ground litter, aerial litter and inflorescences, is especially common on lianas.

Aerial Litter

Tropical forests tend to be characterized by the presence of a considerable amount of what has been referred to as aerial litter, defined herein as dead but still attached portions (e.g., mostly leaves and old inflorescences) of non-woody plants (Stephenson, 2003a). This definition can be expanded to include leaves that have become detached but are still suspended above the ground in some fashion (e.g., being trapped with a group of small lianas). Once aerial litter falls to the ground, it becomes incorporated into ground litter. Although the same type of substrate material is involved, the assemblages of myxomycetes associated with aerial litter are distinctly different from those associated with ground litter. Moreover, aerial litter tends to be a much more productive microhabitat than ground litter, especially in wet tropical forests (Black et al., 2004). In such forests, myxomycetes seem to be largely displaced from the ground litter microhabitat to the aerial litter microhabitat. Presumably, this is the result of differences in levels of moisture. In a wet tropical forest, substrates on the ground tend to remain relatively moist, often with an actual film of water present. This is not an especially good situation for myxomycetes, since moisture promotes the growth of fungi. The latter almost surely is a major constraint for the growth and development of myxomycetes, since fungal hyphae can quickly colonize the fruiting bodies of any myxomycete that appear on ground litter, which limits their ability to produce viable spores. In contrast, even after a period of rain, aerial litter quickly dries out, which apparently provides more favorable conditions for myxomycetes. In one study carried out on an isolated tropical island, aerial litter collected from areas of the forest that were more exposed to the wind were characterized by higher species richness and greater numbers of myxomycetes than either ground litter and other types of aerial litter from more protected forests (Rojas and Stephenson, 2008b). There is little question that samples of aerial litter can be amazingly productive when placed in moist chamber cultures for the isolation of myxomycetes.

Long-stalked members of the genus *Didymium*, including *Didymium iridis*, tend to be exceedingly common on aerial litter. Three other species likely to be encountered in this microhabitat are *Lamproderma scintillans*, *Physarum compressum* and *P. melleum*.

Epiphylls

One of the more distinctive features of a tropical forest is the presence of an often extensive assemblage of epiphylls (liverworts and lichens) on the living, often leathery leaves of tropical trees and shrubs. Lücking (1999) reported more than 250 species of lichens from a single locality in a tropical forest in the Amazon Basin, and more than 80 of these were recorded from a single palm leaf. Older leaves that show signs of senescence tend to support the most extensive assemblages of liverworts. Schnittler (2001a) collected samples of such leaves at six localities in Ecuador, Costa Rica and Puerto Rico and used them to prepare a series of moist chamber cultures. These cultures yielded 11 different species, including one species (*Arcyria afroalpina*) not previously known from the Neotropics. Three species (*Arcyria cinerea*, *Didymium iridis* and *D. squamulosum*) were recorded frequently enough to suggest that they are typically associated with this unusual microhabitat. Stephenson (unpubl. data) sampled epiphyll covered leaves collected in a tropical forest in northern Queensland, Australia, and these same three species were recovered. One noteworthy characteristic of the myxomycetes associated with epiphylls is the fact that their fruiting bodies are unusually small, often with only one or a few fruiting bodies occurring on a single leaf.

Canopy Soil

In tropical forests, a mantle of dead organic matter is usually associated with the epiphytes (including both vascular and nonvascular examples) that occur on the branches and to a lesser extent the trunks of trees. This dead organic matter, which has been referred to as "canopy soil" is derived from decaying epiphytes, partially decomposed tree bark, insect frass and intercepted litter (Stephenson and Landolt, 1998). Canopy soil is similar to the soil found on the ground in that it contains many of the same organisms. For example, Stephenson and Landolt (2011) summarized the results obtained from a series of studies of the dictyostelid cellular slime molds (dictyostelids) recorded from the canopy soil microhabitat. They reported that at least 37 species of dictyostelids have been recovered from samples collected at 11 different localities throughout the world. This total includes several species described as new to science, and three of these are not yet known from soil on the ground, which is considered to be the primary microhabitat for dictyostelids. Primary isolation plates prepared for dictyostelids often yield small plasmodia of myxomycetes, and this was the case for some samples of canopy soil. As such, there is clear evidence that at least a few species of myxomycetes are associated with this microhabitat in tropical forests. However, to the best of our knowledge, this has not yet been studied.

Inflorescences of Herbaceous Plants

Large herbaceous monocots are among the more conspicuous plants present in tropical forests, especially areas of forest that have been disturbed. In the Neotropical forests of Central and South America, some of the more common and widespread examples are various species of *Costus* (family Costaceae), *Hedychium* (Zingiberacae), *Calathea* (Marantaceae) and *Heliconia* (Heliconiaceae). All of these plants are characterized by the production of inflorescences made up of dense clusters of flowers usually subtended by sterile and often somewhat fleshy bracts. Schnittler and Stephenson (2002) studied the assemblages of myxomycetes associated with the inflorescences of large herbaceous monocots at a number of localities in Costa Rica, Ecuador and Puerto Rico. They discovered that inflorescences often yielded the fruiting bodies of myxomycetes, either as field collections or in moist chamber cultures in the laboratory. In some instances, relatively large fruitings of such species as *Physarum didermoides* were observed, and a total of 31 species were represented among the 652 specimens recorded during the entire study. Moist chamber cultures prepared with floral parts typically had a pH between 8 and 9, which is higher than that of most substrates from which myxomycetes have been reported. The two authors speculated that the spores of the species of myxomycetes associated with inflorescences (for which they proposed the term "floricolous" myxomycetes) are dispersed by birds that pollinate the flowers or feed upon the fruits that develop from the flowers. Subsequent sampling carried out in other parts of the world, including northern Queensland and Southeast Asia, has indicated that floricolous myxomycetes are not limited to just the Neotropics.

In addition to *Physarum didermoides,* other myxomycetes commonly found on inflorescences are *P. compressum* and *P. superbum.* However, floricolous species also include examples such as *Arcyria cinerea* that tend to be associated with a wide range of different substrates.

Dung

The dung of herbivorous animals often represents an important microhabitat for myxomycetes in some types of ecosystems, most notably deserts and arid grasslands. A few species of myxomycetes are rarely if ever recorded from any other microhabitat. However, dung can be largely discounted as a microhabitat for myxomycetes in tropical forests, simply because it does not persist long enough to allow the growth and development of these organisms.

Ground Litter

The ground litter microhabitat is made up of a complex and heterogeneous mixture of dead and decaying plant material, including leaves, bark fragments, small twigs, fruits, flowers and seeds that are contiguous with the underlying soil microhabitat (Stephenson, 1989; Rollins and Stephenson, 2012). Given this fact, one might expect ground litter to support a species-rich assemblage of myxomycetes, but this does not appear to be the case in tropical forests. This is likely explained by two factors that are not especially conducive for the growth and development of myxomycetes. First, the litter microhabitat often remains waterlogged. Second, it usually decomposes rather quickly. It appears that many species of myxomycetes actually utilize the often abundant aerial litter microhabitats that are subject to periodic drying.

Studies examining the ground litter microhabitat have used the moist chamber culture technique (Stephenson et al., 1999), only collections obtained from the field (Novozhilov et al., 2001), and a combination of field collections and those derived from moist chamber collections. Schnittler and Stephenson (2000) reported that myxomycete species richness within the litter microhabitat in the Neotropics was lower than in temperate forests. Interestingly, Tran et al. (2006) reported data suggesting that myxomycete species richness was comparable, or possibly even greater, in the Old World Tropics of Southeast Asia. This aspect of the myxomycetes associated with the litter microhabitat warrants further study.

Currently, the ground litter microhabitat is still poorly known, but its diverse composition and structural complexity would appear to make it especially favorable for myxomycetes. As already mentioned, the ground litter microhabitat is contiguous with the soil microhabitat (Stephenson and Landolt, 1996). As such, it is not always apparent just which species are restricted largely to ground litter and what other species may be migrating from the soil and using this microhabitat simply for the production of fruiting bodies. It would be worthwhile to examine the patterns of myxomycete occurrence throughout the litter microhabitat in more detail; especially differences in the spatial distribution of species among the various components that make it up (see the section on twigs). As anyone who has processed large numbers of samples of litter in moist chamber cultures will surely notice, fruiting bodies of a particular species (e.g., *Cribraria microcarpa*) are often restricted to small fragments of wood that typically occur in litter. In this context the apparent preferential utilization of decaying fruits by the tropical species *Ceratiomyxa sphaerosperma* should be noted (Novozhilov et al., 2001; Rojas et al., 2008a).

Some of the species commonly appearing on ground litter are just as likely to be encountered in other microhabitats. This is the case for *Arcyria cinerea* (Fig. 6.5), one of the more abundant and widely distributed of all myxomycetes. It is not unusual for numerous fruiting bodies of *Arcyria cinerea* to occur on coarse woody debris, but the species is sometimes just as common on aerial litter, twigs and other types of plant debris. Because the fruiting bodies assigned to this species show considerable morphological variation, it is possible that several different "biotypes" are involved, but this has not yet been subjected to study.

Figure 6.5. Fruiting bodies of *Arcyria cinerea*, a species that has been recorded from virtually all of the microhabitats found in tropical forests (photo by Alain Michaud). Scale bar = 1.0 mm.

Twigs

Twigs are a characteristic component of the ground litter microhabitat, but they are easily extracted and studied independently. Overall, the myxomycetes associated with twigs are poorly known, but it is clear that twigs support a characteristic assemblage of myxomycetes that is different from bark, coarse woody debris and other components of the ground litter microhabitat. It has been suggested that twigs may represent a primary substrate for some less commonly encountered species of myxomycetes such as *Willkommlangea reticulata* (Stephenson et al., 2008b). In perhaps the most comprehensive study of twigs carried out to date (Wrigley de Basanta et al., 2008), reported that those from tropical forest were less productive (23–43% positive cultures) than those from temperate forests (67–85% positive cultures).

Soil

Tropical soils represent a microhabitat for myxomycetes for which virtually nothing is known. However, it is now well established that amoeboid protozoans are consistent and often abundant members of the soil microhabitat, and it has been estimated that roughly 50% of all soil amoebae in temperate forests are myxomycetes (Feest and Madelin, 1985). As such, it is likely that these organisms play important roles in nutrient cycling as members of the detritus food chain. In the relatively few studies that have examined soils, it appears that members of the Physarales are the predominate taxa associated with this microhabitat, with species of *Didymium* especially prominent. This same situation is likely to be the case for tropical soils as well. It has often been noted that plasmodia frequently appear in cultures when samples of soil are plated out on agar media, but these usually do not produce fruiting bodies. However, some success has been obtained with a modified moist chamber culture technique that involves placing an autoclaved substrate over the surface of a soil culture (Rollins and Stephenson, 2012). Stephenson et al. (2011) recently reviewed what is known about myxomycetes in the soil microhabitat.

Summary

Tropical forests are characterized by a complex structure that is reflected in the occurrence of an appreciable number of distinctly different microhabitats known to support assemblages of myxomycetes. Overall, the myxomycetes associated with tropical forests are still not as well known as those of either boreal forests or temperate forests, and this is especially true for particular tropical microhabitats. Especially noteworthy in this regard is the soil microhabitat, for which almost no data exist. Moreover, relatively few studies have been directed towards such microhabitats as twigs, aerial litter and canopy soil. Clearly, the myxomycetes of tropical forests warrant additional study, and our hope is that the information presented in this chapter will provide a starting point for efforts in this direction.

References

Adl, S.M., Simpson, A.G.B., Lane, C.E., Julius, L., Bass, D., Bowser, S.S., Brown, M.W., Burki, F., Dunthorn, M., Hampl, V., Heiss, A., Hoppenrath, M., Lara, E., Gall, L.L., Lynn, D.H., Mcmanus, H., Mitchell, E.A.D., Mozley-Stanridge, A.E., Parfrey, L.W., Pawlowski, J., Rueckert, S., Shadwick, L., Schoch, C.L., Smirnov, A. and Spiegel, F.W. 2012. The revised classification of Eukaryotes. Journal of Eukaryotic Microbiology, 59: 429–493.

Alexopoulos, C.J. 1963. The Myxomycetes II. Botanical Review, 29: 1–77.

Alexopoulos, C.J. 1964. The rapid sporulation of some myxomycetes in moist chamber culture. Southwest Naturalist, 9: 155–159.

Alexopoulos, C.J., Mims, C.W. and Blackwell, M. 1996. Introductory Mycology, 4th ed. John Wiley & Sons, Inc, New York, USA.

Black, D.R., Stephenson, S.L. and Pearce, C.A. 2004. Myxomycetes associated with the aerial litter microhabitat in tropical forests of northern Queensland, Australia. Systematics and Geography of Plants, 74: 129–132.

Blackwell, M. 1984. Myxomycetes and their arthropod associates. pp. 67–90. *In*: Wheeler, Q.D. and Blackwell, M. (eds.). Fungus-Insect Relationships: Perspectives in Ecology and Evolution. Columbia University Press, New York, USA.

Blackwell, M. and Gilbertson, R.L. 1980. Sonoran Desert myxomycetes. Mycotaxon, 11: 139–149.

Cavalier-Smith, T. 2003. Protist phylogeny and the high-level classification of Protozoa. European Journal of Protistology, 39: 338–348.

Collins, O.R. 1980. Apomictic-heterothallic conversion in a myxomycete *Didymium iridis*. Mycologia, 72: 1109–1116.

Collins, O.R. 1981. Myxomycete genetics, 1960–1981. Journal of the Elisha Mitchell Scientific Society, 97: 101–125.

Eliasson, U. 1981. Patterns of occurrence of myxomycetes in a spruce forest in south Sweden. Holarctic Ecology, 4: 20–31.

Everhart, S.E. and Keller, H.W. 2008. Life history strategies of corticolous myxomycetes: the life cycle, plasmodial types, fruiting bodies, and taxonomic orders. Fungal Diversity, 24: 1–16.

Feest, A. and Madelin, M.F. 1985. Mumerical abundance of myxomycetes (myxogastrids) in soils in the West of England. FEMS Microbiology Ecology, 31: 353–360.

Gilbert, H.C. and Martin, G.W. 1933. Myxomycetes found on the bark of living trees. University of Iowa Studies in Natural History, 15: 3–8.

Harkönen, M. 1977. Corticolous myxomycetes in three different habitats in southern Finland. Karstenia, 17: 19–32.

Harkönen, M. 1981. Myxomycetes developed on litter of common Finnish trees in moist chamber cultures. Nordic Journal of Botany, 1: 791–794.

Keller, H.W. and Brooks, T.E. 1976. Corticolous myxomycetes V. Observations on the genus *Echinostelium*. Mycologia, 68: 1204–1220.

Ko Ko, T.W., Stephenson, S.L., Hyde, K.D. and Lumyong, S. 2010. Patterns of occurrence of myxomycetes on lianas. Fungal Ecology, 3: 302–310.

Lado, C., Estrada-Torres, A., Stephenson, S.L., Wrigley de Basanta, D. and Schnittler, M. 2003. Biodiversity assessment of myxomycetes from two tropical forest reserves in Mexico. Fungal Diversity, 12: 67–110.

Lindley, L.A., Stephenson, S.L. and Spiegel, F.W. 2007. Protostelids and myxomycetes isolated from aquatic habitats. Mycologia, 99: 504–509.

Lister, A. 1894. A Monograph of the Mycetozoa. Printed by Order of the Trustees, London, UK.

Lücking, R. 1999. Factors influencing the diversity and distribution of foliicolous, lichenized fungi in tropical rain forests. Abstracts of the Sixth International Mycological Congress —Kenes Ltd., Tel Aviv, Israel, 1: 7.

Madelin, M.F. 1984. Myxomycetes, microorganisms and animals: a model of diversity in animal-microbial interactions. pp. 2–33. *In*: Anderson, J.N., Rayer, A.D.A. and Walton, W.H. (eds.). Invertebrate-microbial Interactions. Cambridge University Press, New York, USA.

Maimoni-Rodella, R.C.S. and Gottsberger, G. 1980. Myxomycetes from the forest and the cerrado vegetation in Notucatu, Brazil: a comparative ecological study. Nova Hedwigia, 34: 207–246.

Martin, G.W. and Alexopoulos, C.J. 1969. The Myxomycetes. University of Iowa Press, Iowa City.

Mitchell, D.W. 1980. A Key to the Corticolous Myxomycetes. The British Mycological Society, Cambridge, UK.

Mosquera, J., Lado, C. and Beltrán-Tejera, E. 2000. Morphology and ecology of *Didymium subreticulosporum*. Mycologia, 92: 378–983.

Novozhilov, Y.K., Schnittler, M., Rollins, A.W. and Stephenson, S.L. 2001. Myxomycetes in different forest types in Puerto Rico. Mycotaxon, 77: 285–299.

Olive, L.S. 1975. The Mycetozoans. Academic Press, New York, USA.

Pawlowski, J. and Burki, F. 2009. Untangling the phylogeny of amoeboid protists. Journal of Eukaryotic Microbiology, 56: 16–25.

Rojas, C., Schnittler, M., Biffi, D. and Stephenson, S.L. 2008a. Microhabitat and niche separation in species of *Ceratiomyxa*. Mycologia, 100: 843–850.

Rojas, C. and Stephenson, S.L. 2008b. Myxomycete ecology along an elevational gradient on Cocos Island, Costa Rica. Fungal Diversity, 29: 117–127.

Rollins, A.W. and Stephenson, S.L. 2012. Myxogastrid distribution within the leaf litter microhabitat. Mycosphere, 3: 543–549.

Schnittler, M. 2001a. Foliicolous liverworts as a microhabitat for Neotropical myxomycetes. Nova Hedwigia, 72: 259–270.

Schnittler, M. 2001b. Ecology of myxomycetes from a winter-cold desert in western Kazakhstan. Mycologia, 93: 653–669.

Schnittler, M. and Stephenson, S.L. 2000. Myxomycete biodiversity in four different forest types in Costa Rica. Mycologia, 92: 626–637.

Schnittler, M. and Stephenson, S.L. 2002. Inflorescences of Neotropical herbs as a newly discovered microhabitat for myxomycetes. Mycologia, 94: 6–20.

Shadwick, L.L., Spiegel, F.W., Shadwick, J.D.L., Brown, M.W. and Silberman, J.D. 2009. Eumycetozoa = Amoebozoa?: SSUrDNA phylogeny of protosteloid slime molds and its significance for the amoebozoan supergroup. PLoS One, 4: e6754.

Smith, T. and Stephenson, S.L. 2007. Algae associated with myxomycetes and leafy liverworts on decaying spruce logs. Castanea, 72: 50–57.

Stephenson, S.L. 1988. Distribution and ecology of myxomycetes in temperate forests. I. Patterns of occurrence in the upland forests of southwestern Virginia. Canadian Journal of Botany, 66: 2187–2207.

Stephenson, S.L. 1989. Distribution and ecology of myxomycetes in temperate forests. II. Patterns of occurrence on bark surface of living trees, leaf litter, and dung. Mycologia, 81: 608–621.

Stephenson, S.L. 2003a. Myxomycetes of New Zealand. Fungal Diversity Press, Hong Kong.

Stephenson, S.L. 2003b. Myxomycetes associated with decaying fronds of nikau palm (*Rhopalostylis sapida*) in New Zealand. New Zealand Journal of Botany, 41: 311–317.
Stephenson, S.L., Fiore-Donno, A.-M. and Schnittler, M. 2011. Myxomycetes in soil. Soil Biology and Biochemistry, 43: 2237–2242.
Stephenson, S.L. and Landolt, J.C. 1998. Dictyostelid cellular slime molds in canopy soils of tropical forests. Biotropica, 30: 657–661.
Stephenson, S.L. and Landolt, J.C. 2011. Dictyostelids from aerial "canopy soil" microhabitats. Fungal Ecology, 4: 191–195.
Stephenson, S.L., Landolt, J.C. and Moore, D.L. 1999. Protostelids, dictyostelids, and myxomycetes in the litter microhabitat of the Luquillo Experimental Forest, Puerto Rico. Mycological Research, 103: 209–214.
Stephenson, S.L. and Landolt, J.C. 1996. The vertical distribution of dictyostelids and myxomycetes in the soil/litter microhabitat. Nova Hedwigia, 62: 105–117.
Stephenson, S.L., Schnittler, M., Lado, C., Estrada-Torres, A., Wrigley de Basanta, D., Landolt, J.C., Novozhilov, Y.K., Clark, J., Moore, D.L. and Spiegel, F.W. 2004. Studies of Neotropical mycetozoans. Systematics and Geography of Plants, 74: 87–108.
Stephenson, S.L., Schnittler, M. and Novozhilov, Y.K. 2008a. Myxomycete diversity and distribution from the fossil record to the present. Biodiversity and Conservation, 17: 285–301.
Stephenson, S.L., Urban, L.A., Rojas, C. and McDonald, M.S. 2008b. Myxomycetes associated with woody twigs. Revista Mexicana de Micologia, 27: 21–28.
Stephenson, S.L. and Stempen, H. 1994. Myxomycetes: A Handbook of Slime Molds. Timber Press, Portland, Oregon.
Tesmer, J. and Schnittler, M. 2007. Sedimentation velocity of myxomycete spores. Mycological Progress, 6: 229–234.
Townsend, J.H., Aldrich, H.C., Wilson, L.D. and McCranie, J.R. 2005. First report of sporangia of a myxomycete (*Physarum pusillum*) on the body of a living animal, the lizard *Corytophanes cristatus*. Mycologia, 97: 346–348.
Tran, H.T.M., Stephenson, S.L., Hyde, K.D. and Mongkolporn, O. 2006. Distribution and occurrence of myxomycetes in tropical forests of northern Thailand. Fungal Diversity, 22: 227–242.
Wrigley de Basanta, D. 2000. Acid deposition in Madrid and corticolous myxomycetes. Stapfia, 73: 113–120.
Wrigley de Basanta, D., Lado, C., Estrada-Torres, A. and Stephenson, S.L. 2010. Biodiversity of myxomycetes in subantarctic forests of Patagonia and Tierra del Fuego, Argentina. Nova Hedwigia, 90: 45–79.
Wrigley de Basanta, D., Stephenson, S.L., Lado, C., Estrada-Torres, A. and Nieves-Rivera, A.M. 2008. Lianas as a microhabitat for myxomycetes in tropical forests. Fungal Diversity, 28: 109–125.

CHAPTER 7

Microfungi from Deteriorated Materials of Cultural Heritage

Filomena De Leo[a,*] and *Clara Urzì*[b]

ABSTRACT

Fungi possess a high metabolic versatility, hence are able to colonize several organic and even inorganic surfaces causing irreversible changes in their appearance and intrinsic properties. For this reason they may be considered as the major biodeteriogens of material of commercial and artistic value. The present work offers an overview on the main fungal species, both hyphomycetes and the so-called black meristematic fungi, associated with biodeteriorated materials. The patterns of alteration caused by fungi and the methodology that can be used for the diagnosis and prevention of fungal colonization have also been discussed.

Introduction

Fungi are chemoorganotrophic eukaryotic organisms with simple and versatile nutritional requirements. They present a wide range of adaptations to different climatic conditions being able to grow and/or survive in different environmental conditions such as high temperature or extremely low pH.

Fungi are ubiquitous and thus are able to colonize a wide variety of substrates both, organic and inorganic including building materials. In fact,

Department of Biological and Environmental Sciences, University of Messina, Viale F. Stagno d'Alcontres 31, 98166 Messina, Italy.
[a] Email: fdeleo@unime.it
[b] Email: cl@unime.it
* Corresponding author

these latter contain enough organic matter (dust, soil, airborne particles) to support their growth. They are widespread in nature and mainly abound in soils, and also occur as saprophytes or parasites of plants and animals including humans. Their dispersal in the environment occurs mainly by spores (conidia, sexual spores, chlamydospores, etc.) or hyphal fragments, conidiophores, free or associated with inorganic particles (Comtois, 1990). For this reason, fungi can be considered as one of the most harmful organisms associated with biodeterioration of organic and inorganic materials.

The transport and ultimate settling of fungal propagules on various surfaces/substrates are affected by their physical properties (size, density and shape) and the environmental parameters (velocity and direction of air, relative humidity and temperature) (Urzì et al., 2001). However, a mere presence of fungal propagules does not make it sufficient to consider them as a cause of decay, because only under particular conditions, and when also other climatic factors and pollutants occur, they can cause and/or enhance the process of deterioration. Furthermore, fungal spores after their production and maturation may remain quiescent for a long period. The environmental conditions required for germination of spores are different for each fungal species. So, this period of dormancy may depend on the circumstances, lasting only for a few hours or many years.

Some fungal spores can germinate immediately when they find favorable environmental and climatic conditions (e.g., temperature, humidity, rain, direct or indirect sun exposure, nutrient availability, etc.). Other species, in addition to favorable environmental conditions, require a period of "ageing" and/or of an activation signal such as thermal shock or the presence or application of a chemical substance(s) and so on (Carlile et al., 2001). In fact, it is worth stressing for restoration and conservative purposes, that spores of many fungal species involved in biodeterioration of cultural heritage germinate after chemical activation. These activators include detergents, organic acids, alcohols, etc., commonly used in conservative treatments of valuable objects. Hence, in many cases, a fast fungal colonization of surfaces was observed, during the restoration treatments (Maggi et al., 2009).

Fungi are especially adapted for growth on surfaces and thus they may colonize a given surface by production of a biofilm (Harding et al., 2009). Unfortunately, while there is enough literature on biofilm formation by bacteria, cyanobacteria, algae and even yeasts (Doyle, 1999, 2001), this phenomenon is poorly understood for fungi (Villena and Gutierrez Correa, 2003).

Following the models proposed by Sterflinger and Krumbein (1997), Urzì et al. (2000a) and Harding et al. (2009), it can be stated that the settlement of fungi leading to the formation of fungal biofilms on the valuable surface involves the following steps:

1. propagule contact; airborne hyphae or conidia casually contact the surface (fresco, rock surface or organic material);
2. active attachment to the surface; the propagules reaching the surface are stabilized by attachment structures;
3. filamentous fungi produce explorative hyphae (usually thinner than normal hyphae); their formation is faster and characterized by the production of Extracellular Polymeric Substances (EPS), that reinforce the attachment on the surface;
4. hyphal bundle/aggregate formation; this happens if and where a suitable source of nutrients is found;
5. maturation and development with the production of fruiting bodies, spores and conidia; this step occurs when there are enough nutrients to support the growth of fungi. This process leads to the spread of fungi with consequent aesthetic alteration (biofilm, patina) as well as their dispersal to re-initiate another cycle.
   As an alternative to step 5
6. formation of dormant/resistant structures on the surface; sometimes this occurs when the quest for nutrient sources is unsuccessful due to oligotrophic conditions, or due to rapid exhaustion of nutrient sources before the initiation of maturation stage. This aspect of alteration is characterized by spotted dark colonization.
   Hydrophobins may be involved in increasing the attachment of hyphae to hydrophobic surfaces (Wösten, 2001).

Major groups of fungi involved in the biodeterioration of rocks and stone monuments.

Hyphomycetes

Hyaline (that do not produce and deposit melanin in the hyphae) like *Penicillium, Fusarium* and *Aspergillus* or dematiaceous hyphomycetes like *Alternaria, Ulocladium* and *Cladosporium,* etc. are commonly isolated from cultural heritage artifacts all over the world as in South America (Gaylarde and Gaylarde, 2004), Asia (Sharma and Lanjewar, 2010), North Europe (Braams, 1992; Sterflinger and Prillinger, 2001; Suhiko et al., 2007; Grbic and Vukojevic, 2010) and in the Mediterranean Basin (Sterflinger and Krumbein, 1997; Urzì et al., 2000b; Mosyagin et al., 2009).

Hyphomycetes are characterized by their fast growth in culture and require relatively high water content, mainly eutrophic nutrient conditions and a moderate growth temperature (24–28°C). Their presence is well documented on various kinds of substrata like glass, paper, paintings, frescoes and stones. Their biodeteriorative pattern of alteration is due mainly to aesthetical damage by the production of visible spots (black) and patinas (white, brown, orange, etc.) due to pigment production like melanins and carotenoids (Urzì and Realini, 1998; Urzì et al., 2000a,b; Sterflinger, 2000). They are also able to penetrate into rock material by their hyphal growth and biocorrosive activity, due to excretion of organic acids, or by oxidation of mineral-forming cations, preferably iron and manganese (Urzì and Krumbein, 1994; Warscheid and Braams, 2000; Burford et al., 2003; Gadd, 2010).

Many of these species like *Aspergillus*, *Trichoderma* and *Penicillium* are proven also to cause biochemical deterioration by production of a variety of organic acids like oxalic, citric, succinic, etc. and exoenzymes (proteases, lipases, cellulases, etc.) that contribute to the damage of organic and inorganic substrates (Eckardt, 1985; Braams, 1992; Gomez-Alarcon et al., 1994; Gadd and Raven, 2010).

Hook (1665) was the first scientist who described a fungal attack on organic material (leather). He attributed the fungal attack to microclimatic conditions, susceptibility of the material and improper use of measures taken to destroy this unwanted growth.

Urzì and Krumbein (1994) have also opined similarly. Fortunately, in recent years the approach for the diagnosis and the prevention of fungal attack of organic objects of different ages and natures (Fig. 7.1) has involved

Figure 7.1. Photograph conserved in the Archives of Museo Diocesiano in Palermo with evident attack due to different kind of fungi causing discoloration of the paper by release of pigments.

different methodologies (a multiphasic approach) as discussed below. Besides that in case of libraries and museums, it has also become important to assess the potentials of fungi in causing health risks because it is well known that most of the ubiquitous fungal species such as *Aspergillus* spp., *Stachybotrys chartarum, Thricosporon* spp., *Hortaea werneckii,* etc. can cause systemic infections in human beings through the ingestion/inhalation of spores. Allergic reactions through the release of volatile compounds or by direct contact of contaminated materials are also known (Crook and Burton, 2010; Sterflinger and Pinzari, 2011).

Filamentous fungi are often found on heavily altered stone surfaces and they are very common on the outer rocks and monuments; thus, their amount is indicative of the status of conservation of the monument itself. However, little is still known on their natural role in the underground habitats such as caves and catacombs (Vaughan et al., 2011; Adetutu et al., 2011). In fact, whereas it is widely accepted that fungi play an essential role in the biological weathering of mineral substrates through mechanical separation of particles and excretion of secondary metabolites, the amount of culturable fungi recovered from those environments is quite small (Urzì et al., 2010). Recently (Vaughan et al., 2011), it has been reported that the diversity of fungal species may be greater than reported as shown in speleothem surfaces in Kartchner Caverns in Arizona, USA (Vaughan et al., 2011). However, in the case of heavily disturbed environments due to conservative treatments (application of biocides or consolidants), for example, in case of the well known Cave of Lascaux in France (Bastian and Alabouvette, 2009) or less known Milo's Catacombs in Greece (Pantazidou et al., 1997), an explosion of fungal growth may occur either due to the microclimatic changes and/or due to the disappearance of natural microflora.

Microcolonial Fungi (MCF)

The evidence accumulated during the past few years clearly indicate that the most common fungal colonizers of the rock material are the so-called microcolonial fungi or black meristematic fungi. These terms are used to define the typical pattern of growth of these fungi in natural environment, with colonies of small dimension (about 2–5 mm diameter) and the growth as chain of swollen melanized thick cells without formation of aerial mycelium. They have commonly been isolated from sun-exposed surfaces in the Mediterranean as well as from dry and cold climates (Staley et al., 1982; Anagnostitidis et al., 1992; Urzì and Realini, 1998; Selbmann et al., 2005; Sert et al., 2007a). Their occurrence on the stones is reported not only to cause aesthetical damage to the monuments (Fig. 7.2), as their color changes and black spots develop, but also because these organisms cause crater shaped lesions, chipping and exfoliation of the rock surfaces combined with the loss of materials (Urzì et al., 1995, 2000b; Sterflinger, 2010).

They represent a wide and heterogeneous group of black-pigmented fungi having in common the presence of melanin in the cells (swollen cells, hyphae and/or spores). This group of fungi is also termed "rock-inhabiting fungi" as many of them have exclusively been isolated from rock surfaces. Recently, many species like *Sarcinomyces petricola* (Wollenzien et al., 1997) (Fig. 7.3), *Coniosporium apollinis* and *Coniosporium perforans* (Sterflinger et al., 1997), *Coniosporium uncinatus* (De Leo et al., 1999) (Fig. 7.4), *Coniosporium sümbulii* (Sert and Sterflinger, 2010) *Pseudotaeniolina globosa (*De Leo et al., 2003), *Capnobotryella antalyesis, Capnobotryella erdogani* and *Capnobotryella kiziroglui* (Sert et al., 2007b, 2011) have been described.

Figure 7.2. Marble capital located in the Museum of Messina with alterations described as black patina.

Figure 7.3. Carrara marble statue in the Museum of Messina and the *Sarcinomyces petricola* strains isolated from it. The isolated *Sarcinomyces* strain seen under phase contrast (magnification 400x) is encircled.

Figure 7.4. A small stone fragment from the St. Genis Cathedral in France with evident black clusters from were *Coniosporium uncinatus* was isolated. In the circle, the *Coniosporium* strain seen under phase contrast (magnification 400x).

Color image of this figure appears in the color plate section at the end of the book.

Methodology

Currently, the approach to study the fungi associated with biodeterioration of materials of cultural and other significance have widened its scope. Now the study involves to find out the number and biodiversity of fungal species and their relationships with the material itself in order to:

- understand the step of colonization;
- identify the fungi;
- predict the eventual risk of biodeterioration of the substrate and to
- address the conservative intervention.

A most practical way to understand the ecology and deterioration role of fungi on materials of cultural heritage would be to adapt a multiphasic approach by using microscopic, cultural, biochemical as well as molecular methods. This approach can help to understand the complexity of interactions occurring between the fungi and the substrate, the pattern of colonization, enzymes and pigments released, penetration pattern, the numbers and variety of strains isolated (present colonizers) and species found through molecular methods (past and present colonizers).

However, the success of the approach would be dependent on the steps and care taken while sampling (e.g., using needle, adhesive tape, swabs, scalpel, tweezers). Further, after collection the right laboratory analyses should be undertaken (Urzì et al., 2003a).

Microscopic Analyses

Observations under microscope, aiming to demonstrate the direct presence of fungi in/on the object under study and the relationships with the substrata, should always be carried out. A simple light microscope can successfully be used for this purpose and often no specific preparation is required to observe the samples under light microscope (Fig. 7.5). Due to their dimensions (5–40 µm), the fungi are well evident as fungal units, hyphae or spores. Fungal structures observed under the microscope may not grow in culture perhaps due to their specific nutritional requirements. However, a preliminary identification can still be carried out through the direct observation of the presence of particular reproductive structures such as conidia (Pangallo et al., 2008). A better and magnified visualization of fungi and their relationships with substrata can be done under a scanning electron microscope. A combination of microscopy and molecular techniques (FISH, fluorescent *in situ* hybridization) technique could be useful in the diagnosis of microbial contamination of cultural heritage objects (Urzì et al., 2003b). Sterflinger and Hain (1999) presented a protocol for direct

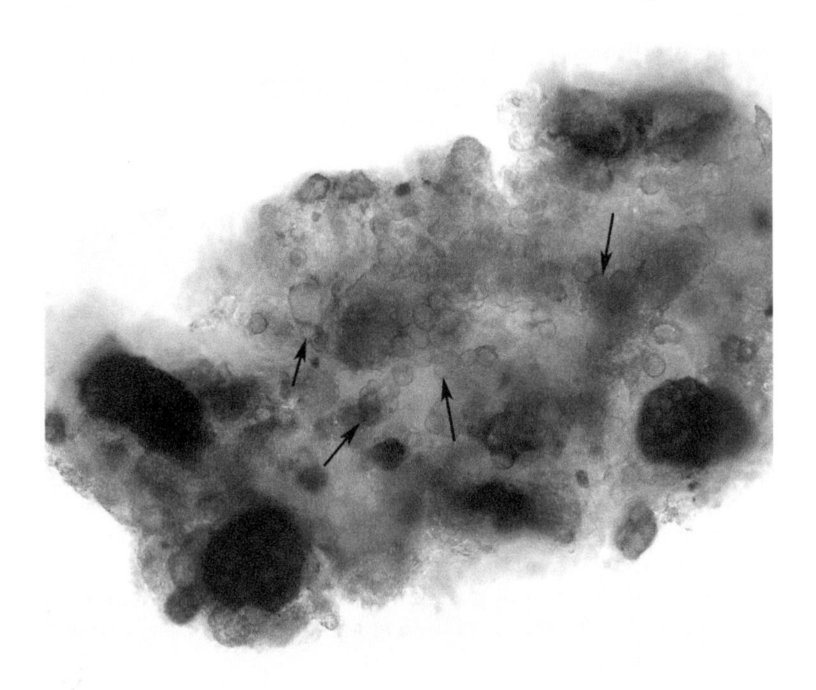

Figure 7.5. Sample taken with adhesive tape (as described by Urzì and De Leo, 2001) from the capital of Fig. 7.2 and observed under a light microscope (magnification 400X). Arrows show some black fungal conidia closely connected to marble flakes.

visualization of meristematic black fungi and black yeasts from a fungal community. This technique allows the identification of single cells of a microbial community by the use of molecular probes specially designed (species- or groups-specific) to be complementary to a region of rRNA genes, and marked with a fluorescent dye. The hybrids can be observed under an epifluorescent microscope or better under a CLSM (Confocal Laser Scanning Microscope).

Cultural Methods

For cultural purposes, it is important to make the right choice of media to reduce the growth of fast growing and spreading fungi, and to prolong the incubation time for slow growing fungi (Urzì et al., 1992) (Fig. 7.6). The identification of fungi based upon the morphology of colony grown on different cultural media and the morphology of reproductive structures is

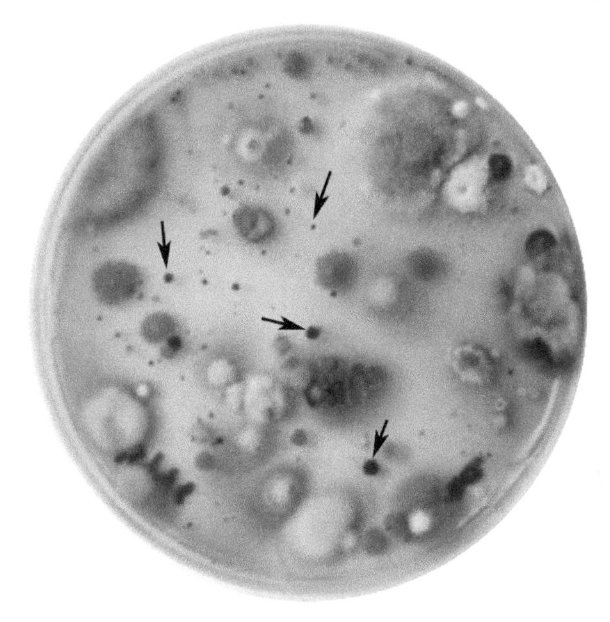

Figure 7.6. Petri dish containing DRBC (Dichlorane Rose Bengal, Chloramphenicol) agar medium with fungal colonies of different kinds and sizes. The medium allows the contemporary growth and thus isolation of fast and slow (arrows) growing fungi.

a time-consuming procedure; but often it is the chosen procedure because the isolated strains can further be tested for their ability to hydrolize specific macromolecules or release extra-cellular pigments in order to demonstrate their active role in biodeterioration.

In some cases, additional biochemical and physiological tests such as assimilation of carbon and nitrogen compounds, production of enzymes, growth at different temperatures and so on, are required to be undertaken in order to characterize eco-physiologically the species.

Molecular Methods

Molecular methods such as DGGE (Denaturing Gradient Gel Electrophoresis) fingerprinting of 18S rDNA genes or better of the ITS region offer the possibility of studying the fungal community from different cultural heritage objects such as paper (Michaelsen et al., 2010; Pangallo et al., 2009), oil paintings (Lopez-Miras et al., 2013) and stone monuments (Piñar and Sterflinger, 2009). If culture independent methods are carried out, the choice of a set of genes to be amplified can be critical. In fact, using DGGE fingerprinting amplification of 18S rRNA genes could lead to a poor identification of fungal communities, while better results can be obtained by amplifying genes of the ITS region (Piñar et al., 2012). In addition, molecular tools more often used in the study of biodeteriogenic fungal communities in/on organic and inorganic substrates help to show a greater biodiversity of fungal species than observed by previous researchers where only cultural methods were used (Piñar et al., 2009, 2012).

Conclusions

The common occurrence of hyphomycetes as well as meristematic fungi on highly biodeteriorated inorganic and organic cultural heritage materials reveals that fungi play a significant role in the decay of both organic and inorganic substrates. Their growth on the material is due to their great adaptation to different conditions, e.g., from oligotrophic to poikilotrophic or rich nutrient conditions, under high Relative Humidity (RH), or under very low humidity conditions (xerophilic). Their colonization on the substrates involves mechanical deterioration through hyphae production as well as the release of molecules of different natures such as pigments, enzymes or/and organic acids and consequently making irreversible changes of the material (biodeterioration "*sensu*" Urzì and Krumbein, 1994). Molecular tools enhance our knowledge concerning the biodiversity of fungi involved in biodeterioration. However, much work still needs to be done for refining the methodology used for the prevention and control of fungi and to conserve the cultural heritage especially those in which the microclimatic parameters are not easy to control (caves, archives, large collections of artifacts stored or displayed in buildings not suitable for conservation).

Dedication

This chapter is dedicated to Wolfgang E. Krumbein, the founder of the first geomicrobiology laboratory in Germany. In his laboratory, several doctorate and postdoctorate students, and guest researchers had the chance to work and the opportunity to discuss and to be enthralled by his enthusiasm and geniality. Clara Urzì was one of several researchers to whom W.E.K. inspired for fruitful mycological investigations.

Acknowledgements

The financial support of the University of Messina through PRA projects of yrs. 2008–2011 is acknowledged. The authors would like to address a special thanks to Mrs. Sherron Collins for her valuable and careful revision of the English text.

References

Adetutu, E.M., Thorpe, K., Bourne, S., Cao, X., Shahsavani, E., Kirby, G. and Ball, A.S. 2011. Phylogenetic diversity of fungal communities in areas accessible and not accessible to tourists in Naracoorte Caves. Mycologia, 103: 959–968.

Anagnostidis, K., Gehrmann, C.K., Gross, M., Krumbein, W.E., Lisi, S., Pantazidou, A., Urzì, C. and Zagari, M. 1992. Biodeterioration of marbles of the Parthenon and Propylaea, Acropolis, Athens—associated organisms, decay and treatment suggestions. pp. 305–325. *In*: Decrouez, D., Chamay, J. and Zezza, F. (eds.). Proceedings of the 2nd International Symposium, Musee d'art et d'histoire, Geneve, CH.

Bastian, F. and Alabouvette, C. 2009. Lights and shadows on the conservation of rock art cave: the case of Lascaux Cave. International Journal of Speleology, 38: 55–60.

Braams, J. 1992. Ecological studies on the fungal microflora inhabiting historical sandstone monuments. PhD thesis, Oldenburg, University, D.

Burford, E.P., Kierans, M. and Gadd, G.M. 2003. Geomycology: fungi in mineral substrata. Mycologist, 17: 98–107.

Carlile, M. J., Watkinson, S.C. and Gooday, W. 2001. The fungi. Academic Press, London, UK.

Comtois, P. 1990. Indoor and mold aerosol. Aerobiologia, 6: 165–176.

Crook, B. and Burton, N.C. 2010. Indoor moulds, sick building syndrome and building related illness. Fungal Biology Reviews, 24: 1–8.

De Leo, F., Urzì, C. and de Hoog, G.S. 1999. Two *Coniosporium* species from rock surfaces. Studies in Mycology, 43: 77–85.

De Leo, F., Urzì, C. and de Hoog, G.S. 2003. A new meristematic fungus, *Pseudotaeniolina globosa*. Antonie van Leeuwenhoek, 83: 351–360.

Doyle, R.J. 1999. Methods in enzymology—Biofilms. Vol. 310. Academic Press, San Diego USA.

Doyle, R.J. 2001. Methods in enzymology. - Microbial growth in Biofilms. Part A. Development and Molecular Biological aspects. Vol. 336. Academic Press, San Diego, USA.

Eckhardt, F.E.W. 1985. Mechanisms of the microbial degradation of minerals in sandstone monuments, medieval frescoes, and plaster. pp. 643–652. *In:* The 5th International Congress on Deterioration and Conservation of Stone, Lausanne, CH.

Gadd, G.M. 2010. Metals, minerals and microbes: geomicrobiology and bioremediation. Microbiology, 156: 609–643.

Gadd, G.M. and Raven, J.A. 2010. Geomicrobiology of eukatyotic microorganisms. Geomicrobiology Journal, 27: 491–519.

Gaylarde, P. and Gaylarde, C. 2004. Deterioration of siliceous stone monuments in Latin America: microorganisms and mechanisms. Corrosion Reviews, 22: 95–415.

Gomez-Alarcon, G., Munoz, M.L. and Flores, M. 1994. Excretion of organic acids by fungal strains isolated from decayed limestone. International Biodegradation & Biodeterioration, 34: 169–180.

Grbic, M.L., Vukojevic J., Simic, G.S., Krizmani, J. and Stupar, M. 2010. Biofilm forming cyanobacteria, algae and fungi on two historic monuments in Belgrade, Serbia. Archives of Biological Sciences, Belgrade, 62: 625–631.

Harding, M.W., Marques, L.L.R., Howard, R.J. and. Olson, M.E. 2009. Can filamentous fungi form biofilms? Trends in Microbiology, 17: 475–479.

Hooke, R. 1665. Micrographia, or some physiological description of minute bodies made by magnifying glasses, with observations and inquires thereupon. Martyn and Allestry, London, UK.

Lopez-Miras, MdM., Martın-Sanchez, I., Yebra-Rodrıguez, A., Romero-Noguera, J., Bolivar-Galiano, F., Ettenauer, J. , Sterflinger, K. and Piñar, G. 2013. Contribution of the Microbial Communities Detected on an Oil Painting on Canvas to Its Biodeterioration. PLoS ONE 8(11): e80198. doi:10.1371/journal.pone.0080198.

Maggi, O., Persiani, A.M., De Leo, F. and Urzì, C. 2009. Structural, functional, and ecological characteristics of the main biodeteriogens. Fungi. pp. 65–70. *In*: Caneva, G., Nugari, M.P. and Salvadori, O. (eds.). Plant Biology for Cultural Heritage. Getty Conservation Institute, Los Angeles CA, USA.

Michaelsen, A., Piñar, G. and Pinzari, F. 2010. Molecular and microscopical investigation of the microflora inhabiting a deteriorated italian manuscript dated from the thirteenth century. Microbial Ecology, 60: 69–80.

Mosyagin, A.V., Knauf, I.V. and Zelenskaya, M.S. 2009. Deterioration of carbonate rocks used for archeological monuments in Tauric Chersonesos (Crimea). Studia Universitatis Babes Bolyai, Geologia, 54: 13–16.

Pangallo, D., Chovanova, K., Šimonovičová, A., De Leo, F. and Urzì, C. 2008. Assessment of the biodeterioration risk of the Ladislav legend fresco in Velka Lomnica (SK) through non-invasive methods. pp. 457–464. *In*: Lukaszewicz, J.W. and Niemcewicz, P. (eds.). Proceedings of the 11th International Congress on Deterioration and Conservation of Stone, Nicolaus Copernicus University Press, Torun, PL. Vol I.

Pangallo, D., Chovanova, K. Simonovicova, A. and Ferianc, P. 2009. Investigation of microbic community isolated from indoor artworks and their environment: identification, biodegradative abilities, and DNA typing. Canadian Journal of Microbiology, 55: 277–287.

Pantazidou, A., Roussomoustakaki, M. and Urzì, C. 1997. The microflora of Milos Catacombs. pp. 321–325. *In*: Sinclair, A. (ed.). Archeological Science. Owbow Books, Park End Place, Oxford, UK.

Piñar, G. and Sterflinger, K. 2009. Microbes and building materials. pp. 463–188. *In*: Comejo, D.N. and Haro, J.L. (eds.). Building Materials: Properties, Performance and Applications. Nova Science Publishers, New York, USA.

Piñar, G., Garcia-Valles, M., Gimeno-Torrente, D., Fernandez-Turiel, G.L., Ettenauer, J. and Sterflinger, K. 2012. Microscopic, chemical and molecular-biological investigation of the decayed medieval stained window glasses of two Catalonian churches. International Biodeterioration & Biodegradation. http://dx.doi.org/10.1016/j.ibiod.2012.02.008.

Selbmann, L., de Hoog, G.S., Mazzaglia, A., Friedmann, E.I. and Onofri, S. 2005. Fungi at the edge of life: cryptoendolithic black fungi from Antarctic deserts. Studies in Mycology, 51: 1–32.

Sert, H.B., Sümbül, H. and Sterflinger, K. 2007a. Microcolonial fungi from antique marbles in Perge/side/Termessos (Antalya/Turkey). Antonie van Leewenhoek, 91: 217–227.

Sert, H.B., Sümbül, H. and Sterflinger, K. 2007b. A new species of *Capnobotryella* from monument surfaces. Mycological Research, 1235–1241.

Sert, H.B. and Sterflinger, K. 2010. A new *Coniosporium* species from historical marble monuments. Mycological Progress, 9: 353–359.

Sert, H.B., Sümbül, H. and Sterflinger, K. 2011. Two new species of *Capnobotryella* from historical monuments. Mycological Progress, 10: 333–339.

Sharma, K. and Lanjewar, S. 2010. Biodeterioration of ancient monument (Devarbija) of chhattisgarh by fungi. Journal of Phytology, 21: 47–49.

Sterflinger, K. and Krumbein, W.E. 1997. Dematiaceous fungi as a major agent for biopitting on Mediterranean marbles and limestones. Geomicrobiology Journal, 14: 219–230.

Sterflinger, K., de Baere, R., de Hoog, G.S., de Wachter, R., Krumbein, W.E. and Haase, G. 1997. *Coniosporium perforans* and *C. apollinis*, two new rock-inhabiting fungi isolated from marble in the Sanctuary of Delos (Cyclades, Greece). Antonie van Leeuwenhoek, 72: 349–363.

Sterflinger, K. and Hain, M. 1999. *In situ* hybrydization with rRNA targeted probes as new tool for the detection of black yeasts and meristematic fungi. Studies in Mycology, 4: 23–30.

Sterflinger, K. 2000. Fungi as geological agents. Geomicrobiological Journal, 17: 97–124.

Sterflinger, K. and Prillinger, H. 2001. Molecular taxonomy and biodiversity of rock fungal communities in an urban environment (Vienna, Austria). Antonie van Leeuwenhoek, 80: 275–286.

Sterflinger, K. 2010. Fungi: their role in deterioration of cultural heritage. Fungal Biology Reviews, 24: 47–55.

Sterflinger, K. and Pinzari, F. 2011. The revenge of time: fungal deterioration of cultural heritage with particular reference to books, paper and parchment. Environmental Microbiology, 1–8.

Suihko, M.L., Alakomi, H.L., Gorbushina, A., Fortune, I. and Saarela, M.M. 2007. Characterization of aerobic bacterial and fungal microbiota on surfaces of historic Scottish monuments. Systematic and Applied Microbiology, 30: 494–508.

Urzì, C., Lisi, S., Criseo, G. and Zagari, M. 1992. Comparazione di terreni per l'enumerazione e l'isolamento di funghi deteriogeni isolati da materiali naturali. Annali di Microbiologia e Enzimologia, 42: 185–193.

Urzì, C. and Krumbein, W.E. 1994. Microbiological impacts on the Cultural Heritage. pp. 107–135. *In*: Krumbein, W.E., Brimblecombe, P., Cosgrove, D.E. and Staniforth, S. (eds.). Durability and Change: The Science, Responsibility, and Cost of Sustaining Cultural Heritage. John Wiley & Sons Ltd., Chichester, UK.

Urzì, C., Wollenzien, U., Criseo, G. and Krumbein, W.E. 1995. Biodiversity of the rock inhabiting microflora with special reference to black fungi and black yeasts. pp. 289–302. *In*: Allsopp, D., Hawksworth, D.L. and Colwell, R.R. (eds.). Microbial Diversity and Ecosystem Function, CAB International, Wallington, USA.

Urzì, C. and Realini, M. 1998. Colour changes of Noto's calcareous sandstone as related to its colonisation by microorganisms. International Biodeterioration & Biodegradation, 42: 45–54.

Urzì, C., De Leo, F., de Hoog, G.S. and Sterflinger, K. 2000a. Recent advances in the molecular biology and ecophisiology of meristematic stone-inhabiting fungi. pp. 3–19. *In*: Ciferri, O., Tiano, P. and Mastromei, G. (eds.). Of Microbes and Art. The Role of Microbial Communities in the Degradation and Protection of Cultural Heritage. Kluwer Academic, New York, USA.

Urzì, C., Salamone, P., De Leo, F. and Vendrell, M. 2000b. Microbial diversity of Greek quarried marbles associated to specific alteration. pp. 35–42. *In*: Monte, M. (ed.). Proceedings of the 8th Workshop Eurocare Euromarble EU496, CNR Editions, Rome, I.

Urzì, C. and De Leo, F. 2001. Sampling with adhesive tape strips: an easy and rapid method to monitor microbial colonization on monument surfaces. Journal of Microbiological Methods, 44: 1–11.

Urzì, C., De Leo, F., Salamone, P. and Criseo, G. 2001. Airborne fungal spores connected with marble colonisation monitored in the terrace of Messina Museum. Aerobiologia, 17: 11–17.

Urzì, C., De Leo, F., Donato, P. and La Cono, V. 2003b. Multiple approaches to study the structure and diversity of microbial communities colonizing artistic surfaces. pp. 187–194. *In*: Saiz-Jimenez, C. (ed.). Molecular Biology and Cultural Heritage. Swets & Zeitlinger, Lisse NL.

Urzì, C., La Cono, V., De Leo, F. and Donato, P. 2003a. Fluorescent *In Situ* Hybridization (FISH) to study Biodeterioration. pp. 55–60. *In*: Saiz-Jimenez, C. (ed.). Molecular Biology and Cultural Heritage, Balkema Publ. Lisse, Amsterdam, N.

Urzì, C., De Leo, F., Bruno, L. and Albertano, P. 2010. Microbial diversity in paleolithic caves: a study case on the phototrophic biofilms of the Cave of Bats (Zuheros, Spain). Microbial Ecology, 60: 116–129.

Vaughan, M.J., Maier, M.R. and Pryor, B.M. 2011. Fungal communities on speleothem surfaces in Kartchner caverns, Arizona, USA. International Journal of Speleology, 40: 65–77.

Villena, G.K. and Gutierrez-Correa, M. 2003. Production of cellulase by *Aspergillus niger* biofilms developed on polyester cloth. Letters in Applied Microbiology, 43: 262–268.

Warscheid, Th. and Braams, J. 2000. Biodeterioration of stone: a review. International Biodeterioration & Biodegradation, 46: 343–368.
Wollenzien, U., de Hoog, G.S., Krumbein, W.E. and Uijthof, J.M.J. 1997. *Sarcinomyces petricola*, a new microcolonial fungus from marble in the Mediterranean basin. Antonie van Leeuwenhoek, 71: 281–288.
Wösten, H.A.B. 2001. Hydrophobins: multipurpose proteins. Annual Review of Microbiology, 55: 625–646.

CHAPTER 8

Fungi in and on Dairy Products

Judit Krisch,[1,*] *József Csanádi*[1] and *Csaba Vágvölgyi*[2]

ABSTRACT

Yeasts and molds in dairy products, having lipolytic and proteolytic enzyme activities can contribute to the development of a desired taste and aroma as secondary starters or, by the same enzyme activity, are responsible for off flavors and off odors as spoilage agents. The most commonly found spoilage yeasts are *Candida* and *Rhodotorula* species, *Yarrowia lipolytica*, and *Geotrichum candidum*. *Candida* species in dairy products can represent opportunistic pathogens causing infections in immune-compromised patients. The main concern about molds in dairy products is their mycotoxin production. Several *Penicillium* and *Aspergillus* species are known mycotoxin producers, even species used as secondary starters. Yeasts and molds in dairy products should undergo risk assessment. Strict hygienic practice in the manufacture process is required to avoid contamination with health-threatened fungi. This chapter embodies these aspects.

Introduction

Milk and dairy products provide a unique environment for microbial growth. Cow's milk contains approximately 3.2% protein, 3.9% lipids and 4.8% lactose as the main carbohydrate (Ray, 2004). In dairy products

[1] Institute of Food Engineering, Faculty of Engineering, University of Szeged, Szeged, Hungary.
[2] Department of Microbiology, Faculty of Informatics and Science, University of Szeged, Szeged, Hungary.
* Corresponding author: krisch@mk.u-szeged.hu

these ingredients are in concentrated form and undergo some degradation processes due to the metabolic activity of primary and secondary starter cultures. Dairy products can be made also from the milk of buffalo, goat and ewe; the main components are the same, although, in different proportions compared to cow's milk. The non-fermented milk products, like cream and butter, and also cheeses have high milk fat content from 20 to 80%. Cheeses are also rich in protein (coagulated casein) and sometimes have a high salt concentration up to 7% (Scott, 1981; Walstra et al., 2006). Fermented dairy products like yogurt, kefir, and koumiss and other probiotic products have a low pH due to the production of organic acids, mainly lactic acid from lactose by lactobacilli in the starter culture. In these special environments, the following properties are required for microbial growth: ability to ferment lactose or to assimilate lactate or citrate, lipolytic and proteolytic enzyme activities, salt and/or acid tolerance, and the ability to grow at low temperatures (Deák, 2007; Flee, 1990; Kure et al., 2004; Pitt and Hocking, 2009). Yeasts and molds in dairy products, having the required properties, have a two-faced behavior: they can contribute to the development of the desired taste and aroma as secondary starters or, by the same enzyme activity, are responsible for off flavors and off odors as spoilage agents. In fermented products and also in cheeses, lactobacilli are used as primary starters for the production of lactic acid, which cause the formation of casein micelle net (clotting, gelation) due to the loss of the negative charge of micelles. Yeasts are used along with lactobacilli for kefir and koumiss production, and can contribute to the taste and texture formation of smear-ripened cheeses (Flee, 1990; Tamine et al., 1999). Strains of the filamentous yeast *Geotrichum candidum* and of the molds of *Penicillium roqueforti* and *P. camemberti* are used in cheese production—but they are also considered as spoilage microorganisms on non mold-ripened cheeses. To make the situation even more complex, non-starter fungi identified on cheeses as contaminants can also contribute to the curing process by their proteolytic and lipolytic activities (Corbo et al., 2001; Minervini et al., 2001; Ropars et al., 2012). First of all in traditionally made cheeses, it may be difficult to distinguish between fungi with ripening and spoilage effect. In this chapter, we will discuss some yeast and filamentous fungi often associated with dairy products and will discuss their effect on food safety and human health.

Dairy Yeasts

Yeasts in Fermented Products, Cream and Butter

The natural yeast flora in raw and pasteurized milk seldom surpasses 10^3 cfu/ml, and will be overgrown by psychrotrophic bacteria when refrigerated. However, when an antibiotic is present in the milk, the yeasts can overgrow (Flee, 1990). In cow, ewe, goat and buffalo milk the most frequent yeast species are *Candida catenulata, C. zeylanoides, Trichosporon cutaneum* and *Yarrowia lipolytica* (Corbo et al., 2001). Their spoilage activity is more pronounced in dairy products than in milk. The main signs of yeast spoilage are gas formation (bulging of the package), off flavors (fruity, bitter or rancid), texture softening, and the formation of white or colored, smooth, creamy colonies on the product surface.

In yogurt, yeasts are the main spoilage microorganisms because of the low pH caused by lactic acid bacteria. To produce visible signs of spoilage (swelling of the package and yeast colonies on the package lid and on the surface of yogurt, yeasty odor and flavor) the number of yeasts has to reach 10^5–10^6 cfu/g. Sources of contamination could be the added ingredients (e.g., fruits, honey, nuts), the production equipment (in case of a non-effective sanitation) or a contaminated starter culture (Fleet, 1990; Mataragas et al., 2011). The most prevalent yeasts in yogurts are *Candida famata* (the anamorph of *Debaryomyces hansenii*) and *Kluyveromyces marxianus* (Fleet, 1990), and *Zygosaccharomyces microellipsoides* (Cappa and Cocconelli, 2001). Other yeasts involved in yogurt contamination are *Torulopsis candida, T. versatilis* (now *Candida versatilis*), *C. intermedia, C. krusei* (Fleet, 1990). From these yeasts only *K. marxianus* is able to ferment lactose.

Cream and butter have high milk fat content so only yeasts with good lipolytic activity are able to grow on these products. The main spoilage agents are *Candida* and *Rhodotorula* species (*C. famata, C. diffluens, R. rubra, R. glutinis*) causing surface discoloration on butter and a foamy appearance and yeasty odor in cream (Fleet, 1990; El Diasty et al., 2007).

Yeasts in Cheeses

The filamentous yeast *Geotrichum candidum,* together with the white mold of *Penicillium camemberti,* is used in the ripening process of soft cheeses like camembert and brie. *G. candidum* grows fast in the first week and builds up a white velvety coat on the surface rind of the cheeses (Fig. 8.1). Later it can also grow inside the cheese. Products of its proteolytic and lipolytic activity are free amino acids and aroma compounds, arising from the degradation of peptides and free fatty acids (Sacristan et al., 2012). Ammonia and volatile sulfides are formed from glutamic and aspartic acid, and from methionine, respectively (Boutrou and Gueguen, 2005). Over ripening of camembert cheeses is considered as spoilage caused by the enzyme activity of *G. candidum* and other yeasts (mainly *D. hansenii*), leading to liquefaction of the cheese interior and pungent smell caused by ammonia.

The most commonly found yeasts in cheeses are *Candida* species (e.g., *C. catenulata, C. zeylanoides, C. famata*), *K. lactis, K. marxianus, D. hansenii* and

Y. lipolytica (Corbo et al., 2001; Fadda et al., 2004; Flee, 1990; Minervini et al., 2001; Vasdinyei et al., 2003; Welthagen et al., 1998). In different types of cheeses the yeast flora is also different. On the surface of unripened cheeses and cottage cheeses, *Y. lipolytica* and *Kluyveromyces* species causes discoloration and rancid flavor. In the interior of Gouda, blue-veined and camembert cheeses, *D. hansenii* is the main species. In goats and other cheeses, *Y. lipolytica* is responsible for production of a brown pigment (Fig. 8.2) by metabolizing the amino acid tyrosine (Carreira et al., 1998, 2001).

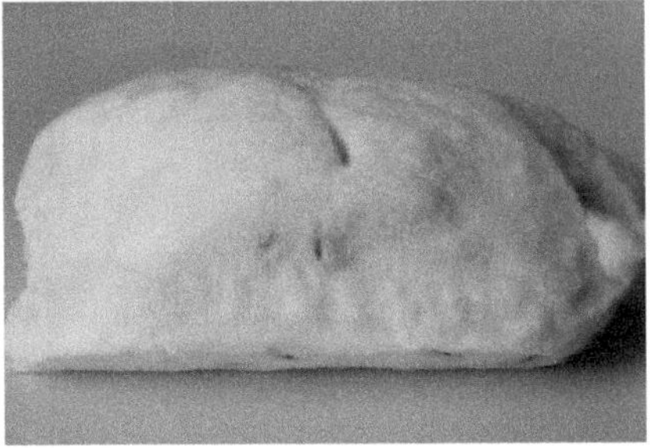

Figure 8.1. Cheeses ripened or spoiled by *Geotrichum candidum*. Left, a ripened camembert cheese; right, an artisanal herb-flavored cheese with undesired white moldy rind.

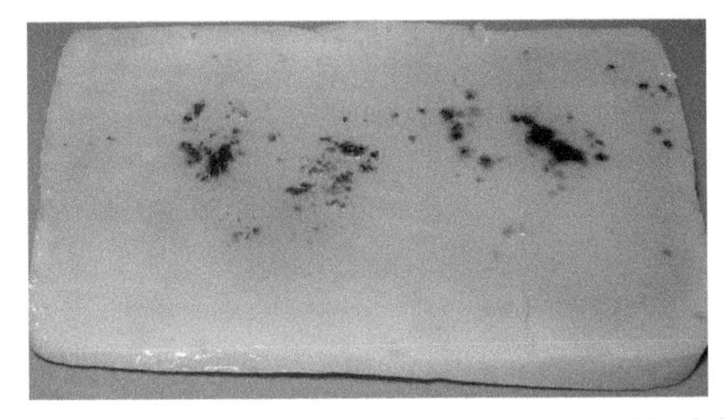

Figure 8.2. Dark pigment producing yeasts on the surface of hard cheese.

Dairy Molds

Dairy molds grow usually on the surface of the products, with the exception of *Penicillium roqueforti*, which can be used as a starter culture in blue-veined cheeses. In this case, the cheese is spiked to let air in the interior and to inoculate it with the mold spores, which in the humid environment will germinate quickly and spread inside to give the characteristic appearance (Ray, 2004; Walstra et al., 2006). In artisanal cheeses the flavoring material (e.g., pepper, herbs and nuts) can cause undesired mold growth inside the cheese, in the case when sufficient air is present (Fig. 8.3).

It seems that in temperate climates the main spoilage molds belong to the *Penicillium* genus, with a frequency of 11–42.6%, while *Aspergillus* and *Mucor* species have a limited occurrence with 0.8–3.5 and 0.4–2.6% frequency, respectively (Kure et al., 2001, 2004; Ropars et al., 2012). In countries with higher average temperature, aspergilli play an important role in the contamination of dairy products with a frequency up to 43.4% (El Diasty, 2007; Gandomi et al., 2009; Saravanakumar et al., 2007). The most frequently isolated spoilage *Penicillium* species are *P. roqueforti*, *P. commune*, *P. palitans*, *P. solitum*, *P. crustosum*, *P. paneum*, *P. brevicompactum*, *P. caseifulvum*, *P. citrinum*, *P. camamberti*, *P. chrysogenum*, *P. nalgiovense* (Bourdichon et al., 2012; Florez et al., 2007; Kure et al., 2001, 2004; Ropers et al., 2012). The source of contamination in cheese factories are the air, floor, walls, wooden shelves in the ageing rooms and the equipment. Therefore, good hygiene is very important in cheese manufacturing. The main *Aspergillus* species

Figure 8.3. *Penicillium roqueforti* in blue-veined cheese and in an artisanal cheese as contaminant from the flavoring pepper fruit.

Figure 8.4. Artisanal cheese with *Penicillium* species on the surface rind.

isolated from dairy products are: *A. niger, A. flavus, A. ochraceus* (El Diasty et al., 2007; Gandomi et al., 2009; Sarvanakumar, 2007). In smear-ripened cheeses, *Fusarium domesticum* and *Fusarium dimerum* can be found on the rind of cheeses in association with yeasts and bacteria (Bachmann et al., 2005; Gandomi, 2009).

Table 8.1. Main characteristics of *Penicillium* species frequently isolated from dairy products.

	P. roqueforti	*P. commune*	*P. camemberti*	*P. paneum*
Colony color on YES CYA	Green Green with brownish center	Green Dark green	White White	Dark green Dark green with brown center
Proteolytic activity	+	+	+	+
Lipolytic activity	+	+	+	n.r.
Starter culture	Yes	No	Yes	No
Flavor production	Yes	Yes	Yes	Yes
Biogenic amine production (bitterness)	n.r.	n.r.	Yes	n.r.
Mycotoxin production	Yes	Yes	Yes	n.r.
References	O'Brien et al., 2008; Florez et al., 2003	Lund et al., 2003	Boutrou et al., 2006	O'Brien et al., 2008

n.r., not reported; YES, yeast extract medium; CYA, Czapek yeast agar.

Interactions Among Dairy Microorganisms

Smear Ripened Cheeses

Smear ripened cheeses are surface ripened cheeses the rind of which is regularly washed with brine containing the mixed microflora necessary for aroma development. The surface microflora of Rebochlon, Tilsit and Limburger type cheeses contains coryneform bacteria and staphylococci together with *D. hansenii* and *G. candidum* as the main yeasts (Mounier et al., 2005, 2008). The interactions are very complex and not really understood. In the study of Mounier et al. (2008) *Y. lipolytica* inhibited the mycelial growth of *G. candidum*, whereas the former yeasts inhibited *D. hansenii* cell viability during the stationary phase. Molds are considered as contaminants on the surface of smear-ripened cheeses with the exception of "Anticollanti", a white mold that prevent the development of stickiness of the smear. This mold was identified as *Fusarium domesticum* and now it is used to prevent the defect of sticky smear (Bachmann et al., 2005).

Mold-ripened Cheeses

During the ripening of white mold cheeses *P. camemberti* and *G. candidum* have a synergistic proteolytic activity. *P. camemberti* tends to produce hydrophobic peptides with a bitter taste but this bitterness is reduced by the aminopeptidases of *G. candidum* by the hydrolysis of these peptides (Boutrou et al., 2006). Although, these fungi can be found in a mixed culture on the surface of the cheeses, they have a slight inhibitory effect of each other's growth (Boutrou and Gueguen, 2005). *G. candidum* caused the total growth inhibition of *P. commune, P. caseifulvum, P. verrucosum, P. discolor, P. solitum, P. coprophilum* and *Aspergillus versicolor* on Camembert cheese at 25°C (Nielsen et al., 1998). *G. candidum* grows faster, competes for nutrients and space, and competitively excludes other molds. In pour-plated experiments, *P. camemberti* had the largest inhibitory effect on *Cladosporium herbarum*, and a high antagonistic effect on *P. roqueforti*. Mixed cultures of *P. camemberti* and *G. candidum*, however, showed a weakened inhibition because of the competition between the starter fungi (Decker and Nielsen, 2005). *G. candidum* also inhibited the growth of the pathogenic bacterium *Listeria monocytogenes* due to organic acid production (Dieuleveux et al., 1998). It seems that dairy molds have competitive and antagonistic interactions with each other and also with other microorganisms, and these interactions can help keep pathogenic or spoilage agents away.

Other Dairy Products

D. hansenii was effective against the dairy molds *Aspergillus* sp., *Byssochlamys fulva, B. nivea, Cladosporium* sp., *Eurotium chevalieri, P. candidum* and *P. roqueforti* in yogurt and on cheese surface at 30 and 20°C. The growth inhibition effect was associated with the competition for nutrients and also with the production of mycocin, a secondary metabolite which inhibits cell wall synthesis in molds. However, the initial mold count was critical to the effectiveness of *D. hansenii* as a biocontrol agent (Liu and Tsao, 2009).

Food Safety Assessment of Dairy Fungi

Dairy products have been manufactured from ancient times and starter cultures have a long "history of use". The consumption of these products over several generations provides reasonable certainty that there are no harmful effects on humans. In the United States the status Generally Recognized As Safe (GRAS) was introduced for the use of substances which are shown to be safe under the conditions of its intended use. In the European Union in

2007, the European Food Safety Authority (EFSA) introduced "Qualified Presumption of Safety" (QPS) for the safety assessment of microorganisms used in food and feed (Bourdichon et al., 2012).

Risk Assessment of Yeasts

Yeasts (especially *Candida* species) in dairy products can represent opportunistic pathogens causing infections in immunocompromised and hospitalized patients. However, the most common pathogenic *Candida* species were not found in cheese. New emerging pathogens like *D. hansenii* and *Y. lipolytica* can cause rare infections in the above mentioned patients which are generally mild and easily treated. "It seems that it is more the exposure to high doses of yeast than the identity of the species or strain that is associated with infection. As such yeasts in cheese cannot be considered to constitute a risk for healthy individuals" (Jacques and Casagerola, 2008). The filamentous yeast *G. candidum* was proposed for QPS status because "the risk of developing an infection due to *G. candidum* in connection with its technological use and consumption of dairy products is virtually nil" (Pottier et al., 2008). Molecular methods in the identification of yeasts would help better distinguish between pathogenic and non-pathogenic yeast strains.

Risk Assessment of Molds

The main concern about molds in dairy products is their mycotoxin production. Several *Penicillium* and *Aspergillus* species are known mycotoxin producers, even species used as secondary starters (Table 8.2).

The most important mycotoxins produced by *P. roqueforti* are roquefortine and PR toxin. Small amounts of both toxins were isolated from blue-veined cheeses but these amounts were too low to cause any adverse effects in animal experiments. PR toxin is not stable in cheese and *P. roqueforti* has a history of safe use without noted reports of adverse human health

Table 8.2. Mycotoxin producing molds found in dairy products.

Species	Mycotoxin	References
P. roqueforti	roquefortine, PR toxin, isofumigaclavine, myco-phenolic acid, penicillic acid	Sengun et al., 2008
P. commune	cyclopiazinic acid	Lund et al., 2003
A. versicolor	sterigmatocystin	Sengun et al., 2008
Aspergillus spp.	aflatoxin M1	Sengun et al., 2008
Penicillium spp.	citrinin, patulin	Sengun et al., 2008
A. ochraceus	ochratoxin	Saravanakumar et al., 2007

effects (US Environmenal Protection Agency, 1997). The best way to prevent mycotoxins in dairy products is the prevention of mold contamination. There are several methods from cold storage to Modified Atmosphere Packaging (MAP), or the use of artificial or natural preservatives to limit the growth of molds in dairy products (Singh et al., 2011; Smith-Palmer et al., 2001; Taniwaki et al., 2001). Good sanitation of the production equipment and strict hygienic practice in the whole manufacture process is required to avoid mycotoxin producing mold contamination. Artisanal dairy products with spontaneous mold growth also should undergo risk assessment, and mold contaminated products in households should not be consumed (Sengun et al., 2008).

Conclusion

Fungi, yeasts and molds are often found in dairy products either as secondary starters or as natural microflora. Their spoilage activity, especially in case of yeasts, can cause visible effects only when the number of cells reached a certain level. Non-starter fungi can also contribute to the development of the desired taste and aroma in dairy products, but undesired growth of molds should be handled as contamination because of the possibility of mycotoxin production.

References

Bachmann, H.P., Bobst, C., Butikofer, U., Casey, M.G., Torre, M.D., Frohlich-Wyder, M. and Furst, M. 2005. Occurrence and significance of *Fusarium domesticum* alias Anticollanti on smear-ripened cheeses. LWT, 38: 399–407.

Bourdichon, F., Casaregola, S., Farrokh, C., Frisvad, J.C., Gerds, M.L., Hammes, W.P., Harnett, J., Huys, G., Laulund, S., Ouwehand, A., Powell, I.B., Prajapati, J.B., Seto, Y., Schure, E.T., Van Boven, A., Vankerckhoven, V., Zgoda, A., Tuijtelaars, S. and Hansen, E.B. 2012. Food fermentations: Microorganisms with technological beneficial use. Int. J. Food Microbiol., 154: 87–97.

Boutrou, R., Aziza, M. and Amrane, A. 2006. Enhanced proteolytic activities of *Geotrichum candidum* and *Penicillium camembertii* in mixed culture. Enzyme Microbial. Technol., 39: 325–331.

Boutrou, R. and Gueguen, T.M. 2005. Interests in *Geotrichum candidum* for cheese technology. Int. J. Food Microbiol., 102: 1–20.

Cappa, F. and Cocconcelli, P.S. 2001. Identification of fungi from dairy products by means of 18S rRNA analysis. Int. J. Food Microbiol., 69: 157–160.

Carreira, A., Ferreira, L.M. and Loureiro, V. 2001. Brown pigments produced by *Yarrowia lipolytica* result from extracellular accumulation of homogentisic acid. Appl. Environ. Microbiol., 67: 3463–3468.

Carreira, A., Paloma, L. and Loureiro, V. 1998. Pigment producing yeasts involved in the brown surface discoloration of ewes' cheese. Int. J. Food Microbiol., 41: 223–230.

Corbo, M.R., Lanciotti, R., Albenzio, M. and Sinigaglia, M. 2001. Occurrence and characterization of yeasts isolated from milks and dairy products of Apulia region. Int. J. Food Microbiol., 69: 147–152.

Deák, T. 2007. Handbook of food spoilage yeasts. CRC Press, Boca Raton, US.
Decker, M. and Nielsen, P.V. 2005. The inhibitory effect of *Penicillium camemberti* and *Geotrichum candidum* on the associated funga of white mould cheese. Int. J. Food Microbiol., 104: 51–60.
Dieuleveux, V., Lemarinier, S. and Guéguen, M. 1998. Antimicrobial spectrum and target site of D-3-phenyllactic acid. Int. J. Food. Microbiol., 40: 177–183.
El-Diasty, E.M. and Salem, R.M. 2007. Incidence of lipolytic and proteolytic fungi in some milk products and their public health significance. J. Appl. Sci. Res., 3: 1684–1688.
Fadda, M.E., Mossa, V., Pisano, M.B., Deplano, M. and Cosentino, S. 2004. Occurrence and characterization of yeasts isolated from artisanal Fiore Sardo cheese. Int. J. Food Microbiol., 95: 51–59.
Flee, G.H. 1990. Yeasts in dairy products. J. Appl. Bact. 68: 199–211.
Florez, A.B., lvarez-Martına, P.A., Lopez-Dıaz, T.M. and Mayo, B. 2007. Morphotypic and molecular identification of filamentous fungi from Spanish blue-veined Cabrales cheese, and typing of *Penicillium roqueforti* and *Geotrichum candidum* isolates. Int. Dairy J., 17: 350–357.
Gandomi, H., Misaghi, A., Basti, A.A., Bokaei, S., Khosravi, A., Abbasifar, A. and Javan, A.J. 2009. Effect of *Zataria multiflora* Boiss. essential oil on growth and aflatoxin formation by *Aspergillus flavus* in culture media and cheese. Food Chem. Toxicol., 47: 2397–2400.
Jacques, N. and Casaregola, S. 2008. Safety assessment of dairy microorganisms: The hemiascomycetous yeasts. Int. J. Food Microbiol., 126: 321–326.
Kure, C.F., Skaar, I. and Brendehaug, J. 2004. Mould contamination in production of semi-hard cheese. Int. J. Food Microbiol., 93: 41–49.
Kure, C.F., Wasteson, Y., Brendehaug, J. and Skaar, I. 2001. Mould contaminants on Jarlsberg and Norvegia cheese blocks from four factories. Int. J. Food Microbiol., 70: 21–27.
Liu, S.-Q. and Tsao, M. 2009. Biocontrol of dairy moulds by antagonistic dairy yeast *Debaryomyces hansenii* in yoghurt and cheese at elevated temperatures. Food Control, 20: 852–855.
Lund, F., Bech Nielsen, A. and Skouboe, P. 2003. Distribution of *Penicillium commune* isolates in cheese dairies mapped using secondary metabolite profiles, morphotypes, RAPD and AFLP fingerprinting. Food Microbiol., 20: 725–734.
Mataragas, M., Dimitriou, V., Skandamis, P.N. and Drosinos, E.H. 2011. Quantifying the spoilage and shelf-life of yoghurt with fruits. Food Microbiol., 28: 611–616.
Minervini, F., Montagna, M.T., Spilotros, G., Monaci, L., Santacroce, M.P. and Visconti, A. 2001. Survey on mycoflora of cow and buffalo dairy products from Southern Italy. Int. J. Food Microbiol., 69: 141–146.
Mounier, J., Gelsomino, R., Goerges, S., Vancanneyt, M., Vandemeulebroecke, K., Hoste, B., Scherer, S., Swings, J., Fitzgerald, G.F. and Cogan, T.M. 2005. Surface microflora of four smear-ripened cheeses. Appl. Environ. Microbiol., 71: 6489–6500.
Mounier, J., Monnet, C., Vallaeys, T., Arditi, R., Sarthou, A.-S., Helias, A. and Irlinger, F. 2008. Microbial interactions within a cheese microbial community. Appl. Environ. Microbiol., 74: 172–181.
Nielsen, M.S., Frisvad, J.C. and Nielsen, P.V. 1998. Protection by fungal starters against growth and secondary metabolite production of fungal spoilers of cheese. Int. J. Food Microbiol., 42: 91–99.
O'Brien, M., Egan, D., O'Kiely, P., Forristal, P.D., Doohan, F.M. and Fuller, H.T. 2008. Morphological and molecular characterisation of *Penicillium roqueforti* and *P. paneum* isolated from baled grass silage. Mycol. Res., 112: 921–932.
Pitt, J.I. and Hocking, A.D. 2009. Fungi and food spoilage. Springer, New York, US.
Pottier, I., Gente, S., Vernoux, J.-P. and Guéguen, M. 2008. Safety assessment of dairy microorganisms: *Geotrichum candidum*. Int. J. Food Microbiol., 126: 327–332.
Ray, B. 2004. Fundamental Food Microbiology. CRC Press, Boca Raton, USA.
Ropars, J., Cruaud, C., Lacoste, S. and Dupont, J. 2012. A taxonomic and ecological overview of cheese fungi. Int. J. Food Microbiol., 155: 199–210.

Sacristán, N., González, L., Castro, J.M., Fresno, J.M. and Tornadijo, M.E. 2012. Technological characterization of *Geotrichum candidum* strains isolated from a traditional Spanish goats' milk cheese. Food Microbiol., 30: 260–266.

Saravanakumar, K., Kumaresan, G. and Sivakumar, K. 2007. Incidence and characterization of ochratoxigenic moulds in Curd (Dahi). Res. J. Agric. Biol. Sci., 3: 818–820.

Scott, R. 1981. Cheese making practice. Applied Science Publisher LTD, Ripple Road, Barkig, Essex, England.

Sengun, I.Y., Yaman, D.B. and Gonul, S.A. 2008. Mycotoxins and mould contamination in cheese: a review. World Mycotox. J., 1: 291–298.

Singh, G., Kapoor, I.P.S. and Singh, P. 2011. Effect of volatile oil and oleoresin of anise on the shelf life of yogurt. J. Food Proc. Preserv., 35: 778–783.

Smith-Palmer, A., Stewart, J. and Fyfe, L. 2001. The potential application of plant essential oils as natural food preservatives in soft cheese. Food Microbiol., 18: 463–470.

Tamine, A.Y. and Robinson, R.K. 1999. Yoghurt Science and Technology. CRC Press, Boca Raton, USA.

Taniwaki, M.H., Hocking, A.D., Pitt, J.I. and Fleet, G.H. 2001. Growth of fungi and mycotoxin production on cheese under modified atmospheres. Int. J. Food Microbiol., 68: 125–133.

US Environmenal Protection Agency. 1997. *Penicillium roqueforti*. Final Risk Assessment http://epa.gov/oppt/biotech/pubs/fra/fra008.htm.

Vasdinyei, R. and Deák, T. 2003. Characterization of yeast isolates originating from Hungarian dairy products using traditional and molecular identification techniques. Int. J. Food Microbiol., 86: 123–130.

Walstra, P., Wouters, J.T.M. and Geurts, T.J. 2006. Dairy Science and technology. Taylor and Francis Group, Boca Raton, London, New York.

Welthagen, J.J. and Viljoen, B.C. 1998. Yeast profile in Gouda cheese during processing and ripening. Int. J. Food Microbiol., 41: 185–194.

CHAPTER 9

Coprophilous Fungi—A Review and Selected Bibliography

J.K. Misra,[1,*] *Shubha Pandey,*[1] *Awanish Kumar Gupta*[1] and *Sunil K. Deshmukh*[2]

ABSTRACT

Dung provides a unique environment, both physical and chemical, to a variety of microorganisms, insects and other invertebrates. Fungi thrive well on this rich substratum. Almost all groups of fungi growing on dung from different parts of the world are known, mostly from the temperate regions. Comparatively, there are limited reports on these fungi from the tropical parts of the world. Here we have tried to consolidate the available information on the taxonomy, ecology and other aspects of coprophilous fungi besides describing the results of the investigations of two of us (Shubha Pandey and Awanish Kumar Gupta) who have studied the dung fungi of some herbivores, both domesticated and wild. The need for further research on the applied aspects of dung fungi as well as surveys from the geographical areas that have not been investigated adequately have also been emphasized.

[1] Department of Botany, Saroj Lalji Mehrotra Bharatiya Vidya Bhavan Girls Degree College, 6, Vineet Khand, Gomti Nagar, Lucknow-226 010 India.

[2] Piramal Enterprises Limited, 1, Nirlon Complex, Off Western Express Highway, Goregaon (East).

Email: Sunil.deshmukh@piramal.com

* Corresponding author: jitrachravi@gmail.com

Introduction

Dung, a marvellous resource, is a world of its own fauna and flora comprising—annelids, arthropods, bacteria, fungi, nematodes, platyhelminths and protozoa. Dung has differently been referred to/ named as apples, bowel movements, caca, chips, crap, dirt, droppings, do-do, faeces, frass, guano, pats, pellets, poop, scats and so on. Even fossilized faeces have a name—coprolites (Seifert et al., 1983). In different parts of India, it is referred to as *Gobar* in Hindi that mainly means the dung of a cow. This unique natural substrate contains a large quantity of fibrous plant material, both coarse and fine-textured, depending on the type and nature of animal that it belongs to. Although dung has little available protein, it is rich in water-soluble vitamins, growth factors, mineral ions, carbohydrates, and is also high (4%) in nitrogen (Webster, 1970; Lodha, 1974). Furthermore, the physical structure, pH and varying moisture content of the dung make it more suitable for the growth of microorganism including a variety of fungi. The fungi growing in or on the dung develop certain features to adapt to the dung-environment. They also have developed a cyclic relationship between dung-herbage-animal gut-dung in the course of evolution (Ingold, 1953; Webster, 1970; Lodha, 1974). These fungi also play an important role in the decomposition of dung, the carbon flow and ecosystem energetics (Angel and Wicklow, 1971). As mentioned above, dung harbours a bizarre group of microorganism. However, this chapter aims to consolidate available information, present a selected bibliography on coprophilous fungi (except myxomycetes, chytrids, bacteria and yeasts), and to highlight the need for further research on the fungi that inhabit such an excellent substratum—the dung.

Methodology

Collection of samples: The dung samples, both fresh and dry, can be collected with the help of sterilized spatula or forceps and kept in paper bags. Depending on the object of study, the collection and isolation of a specific fungal group or groups, the collection techniques may be a bit different requiring special care to get the desired results. However, for observing and isolating the coprophilous fungi in general, that mainly and commonly include members of zygomycetes, ascomycetes, and basidiomycetes, fresh dung samples are preferred, but dried samples can also be collected and stored in paper bags at low temperatures for a longer period of time. These aspects can be found in greater details in Krug et al. (2004).

For observing/isolating fungi from the samples, a teaspoon full of fresh dung is placed onto sterilized moist Whatman's filter paper in a Petri dish and incubated in the laboratory in triplicates at room temperature or at the desired temperature(s) and light or dark conditions. The Petri dishes so incubated are examined weekly up to a month at definite time intervals under a stereo-binocular microscope. The growing fungi/fruiting bodies are picked up with a sterilized needle and placed onto the microscopic glass slides, covered with cover slip, and examined before and after the infiltration of lacto-phenol cotton-blue under a compound microscope. The fungal material is photographed using the camera attached to a microscope. The photographs of fruiting bodies/habits should also be taken. Freshly collected samples should also be observed for any mycelia/ fruiting structures present.

Culturing the fungi directly from the dung samples using Dilution plate method was not found very encouraging by Pandey (2009) and Gupta (2010), but they could isolate the fungi through moist chamber incubation and then culturing them on appropriate media.

Zygomycetes, which are noticed first on the dung surface in the moist chamber, should be isolated using a sterilized fine needle to pick-up spores/ sporangia that can be transferred to suitable culture media such as malt extract yeast-extract agar, V8-juice agar and potato-dextrose agar. Many other media have also been used for isolating these fungi (Benny and Benjamin, 1975; Krug et al., 2004).

Ascomycetous genera develop fruiting bodies in moist chamber. They should be picked-up with a sterilized fine needle while observing under a binocular microscope and placed onto microscope slides to make preparations for direct observation or placed onto Petri dishes containing appropriate media for them to grow.

Basidiomycetes develop their fructifications in moist chamber. Fructifications should be studied in addition to basidia and basidiospores to ascertain their identity.

Since there are no monographs on coprophilic fungi, therefore, respective literature (monographs and publications) of particular groups help in the identification of fungi observed on dung (Sanwal, 1953a,b; Denison, 1964; Arx and Muller, 1975; Dennis, 1954, 1968a,b, 1978; Fennell and Raper, 1955; Ames, 1961, 1963; Lodha, 1962, 1964, 1971; Batra and Batra, 1963; Subramanian and Lodha, 1964, 1968; Ellis, 1966, 1971, 1976, 1988; Narendra, 1973a,b; Arx, 1973, 1981; Hanlin, 1973, 1997, 1998; Narendra and Rao, 1976; Raper and Fennel, 1977; Subramanian and Chandrashekara, 1977; Samson, 1979; Kaushal and Thind, 1983; Seifert et al., 1983; Bell, 1983, 2005; Arx et al., 1984, 1986; Kaushal et al., 1985; Klich and Pitt, 1985; Ellis and Ellis, 1988a,b; Klich, 1993, 2002a; Arora, 1999; Kiffer and Morelet, 2000; Webster and Weber, 2000).

Floristic Studies

Dung supports all groups of fungi—Zygomycetes, Ascomycetes and the Basidiomycetes. The researches on fungi from dung have mainly focussed on the dung of herbivores, however, reports are there to indicate that the dung of carnivores also support fungal growth.

We know today almost 52 genera of Zycomycetes (Table 9.1), 169 of Ascomycetes (Table 9.2), 33 of Basidiomycetes (Table 9.3), and many of Deuteromycetes being associated with one or the other type of dung as a substrate to grow. Some may be represented by a few or many species and this has not been detailed here.

Literature available on fungi from the dung of various animals mainly describes the fungi recovered from different parts of the world, particularly from temperate regions (Outlemans, 1882; Massee and Salmon, 1901, 1902a,b; Cain, 1934, 1950, 1956a,b, 1957, 1961b, 1962; Meyer and Meyer, 1949; Juniper, 1954, 1957; Tubaki, 1954; Cain and Farrow, 1956; Dade, 1957; Dring, 1959; Lundqvist, 1960, 1964a,b; Watling, 1963; Harper and Webster, 1964, 1966; Kar and Pal, 1968; Richardson and Watling, 1968, 1969, 1997; Seth, 1968a, 1995; Fakirova, 1969; Webster, 1970; Otani and Kanzawa, 1970;

Table 9.1. Genera of Zygomycetes known from dung.

1.	*Absidia*	27.	*Mycotypha*
2.	*Actinomucor*	28.	*Parasitella*
3.	*Backusella*	29.	*Phascolomyces*
4.	*Basidiobolus*	30.	*Phycomyces*
5.	*Benjaminiella*	31.	*Pilaira*
6.	*Blakeslea*	32.	*Pilobolus*
7.	*Chaenophora*	33.	*Piptocephalis*
8.	*Chaetocladium*	34.	*Pirella*
9.	*Circinella*	35.	*Protomycocladus*
10.	*Cokeromyces*	36.	*Radiomyces*
11.	*Coemansia*	37.	*Reticulocephalis*
12.	*Conidiobolus*	38.	*Rhizomucor*
13.	*Cunninghamella*	39.	*Rhizopus*
14.	*Dichotomocladium*	40.	*Rhopalomyces*
15.	*Dimargaris*	41.	*Sigmoideomyces*
16.	*Dispira*	42.	*Spirodactylon*
17.	*Dissophora*	43.	*Spiromyces*
18.	*Ellisomyces*	44.	*Stylopage*
19.	*Fennellomyces*	45.	*Syncephalastrum*
20.	*Gilbertella*	46.	*Syncephalis*
21.	*Helicocephalum*	47.	*Teighemiomyces*
22.	*Helicostylum*	48.	*Thamnidium*
23.	*Kickxella*	49.	*Thamnocephalis*
24.	*Micromucor*	50.	*Thamnostylum*
25.	*Mortierella*	51.	*Utharomyces*
26.	*Mucor*	52.	*Zychaea*

Table 9.2. Genera of Ascomycetes known from dung.

1.	*Achaetomium*	49.	*Delitschia*	97.	*Mycoarcticum*
2.	*Amauroascus*	50.	*Dennisiopsis*	98.	*Mycorhynchidium*
3.	*Anopodium*	51.	*Dictyocoprotus*	99.	*Myxotrichum*
4.	*Aphanoascus*	52.	*Emblemospora*	100.	*Nectria*
5.	*Apiosordaria*	53.	*Emericella*	101.	*Neocosmospora*
6.	*Apiospora*	54.	*Emericellopsis*	102.	*Neosartorya*
7.	*Apodospora*	55.	*Enterocarpus*	103.	*Neurospora*
8.	*Apodus*	56.	*Eoterfezia*	104.	*Nigrosabulum*
9.	*Arachniotus*	57.	*Eremomyces*	105.	*Ochotrichobolus*
10.	*Arachnomyces*	58.	*Eupenicillium*	106.	*Onygena*
11.	*Arniella*	59.	*Eurotium*	107.	*Orbicula*
12.	*Arnium*	60.	*Farrowia*	108.	*Orbilia*
13.	*Arthroderma*	61.	*Faurelina*	109.	*Periamphispora*
14.	*Ascobolus*	62.	*Fimaria*	110.	*Petriella*
15.	*Ascocalvatia*	63.	*Fimetariella*	111.	*Peziza*
16.	*Ascodesmis*	64.	*Gelasinospora*	112.	*Phomatospora*
17.	*Ascophanus*	65.	*Guilliermondia*	113.	*Pleophragmia*
18.	*Ascotricha*	66.	*Gymnascella*	114.	*Pleospora*
19.	*Ascozonus*	67.	*Gymnoascoideus*	115.	*Pleuroascus*
20.	*Auxarthron*	68.	*Gymnoascus*	116.	*Podosordaria*
21.	*Bombardia*	69.	*Hamigera*	117.	*Podospora*
22.	*Bombardioidea*	70.	*Hapsidomyces*	118.	*Polytolypa*
23.	*Boothiella*	71.	*Hapsidospora*	119.	*Poronia*
24.	*Boubovia*	72.	*Heleococcum*	120.	*Protoventuria*
25.	*Bulbithecium*	73.	*Hypocopra*	121.	*Preussia*
26.	*Byssochlamys*	74.	*Idophanus*	122.	*Pseudallescheria*
27.	*Byssonectria*	75.	*Jugulospora*	123.	*Pseudascozonus*
28.	*Caccobius*	76.	*Kathistes*	124.	*Pseudeurotium*
29.	*Camptosphaeria*	77.	*Kernia*	125.	*Pseudogymnoascus*
30.	*Cercophora*	78.	*Khuskia*	126.	*Pseudombrophila*
31.	*Chadefaudiella*	79.	*Klasterskya*	127.	*Pteridiosperma*
32.	*Chaetomidium*	80.	*Kuehmiella*	128.	*Pyxidiophora*
33.	*Chaetomium*	81.	*Lanzia*	129.	*Ramgea*
34.	*Chaetopreussia*	82.	*Lasiobolus*	130.	*Renispora*
35.	*Chalazion*	83.	*Lasiobolidium*	131.	*Rhytidospora*
36.	*Cheilymenia*	84.	*Lasiosphaeria*	132.	*Roumegueriella*
37.	*Cleistoiodophanus*	85.	*Leptokalpion*	133.	*Rutstroemia*
38.	*Cleistothelebolus*	86.	*Leuconeurospora*	134.	*Saccobolus*
39.	*Collematospora*	87.	*Leucosphaerina*	135.	*Schizothecium*
40.	*Coniochaeta*	88.	*Leucothecium*	136.	*Scutellinia*
41.	*Coprobia*	89.	*Lophotrichus*	137.	*Selinia*
42.	*Coprobolus*	90.	*Martininia*	138.	*Semidelitschia*
43.	*Copromyces*	91.	*Melanospora*	139.	*Shanorella*
44.	*Coprotiella*	92.	*Microascus*	140.	*Sordaria*
45.	*Coprotinia*	93.	*Microthecium*	141.	*Sphaeronaemella*
46.	*Coprotus*	94.	*Microthyrium*	142.	*Sphaerodes*
47.	*Corynascus*	95.	*Monascus*	143.	*Sphaeronaemella*
48.	*Ctenomyces*	96.	*Mycoarchis*	144.	*Sporormia*

Table 9.2. contd....

Table 9.2. contd.

145. *Sporormiella*	158. *Tripterosporella*
146. *Spororminula*	159. *Tripterospora*
147. *Strattonia*	160. *Uncinocarpus*
148. *Subramaniula*	161. *Unguiculella*
149. *Syspastospora*	162. *Viennotidia*
150. *Talaromyces*	163. *Wawelia*
151. *Thecotheus*	164. *Westerdykella*
152. *Thelebolus*	165. *Xylaria*
153. *Thielavia*	166. *Xynophila*
154. *Triangularia*	167. *Zopfiella*
155. *Trichobolus*	168. *Zygopleurage*
156. *Trichodelitschia*	169. *Zygospermella*
157. *Trichophaea*	

Table 9.3. Genera of Basidiomycetes known from dung.

1. *Agaricus*	18. *Naucoria*
2. *Agrocybe*	19. *Panaeolus*
3. *Anellaria*	20. *Parsaola*
4. *Bolbitius*	21. *Phaeogalera*
5. *Byssocorticium*	22. *Pholiotina*
6. *Clitocybe*	23. *Platygloea*
7. *Clitopilus*	24. *Psathyrella*
8. *Conocybe*	25. *Pseudoclitocybe*
9. *Copelandia*	26. *Psilocybe*
10. *Coprinellus*	27. *Schizostoma*
11. *Coprinopsis*	28. *Sebacina*
12. *Coprinus*	29. *Sphaerobolus*
13. *Crucibulum*	30. *Stropharia*
14. *Cyathus*	31. *Tephrocybe*
15. *Cystobasidium*	32. *Volvariella*
16. *Laetisaria*	33. *Weraroa*
17. *Lepista*	

Udagawa and Takada, 1971; Larsen, 1971; Furuya and Udagawa, 1972, 1973, 1976, 1977; Richardson, 1972, 1998a,b, 1999, 2000; Yang, 1972; Lundqvist and Fakirova, 1973; Saniel and Alma, 1974; Bell, 1983, 2005; Dickinson and Underhay, 1977; Jeng and Krug, 1977; Liou and Chen, 1977; Aas, 1978; Nagy and Harrower, 1979; Udagawa and Muroi, 1979; Piontelli et al., 1981; Booth, 1982; Muroi and Udagawa, 1984; Barrasa et al., 1985; Cribb, 1988, 1989, 1991, 1992, 1994, 1996a,b, 1997, 1998, 1999a,b; Dissing, 1989, 1992; Lorenzo, 1989, 1992; Eberson and Eicker, 1992, 1997; Wang, 1993, 1995, 1996, 1999; Prokhorov, 1994; Bell and Mahoney, 1995; Krug and Jeng, 1995; Weber and Webster, 1997, 1998; Richardson, 1998; Spooner and Butterfill, 1999; Barr, 2000; Krug et al., 2004).

After the year 2000, a number of papers have appeared that describe the fungal diversity of dung fungi from different part of the world. Pubications particularly those of Mike L. Richardson are of greater significance as he has described the biodiversity of these fungi and touched upon their ecological aspects as well (Richardson, 2001a,b, 2002a,b, 2004, 2005, 2006, 2007, 2008a,b, 2011; Watling and Richardson, 2010; Farouq et al., 2012; Mungai et al., 2011, 2012).

Richardson reported 81 species of coprophilous fungi from Iceland (Richardson, 2004). There were almost 50% of the species that were new records for Iceland. He observed only two genera of zygomycetes (*Pilaira* and *Pilobolus*) while ascomycetous genera were 22 in number, dominating the mycoflora. Basidiomycetes observed were—*Coprinus, Panaeolus, and Psilocybe*. His latest addition to the coprophilous mycota of Iceland recorded 19 species new to Iceland (Richardson, 2011). From Faroe Island, he collected a total of 20 samples of dung of hare and sheep and could observe 59 fungi in moist chamber. The fungi from the two dung types were different, although the samples came from the same habitat (Richardson, 2005). From Falkland Islands, Watling and Richardson (2010) recorded 97 fungi from the dung samples of sheep, cattle, horse, rabbit, and goose. In this study also genera of ascomycetes (23 in number) dominated the fungal flora followed by the genera of basidiomycetes (11 in number) and only three genera were that of zygomycetes.

As compared to the temperate regions, however, not much has been done for these fungi from the tropical and subtropical parts of the world (Mahju, 1933a,b; Ginai, 1936a; Lodha, 1962, 1964a,b, 1971; Subramanian and Lodha, 1964, 1968, 1975; Ahmed and Asad, 1971; Abdulla et al., 1971, 1999; Saxena and Mukerji, 1973; Jain and Cain, 1973; Iyer et al., 1973; Narendra, 1973a,b, 1974; Narendra and Rao, 1976; Waritch, 1976; Subramanian and Chandreshekara, 1977; Singh and Mukerji, 1979; Kaushal et al., 1985; Swofford, 1993; Gene et al., 1993; Manoch et al., 2000; Elshafie, 2005). Perusal of available literature indicates that the earlier contributions regarding the coprophilous fungi of Indian sub-continent came from the then Punjab University Lahore (now in Pakistan) by Mahju (1933) and Ginai (1935, 1936). Mahju had studied the dung of six animals—rabbit, sambhar, horse, goat, buffalo and sheep of the Zoological Garden, Lahore. He described 29 species belonging to 29 genera. While Ginai's further contributions to the knowledge of these fungi was the addition of 48 species belonging to 26 genera. He studied the dung of five other animals—cow, neoga, camel, zebra and donkey as well as that of buffalo.

Two of us Pandey (2009) and Gupta (2010), studied the dung flora of four herbivores-buffalo, cow, goat, and horse and wild but domesticated inmates—deer, elephant, giraffe, and zebra of the Prince of Wales

Zoological Garden, Lucknow, respectively. Pandey (2009) recovered 104 fungal forms belonging to the phyla Zygomycota, Ascomycota, Basidiomycota, and Deuteromycota in varying numbers. Out of these, 16 belonged to Zygomycota, 52 to Ascomycota, four to Basidiomycota and the rest 32 to the asexual fungal group—Deuteromycetes. Buffalo dung alone yielded a total of 53 fungal forms, though some of them were also common in other animals also. Out of 53 forms recovered, 11 belonged to Zygomycota, 26 to Ascomycota, three to Basidiomycota, and 13 to Deuteromycetes. The cow dung yielded a total of 61 fungal forms out of which, 10 belonged to Zygomycota, 28 to Ascomycota, two to Basidiomycota and 21 to Deuteromycetes. From the goat dung a total of 44 fungal forms were observed that belonged to Zygomycota (nine), Ascomycota (20), Basidiomycota (one) and 14 to the Deuteromycetes. Comparatively little number of fungi (25) was recovered from the dung of horses. Out of those 25 fungal forms, seven belonged to Zygomycota, 11 to Ascomycota, three to Basidiomycota, and four to Deuteromycetes.

As is evident from the data of her studies (Pandey, 2009), members of the Phylum Ascomycota dominated the dung flora of the herbivores studied followed by Deuteromycetes, Zygomycota, and Basidiomycota which were lowest in the number. Specifically, nine species of *Ascobolus*, two species of *Ascodesmis*, six species of *Chaetomium*, two species of *Lasiobolus*, five species of *Podospora*, and three species of *Saccobolus* were isolated. Among the Basidiomycota, five species of *Coprinus* and two unknown forms were observed. In the Deuteromycetes, five species of *Alternaria*, three species of *Arthrobotrys*, six species of *Aspergillus*, and two species of *Cephaliophora* were observed. In Zygomycota two species of *Circinella*, three species of *Mucor*, five species of *Pilobolus* and a single genus *Rhophalomyces elegans* were observed. She claimed five new records of fungi for the dung substrate and 17 new for Indian fungi. Her results are in conformity with those of other workers.

Gupta (2010) observed a total of 52 fungal forms from the dung of four domesticated wild herbivores—deer, elephant, giraffe and zebra. Out of 52 fungal forms, all groups were in varying numbers as indicated in parentheses against them, Zygomycota (seven), Ascomycota (26) Basidiomycota (one) and the asexual fungal group—Deuteromycetes (18).

Deer dung alone yielded a total of 30 fungal forms, though some of them were common to other animals also. Out of 30 forms recovered, five belonged to Zygomycota, 15 to Ascomycota, one to Basidiomycota, and nine to Deuteromycetes. Comparatively little number of fungi (13) were recovered from the dung of elephant. Out of those 13 fungal forms, three belonged to Zygomycota, five to Ascomycota, one to Basidiomycota and four to Deuteromycetes.

From the giraffe dung a total of 27 fungal forms were observed that belonged to Zygomycota (six), Ascomycota (13), Basidiomycota (one) and seven to the Deuteromycetes. The zebra dung yielded a total of 15 fungal forms. Of which, three belonged to Zygomycota, seven to Ascomycota, five to Deuteromycetes and none to Basidiomycota. The Phylum Ascomycota dominated the dung flora of the herbivores studied followed by Deuteromycetes, Zygomycota and Basidiomycota. Basidiomycetous fungi were recorded in lesser numbers.

Specifically, seven species of *Ascobolus*, only one species of genera like *Ascodesmis sphaerospora, Coniochaeta discospora, Lasiobolus microsporus, Lophotrichus ampullus, Sordaria fimicola, Sphaeronaemella fimicola, Thelebolus microsporus, Thielavia ampullata, Tripterospora erostrata,* and four species of *Chaetomium,* two species of *Iodophanus,* three species of *Podospora* and two species of *Saccobolus* were isolated.

Among the Basidiomycota, a single genus *Coprinus astramentarius* was observed. In Zygomycota, one species each of *Circinella simplex* and *Cokeromyces* sp., three species of *Mucor,* and two species of *Pilobolus* were observed. In Deuteromycetes one species of genera like *Alternaria* sp., *Arthrobotrys thaumasia, Dactylaria brochophaga, Dactylella bembicodes, Didymostilbe* sp., *Fusarium* sp., *Gonytrichum* sp., *Isaria brachiata, Memnoniella echinata, Microsporum gypseum, Oedocephalum* sp., *Phialocephala phycomyces, Stachybotrys* sp., *Stysanus* sp., *Trichurus spiralis* and three species of *Aspergillus* were observed. Gupta (2010) in this study claimed 42 new records of fungi for dung substrate in India for wild domesticated animals studied—23 new records are for the deer dung; 10 for elephant dung; all 27 forms recovered from giraffe dung are new records; and 13 are new records for zebra dung.

Ecological Studies

As compared to the floristic studies, there is not much work on the ecological aspects of dung fungi. But there is information on the flora of fungi from different dung types (dung of different animals), and quite significant information is in print in this regard (Abdullah, 1982; Masunga et al., 2006; Watling and Richardson, 2010). There is a large amount of information available for the diversity and species richness of coprophilous fungi through the publication of Richardson (2001, 2007a, 2008a) and later on Watling and Richardson (2010) have concluded through their studies that, "there is a latitudinal gradient of increasing species richness of coprophilous fungi with decreasing latitude."

Succession of Fungi

Succession has variously been defined by plant ecologists. The gradual replacement of one type of plant community by the other is referred to as plant succession. According to Odum (1963), "plant succession is an orderly process of community change in a unit area". Clements (1916) defined the plant succession in a very simple manner and according to him, "succession is a natural process by which the same locality becomes successively colonised by different groups of communities". Thus, succession is a complex universal process which begins, develops, and finally stablizes at the climax stage. Succession is generally progressive and, therefore, it brings about progressive changes in the habitat(s). It is well established that fungi on a specific substrate show succession (Brown and Graff, 1931; Hering, 1965; Webster, 2000). Coprophilous fungi are no exception. There are studies that indicate that a variety of fungi appear one after the other on the dung (Watling, 1963; Harper and Webster, 1964; Nickolson et al., 1966; Mitchell, 1970; Larsen, 1971; Lodha, 1974; Wicklow and Moore, 1974; Bell, 1975; Underhay and Dickinson, 1978; Nagy and Harrower, 1979; Yocom and Wicklow, 1980; Kuthubutheen and Webster, 1986a; Ebershon and Eicker, 1997; Richardson, 2002).

Pandey (2009) while screening the fungi from the dung of four herbivores-buffalo, cow, goat and horse also did some ecological studies that include phenology of the fungal flora of the dung and its correlation with the physicochemical factors like pH, sodium, potassium, electrical conductivity and moisture content of the dung; succession of fungi on dung; storage-time effect on fungal flora of the dung; and influence of habitat and the colour of two herbivores (buffalo and cow) on their dung flora.

As regards the phenology of fungi of dungs studied by Pandey (2009), varying numbers of fungi from different dung types were observed in different months of the year of isolation (Fig. 9.1). The physicochemical factors like the air temperature showed negative and significant to highly significant correlation with the fungi recovered from buffalo and cow dung, respectively ($r = -0.649, -0.887$). Surprisingly, air temperature did not show any correlations with the dung fungi recovered from goat and horse. The relative humidity of air had, however, also not shown any correlations with the number of fungal forms isolated from any of the four dung types. In the case of buffalo, pH showed negative and highly significant correlation ($r = -0.727$) with the numbers of fungi. Sodium had positive and significant correlation ($r = 0.615$) with the number of fungi recovered from buffalo and cow ($r = 0.571$), respectively, but it did not show any correlations in the case of goat and horse dung. Interestingly, potassium did not exhibit any correlations with any of the dung types studied.

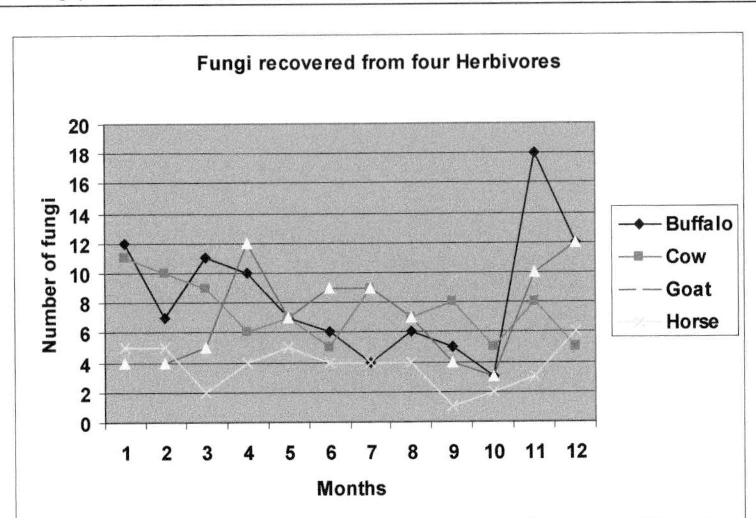

Figure 9.1 Number of fungi observed in different months from dung of four herbivores.

Similar phenological study, done for dung fungi of deer, elephant, giraffe, and zebra by Gupta (2010), also showed variations in the number of fungi recovered from the dung of animals under study in different months of the year of survey. Correlation co-efficient worked out for the fungi and physicochemical factors of the dung studied showed varying relationships. Surprisingly, air temperature did not show any correlation with the number of fungi recovered from any of the dungs-deer, elephant, giraffe, and zebra. Relative humidity, however, showed negative and significant correlation with the fungi recovered from giraffe dung ($r = -0.530$), but no correlation with the number of fungi from deer, elephant, and zebra dung. Potassium showed negative and significant correlation with fungi recovered from giraffe dung ($r = -0.571$), but no correlation in other cases. Sodium showed no correlation with fungi recovered from any of the dung types.

To investigate successional changes in types and numbers of fungi of dung, Pandey (2009) collected dungs in large quantities. She observed fungi in moist chamber at room temperature and recorded some data on moisture content, pH, sodium, potassium and electrical conductivity, both of the fresh dung as well as of those that she left in the earthen pots in the college garden for 12 months. The observations for the left-overs in the garden were made at the interval of three months upto 12 months. Air temperature and atmospheric humidity data of the day of collection were also recorded.

Fresh dungs of the animals under study yielded greater numbers of fungal forms than those that were left in natural environment upto 12 months and screened periodically at the interval of three months. The decrease in numbers for all types of dungs was evident as compared to the fresh dung. Besides decrease in the number of fungal forms, there was obvious change in the groups of fungi that were recovered indicating succession. The fresh dung of buffalo resulted into the isolation of 16 fungal forms dominated by seven zygomycetous ones that in later screening at 3–12 months did not appear at all, but other groups of fungi did appear. In the case of the cow, the fresh dung yielded 14 fungal forms including six zygomycetous fungi whose numbers declined by 50% at three months of screening and then after none of this group could be observed. However, asco- and basidiomycetous forms were recovered. Fresh dung of goat and horse yielded five and seven fungal forms, respectively. In both the cases the zygomycetous fungi were three in number but their number declined after three months. Other fungi were, however, recovered upto 12 months.

The physicochemical factors that were estimated for all the fresh dungs and the ones incubated behaved differently. In all cases, the pH of the fresh dung did not vary much rather it ranged from 8.2–8.3, whereas after incubation the pH shifted towards acidic via neutral. The moisture content of fresh dung for all cases was greater but declined after incubation. Electrical conductivity recorded for the dungs showed variation from fresh to incubated dungs in all cases. Sodium, in all the cases, showed a varying trend. In the case of buffalo, it increased from fresh to incubated ones upto 12 months, while for cow, goat and horse it decreased periodically. Potassium in all cases showed increasing trend from fresh to incubated ones upto 12 months.

The colour and texture of the fresh as well as dungs that were periodically screened for fungal forms have also been observed. The fresh buffalo dung was more watery and yellowish brown but after incubation, the moisture level as well as colour changed. The fresh cow dung was comparatively solid, less watery and dark brownish in colour, but it also changed over time. The fresh goat pellets appeared smooth, less moist or almost dry and blackish in colour. After three months and onwards, the shape of pellets changed and they got mineralized and amorphous. The horse dung, while fresh, looked solid and yellowish in colour but this too upon incubation became loose, mineralized and amorphous.

Thus, the gradual decrease in hydrogen ion concentration and number of fungi has been observed over time. The dung flora declines with the increase in the period of incubation. This indicates that many dung fungi tend to grow when the pH of the medium is alkaline and not acidic. At acidic level, only a few or no fungi could be recovered. That moisture/ humidity does play an important role on the growth and sporulation of fungi had been observed by Pandey (2009) during her study. Dung with high moisture supported more fungi than drier ones. This supports the finding of Kuthubutheen and Webster (1986a,b). Gupta (2010) also found almost similar results for his studies done for the dung of domesticated animals of the zoo. The fact, that chemistry of the dung might not be affecting the succession of the fungal flora, should not be ignored. This definitely needs further investigations in the laboratory.

The decrease in numbers and types of fungal forms in/over all dungs types studied is a natural phenomenon and indicates sucession that has also been observed by other workers who studied the dung fungal flora (Harper and Webster, 1964; Lodha, 1974; Ebersohn and Eicker, 1997; Webster, 2000; Richardson, 2002). Several logical reasons have been advanced by the investigators to substantiate this change in the fungal flora of the dungs over time and those are: nutritional; varying time of sporulation and formation of fruiting bodies; synergism; antagonism and insect predation. On the basis of nutrition, it was assumed that zygomycetous fungi due to their faster growth consume readily available carbon sources of the substrate. They grow vigorously, sporulate heavily and disappear giving way to those fungi that take time to grow, sporulate and form fruiting bodies, and are able to consume more complex carbon sources such as hemicellulose and cellulose—the asco- and finally the basidiomycetous fungi that have capability to exploit both cellulose and lignin. But this hypothesis was found not very sound when experimentally tested. It was found that the fairly accessible carbon source—the sugars that were supposed to be consumed by the zygomycetous fungi, before they disappear from the substrate, remain in the substrate and are not fully exhausted. This led the advancement of another thought that different fungi require varying period of time to accumulate enough food reserves before fruiting. Sporangiophores of Zygomycetes develop in shorter time while the fruiting bodies of Asco- and Basidiomycetes take a longer time. Thus, it can be assumed that there exists a variety of spores in the dung and they might have been triggered to sporulate while passing through the mammalian gut. This hypothesis, though logical and convincing, has not taken into account some other important aspects that cause succession in the fungal forms—the antagonism. Antagonism does exist in the community of fungi at a given substrate—the dung in the present case. It has been reported that the hyphae of *Coprinus* are extremely antagonistic to the hyphae of many other coprophilous fungi (Kendrick, 1992; Webster, 2000). So also the other fungi like *Stilbella* that inhibits growth of many coprophilous fungi (Ikediugwa and Webster, 1970b; Singh and Webster, 1973; Webster, 2000). That synergism may also play a role in fungal succession cannot be ignored (Page, 1960). However, it needs to be tested more elaborately for many fungi. Mycoparasitic and nematophagus fungi may also play significantly in the process of succession. Pandey's (2009) and Gupta's (2010) results presented here are in agreement with what has been hypothesized by earlier workers.

Harper and Webster (1964) studied experimentally the succession of fungi on rabbit pellets from three locations in England and concluded that the succession was related more to the time required for different species to mature and fruit. According to them, it is not the dormancy or growth rate that were responsible and under experimental conditions the sequence of fruiting was quite consistent, and events of fruiting during experimentation and nature were similar, whereas Larsen (1971) found that the time to fruiting was variable for any particular species. She observed that *Pilobolus* and other zygomycetes appear within 2–4 days, discomycetes in the second week and pyrenomycetes and basidiomycetes appear one week later. This variation according to her may be due to the passage of food through gut (1–10 days). Wicklow and Moore (1974) reported that temperature affects fruiting of fungi. The widely accepted concept of the appearance of fungi on dung is that the sugar loving fungi appear first followed by ascomycetes and basidiomycetes but Lodha (1974) favoured the explanation that it is time taken by each fungus to produce its fruiting bodies, rather than nutrients availability that result in succession of fungi on dung.

Webster (1970) put forth many aspects of succession while reviewing the biology of dung fungi in the Presidential address to the British Mycological Society. The review by Wicklow (1992) is also quite in detail. On the basis of a large collection, information on sequence of fruiting body appearance is available (Richardson, 2001, 2002). Richardson's observation, both from the field and in incubation chambers, are in conformity with the widely accepted sequence of fruiting of coprophilous fungi, that is, Zygomycetes appear first followed by apothecial fungi and other pyrenomycetes, and lately the basidiomycetes. Different fungi taking varying time to appear may be due to various factors such as their metabolic activity (Wood and Cooke, 1987); ability to grow under near anaerobic conditions in the gut (Brewer et al., 1972); antagonism (Harper and Webster, 1964; Ikediugwu and Webster, 1970); varying ability to use more complex substances that appear late during the process of decomposition of dung (Lodha, 1974). However, we still lack information about the chemical conditions that might be responsible for succession of fungi in dung (Richardson, 2002). Experiments similar to that of Wood and Cooke (1984, 1987) and Safar and Cooke (1988a,b) need to be done for more coprophilous fungi to have a clearer picture of succession and find out as to which of the late fruiting fungi in dung really use lignin.

To assess the life span or longevity of spores of dung fungi in/on the stored dung, Pandey (2009) stored, all the four types of dung after air drying, in the laboratory at room temperature for 45 months. Screening of fungi was done after storage of 36 months at the interval of three months upto 45 months under moist chamber. A total of 15 fungal forms were recovered from the stored dung of buffalo. The number of fungi decreased with increase in the time period of storage, i.e., from the samples stored for 45 months, she could recover four forms as against five forms from the samples that were stored for 36 months. In the case of cow, a total of 12 fungal forms were recovered. In this case too, the trend remained the same as in the case of buffalo. From goat dung a total of 15 forms were recovered while from the dung of horse only four forms could be recovered. The trends in both these cases remained the same as for the other two dung types. She could observe the initiation of the development of the genus *Pilobolus* of zygomycetes indicating thereby that the spores of this fungus can survive upto 45 months but the fungus failed to develop fully. She assigned no reason to her observation. However, the lack of nutrition can be a more probable reason. In buffalo dung, 15 fungal forms were observed, out of which nine were those of ascomycetes that are known to have thicker-walled spores. Also in the case of cow dung, eight genera belonging to ascomycetes were recovered. In cases of goat and horse dungs, the ascomycetous genera recovered were nine and one, respectively. Significantly, no basidiomycetous fungi could be recovered from any of the stored dungs. This appears logical because the possibility of survival of basidiospores can not be high as they are thin-walled spores that readily germinate or perish. Thus, this preliminary study indicates that ascospores can withstand stresses of the environment and, therefore, their surviability is greater than other fungi. However, laboratory experiments to further strengthen this observation are warranted. Gupta (2010) also had similar observations for the stored dungs he studied.

It is been noted that the territory on which the animals graze affect the diversity of fungi of their dung (Angel and Wicklow, 1975). This made the basis of a study by Pandey (2009) to find out whether this concept stands true for Indian conditions also. In the urban areas of India, the animals have very restricted grazing areas and mostly remain with the owner who feeds them. On the other hand, in the rural areas the animals get enough places/areas to loiter and graze upon besides what they get as feed from their owners. Whether the colour of the cows—red, black, and white can affect the coprophilous fungi has also been tested by Pandey (2009). What she did was that she aseptically collected 50 samples each of buffalo and cow dung in sterilized bamboo-paper bags at random intervals from rural and urban areas of various districts of Uttar Pradesh including Lucknow, and screened them for fungi under a moist chamber. She also randomly collected dung and screened for fungi in a similar manner 25 samples of cows of three colours—red (five), black (10) and white (10). A total of 18 forms were recovered from the buffalo dung of rural and urban areas. Greater numbers of fungi (14) were recovered from urban areas in comparison to rural areas. From the dung of cow, a total of 13 fungal forms were recovered, both from rural and urban areas. From both, the areas, in buffalo and cow dung, ascomycetous fungi were the dominant ones. Only one asexual fungal form was recovered from buffalo dung of an urban area but in case of cow no fungal form of this group appeared.

As regards the colour of the cow and the number of fungi that could be recovered from their dung, it was found that out of a total of 14 fungal forms recovered, red cow's dung harboured six, black had 10 and from the dung of white cows eight fungi were recovered. Ascomycetous fungi were the only ones that were recovered. No asexual fungal forms were recovered.

Data from Pandey (2009) indicated that the habitats, rural and urban, were not appreciably different. In the rural areas where greater area for grazing is expected and consequently more fungi in the dung did not come true for her study. Similarly, in case of the colour of cow, the dung flora also did not provide any conclusive results; hence she advised further work with greater number of samples and wider area of coverage of sample collection to reach to any definite conclusions.

Applied aspects: Coprophilous fungi are known to produce secondary metabolites (Ridderbusch et al., 2004; Weber et al., 2005). However, many novel bioactive secondary metabolites from dung fungi still need discovery. In a recent review Bills et al. (2013) have very elaborately dealt with this aspect and would be profitable to read.

Future Research

1. Why study coprophilous fungi? The study of these fungi would be rewarding because, inspite of their wider distributional records, they have not been adequately searched for all over the world. From the tropical climate, however, their search is very limited whereas we can expect much more diverse mycobiota than what we know today.
2. The dung fungi are easy to handle and can be grown with very little effort in the laboratory for demonstration to the students of mycology and learn both biodiversity and ecology of these fungi that belong to all phyla/groups of fungi.
3. Variations in the mycoflora can easily be noted with age of the dung, change in its moisture content, pH and other chemicals, etc. present in the dung.
4. Many of the coprophilous fungi produce potentially useful compounds. We know little about these compounds; hence more research is warranted.

References

Aas, O. 1978. Two new coprophilous species of *Saccobolus*. Norwegian J. Bot., 25: 65–68.
Aas, O. 1983. The genus *Copronus* (Pezizales) in Norway. Nordic J. Bot., 3: 253–259.
Abbott, T.P. and Wicklow, D.T. 1984. Degradation of lignin by *Cyathus* species. Appl. Environ. Microbiol., 47: 585–587.
Abdulla, S.K., Al-Saadoon and Gurro, J. 1999. New and interesting coprophilous Ascomycetes from Iraq. Nova Hedwigia, 69: 211–216.
Abdulla, S.K., Ismail, A.L.S. and Rattan, S.S. 1976. New and interesting coprophilous fungi from Iraq. Nova Hedwigia, 28: 241–251.
Abdullah, S.K. 1982. Coprophilous mycoflora on different dung types in southern desert of Iraq. Sydowia, 35: 1–5.
Abdullah, S.K. and Rattan, S.S. 1978. *Zygopleurage, Tripterosporella* and *Podospora* (Sordariaceae: Pyrenomycetes) in Iraq. Mycotaxon, 7: 102–116.
Agnihothrudu, V. 1958. Notes on fungi from North-east India. I. A new genus of Tuberculariaceae. Mycologia, 50: 570–579.
Ahmad, S. 1949. A contribution to the fungus flora of Pakistan and India. Indian Phytopath., 2: 11–16.
Ahmed, S.I. and Asad, F. 1971. Coprophilous fungi of West Pakistan. IV. Pak. V. Sci. Inc. Res., 14: 504–506.
Ahmed, S.I. and Cain, R.F. 1972. Revision of the genera *Sporormia* and *Sporormiella*. Can. J. Bot., 50: 419–477.
Ahmed, S.I., Ismail, A.L.S. and Abdullah, S.K. 1970. Contribution to the fungi of Iraq. II (Coprophilous fungi). Bull. Biol. Res. Centre (Baghdad), 5: 1–16.
Ahmed, S.I., Ismail, A.L.S. and Abdullah, S.K. 1971. Contribution to the fungi of Iraq. III (Coprophilous fungi). Bull. Coll. Sci, Univ. Basrah, 2: 1–16.
AI-Musallam, A. 1980. Revision of the black *Aspergillus* species. Ph.D. Thesis. Utrecht, Netherlands: University of Utrecht.
Ajello, L. 1959. A new *Microsporum* and its occurrence in soil and on animals. Mycologia, 51: 69–76.
Ames, L.M. 1961. A monograph of the Chaetomiaceae. US Army Res. Dev. Ser., No. 2: 1–125.
Ames, L.M. 1963. A monograph of Chaetomiaceae. United States Army Research and Development Series, 2: 1–65.
Angel, K. and Wicklow, D.T. 1975. Relationships between coprophilous fungi and faecal substrates in a Colorado grassland. Mycologia, 67: 63–74.
Angel, K. and Wicklow, D.T. 1983. Coprophilous fungal communities in semi-arid to mesi grassland. Can. J. Bot., 61: 594–602.
Angel, K. and Wicklow, D.T. 1974. Decomposition of rabbit feces: An indication of the significance of the coprophilous microflora in energy flow schemes. J. Ecol., 62: 429–437.
Aolm, L. 1959. Some comments on the ascocarps of the Pyrenomycetes. Mycologia, 50: 777–788.
Arenal, F., Platas, G. and Pelaez, F. 2004. Variability of spore length in some species of the genus *Preussia* (*Sporormiella*). Mycotaxon, 89: 137–151.
Arx, J.A. von. 1973. Ostiolate and nonostiolate pyrenomycetes. Proc. Konin. Nederlandse Akad. van Wetenschappen, Series C, 76: 289–296.
Arx, J.A. von. 1975a. On *Thielavia angulata* and some recently described *Thielavia* species. Kavaka, 3: 33–36.
Arx, J.A. von. 1975b. On *Thielavia* and some similar genera of Ascomycetes. Stud. Mycol., 8: 1–31.
Arx, J.A. von. 1981a. The genera of fungi sporulating in pure culture, 3rd ed. J. Cramer.

Arx, J.A. von. 1981b. On *Monila sitophila* and some families of Ascomycetes. Sydowia, 34: 13–29.

Arx, J.A. von and Muller, E. 1975. A re-evaluation of the bitunicate Ascomycetes with keys to families and genera. Stud. Mycol., 9: 1–159.

Arx, J.A. von and Storm, P.K. 1967. Uber einige aus dem Erdboden isolierte, zu . *Sporormia, Preussia* und Westerdykella gehörende Ascomyceten. Persoonia, 4: 407–415.

Arx, J.A. von and Muller, E. 1954. Die Gattungen der amerosporen Pyrenomyceten. Beitr. Kryptogamenfl. Schweiz., 11: 1–434.

Arx, J.A. von, Dreyfuss, M. and Muller, E. 1984. A revaluation of *Chaetomium* and the Chaetomiaceae. Persoonia, 12: 169–179.

Arx, J.A. von, Figueras, M.J. and Guarro, J. 1988. Sordariaceous Ascomycetes without ascospore ejaculation. Nova Hedwigia, 94: 1–104.

Arx, J.A. von, Guarro, J. and Figueras, M.J. 1986. The ascomycete genus *Chaetomium*. Nova Hedwigia Beih, 84: 1–162.

Bachman, M. 1908. Discomycetes in the vicinity of Oxford, Ohio. Contributions from the Botanical Laboratory of Miami University I. Proc. Ohio State Acad. Sci., 5: 57–59.

Baijal, U. and Mehrotra, B.S. 1965. Species of *Mucor* from India—II. Sydowia, 19: 204–212.

Banier, G. 1910. Monographie des *Chaetomidium* et des *Chaetomium*. Bull. Soc. Mycolog. France, 25: 191–237.

Baral, H.O. 1994. Comments on Outline of the ascomycetes—1993. Syst. Ascomycetum, 13: 113–128.

Barnes, R.F.W., Asamoah-Boateng, B., Majam, J.N. and Agyei-Ohemeng, J. 1997. Rainfall and the population dynamics of elephant dung piles in the forests of southern Ghana. African J. Ecol., 35: 39–52.

Barr, M.E. 1970. Some amerosporous ascomycetes on Eriaceae and Empertraceae. Mycologia, 62: 377–394.

Barr, M.E. 2000. Notes on coprophilous bitunicate ascomycetes. Mycotaxon, 76: 105.

Barrasa, J.M., Lundqvist, N. and Moreno, G. 1986. Notes the genus *Sordaria* in Spain. Persoonia, 13: 83–88.

Barrasa, J.M., Solans, M.J. and Moreno, G. 1985. Stratt dissimills (Sordariales), Una nueva especie coprofila. Intl. J. Mycol. Lichenol., 2: 75–84.

Batra, L.R. and Batra, S.W.T. 1963. Indian Discomycetes. Univ. Kansas Sci. Bull., 44: 109–256.

Bell, A. 1975. Fungal succession on dung of the Brushtailed Opossum in New Zealand. New Zealand J. Bot., 13: 437–462.

Bell, A. 1983. Dung Fungi. An Illustrated Guide to Coprophilous Fungi in New Zealand. Victoria University Press, Private Bag Wellington, 88.

Bell, A. 2005. An Illustrated Guide to the Coprophilous Ascomycetes of Australia. CBS Biodiversity Series 3. Central bureau voor Schimmelcultures, Utrecht, The Netherlands.

Bell, A. and Mahoney, D.P. 1995. Coprophilous fungi in New Zealand. I. *Podospora* species with swollen agglutinated perithecial hairs. Mycologia, 87: 375–396.

Bell, A. and Mahoney, D.P. 1996. Perithecium development in *Podospora tetraspora* and *Podospora vesticola*. Mycologia, 88: 163–170.

Benjamin, R.K. 1963. Addenda to 'The merosporangiferous Mucorales'. Aliso, 5: 273–288.

Benjamin, R.K. 1949. Two new species representing a new genus of the Chaetomiaceae. Mycologia, 41: 347–354.

Benjamin, R.K. 1959. The merosporangiferous Mucorales. Aliso, 4: 321–433.

Benjamin, R.K. 1961. Addenda to 'The merosporangiferous Mucorales'. Aliso, 5: 11–19.

Benny, G.L. and Bernjamin, R.K. 1975. Observations on Thamnidiaceae (Mucorales). New taxa, new combination, and notes on selected species. Aliso, 8: 301–351.

Benny, G.L. and Bernjamin, R.K. 1976. Observations on Thamnidiaceae (Mucorales) II. *Chaetocladium, Cokeromyces, Mycotypha* and *Phascolomyces*. Aliso, 8: 391–424.

Bessette, A.E., Miller, O.K., Jr., Bessette, A.R. and Miller, H.H. 2000. Mushrooms of North America. Everbest Printing Company, Ltd., Hong Kong.

Bezerra, J.L. and Kimbrough, J.W. 1975. The genus *Lasiobolus* (Pezizales, Ascomycetes). Can. J. Bot., 53: 1206–1299.

Bills, Gerald F., Gloer, James B. and An, Zhiqiang. 2013. Coprophilous fungi: antibiotic discovery and functions in an underexplored arena of microbial defensive mutualism. Curr. Opinion Microbiol., 16: 1–17.

Bisby, G.R. 1938. The Fungi of Manitoba and Saskatchewan. Nat. Res. Coun. Can. Bull., 20: 1–189.

Bisby, G.R. 1943. *Stachybotrys*. Trans. Brit. Mycol. Soc., 26: 133–143.

Boedijn, K.B. 1927. Uber, *Rhopalomyces elegans* Corda. Ann. Mycol., 25: 161–166.

Booth, C. 1957. Studies of Pyrenomycetes. I. Four species of *Chaetosphaeria*, two with *Catenularia* conidia. II. *Melanopsamma pomiformis* and its *Stachybotrys* conidia. Commonw. Mycol. Inst. Mycol. Pap., 68: 1–27.

Booth, C. 1959. Studies of Pyrenomycetes: IV. *Nectria* (Part I). Commonw. Mycol. Inst. Mycol. Pap., 73: 1–115.

Booth, C. 1961. Studies of Pyrenomycetes: VI. *Thielavia*, with notes on some allied genera. Commonw. Mycol. Inst. Mycol. Pap., 83: 1–17.

Booth, C. 1971. The genus *Fusarium*. Commonw. Mycol. Inst., Kew.

Booth, T. 1982. Taxonomic notes on coprophilous fungi of the Areti: Churchill, Resolute Bay and Devon Island. Can. J. Bot., 60: 115–1125.

Breitenbach, B.K.J. and Kranzlin, F. 1984. Fungi of Switzerland. Vol. 1. Ascomycetes. Verlag Mykologia, Luzern.

Brewer, D., Duncan, J.M., Safe, S. and Taylor, A. 1972. Ovine ill-thrift in Nova Scotia. IV. The Survival at low oxygen partial pressure of fungi isolated from the contents of the ovine rumen. Can. J. Microbiol., 18: 119–128.

Brown, W.H. and Graff, P.W. 1931. Factors influencing succession on dung cultures. Phil. J. Sci., C. Bot., 8: 21–29.

Buchwald, N.F. 1958. Fungi Imperfecti of the Faeroes. F.H. Moller: Fungi of the Faeroes. Part II. 139–221.

Buller, A.H.R. 1934. The biology and taxonomy of *Pilobolus* X. *In*: Researches on Fungi, Vol. 6 p. 1–224. Longmans, Green & Co., London.

Cacialli, G., Caroti, V. and Doveri, F. 1996a. Contributo allo studio dei funghi fimicoli. II. Coprinaceae: *Panaeolus fimicola* (Fries) Quélet e *Panaeolus dunensis* Bon & Courtecuisse. Micologia Italiana, XXV, 2: 49–56.

Cacialli, G., Caroti, V. and Doveri, F. 1996b. Contributo allo studio dei funghi fimicoli. III. Basidiomycetes: *Coprinus marculentus* Britzelmayr e *Coprinus xenobius* Orton. Micologia Italiana XXV, 3: 69–73.

Cacialli, G., Caroti, V. and Doveri, F. 1996c. Contributo allo studio dei funghi fimicoli—VI. Ascomycetes: *Sordaria humana* (Fuckel) Winter e *Sporormiella minima* (Auerswald) Ahmed & Cain. Pagine di Micologia, 5: 40–49.

Cacialli, G., Caroti, V. and Doveri, F. 1996d. Contributo allo studio dei funghi fimicoli—VII. Pezizales: *Lasiobolus cuniculi* Velenovský e *Cheilymenia pulcherrima* (Crouan & Crouan) Boudier. Pagine Micologia, 5: 50–59.

Cacialli, G., Caroti, V. and Doveri, F. 1996e. Contributo allo studio dei funghi fimicoli—IX. Agaricales: *Coprinus miser* (P. Karst.) P. Karst. e *C. cothurnatus* Godey. Documents Mycologiques, Tome XXVI, fasc. 102: 59–66 + tav.

Cacialli, G., Caroti, V. and Doveri, F. 1996f. Contributo allo studio dei funghi fimicoli—X. *Conocybe siliginea* (Fries: Fries) Kühner, ss. str. Moser: tentativo di risolvere un rompicapo nomenclaturale. Boll. Svizzero Micologia, 74, 11: 219–229.

Cacialli, G., Caroti, V. and Doveri, F. 1996g. Contributo allo studio dei funghi fimicoli—XI. Agaricales: *Psilocybe semilanceata* (Fries: Fries) Kummer e *Pholiotina coprophila* (Kühner) Singer. Funghi e Ambiente, 72: 5–16.

Cacialli, G., Caroti, V. and Doveri, F. 1996h. Contributo allo studio dei funghi fimicoli—XII. Pyrenomycetes: *Sordaria fimicola* (Roberge) Ces. & De Not. e *Arnium arizonense* (Griffiths) Lundq. & Krug. Informatore Bot. Italiano, 28(1): 78–90.

Cacialli, G., Caroti, V. and Doveri, F. 1996i. Contribución al estudio de los hongos coprófilos en Italia. XVI. Discomycetes: el Género *Coprotus*. Bole. Soci. Micológica Madrid, 21: 113–125.
Cacialli, G., Caroti, V. and Doveri, F. 1996j. Contributo allo studio dei funghi fimicoli—XIX. Prima segnalazione in Italia di *Coprotus leucopocillum*. Micologia Vegeta. Mediter., 11(1): 3–6.
Cacialli, G., Caroti, V. and Doveri, F. 1996k. Contributo allo studio dei funghi fimicoli—XXI. Agaricales: *Psathyrella prona* (Fr.) Gillet e *Psilocybe inquilina* (Fr.: Fr.) Bres. Boll. Circolo Micologico "G. Carini", 31: 19–27.
Cacialli, G., Caroti, V. and Doveri, F. 1997a. Note su alcune Bolbitiaceae Singer. Contributo allo studio dei funghi fimicoli—I. Rivista Micologia, XL(2): 109–123.
Cacialli, G., Caroti, V. and Doveri, F. 1997b. Contributo allo studio dei funghi fimicoli—VIII. Discomycetes: *Ascobolus albidus* Crouan & Crouan e *Thecotheus cinereus* (Crouan & Crouan) Chenantais. Funghi e Ambiente, 73: 5–15.
Cacialli, G., Caroti, V. and Doveri, F. 1997c. Contributo allo studio dei funghi fimicoli—XIII. *Coprinus curtus* Kalchbrenner e Coprinus utrifer Watling. Pagine Micologia, 7: 61–71.
Cacialli, G., Caroti, V. and Doveri, F. 1997d. Notes on some *Podospora* with agglutinated hairs. Contribution to the study of fimicolous fungi—XVII. Documents Mycologiques, Tome XXVIII, fasc., 104: 41–52.
Cacialli, G., Caroti, V. and Doveri, F. 1997e. Contributo allo studio dei funghi fimicoli—XXIV. Dove trovare gli altri contributi. Pagine Micologia, 7: 75–80.
Cacialli, G., Caroti, V. and Doveri, F. 1997f. *Peziza perdicina* e *Peziza moravecii*: una sola entità? Funghi e ambiente, 74-75: 39–40.
Cacialli, G., Caroti, V. and Doveri, F. 1997g. *Laccaria affinis* fo. *macrocystidiata* con cheilocistidi di forma bizzarra. Pagine Micologia, 8: 65–68.
Cacialli, G., Caroti, V. and Doveri, F. 1997h. Contributo allo studio dei funghi fimicoli—V (ut XXIV) Il Genere *Lasiobolus* in Italia. Boll. Circolo Micologico "G.Carini", 34: 31–40.
Cacialli, G., Caroti, V. and Doveri, F. 1997i. Studio su *Laccaria laccata* var. *moelleri* e sul "complesso pallidifolia". Bolle. Gruppo Micologico G. Bresadola, Nuova Serie, 40(2-3): 123–132.
Cacialli, G., Caroti, V. and Doveri, F. 1998a. Mise à jour concernant les Ascobolaceae et les Thelebolaceae d'Italie. Contribution à l'étude des champignons fimicoles—XXV. Documents Mycologiques, Tome XXVIII, fasc., 109-110: 33–65.
Cacialli, G., Caroti, V. and Doveri, F. 1998b. Contributo allo studio dei funghi fimicoli—XIV. *Ascobolus mancus* (Rehm) van Brummelen, una specie rara appartenente alla sezione Dasyobolus (Saccardo) van Brummelen. Bolle. Svizzero Micologia, 76, 3: 130–135.
Cacialli, G., Caroti, V. and Doveri, F. 1999a. Sul genere *Pilobolus* Tode 1784. Contributo allo studio dei funghi fimicoli—XXIX. Funghi e Ambiente, 78–79, 1998–1999: 13–25.
Cacialli, G., Caroti, V. and Doveri, F. 1999b. Contributo allo studio dei funghi fimicoli—IV. *Didymium squamulosum* (Alb. & Schwein.) Fr. e *Perichaena corticalis* (Batsch) Rostaf. Acta Bot. Malacitana, 23: 201–204.
Cacialli, G., Caroti, V. and Doveri, F. 1999c. Contributo allo studio dei funghi fimicoli—XX. *Ascobolaceae*: *Saccobolus minimus* Velen. e *Saccobolus depauperatus* (Berk. & Br.) E.C. Hansen. BSM, 6: 292–301.
Cacialli, G., Caroti, V. and Doveri, F. 1999d. *Coprinus fimicoli* in Italia. Contributo allo studio dei funghi fimicoli—XXXI. *In*: Contributio ad Cognitionem Coprinorum. A.M.B. Fondazione Centro Studi Micologici.
Cacialli, G., Caroti, V. and Doveri, F. 2000a. Ammerkungen zu *Podospora setosa* (Winter) Niessl und taxonomische Überlegungen über die Lasiosphaeriaceae Nannf. emend. Lundqv. Beiträge zum Studium mistbewohnender Pilze—XVIII. Zeitschrift Mykolo., 66: 95–100.
Cacialli, G., Caroti, V. and Doveri, F. 2000b. *Poronia erici* Lohmeyer & Benkert 1988: primo ritrovamento italiano di una rara specie. Contributo allo studio dei funghi fimicoli—XXXV. Funghi e Ambiente, 83: 5–12.

Cacialli, G., Caroti, V. and Doveri, F. 2002. Aggiornamento della monografia Contributio ad Cognitionem Coprinorum. Mycol. Monten., 5: 121–129.
Cailleux, R. 1971. Recherches sur la mycoflore coprophile Centrafricaine. Les genres *Sordaria, Gelasinospora, Bombardia* (Biologie-Morphologie- Systematique); Ecologie. Bull. Soc. Mycol. Fr., 87: 461–626.
Cain, R.F. 1934. Studies of coprophilous Sphaeriales in Ontario. Univ. Toronto studies, Biol. Ser., 38: 1–126.
Cain, R.F. 1950. Studies of coprophilous Ascomycetes I. *Gelasinospora.* Can. J. Bot., 34: 566–575.
Cain, R.F. 1956a. Studies of coprophilous Ascomycetes II. *Phaeotrichum,* a new cleistocarpous genus in a new family, and its relationships. Can. J. Bot., 34: 575–687.
Cain, R.F. 1956b. Studies of coprophilous Ascomycetes IV. *Tripterospora,* a new cleistocarpous genus in a new family. Can. J. Bot., 34: 699–713.
Cain, R.F. 1957. Studies of the coprophilous Ascomycetes VI. Species from the Hudson Bayaria. Can. J. Bot., 35: 255–268.
Cain, R.F. 1961. Studies of the coprophilous Ascomycetes VII. *Preussia.* Can. J. Bot., 39: 1633–1699.
Cain, R.F. 1962. Studies of the coprophilous Ascomycetes VIII. New species of *Podospora.* Can. J. Bot., 40: 447–490.
Cain, R.F. and Farrow, W.M. 1956. Studies of the coprophilous Ascomycetes III. The genus *Triangularia.* Can. J. Bot., 34: 689–697.
Cannon, P.F. and Hawksworth, D.L. 1984. A revision of the genus *Neocosmospora* (Hypocreales). Trans. Brit. Mycol. Soc., 82: 673–688.
Caretta, G., Piontelli, E., Savino, E. and Bulgheroni, A. 1998. Some coprophilous fungi from Kenya. Mycopathologia, 142: 125–134.
Chivers, A.H. 1912. The Ascomycete Genus *Chaetomium* Arx, Guarro and Fugueras (eds.). Strauss offsetdruck gmbh, Hirschberg 2. 162.
Chivers, A.H. 1915. A monograph of the genus *Chaetomium* and *Ascotricha.* Memoires Torrey Bot. Club, 14: 155–240.
Clements, F.E. 1916. Plant Succession, an analysis of the development of vegetation.
Cribb, A.B. 1988. Fungi on kangaroo dung. *In*: Scott, G. (ed.). Lake Broadwater—The natural History of an Inland Lake and its Enviroments, Darling Downs Institute Press, Toowoomba, Queensland.
Cribb, A.B. 1989. Fungi on Wombat dung in Tasmania. Queensland Natural., 29: 114–115.
Cribb, A.B. 1991. Fungi on dung of an Emu. Queensland Natural., 30: 130–134.
Cribb, A.B. 1992. Fungi on dung of Carpet python. Queensland Natural., 30: 130–134.
Cribb, A.B. 1994. New recorded fungi on dung from Cape York Peeninsula, Queensland. Proc. Royal Soc. Queensland, 104: 19–24.
Cribb, A.B. 1996a. A species of *Xylaria* from Cattle dung in Queensland. Queensland Natural., 34: 22–24.
Cribb, A.B. 1996b. Two dung inhabiting fungi newly recorded from Queensland. Queensland Natural., 34: 62–63.
Cribb, A.B. 1997. Two coprophilous fungi on Koala dung. Queensland Natural., 35: 91.
Cribb, A.B. 1998. Ascomycetes fungi developing on macropad dung from the Musselbrook Reserve Scientific Study. Royal Geogra. Soc. Queensland, 67–69.
Cribb, A.B. 1999a. The fungus *Chaetomium murorum* on wombat dung. Queensland Natural., 37: 22–26.
Cribb, A.B. 1999b. The fungus *Chaetomium homopilatum* on dung of a carpet Python. Queensland Natural., 37: 55–56.
Dade, H.A. 1957. Coprophilous fungi. J. Quekett Microsco. Club Ser., 44: 396–406.
Denison, W.C. 1964. The genus *Cheilymenia* in North America. Mycologia, 56: 718–737.
Dennis, R.W.G. 1954. Some inoperculate discomycetes of tropical America. Kew Bull., 2: 289–348.
Dennis, R.W.G. 1968a. British Ascomycetes. J. Cramer, Liechtenstein.

Dennis, R.W.G. 1968b. British cupfungi and their allies. An Introduction to Ascomycota. J. Cramer, Liechtenstein.

Dennis, R.W.G. 1978. British Ascomycetes, 2nd edition. J. Cramer, Liechtenstein.

Dickinson, C.H. and Underhay, V.H.S. 1977. Growth of fungi in Cattle dung. Trans. Brit. Mycol. Soc., 69: 473–477.

Dickinson, C.H., Underhay, V.H.S. and Ross, V. 1981. Effect of season, soil fauna and water content on the decomposition of cattle dung pats. New Phytologist, 88: 129–141.

Dissing, H. 1989. Four New coprophilous species of *Ascobolus* and *Saccobolus* from Greenland (Pezizales). Opera Bota., 100: 43–50.

Dissing, H. 1992. Notes on the coprophilous Pyrenomycetes *Sporormia fimetaria*. Persoonia, 14: 389–394.

Doveri, F., Caroti, V. and Cacialli, G. 1997. An exceptional finding from horse dung: *Enterocarpus grenotii*. Contribution to the study of fimicolous fungi—XXVI. Bollettino del Gruppo Micologico G. Bresadola, Nuova Serie, 40(2-3): 187–190.

Doveri, F., Cacialli, G. and Caroti, V. 2000a. A guide to the classification of fimicolous pyrenomycetes ss, lato from Italy. Contribution to the study of fimicolous fungi XXXIII. pp. 603–705. *In*: Micologia 2000 A.B.M. Fondazione Centro Studi Micologici.

Doveri, F., Cacialli, G. and Caroti. V. 2000b. Guide pour identification des Pezizales fimicoles d'Italie. Contribution a I'etude des champignons fimicoles XXXII. Docum. Mycologiq., (117-118): 3–97.

Doveri, F. 2004. *Podospora alexandri*, una nuova specie fimicola dall'Italia. Rivista di Micologia, XLVII(3): 211–221.

Doveri, F. 2005. *Sporormiella hololasia*, a new hairy species from Italy. Rivis. Micol., XLVIII(1): 31–41.

Doveri, F., Granito and Lunghini. 2005. Nuovi ritrovamenti di *Coprinus* s.l. fimicoli in Italia. Rivis. Micol., XLVIII(4): 319–340.

Doveri, F. 2006. Nuove segnalazioni di *Onygenales* coprofile dall'Italia. Rivis. Micol., XLIX: 245–266.

Doveri, F. 2007. An updated key to coprophilous Pezizales and Thelebolales in Italy. Mycol. Monten., 10: 55–82.

Doveri, F. 2008a. Aggiornamento sul genere *Chaetomium* con descrizione di alcune specie coprofile, nuove per l'Italia-An update on the genus *Chaetomium* with descriptions of some coprophilous species, new to Italy. Pagine di Micologia, 29: 1–60.

Doveri, F. 2008b. A bibliography of *Podospora* and *Schizothecium*, a key to the species, and a description of *Podospora dasypogon* newly recorded from Italy. Pagine di Micologia, 29: 61–159.

Doveri, F. and Coué, B. 2008a. Une nouvelle variété de *Schizothecium* coprophile de France. A new variety of coprophilous *Schizothecium* from France. Documents Mycologiques, 34(135-136): 1–14.

Doveri, F. and Coué, B. 2008b. *Sporormiella minutisperma*, une nouvelle espèce coprophile, récoltéee en France. Bull. trim. Dauphiné-Savoie, 191: 39–44.

Doveri, F. 2010. Occurrence of coprophilous *Agaricales* in Italy, new records, and comparisons with their European and extraeuropean distribution. Mycosphere, 1: 103–140.

Doveri, F. 2010. A new variety of *Tripterosporella* from dung—an opportunity to recombine the cleistothecial pyrenomycete *Cercophora heterospora* into the genus *Tripterosporella*. Bull. Mycol. Bot. Dauphiné-Savoie, 196: 49–55.

Doveri, F., Sabrina Sarrocco, Susanna Pecchia, Maurizio Forti and Giovanni Vannacci. 2010. *Coprinellus mitrinodulisporus*, a new species from chamois dung. Mycotaxon, 114: 351–360.

Doveri, F. 2011. Addition to "Fungi Fimicoli Italici": An update on the occurrence of coprophilous Basidiomycetes and Ascomycetes in Italy with new records and descriptions. Mycosphere, 2: 331–427.

Doveri, F., Susanna Pecchia, Mariarosaria Vergara, Sabrina Sarrocco and Giovanni Vannacci. 2012. A comparative study and relationship with Onygenales of *Neogymnomyces virgineus*, a new keratinolytic species from dung. Fung. Divers., 52: 13–34.

Doveri, F. 2012a. An exceptional find on rabbit dung from Italy: third record worldwide of *Ascobolus perforatus*. Mycosphere, Doi 10.5943/mycosphere/3/1/3/ 29.

Doveri, F. 2012b. Coprophilous discomycetes from the Tuscan archipelago (Italy). Description of two rare species and a new *Trichobolus*. Mycosphere, 3: 503–522, Doi 10.5943/mycosphere/3/4/13.

Doveri, F. 2013. An additional update on the genus *Chaetomium* with descriptions of two coprophilous species, new to Italy. Mycosphere, 4: 820–846, Doi 10.5943/mycosphere/4/4/17.

Doveri, F. 2014. An update on the genera *Ascobolus* and *Saccobolus* with keys and descriptions of three coprophilous species, new to Italy. Mycosphere, 5: 86–135, Doi 10.5943/mycosphere/5/1/4.

Doveri, F. and Sarrocco, S. 2013. *Sporormiella octomegaspora*, a new hairy species with eight–celled ascospores from Spain. Mycotaxon, 123: 129–140.

Doveri, F., Sarrocco S. and Vannacci, G. 2013. Studies on three rare coprophilous plectomycetes from Italy. Mycotaxon, 124: 279–300.

Drechsler, C. 1937. Some Hyphomycetes that prey on free-living terricolous nematodes. Mycologia, 29: 446–552.

Drechsler, C. 1940. Three new Hyphomycetes preying on free-living terricolous nematodes. Mycologia, 32: 448–470.

Drechsler, C. 1941. Some Hyphomycetes parasitic on free-living terricolous nematodes. Phytopathol., 31: 773–802.

Drechsler, C. 1946. A clamp-bearing fungus parasitic and predaceous on nematodes. Mycologia, 38: 1–23.

Drechsler, C. 1950. Several species of *Dactylella* and *Dactylaria* that capture free living nematodes. Mycologia, 42: 1–79.

Drechsler, C. 1957. A nematode-capturing Phycomycete forming chlamydospores terminally on lateral branches. Mycologia, 49: 364–388.

Dring, V.J. 1959. *Phymatotrichum fimicola* sp. nov. A coprophilous Hyphomycetes. Trans. Br. Mycol. Soc., 42: 06–408.

Duddington, C.L. 1951. The ecology of Predacious fungi. Trans. Br. Mycol. Soc., 34: 322–331.

Ebersohn, C. and Eicker, A. 1992. Coprophilous fungal species composition and species diversity on various dung substrates of African game animals. Bot. Bull. Acad. Sin., 33: 85–95.

Ebersohn, C. and Eicker, A. 1997. Determination of the coprophilous fungal fruit body successional phases and the delimitation of species association classes on dung substrates of African game animals. Bot. Bull. Acad. Sin., 38: 183–190.

Ellis, J.B. and Everhart, B.M. 1892. The North American Pyrenomycetes. Publ. by the authors. Newfield, New Jersey. 793 pp. + 41 plates.

Ellis, J.J. 1963. A study of *Rhopalomyces elegans* in pure culture. Mycologia, 55: 183–198.

Ellis, M.B. 1966. Dematiaceous Hyphomycetes.VII: *Curvularia*, *Brachysporium*, etc. Mycol. Pap. Commonw. Mycol. Inst., 106: 57.

Ellis, M.B. 1971. 'Dematiaceous Hyphomycetes'. Commonw. Mycol. Inst., Kew.

Ellis, M.B. 1976. 'More Dematiaceous Hyphomycetes'. Commonw. Mycol. Inst., Kew.

Ellis, M.B. 1988. 'Microfungi on Miscellaneous Substrates'. pp. 100–158. *In*: Ellis, M.B. and Ellis, J.P. (eds.). An Identification Handbook, University Press, Cambridge.

Ellis, M.B. and Ellis, J.P. 1988. Microfungi on Miscellaneous Substrates. Croom Helm, London.

Elshafie, A.E. 2005. Coprophilous mycobiota of Oman. Mycotaxon, 93: 355–357.

Eriksson, O.E. and Hawksworth, D.L. 1993. Outline of the ascomycetes. Syst. Ascomyc., 12: 51–257.

Fakirova, V.I. 1969. Research into the coprophilous Ascomycetes in Bulgaria III. (English translation), Bulgarische Akad. Wisseneschaften, 19: 199–210.

Farouq, Ahmadu Ali, Abdullah, Dzulkefly Kuang, Foo Hooi-Ling, and Abdullah Norhafizah 2012. Isolation and characterization of coprophilous cellulolytic fungi from Asian elephant (*Elephas maximus*) dung. J. boil., Agri Healthcare, 2: 44–51.

Fennell, D.I. and Raper, K.B. 1955. New species and varieties of *Aspergillus*. Mycologia, 47: 68–69.

Furuya, K. and Udagawa, S. 1972. Coprophilous Pyrenomycetes from Japan I. J. Gen. Appl. Microbiol., 18: 433–454.

Furuya, K. and Udagawa, S. 1973. Coprophilous Ascomycetes from Japan III. Trans. Mycol. Soc. Japan, 14: 7–30.

Furuya, K. and Udagawa, S. 1976. Coprophilous Pyrenomycetes from Japan IV. Trans. Mycol. Soc. Japan, 17: 248–261.

Gams, W. and Holubová-Jechová, V. 1976. *Chloridium* and some other dematiaceous Hyphomycetes growing on decaying wood. Stud. Mycol., 13: 1–99.

Gams, W., Fisher, P.J. and Webster, J. 1984. *Onychophora*, a new genus of phialidic hyphomycetes from dung. Trans. Brit. Mycol. Soc., 82: 174–177.

Gamundi, I.J. 1972. Discomycetes de Tierra del Fuego I. Especies nuevas o criticas del genero *Cheilymenia* (Humariaceae). Bolet. Socied. Argentina Botan., 14: 167–176.

Gene, J., Elshafie, A.I. and Guarro, J. 1993. Two coprophilous fungi Pezizales from the Sultanate of Oman. Mycotaxon, 46: 275–284.

Ginai, M.A. 1936. Further contribution to our knowledge of Indian coprophilous fungi. J. Indian Bot. Soc., 25: 269–284.

Griffin, D.M. 1977. Water potential and wood decay fungi. Ann. Rev. Phytopathol., 15: 319–329.

Grove, W.B. 1936. A Handsome Hyphomycete. Ann. Mycol., 34: 106–107.

Gupta, Awanish Kumar. 2010. Studies on fungi from the dung of some herbivorous inmates of Prince of Wales Zoological Garden, Lucknow. Ph. D. thesis, University of Lucknow, Lucknow, 178 p.

Hanlin, R.T. 1973. Key to the Families, Genera and Species of the Mucorales. Translated from German, Verlag Von J. Cramer, 3301, Lehre, Germany.

Hanlin, R.T. 1997. Illustrated Genera of Ascomycetes-II. The APS Press, USA. 258.

Hanlin, R.T. 1998. Illustrated Genera of Ascomycetes-I. The APS Press, USA. 263.

Harper, J.E. and Webster, J. 1964. An experimental analysis of the coprophilous fungus succession. Trans. Brit. Mycol. Soc., 47: 511–530.

Harrower, K.M. and Nagy, L.A. 1979. Effects of nutrients and water stress on growth and sporulation of coprophilous fungi. Trans. Brit. Mycol. Soc., 72: 459–462.

Helsel, E.D. and Wicklow, D.T. 1978. Arthropod colonization of pre-aged as compared to fresh faeces. Can. Entomol., 110: 217–222.

Helsel, E.D. and Wicklow, D.T. 1979. Decomposition of rabbit faeces: Role of the sciarid fly *Lycoriella mali* Fitch in energy transformations. Can. Entomol., 111: 213–217.

Herrera, José, Poundel, Ravin, Khidir and Hana, H. 2011. Molecular characterization of coprophilous fungal communities reveals sequences related to root-associated fungal endophytes. Microb. Ecol., 61: 239–244.

Hering, T.R. 1965. Succession of fungi in the litter of a Lake District oakwood. Trans. Brit. Mycol. Soc., 48: 391–408.

Hesseltine, C.W., Whitehill, A.R., Pidacks, C., Tenhagen, M., Bohonos, N., Hutchings, B.L. and Williams, J.H. 1953. Coprogen, a new growth factor in dung required by *Pilobolus*. Mycologia, 45: 7–19.

Ikediugwa, F.E.O. and Webster, J. 1970a. Antagonism between *Coprinus heptemerous* and other coprophilous fungi. Trans. Br. Mycol. Soc., 54: 181–204.

Ikediugwa, F.E.O. and Webster, J. 1970b. Hyphal interference in a range of coprophilous fungi. Trans. Br. Mycol. Soc., 54: 205–210.

Ikediugwa, F.E.O., Dennis, C. and Webster, J. 1970. Hyphal interference by *Peniophora gigantean* against *Heterobasidion annosum*. Trans. Br. Mycol. Soc., 54: 307–309.

Iyer, R., Ghosh, S.K. and Sarbhoy, A.K. 1973. Studies on coprophilous fungi—I. Proc. Indian Natl. Sci. Acad. Part B: Bio. Sci., 39: 199–202.

Jain, K. and Cain, R.F. 1973. *Mycoarctium*, a new coprophilous in Thelebolaceae. Can. J. Bot., 51: 305–307.

Jeng, R.S. and Krug, J.C. 1977. New records and new species of coprophilous Pezizales from Argentina and Venezuela. Can. J. Bot., 55: 2987–3000.

John, R. and Hessetline, C.W. 1952. A survey of Mucorales. Trans. N. Y. Acad. Sci., 14: 210–4.

Juniper, A.J. 1954. Some Predacious fungi occurring in dung. II. Trans. Br. Mycol. Soc., 37: 171–175.

Juniper, A.J. 1957. Dung as a source of predacious fungi. Trans. Br. Mycol. Soc., 40: 346–348.

Kar, A.K. and Pal, K.P. 1968. Some coprophilous Discomycetes of Eastern Himalaya (India). Mycologia, 60: 1086–1092.

Kaushal, S.C. and Thind, K.S. 1983. Western Himalayan species of *Ascobolus*. J. Indian Bot. Soc., 62: 16–24.

Kaushal, S.C., Kaushal, R. and Bhatt, R.P. 1985. Addtions to Indian dung fungi. J. Indian Bot. Soc., 66: 116–188.

Kendrick, B. 1992. The Fifth Kingdom. Second edition. Mycologue Publications, Canada. 406.

Khare, K.B. 1976. Two *Ascobolus* species from India. Curr. Sci., 45: 385–386.

Kiffer, E. and Morelet, M. 2000. The Deuteromycetes: Mitosporic Fungi Classification and Generic keys. Science Publishers, Inc. USA.

Kimbrough, J.W., Luckn-Allen, E.R. and Cain, R.F. 1972. North American species of *Coprotus* (Thelebolaceae: Pezizales). Can. J. Bot., 50: 957–971.

Klich, M.A. 1993. Morphological studies of *Aspergillus* section Versicolores and related species. Mycologia, 85: 100–107.

Klich, M.A. 2002a. Identification of common *Aspergillus* species. Centraalbureau voor Schimmelcultures, Netherlands.

Klich, M.A. 2002b. Biogeography of *Aspergillus* species in soil and litter. Mycologia, 94: 21–27.

Klich, M.A. and Pitt, J.I. 1985. The theory and practice of distinguishing species of the *Aspergillus flavus* group. pp. 211–220. *In*: Samson, R.A. and Pitt, J.I. (eds.). Advances in Penicillium and Aspergillus Systematics. Plenum Press, New York.

Kozakiewicz, Z. 1989. *Aspergillus* species on stored products. Mycol. Pap., 161: 1–188.

Krug, J.C. 1971. Some new records of Ascomycetes from Scotland. Trans. Bot. Soc. Edinburgh, 41: 197–199.

Krug, J.C. and Jeng, R.S. 1995. A new coprophilous species of *Podosordaria* from Venezuela. Can. J. Bot., 73: 65–69.

Krug, J.C., Udagawa, S. and Eng, R.S. (1983). The genus *Apiosordaria*. Mycotaxon, 17: 533–549.

Krug, J.C., Benny, G.L. and Keller, H.W. 2004. Corophilous fungi. pp. 467–499. *In*: Mueller, G., Bills, G.F. and Foster, M.S. (eds.). Biodiversity of Fungi, Inventory and Monitoring Methods. Elsevier Academic Press, Amsterdam.

Kuthubutheen, A.J. and Webster, J. 1986a. Water availability and the coprophilous fungus succession. Trans. Br. Mycol. Soc., 86: 63–76.

Kuthubutheen, A.J. and Webster, J. 1986b. Effects of water availability on germination, growth and sporulation of coprophilous fungi. Trans. Br. Mycol. Soc., 86: 77–91.

Lange, M. and Hora, F.B. 1963. Collins Guide to Mushrooms and Toadstools. Collins, London.

Larsen, K. 1971. Danish endocoprophilous fungi, and their sequence of occurrence. Bota. Tidsskrif, 66: 1–32.

Liou, S.C. and Chen, Z.C. 1977. Preliminary studies on coprophilous discomycetes in Taiwan. Taiwania, 22: 44–58.

Lodha, B.C. 1962. Studies on coprophilous fungi. J. Indian Bot. Soc., 41: 121–140.
Lodha, B.C. 1963. Notes on two species of *Trichurus*. J. Indian Bot. Soc., 42: 135–142.
Lodha, B.C. 1964a. Studies on coprophilous fungi—I. *Chaetomium*. J. Indian Bot. Soc., 43: 121–140.
Lodha, B.C. 1964b. Studies on coprophilous fungi—II. Anto. von Leeuwen., 30: 163–167.
Lodha, B.C. 1971. Studies on coprophilous fungi—IV. Some cleistothecial ascomycetes. J. Indian Bot. Soc., 50: 196–208.
Lodha, B.C. 1974. Decomposition of digested litter. pp. 213–241. *In*: Dickison, C.H. and Pugh, G.J.F. (eds.). Biology of Plant Litter Decomposition, Vol. 1. Academic Press, London.
Lorenzo, L.M. 1989. Navedades para flora de pyrenomycetes "Sensulato" (Ascomycotina) coprofilos de la Argentinas Bolet. Socied. Argentina Botan., 26: 35–38.
Lorenzo, L.M. 1992. Contribution al estudio de Pyrenomycetes "Sensulato" (Ascomycotina) coprofilos de la Parque Nacional Nahuel Huapi (Argentina) III. Boletin de la Sociedad Argentina de Botanica, 28: 183–193.
Lundqvist, N. and Fakirova, V.I. 1973. *Cercophora brevifila*, a new species of coprophilous Ascomycetes. Compts rendus Acade. Bulgare Scien., 26: 1395–1398.
Lundqvist, N. 1960. Coprophilous Ascomycetes from northern Spain. Swensk Botan. Tidskrift, 54: 523–529.
Lundqvist, N. 1964a. *Fimetariella*, a new genus of coprophilous Pyrenomycetes. Botan. Notiser, 117: 228–248.
Lundqvist, N. 1964b. *Anopodium*, a new genus of coprophilous Pyrenomycetes with apically pedicellate spore. Botan. Notiser, 117(4): 355–365.
Lundqvist, N. 1967. On spore ornamentation in the Sordariaceae, exemplified by the new cleistocarpous genus *Copromyces*. Arkiy Botan., 2: 327–337.
Lussenhop, J. and Wicklow, D.T. 1985. Interaction of competing fungi with fly larvae. Microb. Ecol., 11: 175–182.
Lussenhop, J., Kumar, R., Wicklow, D.T. and Lloyd, J.E. 1980. Insect effects on bacteria and fungi in cattle dung. Oikos, 34: 54–58.
Lussenhop, J., Wicklow, D.T., Kumar, R. and Lloyd, J.E. 1982. Increasing the rate of cattle dung decomposition by nitrogen fertilization. J. Range Management, 35: 249–250.
Mahju, N.A. 1933. A contribution to our knowledge of Indian coprophilous fungi. J. Indian Bot. Soc., 12: 153–166.
Mains, E.B. 1958. North American entomogenous species of *Cordyceps*. Mycologia, 50: 169–222.
Manoch, L.C., Chana, P., Athipuayakom and Somrith, A. 2000. Noteworthy Ascomycota from Thailand, with special emphasis on Coprophilous Pyrenomycetes and Discomycetes. Abstracts. BMS/Tropical Mycology Symposium. Liverpool John Moores University, Liverpool. UK. 25–29 April.
Masunga, Gaseitsiwe S., Andresen, øystein, Taylor, Joanne E., Dhillon, Shivcharn. 2006. Elephant dung decomposition and coprophilous fungi in two habitats of semi-arid Botswana. Mycol. Res., 110: 1214–1226.
Massee, G. and Salmon, E.S. 1901. Records on coprophilous fungi II. Ann. Bot., 15 (old series): 313–357.
Massee, G. and Salmon, E.S. 1902a. Researchers on coprophilous fungi. Ann. Bot., 16 (old series): 57–93 (16: 62).
Massee, G. and Salmon, E.S. 1902b. Researches on coprophilous fungi from India. Curr. Sci., 37: 245–248.
McGranaghan, P., Davies, J.C., Griffith, G.W., Davies, D.R. and Theodorou, M.K. 1999. The survival of anaerobic fungi in cattle faeces. FEMS Microbiol. Ecol., 29: 293–300.
Meculloch, J.S. 1977. New species of nematophagous fungi from Queensland. Trans. Brit. Mycol. Soc., 68: 173–179.
Mehrotra, B.S. and Mehrotra, M.D. 1962. A peculiar new Hyphomycetes from Cow—dung. Sydowia, 16: 212–214.

Meyer, S.L. and Meyer, V.G. 1949. Some coprophilous Ascomycetes from Panama. Mycologia, 41: 594–600.

Minter, D.W. and Webster, J. 1983. *Wawelia octospora* sp. nov., Xerophilous and coprophilous member of Xyalariaceae. Trans. Brit. Mycol. Soc., 80: 370–373.

Mirza, J.H. 1963. Classification of coprophilous ascomycetes: The genus *Podospora*. Ph.D. thesis. University of Toronto.

Mirza, J.H. and Ahmed, S.I. 1970. A new species of *Podospora* from Pakistan. Mycologia, 62: 1003–1007.

Mirza, J.H. and Khan, R.S. 1979. Studies on coprophilous ascomycetes of Pakistn III. *Semidelitschia tetraspora* sp. nov. Pakistan J. Bot., 11: 99–101.

Mitchell, D.I. 1970. Fungus succession on dung of South African Ostrich and Angora Goat. J. S. Afr. Bot., 36: 191–198.

Moravec, J. 1968. Pripevek k pozhani operkulanich diskomycetu rodu *Cheilymenia* Boud. Ceska Mykologie, 22: 32–41.

Moravec, J. 1984. Two new species of *Coprobia* and taxonomic remarks on the genera *Cheilymenia* and *Coprobia* (Discomyeetes, Pezizales). Ceska Mykologie, 38: 146–155.

Moravec, J. 1988. *Cheilymenia fraudans* and remarks on genera *Cheilymenia* and *Coprobia*. Mycotaxon, 31: 483–489.

Moravec, J. 1989a. *Cheilymenia megaspore* comb. nov., a new combination in the genus *Cheilymenia* (Discomycetes, Pezizales, Pyronemataceae). Mycotaxon, 35: 65–69.

Moravec, J. 1989b. A taxonomic revision of the genus *Cheilymenia* l. Species close to *Cheilymenia rubra*. Mycotaxon, 36: 169–186.

Moravec, J. 1990a. A taxonomic revision of the genus *Cheilymenia* 2. *Cheilymenia asteropila* spec. nov. and *C. gemella* comb. nov. Mycotaxon, 37: 463–470.

Moravec, J. 1990b. Taxonomic revision of the genus *Cheilymenia* 3. A new generic and infrageneric classification of *Cheilymenia* in a new emendation. Mycotaxon, 38: 459–484.

Moravec, J. 1993. Taxonomic revision of the genus *Cheilymenia* 5. The section *Cheilymenia*. Ceska Mykologie, 47: 7–37.

Morrison, F.B. 1959. Feeds and Feeding. Morrison Publishing Co. Clinton, Iowa.

Mungai, P.G., Njogu, J.G., Chukeatirote, E. and Hyde, K.D. 2012. Studies of coprophilous ascomycetes in Kenya—*Ascobolus* species from wildlife dung. Curr. Res. Environ. Appl. Mycol., 2: 1–16; DOI 10.5943/cream/2/1/1.

Mungai, P., Hyde, K.D., Cai, L., Njogu, J. and Chukeatirote, E. 2011. Coprophilous ascomycetes of northern Thailand. Curr. Res. Environ. Appl. Mycol., 1: 135–159, DOI 10.5943/cream/1/2/2.

Muroi, T. and Udagawa, S. 1984. Some coprophilous Ascomycetes from Chile. pp. 161–167. *In*: Inoue, H. (ed.). Studies on Cryptogams in southern Chile.

Nag Raj, T.R. 1979. Miscellaneous microfungi III. Can J. Bot., 57: 2489–2496.

Nagy, L.A. and Harrower, K.M. 1979. Analysis of two southern Hemisphere coprophilous fungus succession. Trans. Brit. Mycol. Soc., 72: 69–74.

Narendra, D.V. 1973a. Studies into coprophilous fungi of Maharashtra. Two rare and interesting ascomycetes. J. Univ. Bombay Sci., Vol. XLII. No. 69.

Narendra, D.V. 1973b. Studies into coprophilous fungi of Maharashtra III. *Sporormia* and *Sporomeilla*. Two rare ascomycetes from sheep dung. Nova Hedwigia, 24: 481–486.

Narendra, D.V. 1974. Studies of coprophilous fungi of Maharashtra—I. J. Biol. Sci., 17: 103–107.

Narendra, D.V. and Rao, V.G. 1976. Studies into coprophilous fungi of Maharashtra (India)—V. Nova Hedwigia, 27: 631–645.

Nickolson, P.B., Bocock, K.L. and Heal, O.W. 1966. Studies on the decomposition of the faecal pellets of a millipede (*Glomeris marginata* Villers). J. Ecol., 54: 755–766.

Nusrath, M. 1977. Studies on the ecological distribution of coprophilous fungi. Geobios, 54: 202–204.

Obrist, W. 1961. The genus *Ascodesmis*. Can. J. Bot., 39: 943–953.

Odum, E.P. 1963. Ecology: Holt Reinhart and Winston Inc.

Otani, Y. and Kanzawa, S. 1970. Notes on coprophilous Discomycetes in Japan II. Trans. Mycol. Soc. Japan, 24: 43–48.

Outlemans, C.A. 1882. Notiz uber einige eue Fungi Coprophili. Nova Hedwigia, 21: 161–166.

Pandey, Shubha 2009. Studies on fungi growing on the Dung of some herbivores. Ph. D. thesis, University of Lucknow, Lucknow, 207.

Parle, J.N. 1963. A microbiological study of earthworm casts. J. Gen. Microbiol., 31: 13–22.

Petch, T. 1930. Notes on entomogenous fungi. Trans. Brit. Mycol. Soc., 16: 55–75.

Pfister, D.H. 1994. *Orbillia fimicola*, a nematophagous discomycete and its *Arthrobotrys* anamorph. Mycologia, 86: 451–453.

Piontelli, E., Toro Santa-Maria, M.A. and Caretta, G. 1981. Coprophilous fungi of horse. Mycopathologia, 74: 89–105.

Prokhorov, V.P. 1994. The key to the genera of coprophilous discomycetes. Micol. Fitopatol., 28: 20–23.

Rai, J.N., Mukerji, K.G. and Tewari, J.P. 1961. A new *Helicostylum* from Indian soils. Can. J. Bot., 39: 1281–1285.

Raper, K.B. and Fennell, D.I. 1977. The Genus Aspergillus. Williams & Wilkins, Baltimore, USA.

Richardson, M.J. 1972. Coprophilous Ascomycetes on different dung types. Trans. Brit. Mycol. Soc., 58: 37–48.

Richardson, M.J. 1998a. New and interesting records of coprophilous fungi. Bot. J. Scotl., 50: 161–175.

Richardson, M.J. 1998b. *Coniochaeta polymegasperma* and *Delitschia trichodelitschiodes*, two new coprophilous asecomycetes. Mycol. Res., 102: 1038–1040.

Richardson, M.J. 2001a. Diversity and occurrence of coprophilous fungi. Mycol. Res., 105: 387–402.

Richardson, M.J. 2001b. Coprophilous fungi from Brazil. Brazilian Arch. Biol. Technol., 44: 283–389.

Richardson, M.J. 2002. The Coprophilous succession. Fung. Diver., 10: 101–111.

Richardson, M.J. 2004a. Coprophilous fungi from Iceland. Acta Bot. Isl., 14: 77–103.

Richardson, M.J. 2004b. Coprophilous fungi from Morocco. Bot. J. Scotl., 56: 147–162.

Richardson, M.J. 2005. Coprophilous fungi from the Faroe Islands. Fróöskaparrit, 53: 67–81.

Richardson, M.J. 2006. A new species of *Coniochaeta* from Perthshire. Bot. J. Scot., 58: 105–107.

Richardson, M.J. 2007a. The distribution and occurrence of coprophilous Ascobolaceae. Mycol. Montenegrena, 10: 211–277.

Richardson, M.J. 2007b [2006]. New records of fungi from Orkney and Shetland. Bot. J. Scotland, 58: 93–104.

Richardson, M.J. 2008a. Records of French coprophilous fungi. Cryptogamie Mycol., 29: 157–177.

Richardson, M.J. 2008b. Records of coprophilous fungi from the Lesser Antilles and Puerto Rico. Caribb. J. Sci., 44: 206–214.

Richardson, M.J. 2011. Additions to the Coprophilous Mycota of Iceland. Acta Bot. Isl., 15: 23–49.

Richardson, M.J. and Watling, R. 1968. Keys to Fungi on dung. Bull. Br. Mycol. Soc., 2(1): 18–43.

Richardson, M.J. and Watling, R. 1969. Keys to Fungi on dung. Bull. Br. Mycol. Soc., 3(2): 86–88, 121–124.

Richardson, M.J. and Watling, R. 1997. Keys to Fungi on Dung. Stourbridge, UK: Brit. Mycol. Soc., London.

Ridderbusch, D.C., Weber, R.W.S., Anke, T. and Sterner, O. 2004. Tulasnein and podospirone from the coprophilous xylariaceous fungus *Podosordaria tulasnei*. Z. Naturforsch, 59(C): 379–383.

Safar, H.M. and Cooke, R.C. 1998a. Exploitation of faecal resource units by coprophilous Ascomycotina. Trans. Brit. Mycol. Soc., 90: 593–599.
Safar, H.M. and Cooke, R.C. 1998b. Interaction between bacteria and coprophilous Ascomycotina and a *Coprinus* species on agar and in copromes. Trans. Brit. Mycol. Soc., 91: 73–80.
Samson, R.A. 1979. A compilation of the *Aspergilli* described since 1965. Studies in Mycology, 18: 1–38.
Saniel, L.S. and Alma-Cayabyab, V. 1974. Coprophilous fungi. Phillipp. Soc. Microbiol., 3: 126 (abstr.).
Sanwal, B.D. 1953a. Contribution towards our knowledge of the Indian Discomycetes-I. Operculate Discomycetes. Sydowia, 7: 191–199.
Sanwal, B.D. 1953b. Contribution towards our knowledge of the Indian Discomycetes II. Two new Operculate Discomycetes. Sydowia, 7: 200–205.
Sarbhoy, A.K., Agarwal, D.K. and Varshney, J.L. 1986. Fungi of India-1977–1981. Associated Publishing Company, New Delhi, India.
Saxena, A.S. and Mukerji, K.G. 1973. Fungi of Delhi—XVII. Three unrecorded coprophilous Ascomycetes. Cska. Mykol., 27: 165–168.
Seifert, K., Kendrick, B. and Murase, G. 1983. A key to Hyphomycetes on dung. University of Waterloo Biology Series No. 27. Ontario, Canada, 62: 1–59.
Seth, H.K. 1968. Coprophillic Ascomycota from Germany. Nova Hedwigia, 16: 495–499.
Seth, H.K. 1995. Corophilic fungi associated with the seeds of the wild plants and their possible role in biological control. J. Indian Bot. Soc., 74: 355–356.
Shoemaker, R.A. and Kokko, E.G. 1977. *Trichurus spiralis*. Fungi Canadenses: 100.
Singh, N. and Webster, J. 1972. Effect of coprophilous species of *Mucor* and bacteria on sporangial production of *Pilobolus*. Trans. Brit. Mycol. Soc., 59: 43–49.
Singh, N. and Mukerji, K.G. 1979. Studies on coprophilous fungi. I. Some rare records. J. Indian Bot. Soc., 58: 163–167.
Singh, N. and Webster, J. 1973. Antagonism between *Stilbella erythrocephala* and other coprophilous fungi. Trans. Brit. Mycol. Soc., 61: 489–495.
Singh, N. and Webster, J. 1976. Effect of dung extracts on the fruiting of *Pilobolus* species. Trans. Brit. Mycol. Soc., 67: 377–379.
Soman, A.G., Gloer, J.B., Koster, B. and Malloch, D. 1999. Sporovexins A-C and a new preussomerin analog: Antibacterial and antifungal metabolites from the coprophilous fungus *Sporormiella vexans*. J. Nat. Prod., 62: 659–661.
Speare, A.T. 1920. On certain entomogenous fungi. Mycologia, 12: 62–76.
Spooner, B.M. and Butterfill, G.B. 1999. Coprophilous discomycetes from the Arores. Kew Bull., 54: 541–560.
Subramanian, C.V. and Chandrashekara, K.V. 1977. *Beejasamuha* and *Sutravarana*, two new genera of coprophilous Hyphomycetes. Can. J. Bot., 55: 245–253.
Subramanian, C.V. and Lodha, B.C. 1964. Four New Coprophilous Hyphomycetes. Anton. von Leeuwen., 30: 317–330.
Subramanian, C.V. and Lodha, B.C. 1968. Two interesting coprophilous fungi from India. Curr. Sci., 37: 245–248.
Subramanian, C.V. and Lodha, B.C. 1975. A study of fungus flora developing on steamed and unsteamed dung. Kavaka, 3: 135–142.
Swofford, D.L. 1993. Two interesting coprophilous fungi from India. Curr. Sci., 37: 245–248.
Sydow, H.P. and Butler, E.J. 1911. Fungi Indiae Orientalis Pars. III. Ann. Mycol., 9: 372–421.
Tandon, R.N. 1968. Mucorales of India. United India Press New Delhi.
Thind, K.S. and Batra, L.R. 1957. The Pezizaceae of the Mussorie hills-IV. J. Indian Bot. Soc., 36: 428–438.
Thind, K.S. and Warattch, K.S. 1971. The Pezizales of India XI. Indian J. Mycol. Pl. Pathol., 1: 36–50.
Trappe, J.M. 1976. Notes on Japanese hypogeous Ascomycetes. Trans. Mycol. Soc. Japan, 17: 209–217.

Tubaki, K. 1954. Studies on the Japanese Hyphomycetes. (I) coprophilous group. Nagaoa, 4: 1–20.

Udagawa, S. 1980. New or noteworthy Ascomycetes from Southeast Asian soil l. Trans. Mycol. Soc. Japan, 21: 17–34.

Udagawa, S. and Muroi, T. 1979. Coprophilous pyrenomycetes from Japan V. Trans. Mycol. Soc. Japan, 20: 453–468.

Udagawa, S. and Takada, M. 1971. Soil and coprophilous microfungi. Bull. Nat. Sci. Mus. (Tokyo), 14: 501–515.

Uljé, C.B., Doveri, F. and Noordeloos, M.E. 2000. Additions to *Coprinus* subsection Lanatuli. Persoonia, 17: 465–471.

Uljé, C.B., Gennari, A., Doveri, F., Cacialli, G. and Caroti, V. 1998. First report of *Coprinus spadiceisporus* in Europe. Persoonia, 16: 537–540.

Ulloa, M. and Hanlin, R.T. 2000. Illustrated Dictionary of Mycology. The American Phytopathological Society, St. Paul, Minnesota.

Underhay, V.H.S. and Dickinson, C.H. 1978. Water, mineral and energy fluctuations in decomposing cattle dung pats. J. Brit. Grassland Soc., 33: 189–196.

Upadhyay, H.P. 1973. *Helicostylum* and *Thaminostylum* (Mucorales). Mycologia, 65: 733–751.

Wang, Y.Z. 1993. Notes on coprophilous Discomycetes from Taiwan. I. Bull. Nat. Muse. Natu. Sci., 4: 113–123.

Wang, Y.Z. 1995. Notes on coprophilous Discomycetes from Taiwan. II. Bull. Nat. Muse. Natu. Sci., 5: 147–152.

Wang, Y.Z. 1996. Notes on coprophilous Discomycetes from Taiwan. III. Bull. Nat. Muse. Natu. Sci., 7: 131–136.

Wang, Y.Z. 1999. The coprophilous Discomycetes of Taiwan. Bull. Nat. Muse. Natu. Sci., 12: 49–74.

Waritch, K.S. 1976. A contribution to the knowledge of coprophilous Pezizales of India. Sydowia, 29: 1–9.

Watling, R. 1963. The fungus succession on hawk pellets. Trans. Br. Mycol. Soc., 46: 81–90.

Watling, R. and Richardson, M.J. 2010. Coprophilous fungi of the Falkland islands. Edinburgh J. Bot., 67: 399–423.

Weber, R.W.S. and Webster, J. 1997. The coprophilous fungus *Sphaeronaemella fimicola*, a facultative mycoparasite. Mycologist, 11: 50–51.

Weber, R.W.S. and Webster, J. 1998. Stimulation of growth and reproduction in *Sphaeronaemella fimicola* by other coprophilous fungi. Mycol. Res., 102: 1055–1061.

Weber, R.W.S., Meffert, A., Anke, H. and Sterner, O. 2005. Production of sordarin and related metabolites by the coprophilous fungus *Podospora pleiospora* in submerged culture and in its natural substrate. Mycol Res., 109: 619–626.

Webster, J. 2000. Some Adavances in Fungal Ecology Over the Past Fifty Years. pp. 1–11. *In*: Bhat, D.J. and Raghukumar, S. (eds.). Ecology of Fungi, Goa University, Goa, India.

Webster, J. 1970. Coprophilous fungi: Presidential address. Trans. Brit. Mycol. Soc., 54: 161–180.

Webster, J. and Weber, E. 2000. Rhizomorphs and perithecial stromata of *Podosordaria tulasnei* (Xylariaceae). Mycologia, 14: 41–44.

Webster, J., Henrici, A. and Spooner, B. 1998. *Orbilia fimicoloides* sp. nov., the teleomorph of *Dactylella* cf *oxyspora*. Mycol. Res., 102: 99–102.

Wicklow, D.I. and Moore, V. 1974. Effect of incubation temperatue on the coprophilous fungus succession. Trans. Brit. Mycol. Soc., 62: 411–415.

Wicklow, D.T. 1981. The coprophilous fungal community: A mycological system for examining ecological ideas. pp. 47–76. *In*: Wicklow, D.T. and Carroll, G.C. (eds.). The Fungal Community—Its Organization and Role in the Ecosystem. Marcel Dekker, New York.

Wicklow, D.T. 1985. *Aspergillus leporis* sclerotia form on rabbit dung. Mycologia, 77: 531–534.

Wicklow, D.T. 1992. The coprophilous fungal community: An experimental system. pp. 715–728. *In*: Carroll, G.C. and Wicklow, D.T. (eds.). The Fungal Community—Its Organization and Role in the Ecosystem, 2nd ed., Marcel Dekker, New York.

Wicklow, D.T., Angel, K. and Lussenhop, J. 1980. Fungal community expression a lagomorph versus ruminat feces. Mycologia, 72: 160–166.

Wicklow, D.T. and Angel, K. 1983. Some reproductive characteristics of coprophilous ascomycetes in three prairie ecosystems. Mycologia, 75: 1070–1073.

Wicklow, D.T. and Angel, K. 1974. A preliminary survey of the coprophilous fungi for a semi-arid grassland in Colorado. US/IBP Tech. Rep. No. 259. Colorado State University, Fort Collins, CO.

Wicklow, D.T. and Hirschfield, B.J. 1979. Evidence of a competitive hierarchy among coprophilous fungal populations. Can. J. Microbiol., 25: 855–858.

Wicklow, D.T. and Malloch, D. 1971. Studies in the genus *Thelebolus*. Temperature optima for growth and ascocarp development. Mycologia, 63: 118–131.

Wicklow, D.T. and Moore, V. 1974. The effect of inoculation temperature on the coprophilous fungal succession. Trans. Brit. Mycol. Soc., 62: 411–415.

Wicklow, D.T. and Yocom, D.H. 1981. Fungal species numbers and decomposition of rabbit faeces. Trans. Brit. Mycol. Soc., 76: 29–32.

Wicklow, D.T. and Yocom, D.H. 1982. Effect of larval grazing by *Lycoriella mali* (Diptera: Sciaridae) on the species abundance of coprophilous fungi. Trans. Brit. Mycol. Soc., 78: 29–32.

Wicklow, D.T. and Zak, J.C. 1983. Viable grass seeds in herbivore dung from a semi-arid grassland. Grass Forage Sci., 38: 25–26.

Wicklow, D.T., Detroy, R.W. and Adams, S. 1980. Differential modification of the lignin and cellulose components in wheat straw by fungal colonists of ruminant dung: Ecological implications. Mycologia, 72: 1065–1076.

Wicklow, D.T., Detroy, R.W. and Jesse, B.A. 1980. Decomposition of lignocellulose by *Cyathus stercoreus* (Schw.) De Toni NRRL 6473, a "white-rot" fungus from cattle dung. Appl. Environ. Microbiol., 40: 169–170.

Wicklow, D.T., Kumar, R. and Lloyd, J.E. 1984. Germination of blue gramma seeds buried by dung beetles. Environ. Entomol., 13: 878–991.

Wicklow, D.T., Langie, R., Crabtree, S. and Detroy, R.W. 1984. Degradation of lignocellulose in wheat straw versus hardwood by *Cyathus* and related species (Nidulariaceae). Can. J. Microbiol., 30: 632–636.

Wood, F.H. 1972. Nematode-trapping fungi from a tussock grassland soil in New Zealand. New Zealand J. Bot., 11: 231–240.

Wood, S.N. and Cooke, R.C. 1984. Use of semi-natural resource units in experimental studies on coprophilous succession. Trans. Brit. Mycol. Soc., 82: 337–339.

Wood, S.N. and Cooke, R.C. 1987. Nutritional competence of *Pilaira anomala* in relation to exploitation of faecal resource. Trans. Brit. Mycol. Soc., 88: 247–255.

Yang, B.Y. 1972. A study on a coprophilous fungus belonging to the genus *Isaria* from Taiwan, Taiwania, 17: 182–189.

Yao, Y.J. and Spooner, B.M. 1996. Notes on British species of *Cheilymenia*. Mycol. Res., 100: 361–367.

Yocom, D.H. and Wicklow, D.T. 1980. Community differentiation along a dune succession: an experimental approach with coprophilous fungi. Ecology, 61: 868–880.

CHAPTER 10

Ubiquitous Occurrence of Thermophilic Molds in Various Substrates

Bijender Singh[1] and *T. Satyanarayana*[2,*]

ABSTRACT

Thermophilic molds have been isolated from various natural habitats including composts, wood chip piles, nesting material of birds and other animals, municipal refuse and several others, and therefore, are ubiquitous in their occurrence and worldwide in their distribution. The warm, humid and aerobic environments in decomposing organic materials provide the basic physiological condition that supports the growth of thermophilic molds. In the natural habitats, thermophilic molds occur as resting propagules or metabolically active vegetative mycelium. The molds that are capable of optimal growth at or beyond 40°C are defined as thermophilic molds. Thermophilic microbes form a diverse group of microorganisms found in various natural and man-made habitats, and among these, some fungi grow at elevated temperatures up to 60°C. This chapter focuses on the wide spread occurrence of thermophilic fungi in various habitats and substrates on the Earth.

[1] Department of Microbiology, Maharshi Dayanand University, Rohtak-124001, Haryana.
[2] Department of Microbiology, University of Delhi South Campus, Benito Juarez Road, New Delhi-110021, India.
* Corresponding author: tsnarayana@gmail.com

Introduction

Temperature is one of the most important ecological factors that affect microbial activities and their distribution (Johri et al., 1999; Maheshwari et al., 2000). Microbial species exist in a great variety of environments with extremes of temperature, pH, chemical contents and/or pressure. This occurrence is due to their genetic and/or physiological adaptations (Cooney and Emerson, 1964; Dix and Webester, 1995; Aguilar, 1996; Stetter, 1999; Johri et al., 1999; Maheshwari et al., 2000; Satyanarayana and Singh, 2004). Of the three domains of life, most of the thermophilic species that have been described belong to Archaea and Eubacteria (Barns et al., 1996). The maximum growth temperature limit for Eukaryota has been recorded as 62ºC (Tansey and Brock, 1978). The fungi that are capable of growth at or above 50ºC and fail to grow at or below 20ºC are considered thermophilic (Cooney and Emerson, 1964; Johri et al., 1999). There are fewer than 50 species of thermophilic fungi which thrive at relatively elevated temperatures (Mouchacca, 1997). These are common in soils and in habitats wherever organic matter heats up. Thermophilic fungi have been isolated from manure, compost, industrial coal mine soils, beach sands, nuclear reactor effluents, Dead Sea valley soils, and desert soils of Saudi Arabia (Redman et al., 1999). In these habitats, thermophiles may occur either as resting propagules or as active mycelia depending on the availability of nutrients and favorable environmental conditions. Generally, there is an inverse relationship between biological diversity and the adaptation required to survive in a specific habitat. Contrary to possible expectations, thermophiles are more frequently isolated from the temperate than the tropical soils (Ellis and Keans, 1981).

Thermophilic fungi are a small assemblage in Eukaryota, which have evolved strategies for growing at elevated temperatures of up to 60 to 62°C. During the last 50–60 years, many species of thermophilic fungi sporulating at 45°C have been reported. The species included in this account are only those which are thermophilic as defined by Cooney and Emerson (1964), "that thermophilic fungi are those capable of growth at or above 50°C and a minimum temperature for growth at or above 20°C". This chapter presents a comprehensive account on the occurrence and distribution of thermophilic fungi in various habitats and substrates. An attempt has also been made to include the methods for the isolation and cultivation of this group of fungi. Much is known about the occurrence of thermophilic fungi from various types of soils and in habitats where decomposition of plant material takes place (Singh and Satyanarayana, 2009). These include: composts, piles of hay, stored grains, wood chip piles, nesting material of birds and animals, snuff, municipal refuse and other accumulations of organic matter wherein the warm, humid and aerobic environment provides the basic physiological conditions for their development. In these habitats, thermophiles may occur either as resting propagules or as active mycelia depending on the availability of nutrients and favorable environmental conditions (Figs. 10.1, 10.2 and 10.3). Soils in tropical countries do not appear to have a higher population of thermophilic fungi than soils in temperate countries as believed earlier. Their widespread occurrence could well be due to the dissemination of propagules from self-heating masses of organic material (Maheshwari et al., 1987). Tansey and Brock (1972) observed that the thermophilic fungi are much more common in acidic thermal habitats than those of neutral to alkaline pH. Thermophilic fungi constitute a heterogeneous physiological group of various genera in the Zygomycetes, Ascomycetes, Deuteromycetes (anamorphic fungi) and Mycelia Sterilia (Johri et al., 1999; Satyanarayana and Singh, 2004).

The occurrence of thermophilic fungi in aquatic sediment of lakes and rivers, as first reported by Tubaki et al. (1974), is mysterious in view of the low temperature (6–7°C) and low level of oxygen (average 10 ppm, <1.0

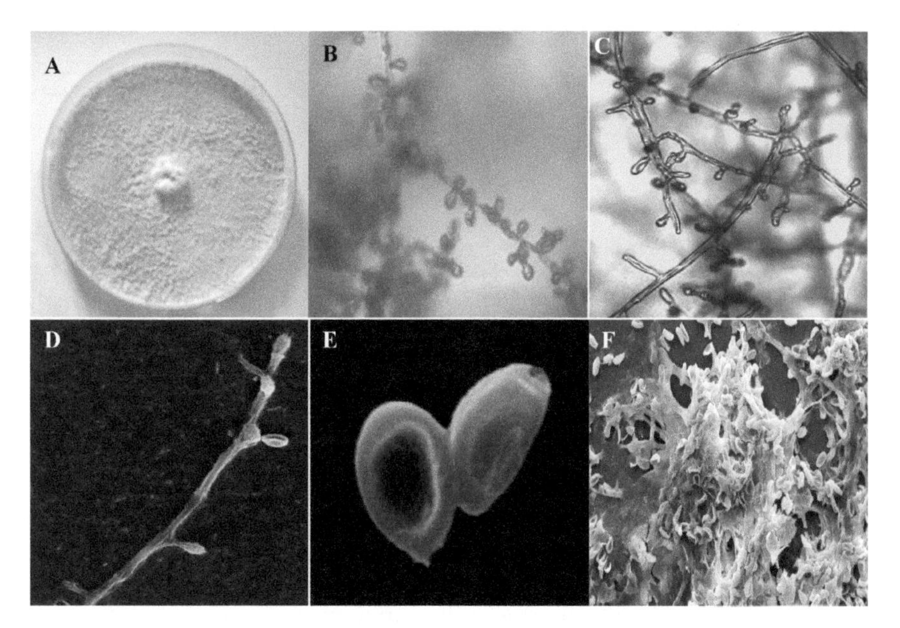

Figure 10.1. Growth and morphology of *Sporotrichum thermophile* (syn. *Myceliopthora thermophila*) under compound and scanning electron microscopes.

A: Petridish showing growth of the mold on YpSs agar
B&C: Morphology of mold showing lateral and terminal conidiospores (100x)
D: Morphology of fungus showing lateral and terminal conidiospores (500x)
E: Shape of conidiospores (500x)
F: Growth of the mold observed on natural substrate showing conidiospores and mycelia (250x)

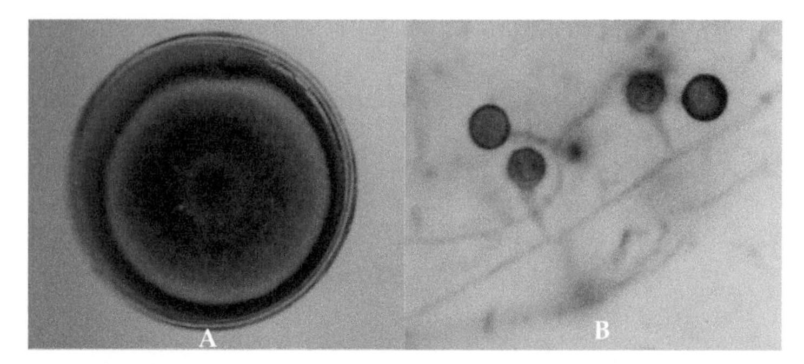

Figure 10.2. Morphology of *Humicola lanuginosa* (syn. *Thermomyces lanuginosus*) observed under compound microscope (A) Petridish showing the growth of the mold on YpSs agar, (B) Micrograph showing the aleuriospores attached on aleuriophores.

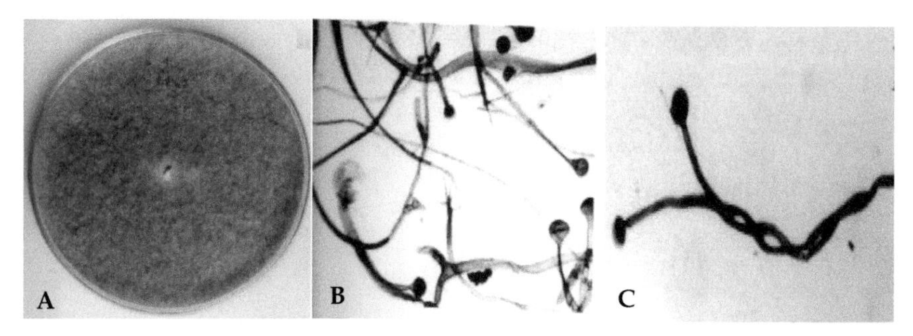

Figure 10.3. Morphology of *Thermomucor indicae-seudaticae* observed under compound microscope (A) Petridish showing the growth of mold, (B) Micrograph showing the full grown sporangium attached to sporangiophores, (C) Micrograph showing the dichotomously branched sporangiophore bearing sporangia.

ppm at a depth of 31 meters) available at the bottom of a lake. A number of thermophilic fungi survive the stresses such as increased hydrostatic pressure, absence of oxygen and desiccation (Mahajan et al., 1986). Undoubtedly, the thermophilic fungi owe their ubiquity and common occurrence in large measure to this special ability to occupy a temperature niche that most other fungi are unable to inhabit. More attempts are, however, needed to provide evidence for the active involvement in the habitats from which the thermophilic fungi are being reported. Cooney and Emerson (1964) published the first modern comprehensive account of the taxonomy, biology and activities of thermophilic fungi in which only 13 species were discussed, and several additional thermophilic fungi have been discovered or redefined since then. A subsequent document of thermophilic fungi was published by Mouchacca (1997) in which he attempted a critical reappraisal of the nomenclatural, and in some cases, also of the taxonomic status of known thermophiles. Latest valid names for all thermophilic fungal taxa have been elaborated by Mouchacca (2000). The original names of some of the taxa have been retained for their taxonomic value.

Geographical Distribution in Various Habitats and Substrates

Thermophilic fungi are world wide in distribution and most species do not show any geographical restriction (Johri et al., 1999; Subrahmanyam, 1999). The tropical sites, however, favor recovery as a consequence of elevated temperature and more competitive microbial environment. Thus, they have been reported from a wide variety of habitats including different soil types and places where self-heating of plant material results in high temperatures (Johri et al., 1999; Subrahmanyam, 1999; Satyanarayana and Singh, 2004). They have been isolated from natural as well as man-made habitats. Most significant natural habitats for saprophytic thermophilic fungi are the decomposing organic materials in which thermogenic conditions result by the activity of microorganisms (Chang and Hudson, 1967; Bergman and Nilsson, 1981). For example, a newly built wood chip pile where temperature rapidly rises to ignition (Feist et al., 1973). Solar heat in the tropics (Bilai, 1984) and moisture content (Festenstein, 1966) are the other factors that significantly influence the development of thermogenic conditions.

Tansey and Jack (1976) and Thakre and Johri (1976) were able to recover several thermophilic fungi from different substrates. Besides these, some non-thermogenic environments like coal and moist soils in Australia, Antarctic and sub-Antarctic soils (Ellis, 1980a,b), aquatic sediments where bottom temperature never exceeds 6 or 7°C and dust on sparingly used library books have also been found to harbor thermophilic fungi (Subrahmanyam, 1999). Salar and Aneja (2006) isolated thermophilic molds from temperate soil of North India. They reported the occurrence of two molds, *Chaetomium senegalense* (Ascomycete) and *Myceliophthora fergusii* for the first time from India. Recently Salar and Aneja (2007) explained the taxonomy and biogeography of thermophilic molds. Extensive collection of thermophilic fungal isolates have been described from man-made habitats such as hay, manure (Crisan, 1959), stored peat (Kuster and Locci, 1964) retting guayule (Cooney and Emerson, 1964), stored grains (Christensen, 1957), mushroom composts (Fergus, 1971; Fergus and Amelung, 1971) and birds' nests (Cooney and Emerson, 1964; Satyanarayana et al., 1977). Habitats of various thermophilic fungi are presented here (Table 10.1).

Table 10.1. Habitats of thermophilic molds.

Source	References
Thermogenic habitats	
Composts	Miehe, 1907; Straatsma et al., 1991; Straatsma and Samson, 1993; Singh, 2007
Soil	Vartaja, 1949; Tansey and Brock, 1973; Morgan-Jones, 1974; Friedman and Galum, 1974; Tansy and Jack, 1976, 1977; Jaitly and Rai, 1982; Chadha et al., 2004; Hassouni et al., 2006; Singh and Satyanarayana, 2006; Salar and Aneja, 2006, 2007; Singh, 2007
Hay	Miehe, 1907
Paddy straw compost	Satyanarayana and Johri, 1984; Singh, 2007
Wheat straw Compost	Chang and Hudson, 1967; Moubasher et al., 1982; Singh, 2007
Mushroom Compost	Fergus, 1964; Fergus et al., 1969, 1971; Straatsma et al., 1991; Wiegant et al., 1992
Municipal Compost	Crisan 1959, 1969; Cooney and Emerson, 1964; Eggins and Mills, 1971; Barnes et al., 1972; Malik and Eggins, 1972; Brown et al., 1974; Mills and Eggins, 1974; Champman et al., 1975; Eriksson and Larsson, 1975; Eriksson and Petersson, 1975; Subrahmanyam et al., 1977; Tansey and Jack, 1977; Matsuo et al., 1977a,b; Matsuo and Yasui, 1985; Gomes et al., 1993
Coal Spoil Tips	Apinis, 1963a; Evans, 1971a,b, 1972; Johri and Thakre, 1975
Stored grains	Christensen, 1957; Clarke et al., 1969; Flanningan, 1970; Mulings and Chesters, 1970; Awao and Mitsugi, 1973; Flanningan, 1974; Mehrotra and Basu, 1975; Taber and Pattit, 1975; Davis et al., 1975
Non-thermogenic habitats	
Cooling towers	Subrahmanyam, 1978
Dust of library books	Subrahmanyam, 1978
Arctic and Antarctic soils	Ellis, 1980a,b
Himalayan soil	Sandhu and Singh, 1981
Human skin	Subrahmanyam et al., 1977
Birds' Nests	Cooney and Emerson, 1964; Satyanarayana et al., 1977
Air	Hughes and Crosier, 1973

Methods of Isolation of Thermophilic Molds

a) Different Media

Various types of media have been used for the isolation of thermophilic molds. The best and most frequently used medium is Yeast extract-phosphate Soluble starch (YpSs) agar medium (Cooney and Emerson, 1964).

1. **Yeast extract starch agar medium:** This medium contains (g/L) starch 15 g, yeast extract 4.0 g, $MgSO_4 7H_2O$ 0.5 g, KH_2PO_4 1.0 g, Rose

Bengal 0.0001 g, trace amount of streptomycin and agar-agar 20 g. The pH of medium is adjusted to 6.2 with 0.1 N HCl/NaOH.

2. **Czapek-Dox agar medium:** This medium contains (g/L) $NaNO_3$ 3.0 g, $MgSO_4$ $7H_2O$ 0.05 g, KCl 0.5 g, $FeSO_4$ 0.01 g, sucrose 30 g and agar-agar 15 g.
3. **Dextrose-peptone-yeast extract Agar:** The medium contains (g/L) glucose 1.0 g, peptone 2.0 g, yeast extract 0.3 g, KH_2PO_4 0.2 g, $MgSO_4$ $7H_2O$ 0.02 g and agar-agar 20 g.
4. **Modified Czapek-Dox medium:** This medium contains (g/L) glucose 2.0 g, L-asparagine 10.0g, KH_2PO_4 1.52 g, KCl 0.52 g, $MgSO_4$ $7H_2O$ 0.52 g, $CuNO_3$ $3H_2O$ trace, $ZnSO_4$ $7H_2O$ trace, $FeSO_4$ $7H_2O$ trace and agar-agar 20g. The medium pH is adjusted to 6.2 (Saxena and Sinha, 1981).

b) Different Methods

Dilution plate technique: Apinis (1963a,b) used this method for the isolation of thermophilic molds. Ten grams of sample are transferred to a flask containing 100 ml sterile water. The contents are shaken on a mechanical shaker for 15 minutes and then serially diluted to obtain 10^{-4}–10^{-5}. From each dilution, 0.5 ml of the sample is transferred to sterile Petri plates containing the yeast extract starch agar medium.

Paired Petri plate technique (Cooney and Emerson, 1964): This method provides moisture and suitable environment for the growth of thermophilic molds. Paired plates are taken and the top plate is fixed with sterile filter paper and the paired plate is sealed with the cellophane tape to prevent loss of moisture. This method gives good results and maximum thermophiles are isolated using this method.

Humid chamber technique (Buxton and Mellanby, 1964): This method is employed especially for isolation of thermophiles from the bird nest materials. Collected bird nest materials are taken and directly placed in a glass chamber which is previously arranged with sterile wet filter paper and sterile glass slide on it. The nest materials are directly placed on sterile glass slides. The internal temperature of the chamber is maintained at 40–45°C. Growth of fungus appearing on the nest material are taken and transferred into sterile yeast extract starch agar slants and checked for thermophilic character.

Warcup's soil plate method: In this method, 2 ml of the sample is placed in a sterile Petri dish and 20 ml of sterile cooled (40°C) Czapek-Dox medium is dispensed. The contents are thoroughly mixed and the plates are incubated at 47 ± 2°C for fungal growth.

Waksman's direct inoculation method: In this method 20 ml of modified Emerson's (YpSs) medium is poured into a sterile Petri dish and is allowed to solidify. Small quantities of the samples are sprinkled over the medium in the dishes and plates are incubated in inverted position at 47 ± 2°C at high humidity.

Occurrence of Thermophilic Molds

The thermophilic molds are present everywhere in nature, that is why they are said to be ubiquitous (Johri et al., 1999; Subrahmanyam, 1999; Satyanarayana and Singh, 2004). They have been isolated from natural as well as man-made habitats. Decomposing organic matter is the main natural habitat of thermophilic molds where thermogenic conditions results in the growth of these molds (Johri et al., 1999; Subrahmanyam, 1999). Solar heat and moisture are the two important factors that result into the development of thermogenic conditions for thermophilic molds. Besides natural habitats, thermophilic molds have also been isolated from man-made habitats such as manure, stored peat, retting guayule, mushroom compost, bird's nests and so on (Johri et al., 1999; Subrahmanyam, 1999).

Thermogenic Habitat

The thermophilic fungi constitute a small physiologically distinct group of about 40 species. Growth of these fungi at high temperature is a rare feature (Johri et al., 1999; Subrahmanyam, 1999). The first thermophilic fungus, *Mucor pusillus,* was isolated from bread and described by Lindt (1886) over a century ago. Tsiklinskaya discovered another thermophilic fungus, *Thermomyces lanuginosus* in 1899 from potato inoculated with garden soil. Miehe (1907) investigated the cause of thermogenesis of stored agricultural produce. He reported four thermophilic fungi viz., *Mucor pusillus, Thermoascus aurantiacus, Thermoidium sulfureum* and *Thermomyces lanuginosus* from self heating hay. Cooney and Emerson (1964) for the first time presented a comprehensive account of thermophilic fungi in which they provided taxonomic descriptions of 13 species of thermophilic fungi known till that time. Mouchacca (1997) reviewed the taxonomic status of thermophilic fungi. The natural habitats with high temperatures result in the growth and development of thermophilic molds. Adopting suitable nutritional conditions, Johri and Thakre (1975) and Tansey and Jack (1976) isolated several thermophilic molds from various substrates. The main and important thermogenic habitats are described below (Johri et al., 1999; Subrahmanyam, 1999):

a. **Composts:** Compost is a natural thermogenic environment where succession of microbes, beginning with mesophilic microbes, takes place (Miehe, 1907). These thermogenic conditions suppress the growth of mesophiles and favor the growth of thermophiles including thermophilic molds. During this composting process, microbes utilize the complex organic matter such as lignocellulose present in paddy straw, wheat straw, municipal solid waste and others. But with the development of recent techniques, thermophilic molds have been utilized for compost preparation for mushroom cultivation (Straatsma et al., 1991; Straatsma and Samson, 1993).
b. **Soil:** Soil is an important habitat for growth and development of thermophilic molds. However, variations are found among the isolates depending upon types of soil, depth, season of the year and the organic matter. Thermophilic molds have been isolated from sun heated mud (Tansey and Brock, 1973), desert soils, rocks (Friedman and Galum, 1974; Singh and Satyanarayana, 2006; Singh, 2007), mangrove soil (Jaitly and Rai, 1982) and forest litter (Morgan-Jones, 1974). Because of solar radiation, top soils in tropical, sub-tropical, and temperate regions also support the growth of thermophilic molds (Vartaja, 1949). Tansey and Jack (1976, 1977) studied the occurrence of thermophilic molds in sun-heated barren soil and grass-shaded and tree-shaded soils, and observed a decrease in occurrence of thermophilic molds due to increase in shade. Several thermophilic fungi were isolated from composting soils and identified as *Rhizomucor pusillus, Scytalidium thermophilum, Melanocarpus albomyces, Chaetomium thermophile* and *Thermomyces lanuginosus* (Chadha et al., 2004). Similarly, Hassouni et al. (2006) isolated thermophilic molds from soil and screened for phytase production.
c. **Hay:** This is the less studied habitat for thermophilic molds. *Thermoascus aurantiacus* and *Malbranchea pulchella* var. *sulfurea* were the first thermophilic molds isolated for self-heated hay by Miehe (1907)
d. **Paddy straw:** This is another substrate which is an excellent habitat for thermophilic molds. Satyanarayana and Johri (1984) extensively studied this substrate and concluded that colonizing ability of thermophilic molds was directly proportional to the concentration of inoculum. *Thermomyces lanuginosus* (Syn. *Humicola lanuginosa*), *Sporotrichum thermophile* (Syn. *Myceliopthora thermophila*) and *Scytalidium thermophilum* (Syn. *Torula thermophila*) are the dominant species of this substrate. Beside these, *Aspergillus fumigatus* showed a strong competition in pure and mixed cultures. Carbon to nitrogen ratio significantly affected the decomposing ability of these molds.

e. **Wheat/rice straw compost:** Chang and Hudson (1967) isolated and studied thermophilic molds from wheat straw compost. The initial population in compost was dominated by mesophilic microbes followed by thermophilic ones due to rise in temperature to around 50°C. The wheat straw compost was dominated by *Penicillium dupontii, Myriococcum albomyces, Thermomyces lanuginosus* and *Sporotrichum thermophile* (Moubasher et al., 1982; Singh, 2007). In the early stage of composting, mostly the mucoraceous members dominate due to their simple nutritional requirements. Johri and Satyanarayana (1983) reported occurrence of *C. thermophile, H. lanuginosa, S. thermophile, Thermoascus aurantiacus* and several other thermophilic fungi in paddy straw compost.

f. **Mushroom compost:** Mushroom composting involves the use of decomposed organic matter for the growth and cultivation of mushrooms. Mushroom composting is a two stage process. In the first phase, there is no growth of thermophilic molds due to high temperature, low oxygen and high CO_2 levels. The conditions are alkaline due to the liberation of ammonia. In the second phase, due to controlled conditions of temperature, oxygen and humidity, the thermophilic molds start growing. Fergus (1964) was the first to report the presence of thermophilic molds in mushroom compost. Later Fergus and his co-workers (1969, 1971) isolated two new thermophilic molds viz. *Papulaspora thermophila* and *Thielavia thermophila*. The role of thermophilic molds in mushroom composting has been extensively investigated. Straatsma et al. (1991) observed that the growth of edible mushroom (*Agaricus bisporus*) was promoted by compounds produced by *Scytalidium thermophilum*. High CO_2 production was correlated with enhanced growth rate of mycelia of edible mushrooms (Wiegant et al., 1992).

g. **Municipal compost:** Municipal waste is mainly composed of substrates rich in lignocellulose. The thermophilic molds degrade these organic compounds that can be used as manure for single cell protein production such as mushrooms (Barnes et al., 1972; Malik and Eggins, 1972; Champman et al., 1975; Eriksson and Larsson, 1975; Eriksson and Petersson, 1975; Tansey and Jack, 1977; Matsuo et al., 1977a,b; Matsuo and Yasui, 1985; Gomes et al., 1993). Some species of thermophilic molds have been shown to degrade plastic substrates (Brown et al., 1974; Eggins and Mills, 1971; Mills and Eggins, 1974). In India, a new genus *Thermomucor* was isolated from municipal compost (Subrahmanyam et al., 1977). *Thermoascus aurantiacus* and *M. thermophila* have been isolated from municipal waste (Cooney and Emerson, 1964; Crisan, 1959, 1969).

h. **Coal Spoil Tips:** Evans (1971a,b) first isolated thermophilic molds from coal spoil tips. These tips provide a suitable environment for the growth and development of thermophiles. *Mucor pusillus, T. lanuginosus, T. aurantiacus* and *A. fumigatus* are the dominant species of thermophilic molds found in coal spoil tips (Apinis, 1963a, 1972). Among these species, *T. aurantiacus* was present in large numbers. *Aspergillus fumigatus* is the dominant species found in Indian coal mine soils (Johri and Thakre, 1975). Besides these, a large number of unique species such as *Mortierella wolfii, Sphaerospora saccata, Talaromyces laycillanum, Acrophialospora fusispora, Penicillium argillaceum* and others are reported (Evans, 1971b).
i. **Stored grains:** A large number of studies have found that stored grains are an important substrate for the occurrence of thermophilic molds (Awao and Mitsugi, 1973; Flanningan, 1970; Mehrotra and Basu, 1975; Mulings and Chesters, 1970; Taber and Pattit, 1975). The temperature of grains rises due to metabolic activity of mesophiles favoring the growth and development of thermophilic molds (Flanningan, 1970). The thermophilic molds are also known to spoil stored grains due to secretion of some toxins (Christensen, 1957; Davis et al., 1975). Besides stored grains, thermophilic molds are also present on aerial parts of crops and freshly harvested grains (Flanningan, 1974). The storage conditions facilitate the development of thermophilic molds such as *Absidia corymbifera, A. fumigatus* and *T. lanuginosus* (Clarke et al., 1969).

Non-thermogenic Habitat

The thermophilic molds occur dominantly in the habitats having high temperature generated due to the metabolic activity of the microbes (Johri et al., 1999; Subrahmanyam, 1999). However, thermophilic molds have been found growing at some places where high temperature does not prevail and such habitats are called non-thermogenic habitats (Johri et al., 1999; Subrahmanyam, 1999). These habitats include red wood, cooling towers, and dust on used books in air conditioned libraries (Subrahmanyam, 1978), arctic and antarctic soils (Ellis, 1980a,b) and Himalayan soil (Sandhu and Singh, 1981). *Thermomucor indicae-seudaticae* was found on human skin (Subrahmanyam et al., 1977; Subrahmanyam, 1999). Thermophilic molds have also been reported in the sediments of rivers where temperature does not exceed 7°C. The occurrence of thermophilic molds in these non-thermogenic habitats has been thought to be due to their migration/transport from the surrounding terrestrial habitats.

Distribution of Thermophilic Molds

The thermophilic molds are able to grow in environmental conditions where temperature rises due to the metabolic activity of microbes or heating by solar radiations (Johri et al., 1999; Subrahmanyam, 1999). Their distribution follows the pattern similar to thermophilic bacteria and actinobacteria on a worldwide basis. Their worldwide occurrence is due to their unique adaptability to high temperature habitats. The presence of self-heated organic matter all around the world, favors the growth and development of thermophilic molds (Johri et al., 1999; Subrahmanyam, 1999). Some of the common thermophilic molds such as *M. pusillus* and *T. lanuginosus* have been isolated as contaminants. Miehe (1907) and Noack (1920) isolated *T. aurantiacus* and *M. pulchella* var. *sulfurea* from self-heated organic materials. Because of their high temperature requirements, tropical regions has been considered as their natural habitats. In contrast, these molds have been isolated from various non-thermogenic environments such as the arctic and antarctic regions as stated earlier. The soil samples from Costa Rica, Honduras, Brazil and some parts of the US, UK and Java have been found as habitats for similar types of thermophilic molds. Maheshwari et al. (1987), in their investigation on geographical distribution of thermophilic molds, reported that less than 50% of known thermophilic molds are found in India. Further studies are, however, needed to show a clear geographical distribution pattern of thermophilic molds.

Conclusions

Thermophilic fungi are ubiquitous in their occurrence, and most of them do not show any geographical restriction. Their main habitat is the thermogenic environments like composts, although there are reports on their occurrence in non-thermogenic environments too. The soil and compost samples from all over the world have been found to harbor thermophilic molds. Further efforts are called for understanding the diversity of thermophilic molds and their metabolites using both culture-dependent and culture-independent approaches.

Acknowledgements

The authors gratefully acknowledge the financial assistance from the University Grant Commission (UGC), Department of Biotechnology (DBT) and Council of Scientific & Industrial Research (CSIR) New Delhi, India while preparing this chapter.

References

Aguilar, A. 1996. Extremophile research in the European Union: from fundamental aspects to industrial expectations. FEMS Microbiol. Rev., 18: 89–92.
Apinis, A.E. 1963a. Occurrence of thermophilous microfungi in certain alluvial soils near Nottingham. Nova Hedgw., 5: 57–78.
Apinis, A.E. 1963b. Thermophilous fungi of costal grasslands. pp. 427–438. *In*: Doeksen, J. and Van der Drift, J. (eds.). The Proceedings of the Colloquium on Soil Fanna, soil microflora and their relationships by soil organisms, North Holland, Amsterdam.
Apinis, A.E. 1972. Thermophilous fungi in certain grasslands. Mycopath. Mycol. Appl., 48: 63–74.
Awao, T. and Mitsugi, K. 1973. Notes on thermophilic fungi of Japan. Trans. Mycol. Soc. Japan, 14: 145–160.
Barnes, T.G., Eggins, H.O.W. and Smith, E.L. 1972. Preliminary stages in the development of a process of the microbial upgrading of waste paper. Int. Biodeterior. Bull., 8: 112–116.
Barns, S.M., Delwiche, C.F., Palmer, J.D. and Pace, N.R. 1996. Perspectives on archaeal diversity, thermophily, and monophyly from environmental rRNA sequences. Proc. Nat. Acad. Sci. USA, 93: 9188–9193.
Bergman, O. and Nilsson, T. 1981. Studies on outside storage of saw mill chips. Inst. For Virkeslara, Skogshogasskolan, Stockholm Res. Note R, 71: pp. 43.
Bilai, T. 1984. Thermophilic micromycetes species from mushroom cellar composts. Microbiol. Zh. (Kiev.), 46: 35–38.
Brown, B.S., Mills, J. and Hulse, J.M. 1974. Chemical and biological degradation of waste plastics. Nature, 250: 161–163.
Buxton, P.A. and Mellanby, K. 1934. The measurement and control of humidity. Bull. Ent. Res., 25: 171–175.
Champman, E.S., Evans E., Jacobelli, M.C. and Logan, A.A. 1975. The cellulolytic and amylolytic activity of *Papulaspora thermophila*. Mycologia, 67: 608–615.
Chang, Y. and Hudson, H.J. 1967. Fungi of wheat straw compost I. Ecological Studies. Trans. Br. Mycol. Soc., 50: 649–666.
Christensen, C.M. 1957. Deterioration of stored grains by fungi. Bot. Rev., 23: 108–134.
Clarke, J.H., Hill, S.T., Niles, E.V. and Howard, M.A.R. 1969. Ecology of microflora of moist barley, barley in sealed silos on farms. Pest. Infest. Res., 1966: 14–16.
Cooney, D.G. and Emerson, R. 1964. Thermophilic fungi, An account of their biology, Activities and Classification. W.H. Freeman and Co. san Fransisco, USA.
Crisan, E.V. 1959. The isolation and identification of thermophilic fungi. Mycologia, 63: 1171–1198.
Crisan, E.V. 1969. The proteins of thermophilic fungi. pp. 32–33. *In*: Grunckel, J.E. (ed.). Current Topics in Plant Science. Academic Press, New York, London, UK.
Davis, N.D., Wagener, R.E., Morgan-Jones, G. and Diener, U.L. 1975. Toxigenic thermophilic and thermotolerant fungi. Appl. Micobiol., 29: 455–457.
Dix, N.J. and Webster, J. 1995. Fungal Ecology. Chapman and Hall, London, UK.
Eggins, H.O.W. and Mills, J. 1971. *Talaromyces emersonii*—a possible biodeteriogen. Int. Biodet. Bull., 7: 105–108.
Ellis, D.H. 1980a. Thermophilic fungi isolated from a heated aquatic habitat. Mycologia, 72: 1030–1033.
Ellis, D.H. 1980b. Thermophilous fungi isolated from some Antarctic and Sub-Antarctic soils. Mycologia, 72: 1033–1036.
Ellis, D.H. and Keane, P.J. 1981. Thermophilic fungi isolated from some Australian soils. Aust. J. Bot., 29: 689–704.

Evans, H.C. 1971a. Thermophilic fungi of coal spoil tips I. Taxonomy. Trans. Br. Mycol. Soc., 57: 241–254.

Evans, H.C. 1971b. Thermophilic fungi of coal spoil tips II. Occurrence and temperature relations. Trans. Br. Mycol. Soc., 57: 255–266.

Feist, W.C., Springer, E.L. and Hajny, G.J. 1973. Spontaneous heating in piled wood chips, contributions of bacteria. TAPPI, 36: 148–151.

Fergus, C.L. 1964. Thermophilic and thermotolerant molds in mushroom compost during peak heating. Mycologia, 56: 267–284.

Fergus, C.L. 1971. The temperature relation and thermal resistance of thermophilic *Populaspora* from mushroom compost. Mycologia, 63: 426–431.

Fergus, C.L. and Amelung, R.M. 1971. The heat resistance of some thermophilic fungi in mushroom compost. Mycologia, 63: 675–679.

Fergus, C.L. and Sinden, J.W. 1969. A new thermophilic fungus in mushroom compost, *Thielavia thermophila* sp. nov. Can. J. Bot., 47: 1635.

Festenstein, G.N. 1966. Biochemical changes during molding of salt heated hay in Dewar flasks. J. Sci. Food Agri., 17: 130–133.

Flanningan, B. 1970. Comparison of seed-borne microflora of barley, oats and wheat. Trans. Br. Mycol. Soc., 55: 267–276.

Flanningan, B. 1974. Distribution of seed-borne microorganisms in naked barley and wheat before harvest. Trans. Br. Mycol. Soc., 62: 51–58.

Friedman, E.I. and Galum, M. 1974. pp. 165–212. *In*: Brown, G.W. (ed.). Desert Biology, Vol. 2, Academic Press London, UK.

Gomes, J., Gomes, I., Kreiner W., Esterbauer, H., Sinner, M. and Steiner, W. 1993. Production of high level of cellulase-free and thermostable xylanase by a wild strain of *Thermomyces lanuginosus* using beechwood xylan. J. Biotechnol., 30: 283–297.

Hassouni, H., smaili-Alaoui, I.M., Gaime-Perraud, I., Augur, C. and Roussos, S. 2006. Effect of culture media and fermentation parameters on phytase production by the thermophilic fungus *Myceliophthora thermophila* in solid state fermentation. Mycol. Apli. Int., 18: 29–36.

Hughes, W.T. and Crosier, J.W. 1973. Thermophilic fungi in the microflora of man and environmental air. Mycopath. Mycol. Appl., 49: 147–152.

Jaitly, A.K. and Rai, J.N. 1982. Thermophilic and thermotolerant fungi from mangrove swamps. Mycologia, 74: 1021–1022.

Johri, B.N. and Thakre, R.P. 1975. Soil amendments and enrichment media in the ecology of thermophilic fungi. Ind. Natl. Sci. Acad., 41: 564–570.

Johri, B.N. and Satyanarayana, T. 1983. Ecology of thermophilic fungi. pp. 349–361. *In*: Mukherji, K.G., Agnihotri, V.P. and Singh, R.P. (eds.). Progress in Microbial Ecology, Print House (India), Lucknow.

Johri, B.N., Satyanarayana, T. and Olsen, J. 1999. Thermophilic molds in biotechnology. Kluwer Academic Publishers, UK.

Kuster, E. and Locci, R. 1964. Studies on peat and peat microorganisms II. Occurrence of thermophilic fungi in peat. Arch. Mikrobiol., 48: 319–324.

Lindt, W. 1886. Mitteilungen über einige neue pathogene Shimmelpilze. Arch. Exp. Pathol. Pharmakol., 21: 269–298.

Mahajan, M.K., Johri, B.N. and Gupta, R.K. 1986. Influence of desiccation stress in xerophilic and thermophilic *Humicola* sp. Curr. Sci., 55: 928–930.

Maheshwari, R., Bharadwaj, G. and Bhat, M.K. 2000. Thermophilic fungi: Their physiology and enzymes. Microbiol. Mol. Biol. Rev., 64: 461–488.

Maheshwari, R., Kamalam, P.T. and Balasubrahamanyam, P.L. 1987. The biogeography of thermophilic fungi. Curr. Sci., 56: 151–155.

Malik, K.A. and Eggins, H.O.W. 1972. Some studies on the effect of pH on the ecology of cellulolytic thermophilic fungi using a perfusion technique. Biologia, 18: 277–279.

Matsuo, M. and Yasui, T. 1985. Properties of xylanase of *Malbranchea pulchella* var. *sulfurea* no. 48. Agric. Biol. Chem., 49: 839–841.

Matsuo, M., Yasui, T. and Kobayashi, T. 1977a. Purification and some properties of β-xylosidase from *Malbranchea pulchella* var. *sulfurea* no. 48. Agric. Biol. Chem., 41: 1593–1599.

Matsuo, M., Yasui, T. and Kobayashi, T. 1977b. Enzymatic properties of β-xylosidase from *Malbranchea pulchella* var. *sulfurea* no. 48. Agric. Biol. Chem., 41: 1601–1606.

Mehrotra, B. and Basu, M. 1975. Survey of microorganism associated with cereal grains and other milling fractions in India, Pt. I, imported wheat. Int. Biodeter. Bull., 11: 56–63.

Miehe, H. 1907. Die Selbsterhitzung des Heus. Eine biologische Studie. Gustav Fischer Verlag, Jena, Germany.

Mills, J. and Eggins, H.O.W. 1974. The biodeterioration of pasticisers by thermophilic fungi. Int. Biodeterior. Bull., 10: 39–44.

Morgan-Jones. 1974. Notes on Hypomycetes V. A new thermophilic species of *Acremonium*. Can. J. Bot., 52: 429–431.

Moubasher, A.H., Hafez, S.I.I., Aboelfattah, H.M. and Moharrarh, A.M. 1982. Fungi of wheat and broad bean straw composts 2, Thermophilic fungi. Mycopathologia, 84: 61–72.

Mouchacca, J. 1997. Thermophilic fungi: biodiversity and taxonomic status. Crypt. Mycol., 18: 19–69.

Mouchacca, J. 2000. Thermophilic fungi and applied research: a synopsis of name changes and synonymies. World J. Microbiol. Biotechnol., 16: 881–888.

Mulings, S.L. and Chesters, C.G.C. 1970. Ecology of fungi associated with moist stored barley grains. Ann. Appl. Biol., 65: 277–284.

Noack, K. 1920. Der Betriebstoffwechsel der thermophilen Pilze. Jahrb. Wiss. Bot., 59: 593–648.

Redman, R.S., Litvintseva, A., Sheehan, K.B., Henson, J.M. and Rodriguez, R.J. 1999. Fungi from geothermal soils in Yellowstone National Park. Appl. Environ. Microbiol., 65: 5193–5197.

Salar, R.K. and Aneja, K.R. 2006. Thermophilous fungi from temperate soils of northern India. J. Agric. Technol., 2: 49–58.

Salar, R.K. and Aneja, K.R. 2007. Thermophilic Fungi: Taxonomy and Biogeography. J. Agric. Technol., 3: 77–107.

Sandhu, D.K. and Singh, S. 1981. Distribution of thermofilous microfungi in forest soils of Darjeeling (Eastern Himalayas). Mycopathologia, 74: 79–81.

Satyanarayana, T. and Johri, B.N. 1984. Thermophilic fungi of paddy straw compost, growth, nutrition and temperature relationships, J. Indian Bot. Soc., 63: 164–170.

Satyanarayana, T., Johri, B.N. and Saksena, S.B. 1977. Seasonal variation in mycoflora of nesting materials of birds with special reference to thermophilic fungi.Trans. Br. Mycol. Soc., 68: 307–309.

Satyanarayana, T. and Singh, B. 2004. Thermophilic molds: diversity and potential biotechnological applications. pp. 87–110. *In*: Gautam, S.P., Sharma, A., Sandhu, S.S. and Pandey, A.K. (eds.). Microbial Diversity: Opportunities and Challenges. Shree Publishers and Distributors, New Delhi, India.

Saxena, R.K and Sinha, U. 1981. L-asparaginase and glutaminase activities in the culture filtrates of *Aspergillus nidulans*. Curr. Sci., 50: 218–219.

Singh, B. 2007. Production, characterization and applications of extracellular phytase of the thermophilic mold *Sporotrichum thermophile* Apinis. PhD. Thesis, University of Delhi South campus, New Delhi, India.

Singh, B. and Satyanarayana, T. 2006. Phytase production by a thermophilic mold *Sporotrichum thermophile* in solid-state fermentation and its application in dephytinization of sesame oil cake. Appl. Biochem. Biotechnol., 133(3): 239–250.

Singh, B. and Satyanarayana, T. 2009. Thermophilic molds in environmental management. pp. 352–375. *In*: Misra, J.K. and Deshmukh, S.K. (eds.). Progress in Mycological Research, Vol. I. Fungi from Different Environments. Environmental Mycology. Science Publishers, USA.

Stetter, K.O. 1999. Extremophiles and their adaptation to hot environments. FEBS Lett., 452: 22–25.

Straatsma, G., Gerrits, J.P.G., Augustin, A.P.A.M., Camp, H.J.M. and Van Griendsven, L.J.L.D. 1991. Growth kinetics of *Agaricus bisporus* mycelium on solid substrate (mushroom compost). J. Gen. Microbiol., 137: 1471–1477.

Straatsma, G., and Samson, R.A. 1993. Taxonomy of *Scytalidium thermophilum*, an important thermophilic fungus in mushroom compost. Mycol. Res., 97: 321–328.

Subrahmanyam, A. 1978. Isolation of thermophilic fungi from dust on books. Curr. Sci., 47: 817–819.

Subrahmanyam, A. 1999. Ecology and distribution. pp. 13–42. *In*: Johri, B.N., Satyanarayana, T. and Olsen, J. (eds.). Thermophilic Molds in Biotechnology. Kluwer Academic Publishers, UK.

Subrahmanyam A., Mehrotra, B.S. and Thirumalacher, M.J. 1977. *Thermomucor*, a new genus of mucorales. Geor. J. Sci., 35: 1–6.

Taber, R.A. and Pattit, R.E. 1975. Occurrence of thermophilic microorganism in peanuts and peanut soil. Mycologia, 67: 157–161.

Tansey, M.R. and Brock, T.D. 1973. *Dactylaria gallopoya*—a cause of avian encephalitis in hot spring effluents, thermal soils and self heated coal waste piles. Nature, 242: 202–205.

Tansey, M.R. and Brock, T.D. 1978. Microbial life at high temperatures: ecological aspects. pp. 159–216. *In*: Kushner, D. (ed.). Microbial Life in Extreme Environments. Academic Press, London, UK.

Tansey, M.R. and Brock, T.D. 1972. The upper temperature limit for eukaryotic organism. Proc. Natl. Acad. Sci. USA, 69: 2426–2428.

Tansey, M.R. and Jack, M.A. 1976. Thermophilic fungi in sun heated soils. Mycologia, 68: 1061–1075.

Tansey, M.R. and Jack, M.A. 1977. Growth of thermophilic fungi in soil *in situ* and *vitro*. Mycologia, 69: 563–578.

Thakre, R.P. and Johri, B.N. 1976. Occurrence of thermophilic fungi in coal mine soils of Madhya Pradesh. Curr. Sci., 45: 271–273.

Tsiklinskaya, P. 1899. Sur les muce´dine´es thermophiles. Ann. Inst. Pasteur (Paris), 13: 500–515.

Tubaki, K., Ito, T. and Matsudu, Y. 1974. Aquatic sediments as a habitat of thermophilic fungi. Ann. Microbiol., 24: 199–207.

Vartaja, O. 1949. High surface soil temperature. Oikos, 1: 6–28.

Wiegant, W.M., Wery J., Britenhmis, E.T. and De Bount, J.A. 1992. Growth promoting effect of thermophilic fungi on mycelium of edible mushroom *Agaricus bisporus*. Appl. Environ. Microbiol., 58: 2644–2659.

CHAPTER 11

Rusts Fungi of Wheat

Huerta-Espino J.,[1,*] *R.P. Singh*[2] and *Alan P. Roelfs*[3]

ABSTRACT

There are three rust diseases on wheat: Stem rust caused by the fungus *P. graminis* Pers. f. sp. *tritici* Eriks. & E. Henn., leaf rust by *Puccinia triticina* Eriksson, *Puccinia tritici-duri* Viennot-B., and stripe rust caused by *Puccinia striiformis* West. f. sp. *tritici* Eriks. & E. Henn. *P. graminis* f. sp. *tritici*, *P. triticina*, *P. tritici-duri* and *P. striiformis* f. sp. *tritici* are macrocyclic heteroecious rusts with uredinia and telia occurring on wheat, and pycnia and aecia occurring on the alternate hosts *Berberis vulgaris* L. (common barberry), *Thalictrum speciosissimum* Loefl. (Meadow rue) *Anchuza italica* L., and *Berberis* spp. (barberry), respectively. All three wheat rusts can cause diseases on a continental level as the urediniospores can be carried away by wind currents for thousands of kilometers. The wheat rusts are also highly variable from region to region and from different epidemiological zones worldwide, as many physiologic races have been described for all three diseases through the use of differential host plants. Among the three wheat rusts, leaf rust is the most common and widespread disease of wheat in Mexico, North America and worldwide, *P. tritici-duri* is important in North Africa and some areas of Spain, Portugal and Italy and yellow or stripe rust is very important in the Middle East and Central Asia. Stem rust on the other hand and particularly the race Ug99 is very important in Africa at the

[1] INIFAP-CEVAMEX, Apdo. Postal 10, 56230, Chapingo, Mexico.
[2] International Maize and Wheat Improvement Center (CIMMYT), Apdo. Postal 6-641, 06600, Mexico, DF, Mexico.
[3] USDA-ARS, Cereal Disease Laboratory, University of Minnesota, St. Paul, MN 55108 (retired).
* Corresponding author: j.huerta@cgiar.org

moment; but is potentially a threat to the wheat production worldwide. Functional alternate hosts are rare or not well documented in most wheat growing areas of the world. Therefore, the rust fungus usually reproduces through continuous cycling of urediniospores on the wheat crop. Stem rust is usually found on the stems and leaf-sheaths of the wheat plant, whereas leaf rust is found mostly on the blades of leaves, and stripe rust on the leaf blades in yellow stripes between the veins, but head infection under heavy inoculum pressure is also common. Inocula production, and greenhouse and field tests for the wheat rusts are similar, but there are a few exceptions regarding temperature and light regimes which have to be considered differently for the individual rusts.

Introduction

The cereal rusts as diseases, reduce the quantity and quality of wheat yields. Losses due to stem, leaf and stripe rusts have been enormous over the years. Despite the great progress made in their control, the rusts are still the most important wheat diseases worldwide. There are three rusts which belong to the genus *Puccinia*, and differ not only in their morphology and life cycle, but also in their environmental conditions for growth. The pathogens *Puccinia graminis* Pers. f. sp. *tritici* Eriks. & E. Henn., *P. triticina* Erikss., and *P. striiformis* Westend. f. sp. *tritici* Eriks. & E. Henn., are widely distributed around the world and all are generally asexually reproduced, but have the capacity to evolve into new virulent forms able to attack previously resistant hosts through mutation. The rust is named for the dry, dusty yellow-red or black spots and stripes (sori or pustules) that erupt through the epidermal tissues of the leaf or stem in a susceptible or moderately susceptible cultivars. In general, the size and surrounding coloration of the rust pustules determine the specific infection types which can be chlorotic flecks, necrotic spots or the infection may result into sporulating pustules of various sizes. The rust fungi are also able to produce large amounts of urediniospores which are effective in long distance dissemination. The rust diseases increase rapidly causing serious losses, making the rusts the most devastating diseases of wheat.

Among the rusts of wheat, stem rust is considered to be the most destructive. It can attack all of the above ground parts of the plant; leaves, sheaths, stem and spikes, including awns and glumes. However, it is more prominent on the stems. In general, it occurs later in the growing season and prefers higher temperatures, depending on the growing environment. Leaf rust is the most common and widely distributed of the wheat rusts. It primarily attacks the leaf blades, and in highly susceptible cultivars the leaf sheath and glumes. Leaf rust causes less damage than stem rust, but it occurs more frequently, thus leaf rust damage tends to be less spectacular than the stem rust; however, it probably results in greater total loss than the other rusts because of its broad adaptation and widespread occurrence (Huerta-Espino et al., 2011).

Yellow or stripe rust in contrast to stem and leaf rust requires relatively lower temperatures. The importance of this rust depends on climate and the degree of resistance of the cultivar from area to area. Because the optimum environments for the three rusts are different, generally one may flourish where the others do not. The three wheat rusts have been controlled by developing resistant cultivars in much of the world. However, resistance in some cases has been short lasting. The wheat resistance and pathogen virulence interact in a typical gene-for-gene relationship (Flor, 1956).

Within some rust species, there is a lot of variability in host range. Plant pathologists have grouped morphologically similar individuals that differ in host range into formae specialis; which are defined according to the ability to attack a particular species. Early in the 20th century, Stakman (1919) discovered that the formae specialis could be further subdivided into physiological races, which differed in their capacity to attack certain wheat cultivars. Physiological races are characterized by the reactions on a set of host differentials carrying different sources of resistance (Singh et al., 2012; McCallum et al., 2012; Wellings et al., 2012).

The classification and characterization of physiological races, have been done by using the pattern of low and high infection types resulting when differential hosts were inoculated with a specific pathogen culture (Stakman et. al., 1962; Mains and Jackson, 1926; Stubbs, 1985). Single gene differential hosts for stem and leaf rusts, have been preferentially used over the original differentials, which often had more than one resistance gene.

Causal Organisms

Wheat rust pathogens are fungi of the class Basidiomycetes, order Uredinales, family Pucciniaceae, and the genus *Puccinia*. These rust fungi are highly specialized biotrophic wheat pathogens with narrow host ranges. Fontana and Tozzetti independently provided the first unequivocal and detailed reports of wheat stem rust in 1767 (Fontana, 1932; Tozzetti, 1952). However, Kislev (1982) reported uredinia, hyphae and germinating urediniospores of a parasite fungus observed on two ancient lemma fragments of wheat and identified as *P. graminis* in Israel dated from 3300 years old. The causal organism of wheat stem rust was named *Puccinia graminis* by Persoon in 1797 (Persoon, 1801). Chester (1946) provided one of the first detailed histories of literature on the rust of wheat. In the early records, wheat leaf rust is not distinguished from stem or stripe rust (Chester, 1946). However, by 1815 de Candolle (1815) had shown that wheat leaf rust was caused by a distinct fungus *Uredo rubigovera*. The pathogen underwent a number of name changes until 1956 when Cummins and Caldwell (1956) suggested *P. recondita*, which has been the generally used nomenclature. Studies by Savile (1984), the virulence analyses on a worldwide basis by Huerta-Espino (1992), and morphological and pathogen genetic studies by Anikster et al. (1997) showed that *P. recondita* is not the causal agent of wheat leaf rust. Currently *P. triticina* has been accepted and preferred. *P. triticina* was used by Mains and Jackson (1926) and has been used in parts of Asia and eastern Europe for many years. In this chapter, *P. triticina* will be used for the leaf rust on wheat (*Triticum aestivum* L.). Although, Gadd (Eriksson and Henning, 1896) first described stripe or yellow rust of wheat in 1777, it was not until 1896 that Eriksson and Henning (1896) showed that stripe rust resulted from a separate pathogen, which they named *P. glumarum*. In 1953, Hylander et al. (1953) revived the name *P. striiformis*.

The Diseases

Of the three rust diseases of wheat caused by these fungi, it is very difficult to define which of them is the most important on a world-wide basis. The recent yellow rust epidemics, caused by *P. striiformis* f. sp. *tritici* in the Middle East and Central Asia, made stripe rust in the area as one of the most feared rust diseases of wheat (Nazari et al., 2010). On the other hand, the appearance of stem or black rust Ug99 race caused by *P. graminis* f. sp. *tritici* with a particular virulence combination which defeated important resistance genes in Africa, its fast evolution and continuous movement to other wheat growing areas of the world is just like the wakening of a furious monster, making it the most devastating disease of the three rusts (Singh et al., 2006, 2008, 2011). From the prevalence and distribution however, the so called leaf or brown rust caused by *P. triticina* is probably the one that causes more yield losses on annual bases (Huerta-Espino et al., 2011).

Stem rust caused by *P. graminis* f. sp. *tritici*, is also known as the black rust or summer rust due to the abundant production of shiny black teliospores, which form in the telium at the end of the season or with unfavorable conditions or when the plant reaches physiological maturity. Stem rust is favored by humid conditions and warmer conditions with temperatures of 15 to 35°C. It is the most devastating of the rust diseases and can cause losses of up to 50% in one month when conditions for its development are favorable. Losses up to 100% can occur with susceptible cultivars.

Leaf rust caused by *P. triticina* occurs on the leaf blades, although leaf sheaths can also be infected under favorable conditions, high inoculum densities and extremely susceptible cultivars. It frequently lacks the abundant teliospore production of stem rust at the end of the season, resulting in a brown leaf lesion rather than a black stem lesion that occurs with stem rust. When leaf rust teliospores are produced, they usually emanate from telia on the lower leaf surfaces, which remain covered by the epidermal cells. The disease develops rapidly at temperatures between 10°C and 30°C. Leaf rust occurs to some extent wherever wheat is grown. Losses in grain yield are primarily attributed to reduced floret set and grain shriveling. In highly susceptible genotypes, florets, tillers and plants can be killed by pre-heading infections.

Stripe or yellow rust, caused by *P. striiformis* f. sp. *tritici*, is principally a disease of wheat grown in cooler climates (2°C to 15°C), which are generally associated with higher elevations, northern latitudes or cooler years. It takes its name from the characteristic stripe of uredinia that produce yellow-colored urediniospores. Because of the disease's early seasonal attack, stunted and weakened plants often occur. Losses can be severe (about 50%) due to shriveled grains and damaged tillers. In extreme situations, stripe rust can cause 100% losses.

General Epidemiology

Rust epidemics develop when susceptible wheat plant, virulent rust fungi and free moisture (dew) exist/together over large areas. Urediniospores of the wheat rusts germinate within one to three hours of contact with free moisture over a range of temperatures depending on the species of rust. Urediniospores are produced in large numbers and can be blown to considerable distances by wind (Hirst and Hurst, 1967; Watson and de Sousa, 1983). However, most urediniospores are deposited close to their source (Roelfs and Martell, 1984) under the influence of gravity. Urediniospores are relatively long-lived and can survive in the field away from host plants for several weeks. They can withstand freezing if their moisture content is lowered from 20 to 30%. Viability rapidly decreases at moisture contents of more than 50%.

Long-distance spread of urediniospores is influenced by latitude and respective wind patterns. In general, spores move west to east due to the winds resulting from the rotation of the earth. At progressively higher latitudes, winds tend to take a more southerly direction in the Northern Hemisphere and a northerly direction in the Southern Hemisphere. Studies in the United States (Roelfs, 1985a) showed spore movements to be from the southwest to northeast, north of 30°N. In the Southern Hemisphere, because most of the wheat growing areas and land masses, in general, are north of 30°S, the movement is more west to east (Luig, 1985). However, over a period of years, barley stripe rust moved south and eastward across South America (Dubin and Stubbs, 1986), and eventually reached the United States (Roelfs et al., 1992; Roelfs and Huerta-Espino, 1994). In most areas studied, spores produced in the upper levels of the crop canopy move into a geographical area where the crop phenology is less advanced.

Stem Rust

Stem or black rust of wheat caused by *P. graminis* at one time, was a feared disease in most wheat regions of the world. The fear of stem rust was understandable because an apparently healthy crop, three weeks before harvest, could be reduced to a black tangle of broken stems and shriveled grains by harvest. In some areas early maturing cultivars were introduced to avoid flowering and grain-filling during hot weather. Early maturing cultivars escape much of the damage caused by stem rust by avoiding the growth period of the fungus. The widespread use of resistant cultivars worldwide has reduced the disease as a significant factor in production. Although changes in pathogen virulence have rendered some resistances ineffective, resistant cultivars have generally been developed ahead of the pathogen. Recent outbreaks of the race Ug99 epidemics in Africa will require more attention in breeding stem rust resistant cultivars.

Epidemiology

The minimum, optimum and maximum temperatures for spore germination are 2°C, 15°C to 24°C, and 30°C, respectively (Hogg et al., 1969) and for sporulation, 5°C, 30°C and 40°C, respectively, for the stem rust which are about 5.5°C higher in each category than for *P. triticina*. Stem rust is more important late in the growing period, on late-sown and maturing wheat cultivars, and at lower altitudes. Spring-sown wheat is particularly vulnerable in the higher latitudes if sources of inoculum are located downwind. Large areas of autumn-sown wheat occur in the southern Great Plains of North America, providing inoculum for the northern spring-sown wheat crop. In warm humid climates, stem rust can be especially severe due to the long period of favorable conditions for disease development when a local inoculum source is available.

Stem rust differs from leaf rust in requiring a longer dew period (six to eight hours are necessary). In addition, many penetration pegs fail to develop from the appressorium unless stimulated by at least 10,000 lux of light for a three-hour period while the plant slowly dries after the dew period. Maximum infection is obtained with 8 to 12 hours of dew at 18°C followed by 10,000+ lux of light while the dew slowly dries and the temperature rises to 30°C (Rowell, 1985). Light is seldom a limiting factor in the field as dews often occur in the morning. However, little infection results when evening dews and/or rains are followed by winds causing a dry-off prior to sunrise. In the greenhouse, reduced light is often the reason for poor rate of infection. The effect of light probably is an effect on the plant rather than the fungus system as urediniospores injected inside the leaf whorl result in successful fungal penetrations without light striking the fungus. Stem rust uredinia occur on both leaf and stem surfaces as well as on the leaf sheaths, spikes, glumes, awns and even grains.

A stem rust pustule (uredinium) can produce about 10,000 urediniospores per day (Katsuya and Green, 1967; Mont, 1970). This is more than the leaf rust, but the infectivity is lower with only about one germling in 10 resulting in a successful infection. Stem rust uredinia, being mostly on the stem and leaf sheath tissues, often survive longer than those of the leaf rust, which are confined more often to the leaf blades. The rate of increase of disease for the two rusts is very similar.

Stem rust urediniospores are rather resistant to atmospheric conditions if their moisture content is moderate (20 to 30%). Long-distance transport occurs annually (800 km) across the North American Great Plains (Roelfs, 1985a), nearly annually (2000 km) from Australia to New Zealand (Luig, 1985) and at least three times in the past 75 years (8,000 km) from East Africa to Australia (Watson and de Sousa, 1983).

Aeciospores can also be a source of inoculum of wheat stem rust. Historically, this was important in North America and northern and eastern Europe. This source of inoculum has generally been eliminated or greatly reduced by removal of the barberry (*Berberis vulgaris* L.) from the proximity of wheat fields. Aeciospores infect wheat similarly to urediniospores infection.

In modern times it is possible to track down the movement of stem rust race Ug99 (Pretorious et al., 2000). Typically, most spores will be deposited close to the source (Roelfs and Martell, 1984); however, long-distance dispersal is well documented with three principal modes of dispersal known to occur. The first mode of dispersal is a single event, extremely long-distance (typically cross-continent) dispersal that results in pathogen colonization of new regions. Dispersion of this type is rare under natural conditions and by nature being inherently unpredictable. Assisted long-distance dispersal, on travelers clothing or infected plant material, is another increasingly important factor for the colonization in new areas by the rust pathogens. More recently, concerns over non-accidental release of plant pathogens as a form of "agricultural bio-terrorism" have arisen. Wheat stem rust has been thought to be as one pathogen of concern (Hugh-Jones, 2002) primarily due to its known ability to cause devastating production losses to a major staple food.

The third mode of dispersal, extinction and recolonization, occurs in areas that have unsuitable conditions for year-round survival. Typically these are temperate areas where hosts are absent during winter or summer. An example of this mechanism is the "*Puccinia* Pathways" of North America, a concept developed by Stakman (1957) in which rust pathogens over-winter in southern US or Mexico and recolonize wheat areas in the Great Plains and further north following the prevailing south-north winds as the wheat crop season progresses.

The most common mode of dispersal for rust pathogens is step-wise range expansion and it occurs over shorter distances, within a country or a region. A good example of this type of dispersal mechanism would include the spread of *Yr9*-virulent race of *P. striiformis* that evolved in eastern Africa and migrated to South Asia through the Middle East and West Asia in a step-wise manner over about 10 years, and caused severe epidemics along its path (Singh et al., 2004b).

Singh et al. (2008) summarized the status of the race Ug99 up to 2007. By then, the presence of Ug99 had been confirmed in Uganda, Kenya, Ethiopia, Sudan and Yemen. Occurrence in Yemen provided evidence that Ug99 was moving towards the Middle East and Asia. Subsequent confirmation of Ug99 (race TTKSK) in Iran, by FAO in 2008 (FAO, 2008), supported these predictions. Positive Iranian Ug99 isolates were actually collected in 2007 from two sites, Borujerd and Hamadan, in northwestern Iran, and underwent extensive testing to confirm the race (Nazari et al., 2009). Given the regular northeasterly airflows out of Yemen (Singh et al., 2008), the possibility that this was a new incursion from Yemen is considered likely. Within Africa, several significant changes have occurred in both the pathogen population and distribution. Virulence to resistance genes, Sr24 and Sr36, was detected in Kenya in 2006 and 2007, respectively (Jin et al., 2008, 2009). Seven variants are now recognized as being part of the "Ug99 race lineage". All are closely related, having nearly identical DNA fingerprints, but differ slightly in their avirulence/virulence phenotypes (Jin et al., 2008; Szabo, 2007; Visser et al., 2010). Pathogen monitoring indicates that Ug99 lineages are rapidly spreading in Africa, as combined virulence to Sr31 and Sr24 (race PTKST) in South Africa was confirmed in 2009 (Pretorius et al., 2010) as well as in Ethiopia. At present, Ug99 (race TTKSK) is the only known race in the lineage confirmed outside of Africa, but the eventual appearance of other variants outside of Africa is considered likely.

Confirmed race data indicate that movements of Ug99 variants are now occurring within Africa. Detection of race TTKSF in South Africa in 2000 and confirmation as part of the Ug99 lineage (Visser et al., 2009) was the first indication of the interchange between East African and southern African stem rust populations. The Sr31-avirulent race TTKSF is presumed to be an exotic introduction into South Africa. Race TTKSP is considered to have evolved locally within South Africa and to have acquired virulence to Sr24 through mutation (Terefe et al., 2010). Southern Africa was considered to be a single epidemiological zone (Saari and Prescott, 1985) and the recent detection of race TTKSF in Zimbabwe supports the spread of common races within the region. Identification of race PTKST, virulent to both Sr31 and

Sr24, during 2009 in South Africa (Pretorius, 2010) was a further indication of the connection between East and southern Africa stem rust populations. Race PTKST was previously detected in Kenya from isolates collected in 2008 and, more recently, from back-dated analysis of isolates collected in Ethiopia in 2007.

Hosts

Wheat, barley, triticale and a few related species are the primary hosts for *P. graminis*. The primary alternate host in nature has been *Berberis vulgaris*, a species native to Europe, although other species have been susceptible in greenhouse tests.

Alternate Hosts/Substrates

The main alternate host for *P. graminis* is *B. vulgaris*, which was spread by humans across the northern latitudes of the Northern Hemisphere. The barberry spread westward with humans and became established as a naturalized plant from Pennsylvania through the eastern Dakotas and southward into north-eastern Kansas. Many species of Berberis, Mahonia and Mahoberberis are susceptible to *P. graminis* (Roelfs, 1985b).

The alternate host was a major source of new combinations of genes for virulence and aggressiveness in the pathogen (Groth and Roelfs, 1982). The amount of variation in the pathogen made breeding for resistance difficult, if not impossible. Of the virulence combinations present one year, many would not reoccur the following year, but many new ones would appear (Roelfs, 1982). The barberry was the source of inoculum (aeciospores) early in the season. Generally, infected bushes were close to cereal fields of the previous season, so inoculum travelled short distances without the loss in numbers and viability associated with long-distance transport. Barberry was a major source of stem rust inoculum in Denmark (Hermansen, 1968) and North America (Roelfs, 1982).

Resistance to *P. graminis* in barberry is reported to result from the inability of the pathogen to directly penetrate the tough cuticle (Melander and Craigie, 1927). *Berberis vulgaris* becomes resistant to infection about 14 days after the leaves unfold. However, infections occur on the berries, thorns and stems, which suggest the toughening of the cuticle, may not be as important as originally thought of. An excellent description of stem rust of cereals including the sexual and asexual stages of the disease, as well as the control was published by Craigie (1957).

Other Hosts

Barley, triticale and an occasional rye plants are infected by wheat stem rust. Wild barley, such as *H. jubatum* L. and rarely *H. pusillum* Nutt., and *Aegilops cylindrica* L., are sometimes infected in the United States, Canada and Mexico (Roelfs, 1986); however, it is thought that the inoculum generally comes from wheat to these grasses rather than vice versa in the US. But, in areas where no wheat stem rust is found, it is possible to find wheat stem rust on the other hosts.

Primary Hosts

T. aestivum and *T. turgidum* var. *durum* L. as cultivated wheat and triticale are the primary hosts of wheat stem rust.

Pathogen

Fontana (1932) made the first known detailed study, including precise drawings, of *P. graminis* in 1767. Persoon named the fungus on barberry *Aecidium berberidis* in 1791 and the form on wheat *P. graminis* in 1794 (Persoon, 1801). DeBary (1866) showed that the two fungi were different stages of a single species. Craigie (1927) made the first controlled crosses between strains of *P. graminis*.

Life Cycle

In most areas of the world, the life cycle (Figs. 11.1 and 11.2) of *P. graminis* f. sp. *tritici* consists of continual uredinial generations (asexual stage). The fungus spreads by airborne urediniospores from one wheat plant to another and from field to field. Primary inoculum may originate locally (endemic) from volunteer plants or be carried long distances (exodemic) by wind and deposited by rain. In North America, *P. graminis* annually moves 2000 km from the southern winter wheat to the most northern spring wheat in 90 days or less and in the uredinial cycle can survive the winter at sea level to at least 35°N. Snow can provide cover that occasionally permits *P. graminis* to survive as infections on winter wheat even at severe sub-freezing temperatures experienced at 45°N (Roelfs and Long, 1987). The sexual cycle seldom occurs except in the Pacific Northwest of the United States (Roelfs and Groth, 1980) and in local areas of Europe (Spehar, 1975; Zadoks and Bouwman, 1985). Although the sexual cycle produces a great genetic diversity (Roelfs and Groth, 1980), it also produces a large number of individuals that are less fit due to avirulence genes (Roelfs and Groth,

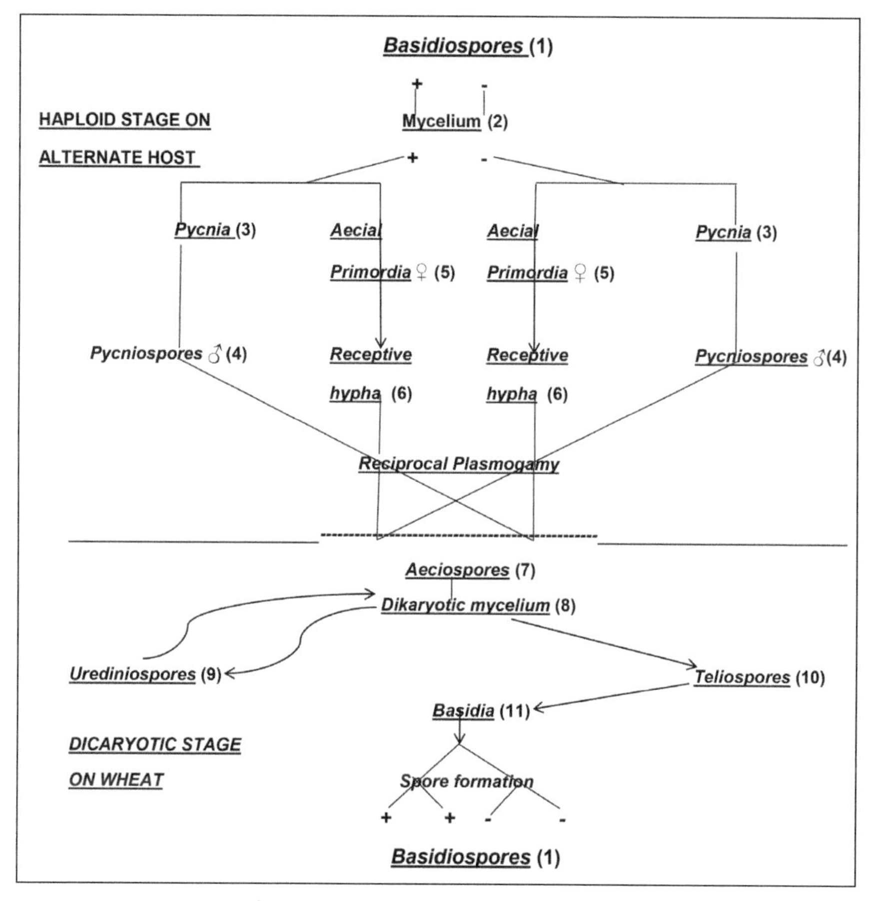

Figure 11.1. Life cycle of *Puccinia graminis.*

1988) and to re-assortment of genes for aggressiveness. *P. graminis* has successfully developed an asexual reproduction strategy that apparently allows the fungus to maintain necessary genes in blocks that are occasionally modified by mutation and selection. There are many diagrams and pictures of the stem rust life cycle; some of them are oversimplified, and some are very detailed. Both a diagram and a drawing of the different stages of the pathogen on the alternate host and on the wheat plant have been given here (Figs. 11.1 and 11.2).

Plus and minus basidiospores (1) [Fig. 11.3b] infecting leaves of the alternate host (Fig 11.3c, *Berberis*) have been shown. The germ tube (2) of the basidiospores produces a sharp peg-like structure that punctures the leaf surface (direct penetration) and then after the fungus begins to grow through the epidermal cells of the barberry leaf and develops mycelia

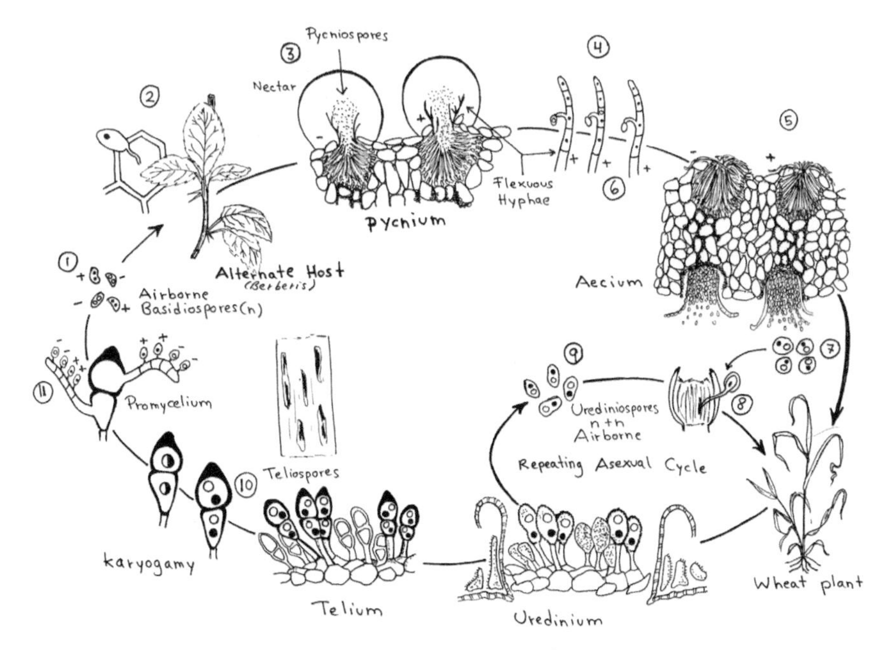

Figure 11.2. Illustrated Life cycle of *P. graminis*.

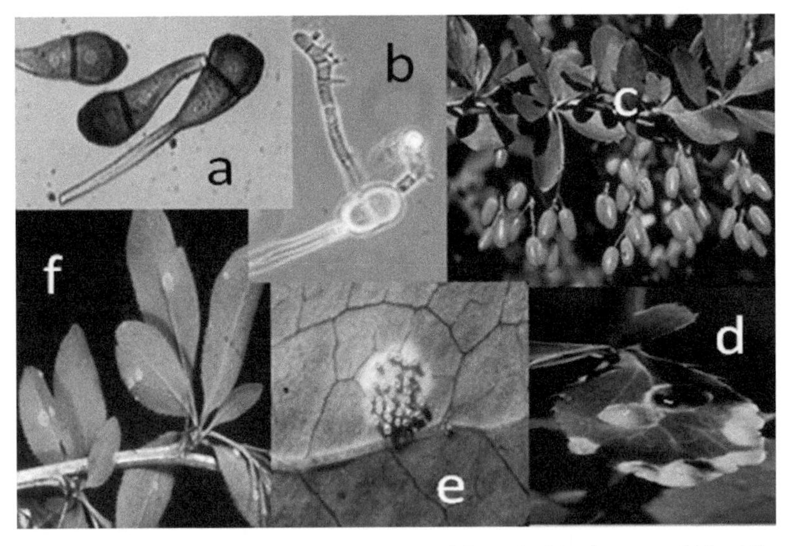

Figure 11.3. Stages and structures of the stem rust life cycle: (a) teliospore, (b) basidia and basidiospore, (c) *Berberis* plant (alternate host), (d) pycnia on barberry, (e) aecia on the lower part of the barberry leaf.

Color image of this figure appears in the color plate section at the end of the book.

within the leaf. The mycelium produces spermatogonia (pycnia) towards the upper surface of the leaf [Fig. 11.3d] (3). New infections may result from the spermatia (pycniospores) produced in the spermatogonia. At the same time, aecial primordial (5) develop on the lower surface of the leaves. Aeciospores are however, only produced after dikaryotization (plasmogamy). The dikaryotization takes place in the following ways: A spermatium (4) fuses with a receptive hypha of a spermatogonium of the opposite mating type (6) or two vegetative hyphae of different matting types may fuse. If a leaf is infected by spores of only one matting type, spermatia of the other matting type should be transferred to it from other leaves by insects. The spermatial nuclei migrate through the mycelium to the basal cells of the aecial primordium and form dikaryons. The spermatogonium can accordingly be considered as the male and the basal cell of the aecial primordium as a female sex organ. Plasmogamy can also occur through somatogamous fusion of *plus* (+) and *minus* (–) mycelia. The dikaryotic basal cell pinches off aeciospore mother cells in a chain-like fashion. They divide into two cells. The apical cell becomes a dikaryotic aeciospore (7), which is disseminated (5) after the aecium opens, while the remaining cells of the pair degenerate. The aeciospore infects the wheat plant by entry of the germ tube through the stomata (8). In the wheat plant, the mycelium produces dikaryotic urediniospores primarily on the lower surfaces of leaves. The urediniospores [Fig. 11.4e] constitute the so-called repeating stage of the rusts

Figure 11.4. Structures present on different parts of the wheat plant: (a) uredinia on stem, (b) uredinia on the head; (c) uredinia on leaf and sheath, (d) uredinia on seedling leaf, (e) urediniospores, (f) telia, and (g) teliospore.

Color image of this figure appears in the color plate section at the end of the book.

since several "harvest" of spores may be produced in one growing season. They form structures called uredinia [Fig 11.4a, b, c, d] (singular uredinium: L. urere = to burn) because of their reddish color. The uredinial cells are formed subepidermally from dikaryotic mycelium originating from the germination of an aeciospore or a urediniospore; these spread the infection on the same plant and to others as the asexual cycle (9). After a period of vegetative growth of the mycelia, the telia [Fig. 11.4f] (singular telium), groups of binucleate cells called teliospores, also termed teleutospores (10) [Fig 11.4g], are produced on the dikaryotic mycelium. The dikaryotic phase comes to an end with karyogamy in each of the two cells of the teliospores (10). The teliospore is a probasidium overwinters. Subsequently each cell of the diploid teliospore germinates when favorable conditions prevail. The promycelium or metabasidium grows out from each cell of the teliospore. The diploid nucleus now migrates into the metabasidium, under goes meiosis and produces four haploid nuclei to form a septate basidium (11) [Fig. 11.3b] from which arise four basidiospores (1).

The repeating asexual cycles then involve urediniospores producing uredinia in about 14-day cycles with optimum conditions. Urediniospore germination starts in one to three hours at optimum temperatures in the presence of free water (dew). The moisture or dew period must last six to eight hours at favorable temperatures for the spores to germinate and produce a germ tube and an appressorium. Visible development will stop at the appressorium stage until at least 10,000 lux (16,000 being optimum) of light are provided. Light stimulates the formation of a penetration peg that enters a closed stoma. If the germling dries out during the germination period, the process is irreversibly stopped. The penetration process takes about three hours as the temperature rises from 18°C to 30°C (Rowell, 1985). The light requirement for infection makes *P. graminis* much more difficult to work within the greenhouse than *P. triticina*. Most likely, light seldom has an effect in the field except when dew periods dissipate before daybreak.

Urediniospores develop in pustules (uredinia) that rupture the epidermis and expose masses of reddish-brown spores (Fig. 11.4a–d). The uredinia are larger than those of leaf rust and are oval-shaped or elongated, with loose or torn epidermal tissue along the margins. The urediniospores are reddish-brown, elliptical to egg-shaped, echinulate structures measuring 24 to 32 µm x 18 to 22 µm (Fig. 11.4e)

As the host matures, the telia (Fig. 11.4f) are produced directly from urediniospore infections or teliospores can be produced in a mature uredinial pustule. The teliospores are dark brown two-celled and somewhat wedge-shaped (Fig. 11.4g). They have thick walls, and measure 40 to 60 µm x 18 to 26 µm. The apical cell is rounded or slightly pointed.

Leaf Rust

Epidemiology

Puccinia triticina can survive the same environmental conditions as the wheat leaf, provided infection but no sporulation has occurred. The fungus can infect with dew periods of three hours or less at temperatures of about 20°C; however, more infections occur with longer dew periods. At lower temperatures, longer dew periods are required, for example, at 10°C a 12-hour dew period is necessary. Few if any infections occur where dew period temperatures are above 32°C (Stubbs et al., 1986) or below 2°C. In some regions of the world, most of the severe epidemics occur when uredinia and/or latent infections survive the winter at some threshold level on the wheat crop, or where spring-sown wheat is the recipient of exogenous inoculum at an early date, usually before heading. Severe epidemics and losses can occur when the flag leaf is infected before anthesis (Chester, 1946). *Puccinia tritici-duri* (Viennot-Bourgin, 1941) has not been intensively studied under controlled conditions but, in general, environmental conditions for infection are likely to be similar. However, the latent period (uredinial) is approximately three to four days longer, and teliospore production starts shortly after initial urediniospore production (Huerta-Espino, 1992). The initial inoculum (aeciospores) is from *A. italica* (Ezzahiri et al., 1992, 1994), and the disease spread is generally limited.

Hosts

Puccinia triticina is primarily a pathogen of wheat, its immediate ancestors and the man-made crop triticale. Recent evidences (Huerta-Espino and Roelfs, 1989) indicated that populations of leaf rust exist in Europe, Asia and Africa that are primarily pathogens of durum wheat. They are all distinct from the population that exists worldwide on bread wheat (Huerta-Espino and Roelfs, 1992).

Alternate Hosts

Allen (1932) demonstrated that *P. triticina* was bipolar and heterothallic rust. The fungus produces its sexual gametes (pycniospores and receptive hyphae) on the alternate host (Fig. 11.5). Most rust researchers have now accepted that *Thalictrum speciosissimum* Loefl. (in the Ranunculaceae family) is the primary alternate host for *P. triticina* in Europe. Alternate hosts seldom, if ever, function in North America (Saari et al., 1968), South America and Australia. The alternate host is considered important at least for recombining virulence factors in part of the Mediterranean area

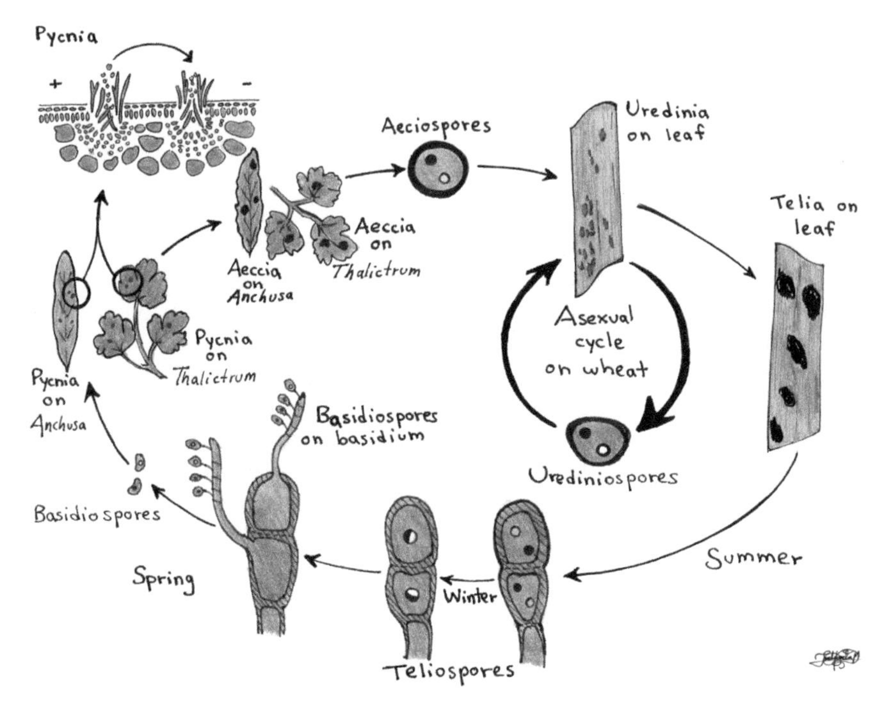

Figure 11.5. Illustrated life cycle of *P. triticina* and *P. tritici-duri*.

Color image of this figure appears in the color plate section at the end of the book.

(d'Oliveira and Samborski, 1964). The importance of the alternate host in generating changes in the pathogen population for virulence combinations and other factors worldwide is unknown.

The primary alternate host of *P. triticina*, including the durum attacking populations, is *T. speciosissimum* (Fig. 11.6f), whereas *A. agregata, A. undulata, Echium glomeratum* and *Lycopsis arvensis* (Boraginaceae) are the alternate hosts for the leaf rusts on wild wheat (*Aegilops* spp.) and rye. However, in the Mediterranean area, a second leaf rust *P. tritici-duri* (Vienot-Bourgin, 1941) on bread and durum wheat has basidiospores that attack *A. italica* [Fig. 11.6g] (Ezzahiri et al., 1992, 1994; Anikster et al., 1997). The alternate host is essential for the survival and spread of *P. tritici-duri*.

The alternate host is infected when the teliospores germinate in the presence of free moisture. Basidiospores (Fig. 11.6e) are produced that are capable of being carried a short distance (a few meters) to infect the alternate hosts. Approximately, seven to ten days following infection, pycnia with pycniospores and receptive hyphae appear. These serve as the gametes, and fertilization occurs when the nectar containing the pycniospores is carried to receptive hyphae of the other mating type by insects, by splashing rain

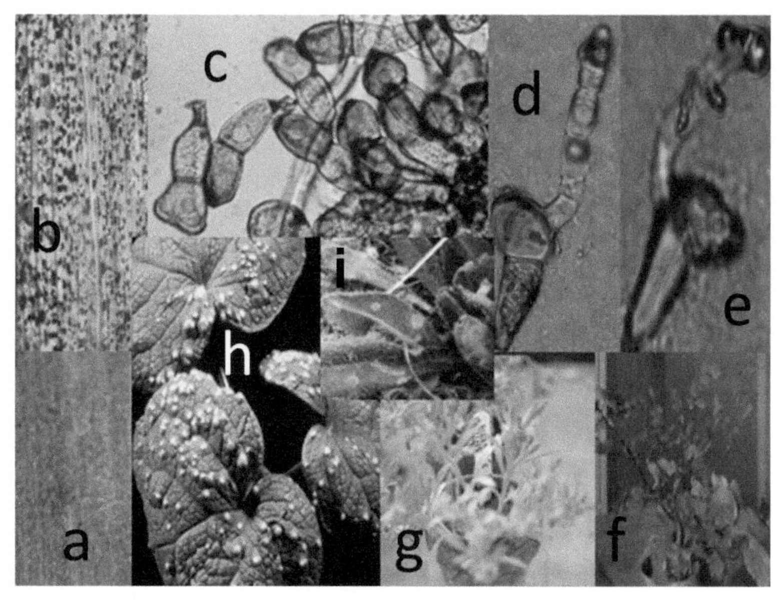

Figure 11.6. Structures and alternate host plants of the life cycle of *P. triticina* and *P. tritici-duri*: (a) Uredinia on wheat leaf, (b) telia, (c) Teliospore, (d) basidia, (e) basidiospore, (f) *Thalictrum* (alternate host), (g) *Anchusa* (alternate host), (h) pycnia on *Thalictrum*, (i) pycnia on *Anchusa*.

Color image of this figure appears in the color plate section at the end of the book.

or by cohesion. The aecial cups appear seven to 10 days later on the lower surface of the leaf, producing aeciospores that are windborne and that cause infection by penetrating the stomata of the wheat leaves.

Other Hosts

P. triticina might attack many species of grasses, but it is unclear which species serve as functional hosts in nature. Many grasses can be infected by artificial inoculation; however, this may not occur in the field. Potential hosts for wheat leaf rusts could be wild or weedy species of the genera *Triticum* and *Aegilops* (now classified as *Triticum*) and the related species of the *Agropyron* complex. In southern Italy, *Agropyron* sp. is also reported to be infected by wheat-*Thalictrum*-infecting rust (Casulli, 1988; Huerta-Espino unpubl. data). The most common non-crop host for wheat leaf rust is volunteer or self-sown wheat. These plants may be in fallow fields, along the edges of fields and roads, as weeds in a second crop and as a cover crop under orchards, along irrigation canals, etc. This is the major source of inoculum throughout much of the world where wheat is autumn or winter sown.

Primary Hosts

The primary host of *P. triticina* is *T. aestivum* L. em. Thell. *T. turgidum* L., is also a host not only in the Mediterranean, the Middle East, Ethiopia, Chile and India, where durum wheat is more extensively cultivated but also in Mexico and the US. Wheat leaf rust would also appear to be a major threat to triticale (X. *Triticosecale* Wittmack), the crop derived from the man-made cross between wheat and rye (Skovmand et al., 1984). *P. tritici-duri* is a pathogen of *T. turgidum* and some very susceptible *T. aestivum* cultivars. This leaf rust is limited to regions around the Mediterranean Sea where *Anchusa italica* is common and where traditional methods of cultivation allow the perennial *Anchusa* to survive in wheat fields (Ezzahiri et al., 1994).

Pathogen

De Candolle (1815) separated wheat leaf rust from other rusts of wheat and called it *U. rubigovera* in 1815. Eriksson and Henning (1894) described the causal organisms of both wheat and rye leaf rust as *P. dispersa*. Eriksson (1894) separated the wheat and rye leaf rust fungi, and the causal organism of wheat leaf rust became *P. triticina*, a name still used in parts of eastern Europe. Mains (1932) placed the causal organism of wheat leaf rust in *P. rubigovera* and established a complex group of 52 formae specialis for the fungus causing wheat leaf rust. *Puccinia recondita* was recommended by Cummins and Caldwell (1956) as the correct designation for the wheat leaf rust pathogen. *P. recondita* f. sp. *tritici* was widely used thereafter (Samborski, 1985). However, in 1984, Savile (1985) stated that *P. triticina* was the binomial for wheat leaf rust and *P. recondita* for rye leaf rust. Recently, an extensive study (Anikster et al., 1997) of morphological characters, alternate host ranges and the intermating of cultures within and among various populations showed that *P. triticina* and *P. recondita* are different organisms. *Puccinia recondita* f. sp. *secalis* and *P. tritici-duri* are more similar and share some ability to attack the same alternate host species.

Life Cycle

Life cycles for *P. triticina* and *P. tritici-duri* and the disease cycle for wheat leaf rust are depicted in Fig. 11.5. The time for each event and frequency of some events like the sexual cycle, wheat cropping season and green-bridge may vary among areas and regions of the world. The events might be very similar to those described to occur in the stem rust (*P. graminis*) life cycle.

The alternate host currently provides little direct inoculum of *P. triticina* to wheat, but may be a mechanism for genetic exchanges between races and perhaps populations. The pathogen survives the period between wheat crops in many areas on a green-bridge of volunteer (self-sown) wheat. Inoculum in the form of urediniospores can be blown by wind from one region to another. The sexual cycle is essential for *P. tritici-duri*. Teliospores can germinate shortly after development, and basidiospore infection can occur throughout the wheat-growing cycle.

The urediniospores (Fig. 11.7f) initiate germination 30 minutes after contact with free water at temperatures of 15° to 25°C. The germ tube grows along the leaf surface until it reaches a stoma; an appressorium is then formed, followed immediately by the development of a penetration peg and a sub-stomatal vesicle from which primary hyphae develop. A haustorial mother cell develops against the mesophyll cell, and direct penetration occurs. The haustorium is formed inside the living host cell in a compatible host-pathogen interaction. Secondary hyphae develop resulting in additional haustorial mother cells and haustoria. In an incompatible host-pathogen response, the haustoria fail to develop or develop at a slower rate. When the host cell dies, the fungus haustorium dies. Depending upon when or how many cells are involved; the host-pathogen interaction will result in a visible resistance response (Rowell, 1981, 1982).

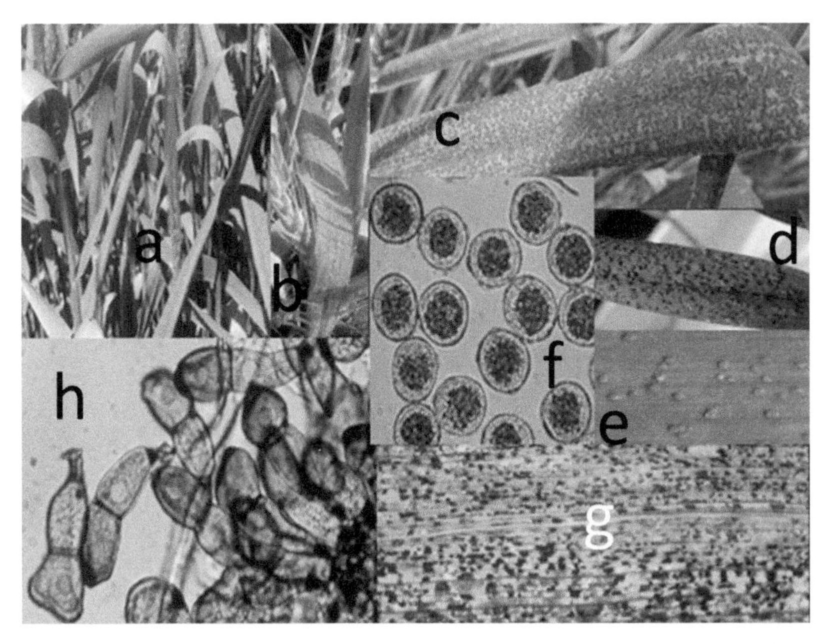

Figure 11.7. Structures on the wheat plant from the life cycle of *P. triticina* and *P. tritici-duri*: (a, b), and (c) Uredinia on flag leaves (field infection), (d, e) uredinia on flag leaf (greenhouse), (f) urediniospores, (g) telia, (h) teliospores.

Color image of this figure appears in the color plate section at the end of the book.

Spore germination to sporulation can occur within a seven- to 10-day period at optimum and constant temperatures. At low temperatures (10°C to 15°C) or diurnal fluctuations, longer periods are necessary. The fungus may survive as insipid mycelia for a month or more when temperatures are near or below freezing. Maximum sporulation is reached in about four days following initial sporulation (at about 20°C). Although the number can vary greatly, about 3,000 spores are produced per uredinium per day. This level of production may continue for three weeks or more if the wheat leaf remains alive that long (Chester, 1946; Stubbs et al., 1986). Uredinial (pustules) are red, oval-shaped and scattered, and they break through the epidermis (Fig. 11.7a–e). Urediniospores are orange-red to dark red, echinulate, spherical and usually measure 20 to 28 µm in diameter (Fig. 11.7f). The teliospores are dark brown, two-celled with thick walls and rounded or flattened at the apex. *Puccinia tritici-duri* differs from *P. triticina* in requiring 10 to 12 days for appearance of urediniospores, and initial teliospore production often occurs within 14 days of initial infection. The uredinia are yellowish-brown and produce fewer urediniospores per uredinia, and within a few days the lesion primarily produces teliospores. Also, *P. tritici-duri* infections are likely to be on the lower leaf surface.

The teliospores of *P. triticina* are formed under the epidermis under unfavorable conditions or senescence and remain with the leaves (Fig. 11.7g). Leaf tissues can be dispersed by wind, animals or humans to considerable distances. Basidiospores are formed and released under humid conditions, which limit their spread. Basidiospores are also hyaline and sensitive to light, further limiting travel to probably 10s of meters. Aeciospores are more similar to urediniospores in their ability to be transported by wind currents, but long-distance transport has not been noted for some reason. *Puccinia tritici-duri* produces abundant teliospores within weeks of initial infection, producing dark secondary rings of uredinia which rapidly turn onto telia around each infection site (Huerta-Espino, 1992).

Stripe Rust

Stripe or yellow rust of wheat caused by *P. striiformis* f. sp. *tritici* can be as damaging as stem rust. However, stripe rust has a lower optimum temperature for development that limits it as a major disease in many areas of the world. Stripe rust is principally an important disease of wheat during the winter or early spring or at high elevations. Stripe rust of wheat may be the cause of stripe rust on barley (Stubbs, 1985). In Europe, a forma specialis of *P. striiformis* has evolved that is commonly found on barley and seldom on any but the most susceptible wheat (Zadoks, 1961). *Puccinia striiformis* f. sp. *hordei* was introduced into South America where it spread across the continent (Dubin and Stubbs, 1986) and was later identified in Mexico and United States (Roelfs et al., 1992; Roelfs and Huerta-Espino, 1994).

Epidemiology

P. striiformis f. sp. *tritici* has the lowest temperature requirements of the three wheat rust pathogens. Minimum, optimum and maximum temperatures for stripe rust infection are 0°C, 11°C and 23°C, respectively (Hogg et al., 1969). *P. striiformis* frequently can actively overwinter on autumn-sown wheat. Most of the epidemiological work has been done in Europe and reviewed by Zadoks and Bouwman (1985) and Rapilly (1979).

In Europe, *P. striiformis* oversummers on wheat (Zadoks, 1961). The amount of over-summering rust depends on the amount of volunteer wheat, which, in turn, is a function of moisture during the off-season. The urediniospores are then blown to autumn-sown wheat. In northwestern Europe, overwintering is limited to uredinial mycelia in living leaf tissues as temperatures of –4°C will kill exposed sporulating lesions. Latent lesions can survive if the leaf survives. In other areas of the world, snow can insulate the sporulating lesions from the cold temperatures, therefore air temperatures below –4°C fail to eliminate the rust lesions. The latent period for stripe rust during the winter can be up to 118 days and is suspected to be as many as 150 days under a snow cover (Zadoks, 1961).

In areas near the equator, stripe rust tends to cycle endemically from lower to higher altitudes and return following the crop phenology (Saari and Prescott, 1985). In more northern latitudes, the cycle becomes longer in distance with stripe rust moving from mountain areas to the foothills and plains.

Due to their susceptibility to ultraviolet light, urediniospores of stripe rust probably are not transported in a viable state as far as those of leaf and stem rusts. Maddison and Manners (1972) found stripe rust urediniospores three times more sensitive to ultraviolet light than those of stem rust. Still, Zadoks (1961) reports that stripe rust was wind-transported in a viable state for more than 800 km. The introductions of wheat stripe rust into Australia and South Africa and barley stripe rust into Colombia were probably aided by humans through jet travel (Dubin and Stubbs, 1986; O'Brien et al., 1980). However, the spread of stripe rust from Australia to New Zealand, a distance of 2,000 km, was probably through airborne urediniospores (Beresford, 1982). Perhaps an average spore of stripe rust has a lower likelihood of being airborne in a viable state over long distances than that of the other wheat rusts, but certainly some spores must be able to survive long-distance transport under special and favorable conditions. There are several examples of the sequential migration of stripe rust. Virulence for the gene *Yr2* (cultivars Siete Cerros, Kalyansona and Mexipak) was first recorded in Turkey and over a period of time was traced to the subcontinent of India and Pakistan (Saari and Prescott, 1985) and may be associated with the weather systems called the "Western Disturbance". As mentioned, barley stripe rust in South America migrated from its introduction point in Colombia to Chile over a period of a few years (Dubin and Stubbs, 1986).

Most areas of the world studied seem to have a local or nearby source of inoculum from volunteer wheat (Line, 1976; Stubbs, 1985; Zadoks and Bouwman, 1985). However, some evidence points to inoculum coming from non-cereal grasses (Hendrix et al., 1965; Tollenaar and Houston, 1967). Future studies of stripe rust epidemiology need to take into account not only the presence of rust on nearby grasses, but also the fact that the rust must occur on the grasses prior to its appearance on cereals. The virulent phenotype must be shown to be the same on both hosts and that it moves from the grass to wheat during the crop season.

Stripe rust epidemics in the Netherlands can be generated by just a single uredinium per hectare surviving the winter if the spring season is favorable for rust development (Zadoks and Bouwman, 1985). Visual detection of a single uredinium per hectare is unlikely, however, as foci develop around the initial uredinium, it becomes progressively easier to detect.

Hosts

P. striiformis is a pathogen of grasses and cereal crops: wheat, barley, triticale and rye. Stripe rust is the only rust of wheat that consistently spreads beyond the initial infection point within the plant.

Alternate Hosts

Before 2009, the alternate host for *P. striiformis* was unknown. Eriksson and Henning (1894) looked for the alternate host among species of the Boraginaceae. Tranzschel (1934) suggested that *Aecidium valerianella*, a rust of valerianella, might be related to *P. striiformis*. Mains (1933) thought that *P. koeleriae* Arth., *P. arrhenatheri* Eriks. and *P. montanensis* Ellis, which have aecidial states on *Berberis* and *Mahonia* spp., could be related to *P. striiformis*. Straib (1937) and Hart and Becker (1939) were unsuccessful in attempts to infect *Berberis*, *Mahonia* and *Valerianella* spp. An alternate host of the rust, *P. agropyri* Ell. & Ev., is *Clematis vitalba*. This rust closely resembles *P. striiformis* thus, Viennot-Bourgin (1934) suggested that the alternate host of stripe rust might occur in the Clematis family. Teliospores readily germinate immediately to produce basidiospores (Wright and Lennard, 1980). Today the mystery has been resolved, and the hypothesis of Mains (1933) was confirmed. Jin et al. (2010) demonstrated successful infection of *Berberis chinensis, B. holstii, B. koreana*, and *B. vulgaris* through inoculum originating from wheat straw that harbored germinating teliospores, confirming that *Berberis* spp. are the alternate hosts of *P. striiformis* [Fig. 11.8] (Jin et al., 2010; Hovmoller et al., 2011). Details of events that occur on the alternate host after infection by the basidiospores of *P. striiformis* onto *Berberis* are not well documented yet, as have been described for *P. graminis* (Craigie, 1927, 1957; Walter, 1961) and *P. triticina* (Jackson and Mains, 1921; Allen, 1931, 1932; Chester, 1946). However, it is expected to follow the same pattern as *P. graminis*; since they have in common *Berberis* as an alternate host.

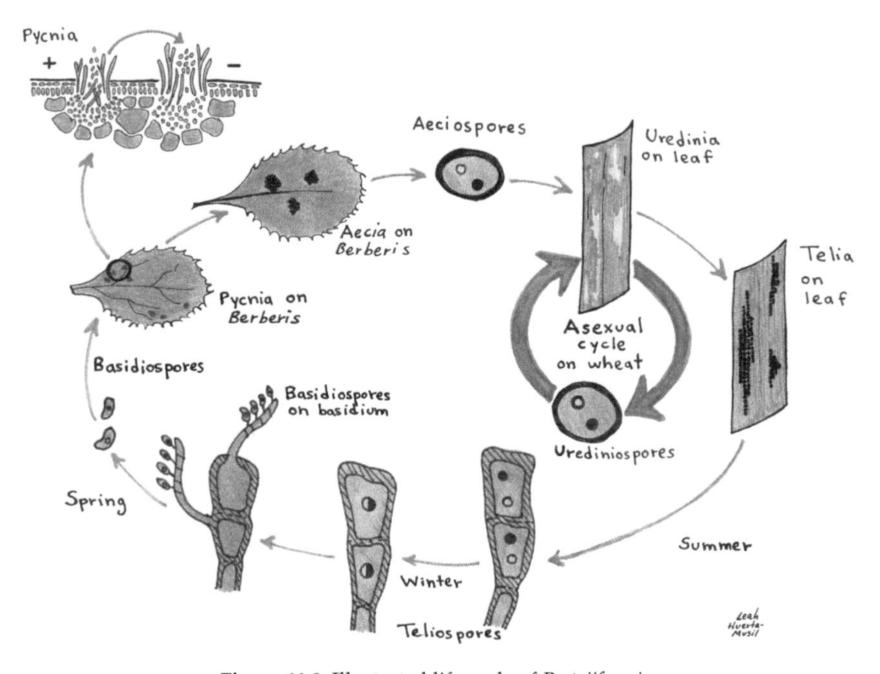

Figure 11.8. Illustrated life cycle of *P. striiformis*.

Color image of this figure appears in the color plate section at the end of the book.

Other Hosts

P. striiformis seems to lack a clearly defined formae specialis that occur with *P. graminis*, and isolates of stripe rust seem to have a wider host range than those of *P. triticina*. Sufficient evidence exists for the separation of the primary wheat attacking form from the barley attacking form (Stubbs, 1985; Zadoks, 1961). *P. striiformis* f. sp. *tritici* attacks primarily bread wheat, durum wheat and triticale. The assumption that stripe rust, which occurs on various grass species, has a similar virulence to that which attacks wheat is probably not valid (Manners, 1960; Tollenaar and Houston, 1967). Likewise, the ability to produce a few uredinia on some plants of a species in greenhouse tests does not prove that the species is a host under field conditions. Furthermore, there is no reason to expect that race-specific resistance does not occur in the other hosts. Many of the existing race-specific genes for resistance have been transferred from species that are other hosts.

Primary Hosts

Triticum spp. and *Triticosecale* (Triticale) are the major hosts for stripe rust *P. striiformis* f. sp. *tritici*.

Pathogen

Gadd first described stripe rust in 1777. It was reported to have caused an epidemic on rye in Sweden in 1794 (Eriksson and Henning, 1896). Schmidt (1827) designated the pathogen as *U. glumarum*; Westendorp designated the stripe rust pathogen of rye as *P. striaeformis* in 1854 (Westendorp, 1854). Eriksson and Henning (1896) chose the name *P. glumarum* in their comprehensive taxonomic work. Hylander et al. (1953) and Cummins and Stevenson (1956) revived the name currently in use, *P. striiformis* West. It is probably desirable to add the forma specialis if it has been determined.

Life Cycle

Puccinia striiformis life cycle is very similar to those of the stem rust fungi. Uredinia develop in narrow, yellow and linear stripes mainly on leaves (Fig. 11.9a, b, d, e, f) and the heads or spikelets (Fig. 11.9g). When the heads are infected, the pustules appear on the inner surfaces of glumes and lemmas (Fig. 11.9h). The urediniospores (Fig. 11.9c) are yellow to orange in color, more or less spherical, echinulate and 28 to 34 µm in diameter. Narrow black stripes are formed on leaves during telia development (Fig. 11.9i). Teliospores are dark brown, two-celled and similar in size and shape to those of *P. triticina* (Fig. 11.9j). Teliospore germinates to produce basidiospores (Fig. 11.10d) and in the presence of the alternate host *Berberis* (Fig. 11.10e), infection might occur through direct penetration as in the case of stem rust, or through the leaf stomata as in the case of leaf rust. Pycnia (Fig. 11.10f) develop on the upper side of the barberry leaf, followed by aecia formation on the lower side of the leaf (Fig. 11.10g). Aecia will in time produce aeciospores which can infect the wheat plant. Under controlled conditions aecia, collected from *Berberis chinensis* and inoculated onto susceptible

Figure 11.9. Structures on the wheat plant from the life cycle of *P. striiformis*: (a, b, d, e and f) Uredinia on flag leaves under field conditions, (c) urediniospores, (g) head infection, (h) spikelet infection, (i) telia on leaf, (j) teliospore.

Color image of this figure appears in the color plate section at the end of the book.

Figure 11.10. Structures and alternate host plants of the life cycle of *P. striiformis*: (a) Uredinia on wheat leaf, (b) telia, (c) Teliospore, (d) basidia and basidiospore, (e) *Berberis* plant (alternate host), (f) pycnia on barberry leaf, (g) aecia on the lower part of the barberry leaf.

Color image of this figure appears in the color plate section at the end of the book.

seedlings of wheat, showed typical yellow striped lesions (uredinia) of the stripe rust of wheat (Jin et al., 2010). The role of the alternate host in the large wheat areas of the world in which stripe rust is important in producing new virulence combinations has to be determined. In the meantime, the asexual stage (urediniospore) plays the most important role in spreading the disease and in the life cycle of the stripe rust of wheat.

Inoculum Production

Studies of the cereal rusts require the increase and preservation of inoculum, which, in most cases, involve urediniospores. For many experiments, inoculum of a particular pathogen phenotype or a particular isolate is needed. In such situations, it is essential to be able to purify and maintain isolates over a period of years. In other cases, larger quantities of inoculum for field inoculation may require multiplication, collection and storage for various periods of time. This is particularly important for breeding programs in which one of the major goals is breeding for rust resistance (Fig. 11.11c).

Figure 11.11. Bulk collection of urediniospores and field infection on the spreader row in a breeding stripe rust nursery.

Color image of this figure appears in the color plate section at the end of the book.

Spore Increase

The usual procedure is to select a susceptible host. A local host line can be used if it is susceptible to the isolate to be increased. Sometimes it is possible to select a host that is susceptible to the isolate to be increased, but resistant to other isolates, eliminating some of the contamination problems. Urediniospores are easily airborne and may be present outside the laboratory, representing a potential source of isolate contamination. To reduce contamination in work areas where rusted plants were grown, wash the area with water before bringing in new plants. To reduce contamination to a minimum, keep the greenhouse clean. Several situations may cause problems in obtaining adequate infection (Rowell, 1985). When spores have been stored dry, a slow rehydration process is required. If seedlings or adult plants are sprayed with water to simulate dew formation, mineral or other contaminants in the water may inhibit spore germination. Pollutants in the air have been reported to reduce infection as well (Melching et al., 1984; Sharp, 1967; Stubbs, 1985). Inoculum can be increased either on seedlings or adult plants. The choice is based primarily on personal preference and local conditions.

Seedling Plants

Generally, inoculate seedlings at 8 to 10 days of age when the primary or seedling leaf is fully expanded in the case of leaf and stem rust (Fig. 11.12). In the case of yellow rust better results are obtained if seedlings are older than 10 days. Following incubation in a dew chamber, move the seedlings to a greenhouse or growth chamber. To prevent contamination by spores from other isolates and urediniospores from outside, isolation of the seedlings is desirable. Isolates maintained on seedlings in single, small pots can be covered with a glass lamp chimney, which has the top covered with a fine mesh cloth to allow heat exchange, but minimizes spore movement. Browder (1971) designed an isolation chamber composed of a chimney and a cap with a space for air exchange. Small plastic-covered cages, 30 x 25 x 20 cm, can be constructed in which single pots or cups can be placed. Some cage designs have the front flap hinged at the top and about 3 cm short of reaching the cage floor to allow air to enter with minimal spore exchange and to allow access for watering and spore collection. It is important to have cages large enough so that the plastic does not come into contact with the inoculated part of the seedling, as some plastics are coated with a phytotoxic substance. If the humidity in the cage is very high, spore viability is affected. If dew formation or guttation drops are present for a long period of time, reinfections may occur. Best results are normally obtained when spores are collected in the afternoon. Plants may be treated

Figure 11.12. Steps in the collection of urediniospores and inoculation in the greenhouse.

Color image of this figure appears in the color plate section at the end of the book.

with maleic hydrazide at a rate of 5 to 10 mg with 50 ml of water per pot (10 cm diameter) at emergence to reduce plant growth and enhance spore production (Rowel, 1985). Flats of any size, thickly sown with wheat, can be used for large increases of inoculum (Fig. 11.13). A 30- x 25-cm flat can produce 5 g of urediniospores which may be collected with a large cyclone collector (Cherry and Peet, 1966).

Adult Plants

Adult plants are also used as hosts for inoculum increase. Inoculum can be collected directly from the field, but it is often a mixture of races, which may not be desirable. Field-collected inoculum is often contaminated with spores of other fungi that can affect subsequent experiments. To increase inoculum on adult plants in the greenhouse, it is important to select a susceptible host, maintain isolation, avoid dew formation in the isolation chamber if possible, and have high sanitation standards to prevent infection or infestation of the host with undesirable diseases or insects. High levels of humidity often result in the development of hyperparasites of the rust

Figure 11.13. Multiplication of *P. graminis*, RTR race (Singh, 1991) in the greenhouse.

Color image of this figure appears in the color plate section at the end of the book.

and lower spore viability of the rust fungi. Contamination most likely takes place before or during the inoculation and incubation processes. Contamination can also occur when isolates are watered or collected. This is particularly important where isolates are maintained for long periods of time on adult plants.

Spore Collection

Urediniospores can be collected in large numbers by tapping a rusted plant over a piece of smooth dry paper or aluminum foil (Browder, 1971). Plastic is unsatisfactory because static electricity usually results in the clinging of spores to the plastic's surface. After collections are dried (20–30% relative humidity), the spores can then be stored in a container. A modification of this method involves tapping the plant directly over a funnel or container. A mini-cyclone collector attached to a vacuum pump allows collecting from a single pustule to more or less one gram of urediniospores (Fig. 11.14b). It is essential to clean and dry the funnel between collections. Precautions should be taken not to tap soil debris or water drops into the spore collection. Collecting aphids also should be avoided.

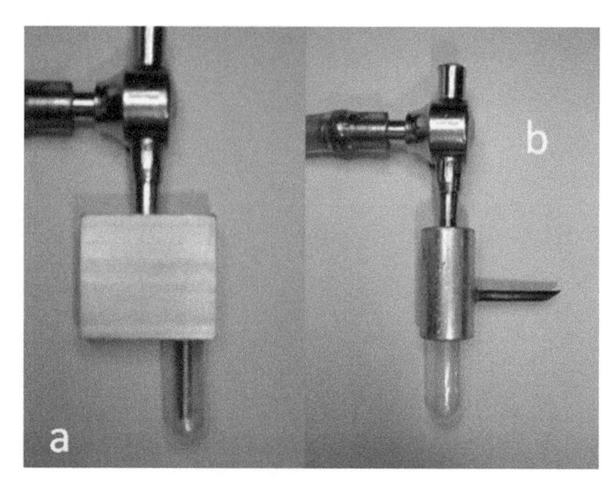

Figure 11.14. (a) Atomizer to inoculate urediniospores in an oil suspension onto seedling and adult plants in the greenhouse. (b) Mini-cyclone to collect single pustules or bulks of urediniospores in the greenhouse.

Spore Storage

There are different methods of spore storage depending on the length of storage time required and the amount of spores involved.

Room Temperature

Urediniospores can be stored at room temperatures for short periods of days (stripe rust), weeks (stem rust), and months (leaf rust) depending on the moisture of spores. The storage time can be increased by drying and maintaining the spores at 20–30% relative humidity over a desiccant.

Refrigeration

After drying the urediniospores, they can be stored at 5–8°C for variable periods of weeks or months depending upon the rust and basic conditions. They must be sealed in an airtight container or kept in a desiccator. This period can be perhaps doubled by storing the spores in non-toxic paraffinic oil or in a partial vacuum desiccator. Urediniospores on dried stem or leaf pieces can be stored for several weeks in a refrigerator.

Vacuum Drying

Vacuum drying of urediniospores in vials makes storage possible for up to 10 years (Sharp and Smith, 1957). The spores can be dried under reduced pressure (40 to 50 Tor). At this reduced pressure, a flame to seal the open end of the vial containing the spores can be used. Vials are generally stored at 5–8°C for long-term storage (longer than one year). This method has been successfully used at CIMMYT where isolates as old as 20 years remain viable. They can also be stored at room temperature for periods of one year or less. After removing the spores from storage, they can be slowly rehydrated over a period of about three hours at 50% relative humidity (Rowel, 1957). However, in cases where dew forms slowly on the plants after inoculation, this extra rehydration step may be unnecessary.

Liquid Nitrogen and Ultra Freezers

Most major laboratories worldwide use a method where urediniospores can be stored for long periods in liquid nitrogen at –196°C (Loegering et al., 1961, 1966). The spores are dried to 20–30% relative humidity and then sealed in. This process however, is less common due to no availability of freezers (–70°C or –80°C). To store inoculum in these types of freezers; one just has to dry the urediniospores at room temperature for 24 to 48 hours, and then place the dried urediniospores in a close container (Fig. 11.15).

Figure 11.15. Leaf rust storage in an ultra-freezer at –57°C, race BBG/BN (Singh et al., 2004a) and other rust in gelatin capsules in plastic and glass containers.

Leaf and stem rust urediniospores stored in these freezers can be used to inoculate directly onto seedlings or adult plants in the greenhouse tests or to inoculate resistance test host and spreader rows in the field. A heat shock in a water bath at 45°C for seven minutes and a pre-germination treatment for three hours in a humid chamber (80% RH) is required. In the case of stripe rust, best results are obtained when the stored urediniospores are first multiplied onto a susceptible host, prior to inoculation of seedlings or adult plants in the greenhouse or field conditions.

Rust Evaluation

Natural Infections

Epidemics of stem rust, leaf rust, and stripe rust are naturally occurring in many wheat growing regions of the world (Singh et al., 2012; Huerta-Espino et al., 2011; Wellings et al., 2012). These epidemics are likely the most common form of disease screening for rust resistance. Urediniospores of *P. graminis*, *P. triticina* and *P. striiformis* are found in abundance in these areas. Moist conditions and dew formation will promote infection of susceptible lines. In many cases irrigation is used to promote disease development.

Artificial Infection—Field Evaluation

Exogenous inoculum can be applied to wheat plants in the field using a variety of methods (Fig. 11.16). Urediniospores can be mixed with carriers such as talcum powder or paraffinic mineral oil such as Soltrol. Urediniospores are very hydrophobic so they donot mix readily with water, however, water based urediniospore suspensions can be injected into elongating wheat stems to infect plants in the field without the need for exogenous moisture. Suspensions of urediniospores in mineral oil can be efficiently applied to plants using sprayers of various kinds. Typically hand held sprayers are used (McIntosh et al., 1995; Roelfs et al., 1992). If needed a large amount of urediniospores can be collected in the field, particularly in dry environments using a cyclone collector with a vacuum and a centripetal force (Roelfs et al., 1992).

Artificial Infection—Greenhouse

Plants can also be artificially inoculated indoors and humidified to result in infection. This can be done either at the seedling or adult plant stages. Urediniospore suspensions in carriers such as water (with surfactant) or

Figure 11.16. Different steps of field inoculation (dew formation is required for successful infection).

mineral oil are applied to the plants which are then humidified, typically for 12–18 hours, after inoculation (Roelfs et al., 1992). High humidity is generated by humidity chambers, which can be either permanent or temporary in design. Specialized inoculators (Fig. 11.11a) have been designed to use compressed air to spray spore suspensions in mineral oil with fine droplet sizes (Browder, 1971). The procedure is essentially the same for both seedling and adult plants.

Disease Symptoms and Signs

The symptoms of stem, leaf and stripe rusts appears as flecks, or tiny white colored spots, approximately 7–10 days after inoculation. These will progress either into small sporulating pustules with chlorosis or necrosis surrounding the pustule in the incompatible or resistant interactions, or to larger sporulating pustules (signs) without chlorosis or necrosis approximately 10–14 days after inoculation (McIntosh et al., 1995; Roelfs et al., 1992). Symptoms and signs of the disease will appear more quickly under higher temperatures than they will under lower temperatures in the case of stem and leaf rust, but stripe rust signs will developed later at 15°C.

Scoring Resistance—Field Evaluation

Typically, field ratings are done when plants become physiologically mature, but can be done at any growth stage depending on the purpose of the evaluation. At maturity of plants, the leaf rust epidemic will have maximum opportunity to increase and infect all plants. Rating normally focuses on the flag and penultimate leaves. The rating consists of two parts, the first is the severity or proportion of the flag leaf covered with leaf rust, expressed as a percentage, the second is the pustule type or types present, expressed as a letter abbreviation. Often the severity is estimated using standardized diagrams such as the Cobb or modified Cobb scale (Peterson et al., 1948) (Fig. 11.17). Pustule type is described using the following scale; R = resistant or miniature uredinia, MR = moderately resistant or small uredinia, MS = moderately susceptible or moderately sized uredinia, S = susceptible or large uredinia (Roelfs et al., 1992). Therefore, a rating of "10MR" would indicate that 10% of the flag leaf was covered by moderately resistant or small uredinia.

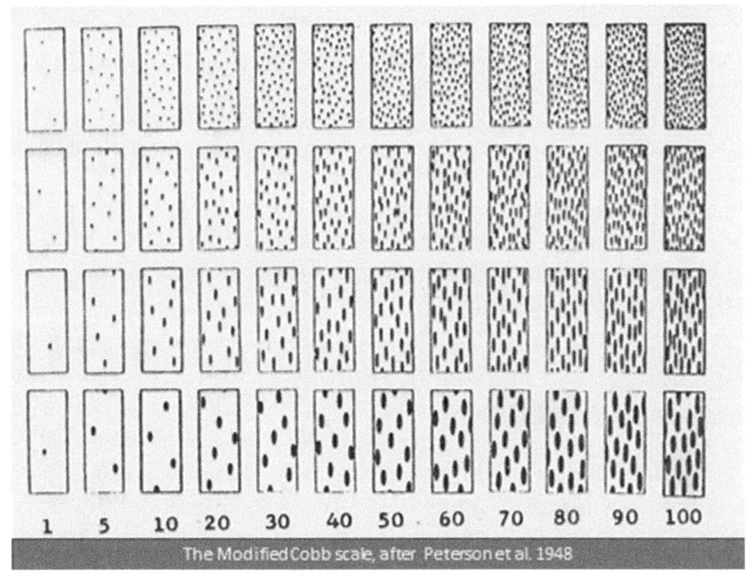

Figure 11.17. Modified Cob scale used to score rust data in the field (Peterson et al., 1946), after Roelfs et al. (1992).

Scoring Resistance—Greenhouse-Evaluation

Seedling evaluation for stem and leaf rust is based on the size of uredinia, the presence of chlorosis and/or necrosis, and the distribution of the pustules on the leaves. It is done approximately 10–14 days after inoculation. A 0–4 scale (Figs. 11.18 and 11.19) is used as described by Roelfs et al. (Roelfs et al., 1992). Normally infection types "0", ";" (fleck), "1", and "2" are considered resistant responses, whereas "3" and "4" are susceptible. Sometimes the uredinias are not consistent over the length of the leaf. Infection type "X" = a random distribution of variable-sized uredinias across a single leaf, "Y" = ordered distribution of uredinias with larger uredinias at the leaf tip and smaller uredinias at the base, and "Z" = ordered distribution of uredinias

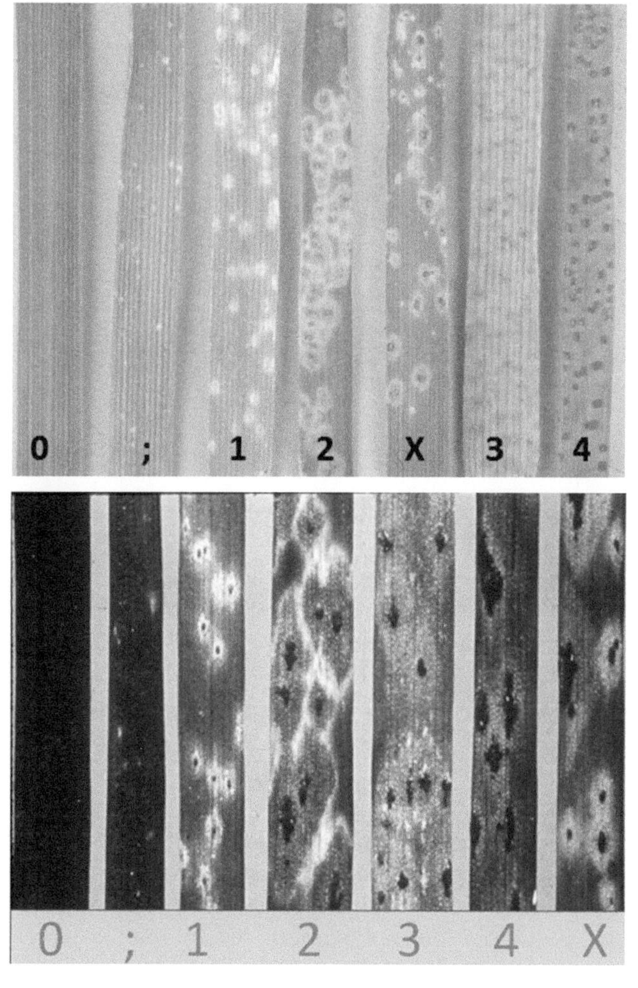

Figures 11.18 and **11.19.** Scale 0–4 to score infection types for leaf and stem rust at seedling stage respectively under greenhouse conditions (Roelfs et al., 1992).

Color image of this figure appears in the color plate section at the end of the book.

with larger uredinias at the leaf base and smaller uredinias at the leaf tip. These pustule types can also be modified using "+" or "–" to indicate larger or smaller uredinias than average for a particular pustule type. For example, a rating of "1+" would indicate slightly larger than average "1" type uredinias (Roelfs et al., 1992). In the case of stripe rust a 0–9 scale (Fig. 11.20) is used (McNeal et al., 1971) Infection types 0 to 6 are considered resistant; whereas 7–9 are susceptible.

Adult plant evaluation in the greenhouse, takes place approximately 14–21 days after inoculation using the same infection type scale as described for seedling evaluation. The leaves and uredinia are generally larger at the adult plant stage but the relative comparison between the different infection types is the same.

Figure 11.20. Scale 0–9 to score seedling infection types for stripe rust in the greenhouse, After McNeal et al. (1971).

Color image of this figure appears in the color plate section at the end of the book.

Concluding Remarks

The rust diseases of wheat caused by the species of the *Puccinia* will remain very important on a world-wide basis. These diseases are essentially clones, originating from dikaryotic urediniospores which can rapidly evolve due to mutations, co-adaptations and somatic recombination which are better fit; carry new virulence combinations, are more aggressive and might have selective advantages over previous races also.

With the changes in the cropping systems and the impact of the global warming the epidemiology of these rust diseases might change dramatically.

An example is the apparently adaptation to warmer temperatures by *P. striiformis* f. sp. *tritici* that make this disease the most feared one, not only in the Middle East, Central Asia, recently in India but also in other wheat growing areas of the world where stripe rust was not an important wheat disease before.

On the other hand, the return of stem or black rust in Europe, along with Ug99 and other stem rust races recently originated in eastern Africa with new virulence combination defeating important resistance genes will challenge scientists to look closely to the epidemiology of the diseases and better understand the role of the alternate hosts, when present, in producing new races or new virulence combinations.

Although the leaf rust is under control; new virulence combination will continue arising so we have to consider that "Rust never sleeps". Better understanding of the rust diseases of wheat, will help breeders of the world to choose the best breeding strategy to achieve durable rust resistant cultivars.

References

Allen, R.F. 1931. Heterothallism in *Puccinia triticina*. Science N. S., 74: 462–463.

Allen, R.F. 1932. A cytological study of heterothallism in *Puccinia triticina*. J. Agr. Res., 44: 733–754.

Anikster, Y., Bushnell, W.R., Eilam, T., Manisterski, J. and Roelfs, A.P. 1997. *Puccinia recondita* causing leaf rust on cultivated wheats, wild wheats, and rye. Can. J. Bot., 75: 2082–2096.

Azbukina, Z. 1980. Economical importance of aecial hosts of rust fungi of cereals in the Soviet Far East. pp. 199–201. *In*: Proc. 5th European and Mediterranean Cereal Rusts Conf., Bari, Rome.

Beresford, R.M. 1982. Stripe rust (*Puccinia striiformis*), a new disease of wheat in New Zealand. Cer. Rusts Bull., 10: 35–41.

Browder, L.E. 1971. Pathogenic specialization in cereal rust fungi, especially *Puccinia recondita* f. sp. *tritici*: Concepts, methods of study and application. U.S. Dep. Agric. Tech. Bull., 1432.51 pp.

Brown, J.K.M. and Hovmøller, M.S. 2002. Aerial dispersal of pathogens on the global and continental scales and its impact on plant disease. Science, 297: 537–541.

Casulli, F. 1988. Overseasoning of wheat leaf rust in southern Italy. pp. 166–168 *In*: Proc. 7th European and Mediterranean Cereal Rusts Conf., 5–9 Sept., Vienna, Italy.

Cherry, E. and Peet, C.E. 1966. An efficient device for the rapid collection of fungal spores from infected plants. Phytopathology, 56: 1102–1103.

Chester, K.S. 1946. The nature and prevention of the cereal rusts as exemplified in the leaf rust of wheat. In Chronica botanica. Walthan, MA, USA, 269 pp.

Craigie, J.H. 1927. Experiments on sex in rust fungi. Nature, 120: 116–117.

Craigie, J.H. 1957. Stem Rust of Cereals. Canada Department of Agriculture. Botany and Plant Pathology Division. Publication 666. Ottawa, Ontario. April 1957.

Cummins, G.B. and Caldwell, R.M. 1956. The validity of binomials in the leaf rust fungus complex of cereals and grasses. Phytopathology, 46: 81–82.

Cummins, G.B. and Stevenson, J.A. 1956. A check list of North American rust fungi (Uredinales). Plant Dis. Rep. Suppl., 240: 109–193.

d'Oliveira, B. and Samborski, D.J. 1964. Aecial stage of *Puccinia recondita* on Ranunculaceae and Boraginaceae in Portugal. pp. 133–150. *In*: First European Brown Rust Conference. Proc. Cereal Rusts Conferences., Cambridge, UK.

DeBary, A. 1866. Neue Untersuchungen uber die Uredineen insbesondere die Entwicklung der *Puccinia graminis* und den Zusammenhang duselben mit *Aecidium berberis*. pp. 15–20. *In*: Monatsber Preuss. Akad. Wiss., Berlin.

De Candolle, A. 1815. Uredo rouille des cereales. *In*: Flora francaise, famille des champignons, p. 83.

Dubin, H.J. and Stubbs, R.W. 1986. Epidemic spread of barley stripe rust in South America. Plant Disease, 70: 141–144.

Dubin, H.J. and Torres, E. 1981. Causes and consequences of the 1976–1977 wheat leaf rust epidemic in northwest Mexico. Ann. Rev. Phytopath., 19: 41–49.

Eriksson, J. 1894. Uber die Spezialisierung des Parasitismus bei dem Getreiderostpilzen. Ber. Deut. Bot. Ges., 12: 292–331.

Eriksson, J. and Henning, E. 1894. Die Hauptresultate einer neuen Untersuchung uber die Getreideroste. Z. Pflanzenkr., 4: 66–73, 140–142, 197–203, 257–262.

Eriksson, J. and Henning, E. 1896. Die Getreideroste. Ihre Geschichte und Natur sowie Massregein gegen dieselben, p. 463. Stockholm, P.A. Norstedt and Soner.

Ezzahiri, B., Diouri, S. and Roelfs, A.P. 1992. The role of the alternate host, *Anchusa italica*, in the epidemiology of *Puccinia recondita* f. sp. *tritici* on durum wheats in Morocco. pp. 69–70. *In*: Zeller, F.J. and Fischbeck, G. (eds.). Proc. 8th European and Mediterranean Cereal Rusts and Mildews Conf. Heft 24 Weihenstephan, Germany.

Ezzahiri, B., Diouri, S. and Roelfs, A.P. 1994. Pathogenicity of *Puccinia recondita* f. sp. *tritici* in Morocco during 1985, 1988, 1990, and 1992. Plant Dis. 78: 407–410.

FAO. 2008. Wheat killer detected in Iran. http://www.fao.org/newsroom/en/news/2008/1000805/index.html (accessed on June 28, 2012).

Flor, H.H. 1956. The complementary Genetic Systems in Flax and Flax Rusts. pp. 29–54. *In*: Demerec, M. (ed.). Advances in Genetics Volume VIII. Academic Press New York, N.Y. USA.

Fontana, F. 1932. Observations on the rust of grain. P.P. Pirone, transl. Classics No. 2. Washington, DC, Amer. Phytopathol. Society (Originally published in 1767).

Groth, J.V. and Roelfs, A.P. 1982. The effect of sexual and asexual reproduction on race abundance in cereal rust fungus populations. Phytopathology, 72: 1503–1507.

Hart, H. and Becker, H. 1939. Beitrage zur Frage des Zwischenwirts fur *Puccinia glumarum* Z. Pflanzenkr. (Pflanzenpathol.) Pflanzenschutz, 49: 559–566.

Hendrix, J.W., Burleigh, J.R. and Tu, J.C. 1965. Oversummering of stripe rust at high elevations in the Pacific Northwest-1963. Plant Dis. Rep., 49: 275–278.

Hermansen, J.E. 1968. Studies on the spread and survival of cereal rust and mildew diseases in Denmark. Contrib. No. 87. Copenhagen, Department of Plant Pathology, Royal Vet. Agriculture College. 206 pp.

Hirst, J.M. and Hurst, G.W. 1967. Long-distance spore transport. pp. 307–344. *In*: Gregory, H. and Monteith, J.L. (eds.). Airborne microbes. Cambridge University Press.

Hogg, W.H., Hounam, C.E., Mallik, A.K., and Zadoks, J.C.1969. Meteorological factors affecting the epidemiology of wheat rusts. WMO Tech. Note No. 99. 143 pp.

Hodson, D.P. 2010. Shifting boundaries: challenges for rust monitoring. pp. 103–118. *In*: McIntosh, R. and Pretorius, Z. (eds.). Proceedings of BGRI 2010Technical Workshop. St Petersburg, Russia. http://www.globalrust.org/db/attachments/about/19/1/BGRI%20oral%20papers%202010.pdf.

Hovmoller, M.S., Sorensen, C.K., Walter, S. and Justesen, A.F. 2011. Diversity of *Puccinia striiformis* on Cereals and Grasses. Annu. Rev. Phytopathol., 49: 197–217.

Huerta-Espino, J. and Roelfs, A.P. 1989. Physiological specialization on leaf rust on durum wheat. Phytopathology, 79: 1218.

Huerta-Espino, J. and Roelfs, A.P. 1992. Leaf rust on durum wheats. *In*: Proc. VIII European and Mediterranean Cereal Rusts and Mildews Conference, 1992, Weihenstephan (Germany). VorträgefürPflanzenzüchtung, 24: 100–102.
Huerta-Espino, J. 1992. Analysis of wheat leaf and stem rust virulence on a worldwide basis. Ph.D. thesis. University of Minnesota, St. Paul, USA.
Huerta-Espino, J., Singh, R.P., German, S., McCallum, B.D., Park, R.F., Chen, W., Bhardwaj, S.C. and Goyeau, H. 2011. Global status of wheat leaf rust caused by *Puccinia triticina*. Euphytica, 179: 143–160.
Hugh-Jones, M.E. 2002. Agricultural bioterrorism. pp. 219–232. *In*: High-Impact Terrorism: Proceedings of a Russian-American Workshop. National Academy Press, Washington DC.
Hylander, N., Jorstad, I. and Nannfeldt, J.A. 1953. Enumeratio uredionearum Scandinavicarum. Opera Bot., 1: 1–102.
Jackson, H.S. and Mains, E.B. 1921. Aecial stage of the orange leaf rust of wheat, *Puccinia triticina* Erikss. Jour. Agr. Res., 22: 151–172.
Jin, Y., Szabo, L.J. and Carson, M. 2010. Century-old mystery of *Puccinia striiformis* life history solved with the identification of Berberis as an alternate host. Phytopathology, 100: 432–435.
Jin. Y, Pretorius, Z.A., Singh, R.P. and Fetch, T., Jr. 2008. Detection of virulence to resistance gene Sr24 within race TTKS of *Puccinia graminis* f. sp. *tritici*. Plant Disease, 92: 923–26.
Jin, Y., Szabo, L.J., Rouse, M.N., Fetch, T., Jr., Pretorius, Z.A., Wanyera, R. and Njau, P. 2009. Detection of virulence to resistance gene Sr36 within the TTKS race lineage of *Puccinia graminis* f. sp. *tritici*. Plant Disease, 93: 367–370.
Katsuya, K. and Green, G.J. 1967. Reproductive potentials of races 15B and 56 of wheat stem rust. Can. J. Bot., 45: 1077–1091.
Kislev, M.E. 1982. Stem rust of wheat 3300 years old found in Israel. Science, 216: 993–994.
Loegering, W.O., Harmon, D.L. and Clark, W.A. 1961. A long term experiment for preservation of urediospores of *Puccinia graminis tritici* in liquid nitrogen. Plant Dis. Rep., 45: 384–385.
Loegering, W.O., Harmon, D.L. and Clark, W.A. 1966. Storage of urediospores of *Puccinia graminis tritici* in liquid nitrogen. Plant Dis. Rep., 50: 502–506.
Line, R.F. 1976. Factors contributing to an epidemic of stripe rust on wheat in the Sacramento Valley of California in 1974. Plant Dis. Rep., 60: 312–316.
Luig, N.H. 1985. Epidemiology in Australia and New Zealand. pp. 301–328. *In*: Roelfs, A.P. and Bushnell, W.R. (eds.). Cereal rusts, Vol. 2, Diseases, Distribution, Epidemiology, and control. Academic Press, Orlando, FL, USA,
Maddison, A.C. and Manners, J.G. 1972. Sunlight and viability of cereal rust urediospores. Trans. Br. Mycol. Soc., 59: 429–443.
Mains, E.B. 1932. Host specialization in the leaf rust of grasses *Puccinia rubigovera*. Mich. Acad. Sci., 17: 289–394.
Mains, E.B. 1933. Studies concerning heteroecious rust. Mycologia, 25: 407–417.
Mains, E.B. and Jackson, H.S. 1926. Physiologic specialization in the leaf rust of wheat, *Puccinia triticina* Erikss. Phytopathology, 16: 89–120.
Manners, J.G. 1960. *Puccinia striiformis* Westend. var. *dactylidis* var. nov. Trans. Br. Mycol. Soc., 43: 65–68.
McCallum, B., Hiebert, C., Huerta-Espino, J. and Cloutier, S. 2012. Wheat leaf rust. pp. 33–62. *In*: Sharma, I. (ed.). Disease Resistance in Wheat. 1. Cabi Plant Protection Series. CAB International. Cambridge, MA, USA.
McIntosh, R.A., Wellings, C.R. and Park, R.F. 1995. Wheat rusts: An atlas of resistance genes. CSIRO Publications, East Melbourne, Australia.
McNeal, F.H., Konzak, C.F., Smith, E.P., Tate, W.S. and Russell, T.S. 1971. A uniform system for recording and processing cereal research data. US. Dept. Agric., Agric. Res. Serv., ARS 34–121.42 pp.

Melander, L.W. and Craigie, J.H. 1927. Nature of resistance of *Berberis* spp. to *Puccinia graminis*. Phytopathology, 17: 95–114.

Melching, J.S., Stanton, J.R. and Koogle, D.L. 1974. Deleterious effects of tobacco smoke on germination and infectivity of spores of *Puccinia graminis tritici* and on germination of spores of *Puccinia striiformis, Pyricularia orzyae*, and *Alternaria* species. Phytopathology, 64: 1143–1147.

Mont, R.M. 1970. Studies of nonspecific resistance to stem rust in spring wheat. M.S. thesis, University of Minnesota. St. Paul, MN, USA.

Nagarajan, S. and Joshi, L.M. 1985. Epidemiology in the Indian Subcontinent. pp. 371–402. *In*: Roelfs, A.P. and Bushnell, W.R. (eds.). The Cereal Rusts Vol. II; Diseases, Distribution, Epidemiology, and Control. Academic Press, Orlando. Fla. USA.

Nagarajan. S. and Singh, D.V. 1990. Long-distance dispersion of rust pathogens. Ann. Rev. Phytopathol., 28: 139–153.

Nazari. K., Mafi, M., Yahyaoui, A., Singh, R.P. and Park, R.F. 2009. Detection of wheat stem rust (*Puccinia graminis* f. sp. *tritici*) race TTKSK (Ug99) in Iran. Plant Disease, 93: 317.

Nazari, K., Yahyaoui, A., Abdalla, O., Nachit, M., Ogbonnaya, F., Brettell, R. and Rajaram, S. 2010. Wheat rust diseases in Central and West Asia and North Africa (CWANA) and breeding for the multiple disease resistance. http://www.abriigenomics.ir/Workshop%20 Presentations/Wheat%20rust%20diseases%20in%20CWANA%20ABRII%20MAY%20 10-12%20K_Nazari%20et%20al.pdf accessed June 24, 2012.

O'Brien, L., Brown, J.S., Young, R.M. and Pascoe, T. 1980. Occurrence and distribution of wheat stripe rust in Victoria and susceptibility of commercial wheat cultivars. Aust. Plant Pathol. Soc. Newsl., 9: 14.

Persoon, C.H. 1801. Synopsis Methodica fungorum Gottingen (Johnson Reprint Corp., New York, 1952).

Peterson, R.F., Campbell, A.B. and Hannah, A.E. 1948. A diagrammatic scale for estimating rust intensity on leaves and stems of cereals. Canadian Journal of Research, 26: 496–500.

Pretorius, Z.A., Singh, R.P., Wagoire, W.W. and Payne, T.S. 2000. Detection of virulence to wheat stem rust resistance gene *Sr31* in *Puccinia graminis* f. sp. *tritici* in Uganda. Phytopathology, 84: 203.

Pretorius, Z.A., Bender, C.M., Visser, B. and Terefe, T. 2010. First report of a *Puccinia graminis* f. sp. *tritici* race virulent to the Sr24 and Sr31 wheat stem rust resistance genes in South Africa. Plant Disease, 94: 784.

Prospero, J.M., Blades, E., Mathieson, G. and Naidu, R. 2005. Interhemispheric transport of viable fungi and bacteria from Africa to the Caribbean with soil dust. Aerobiologia., 21: 1–19.

Rapilly, F. 1979. Yellow rust epidemiology. Ann. Rev. Phytopathol., 17: 59–73.

Roelfs, A.P. 1982. Effects of barberry eradication on stem rust in the United States. Plant Disease, 66: 177–181.

Roelfs, A.P. 1985a. Epidemiology in North America. pp. 403–434. *In:* Roelfs, A.P. and Bushnell, W.R. (eds.). The Cereal Rusts, Vol. 2, Diseases, Distribution, epidemiology, and control. Academic Press. Orlando, FL, USA.

Roelfs, A.P. 1985b. Wheat and rye stem rust. pp. 3–37. *In*: Roelfs, A.P. and Bushnell, W.R. (eds.). The cereal rusts, Vol. 2, Diseases, distribution, epidemiology, and control. Academic Press. Orlando, FL, USA,

Roelfs, A.P. 1986. Development and impact of regional cereal rust epidemics. pp. 129–150. *In*: Leonard, K.J. and Fry, W.E. (eds.). Plant Disease Epidemiology. Vol. 1. MacMillan. New York, NY, USA.

Roelfs, A.P. 1988. Genetic control of phenotypes in wheat stem rust. Ann. Rev. Phytopathol., 26: 351–367.

Roelfs, A.P. and Groth, J.V. 1980. A comparison of virulence phenotypes in wheat stem rust populations reproducing sexually and asexually. Phytopathology, 70: 855–862.

Roelfs, A.P. and Groth, J.V. 1988. *Puccinia graminis* f. sp. *tritici* black stem rust of *Triticum* spp. pp. 345–361. *In*: Sidhu, G.S. (ed.). Advances in Plant Pathology, vol. 6, Genetics of pathogenic fungi. Academic Press. London, UK.
Roelfs, A.P. and Long, D.L. 1987. *Puccinia graminis* development in North America during 1986. Plant Disease, 71: 1089–1093.
Roelfs, A.P. and Martell, L.B. 1984. Uredospore dispersal from a point source within a wheat canopy. Phytopathology, 74: 1262–1267.
Roelfs, A.P., Huerta-Espino, J. and Marshall, D. 1992. Barley stripe rust in Texas. Plant Disease, 76: 538.
Roelfs, A.P. and Huerta-Espino, J. 1994. Seedling resistance in Hordeum to barley stripe rust from Texas. Plant Disease, 78: 1046–1049.
Roelfs, A.P., Singh, R.P. and Saari, E.E. 1992 Rust diseases of wheat. Concepts and Methods of Disease Management. Mexico, D.F.: CIMMYT.
Rotem, J., Wooding, B. and Aylor, D.E. 1985. The role of solar radiation, especially UV, in the mortality of fungal spores. Phytopathology, 75: 510–514.
Rowell, J.B. 1957. Oil inoculation of wheat with spores of *Puccinia graminis tritici*. Phytopathology, 47: 689–690.
Rowell, J.B. 1981. Relation of postpenetration events in Idaed 59 wheat seedling to low receptivity to infection by *Puccinia graminis* f. sp. *tritici*. Phytopathology, 71: 732–736.
Rowell, J.B. 1982. Control of wheat stem rust by low receptivity to infection conditioned by a single dominant gene. Phytopathology, 72: 297–299.
Rowell, J.B. 1985. Controlled infection by *Puccinia graminis* f. sp. *tritici* under artificial conditions. pp. 291–332. *In*: Roelfs, A.P. and Bushnell, W.R. (eds.). The Cereal Rusts, Vol. 1, Origins, specificity, structure, and physiology. Academic Press. Orlando, FL, USA.
Saari, E.E. and Prescott, J.M. 1985. World distribution in relation to economic losses. pp. 259–298. *In*: Roelfs, A.P. and Bushnell, W.R. (eds.). The cereal rusts, Vol. 2, Diseases, distribution, epidemiology, and control. Academic Press.Orlando, FL, USA.
Saari, E.E., Young, H.C., Jr. and Fernkamp, M.F. 1968. Infection of North American *Thalictrum* spp. with *Puccinia recondita* f. sp. *tritici*. Phytopathology, 58: 939–943.
Samborski, D.J. 1985. Wheat leaf rust. pp. 35–59. *In*: Roelfs, A.P. and Bushnell, W.R. (eds.). The cereal rusts, Vol. 2, Diseases, Distribution, Epidemiology, and Control. Academic Press.Orlando, FL, USA.
Savile, D.B.O. 1985. Taxonomy of the cereal rust fungi. pp. 79–112. *In*: Bushnell, W.R. and Roelfs, A.P. (eds.). The cereal rusts, Vol. 1, Origins, specificity, structures, and physiology. Academic Press. Orlando, FL, USA.
Schmidt, J.K. 1827. Allgemeine ökonomisch-technischc Flora oder Abbildungen und Beschreibungen aller in bezug auf Ökonomic und Technologic, merkwürdigen Gewächse, Vol. I, p. 27. Jena, Germany.
Sharp, E.L. 1967. Atmospheric ions and germination of urediospores of *Puccinia striiformis*. Science, f56: 1359–1360.
Sharp, E.L. and Smith, F.G. 1957. Further study of the preservation of *Puccinia* uredospores. Phytopathology, 47: 423–429.
Singh, R.P. 1991. Pathogenicity variation of *Puccinia recondita* f. sp. *tritici* and *P. graminis* f. sp. *tritici* in wheat-growing areas of Mexico during 1988 and 1989. Plant Dis., 75: 790–794
Singh, R.P., Huerta-Espino, J., Pfeiffer, W. and Figueroa-Lopez, P. 2004a. Occurrence and impact of a new leaf rust race on durum wheat in northwestern Mexico from 2001 to 2003. Plant Disease, 88: 703–708.
Singh, R.P., William, H.M., Huerta-Espino, J. and Rosewarne, G. 2004b. Wheat rust in Asia: meeting the challenges with old and new technologies. *In*: New Directions for a Diverse Planet: Proceedings of the 4th International Crop Science Congress. September 26–October 1, 2004. Brisbane, Australia. http://www.cropscience.org.au./icsc2004/symposia/3/7/141_singhrp.htm (accessed on June 24, 2012).
Singh, R.P., Hodson, D.P., Jin, Y., Huerta-Espino, J., Kinyua, M., Wanyera, R., Njau, P. and Ward, R.W. 2006. Current status, likely migration and strategies to mitigate the threat to wheat

production from race Ug99 (TTKS) of stem rust pathogen. CAB Reviews: Perspectives in Agriculture, Veterinary Science, Nutrition and Natural Resources, 1–54.

Singh, R.P., Hodson, D.P., Huerta–Espino, J., Jin, Y., Njau, P., Wanyera, R., Herrera–Foessel, S.A. and Ward, R.W. 2008. Will stem rust destroy the world's wheat crop? Adv. Agron., 98: 271–309

Singh, R.P., Hodson, D.P., Huerta-Espino, J., Jin, Y., Bhavani, S., Njau, P., Herrera-Foessel, S.A., Singh, P.K., Singh, S. and Govindan, V. 2011. The Emergence of Ug99 Races of the Stem Rust Fungus Is a Threat to World Wheat Production. Annu. Rev. Phytopathol., 49: 465–481.

Singh, S., Singh, R.P. and Huerta-Espino, J. 2012. Stem Rust. pp. 18–32. *In*: Sharma, I. (ed.). Disease resistance in Wheat. 1. Cabi Plant Protection Series. CAB International. Cambridge, MA, USA.

Skovmand, B., Fox, P.N. and Villareal, R.L. 1984. Triticale in commercial agriculture: progress and promise. Adv. Agron., 37: 1–45.

Stakman, E.C. 1919. New biologic forms of *Puccinia graminis*. J. Agric. Res., 16: 103–105.

Stakman, E.C. 1957. Problems in preventing plant disease epidemics. American Journal Botany, 44: 259–267.

Stakman, E.C., Stewart, D.M. and Loegering, W.Q. 1962. Identification of physiological races of *Puccinia graminis* var. *tritici*. U.S. Dept. Agric., ARS E617. 53 pp.

Spehar, V. 1975. Epidemiology of wheat rust in Europe. pp. 435–440. *In*: Proc. 2nd Int. Winter Wheat Conf., Zagreb, Yugoslavia.

Straib, W. 1937. Untersuchungen uben dasn Vorkommen physiologischen Rassen der Gelbrostes (*Puccinia glumarum*) in den Jahren 1935–1936 und uber die Agressivitat eninger neuer Formes auf Getreide un Grasern. Arb. Biol. Reichsant. Land. Fortw. Berlin-Dahlem, 22: 91–119.

Stubbs, R.W. 1985. Stripe rust. pp. 61–101. *In*: Roelfs, A.P. and Bushnell, W.R. (eds.). The cereal rusts, Vol. 2, Diseases, Distribution, Epidemiology, and Control. Academic Press. Orlando, FL, USA.

Stubbs, R.W., Prescott, J.M., Saari, E.E. and Dubin, H.J. 1986. Cereal disease methodology manual. pp. 46. CIMMYT. Mexico, DF.

Sturman, A.P., Tyson, P.D. and D'Abreton, P.C. 1997. A preliminary study of the transport of air from Africa and Australia to New Zealand. J. R. Soc. N. Z., 27: 485–98.

Szabo, L.J. 2007. Development of simple sequence repeat markers for the plant pathogenic rust fungus, *Puccinia graminis*. Mol. Ecol. Notes, 7: 92–94.

Tollenaar, H. and Houston, B.R. 1967. A study on the epidemiology of stripe rust, *Puccinia striiformis* West. in California. Can. J. Bot., 45: 291–307.

Terefe, T., Pretorius, Z.A., Paul, I., Mebalo, J., Meyer, L. and Naicker, K. 2010. Occurrence and pathogenicity of *Puccinia graminis* f. sp. *tritici* on wheat in South Africa during 2007 and 2008. S. Afr. J. Plant Soil, 27: 163–167.

Tozzetti, G.T. 1952. V. Alimurgia: True nature, causes and sad effects of the rusts, the bunts, the smuts, and other maladies of wheat and oats in the field. *In*: Tehon, L.R., transl. Phytopathological Classics No. 9, p. 139. St Paul, MN, USA, American Phytopathol. Society (Originally published 1767).

Tranzschel, W. 1934. Promezutocnye chozjaeva rzavwiny chlebov i ich der USSR (The alternate hosts of cereal rust fungi and their distribution in the USSR). pp. 4–40. *In*: Bull. Plant Prot. Ser. 2 (in Russian with German summary).

Viennot-Bourgin, G. 1934. La rouille jaune der graminees. Ann. Ec. Natl. Agric. Grignon Ser., 3. 2: 129–217.

Viennot-Bourgin, G. 1941. Diagnose Latine de *Puccinia tritici-duri* Ann. Ecole Nationale Agric. Grignon, Paris C. Amat. Ser., 2: 146.

Visser, B., Herselman, L., Park, R.F., Karaoglu, H., Bender, C.M. and Pretorius, Z.A. 2010. Characterization of two new *Puccinia graminis* f. sp. *tritici* races within the Ug99 lineage in South Africa. Euphytica. doi:10.1007/s10681-010-0269-x.

Visser, B., Herselman, L. and Pretorius, Z.A. 2009. Genetic comparison of Ug99 with selected South African races of *Puccinia graminis* f. sp. *tritici*. Mol. Plant Pathol., 10: 213–222.

Walter, H. 1961. Einfuhrung in die Phytologie. II. Grundlagen des Pflanzen systems, 2. Aufl. Stuttgart.

Watson, I.A. and de Sousa, C.N.A. 1983. Long distance transport of spores of *Puccinia graminis tritici* in the Southern Hemisphere. *In*: Proc. Linn. Soc. N.S.W., 106: 311–321.

Wellings. C.R., Boyd, L.A. and Chen, X.M. 2012. Resistance to stripe rust in wheat: pathogen biology driving resistance breeding. pp. 63–83. *In*: Sharma, I. (ed.). Disease Resistance in Wheat. 1. Cabi Plant Protection Series. CAB International. Cambridge, MA, USA.

Westendorp, G.D. 1854. Quatrieme notice sur quelques cryptogames recemment decouvertes en Belgique. Bull. L'Academie R. Sci. Belg. Classes Sci., 21: 229–246.

Wright, R.G. and Lennard, J.H. 1980. Origin of a new race of *Puccinia striiformis*. Trans. Br. Mycol. Soc., 74: 283–287.

Zadoks, J.C. 1961. Yellow rust on wheat studies of epidemiology and physiologic specialization. Neth. J. Plant Pathol., 67: 69–256.

Zadoks, J.C. and Bouwman, J.J. 1985. Epidemiology in Europe. pp. 329–369. *In*: Roelfs, A.P. and Bushnell, W.R. (eds.). The cereal rusts, Vol. 2, Diseases, Distribution, Epidemiology, and Control. Academic Press. Orlando, FL, USA.

CHAPTER 12

Current Research on Fungal Pathogens Associated with Rice

Hoai Xuan Truong[a,*] and *Evelyn B. Gergon*[b]

ABSTRACT

Rice is one of the most important crops in the world feeding millions of people at all levels of life. While the world population is continuously increasing, the area of rice production is shrinking. It is, therefore, becoming an urgent necessity to increase rice productivity per unit space and reduce grain losses due to pests and diseases. Fungi, especially those that are seedborne in nature, are one of the major pathogens that can result in tremendous crop losses as high as 50%. Their presence in the field or in/on the seed is only noticed when all factors that support and trigger their growth and development on the plant are there. But at that stage unfortunately, not much can be done to arrest them.

This chapter discusses the recent research efforts worldwide on genetic diversity of some of the important fungal pathogens causing diseases in rice viz., rice blast, sheath blight, brown leaf spot, sheath rot, foot rot, narrow brown leaf spot, stem rot, leaf scald, false smut and head blight.

Philippine Rice Research Institute, Science City of Muñoz, Nueva Ecija, The Philippines.
[a] Email: truongxuan893@yahoo.com
[b] Email: egergon@yahoo.com
* Corresponding author

Introduction

Rice (*Oryza sativa* L.) is the main staple food crop for almost half of the world's population (Xu et al., 2002). Rice is mainly grown in Asia and is currently expanding to other continents like Sub-Saharan Africa. With growing rice requirements, growers and research institutions are compelled to find ways of increasing its productivity using the same area of production amidst the pressure of abiotic stresses owing to climatic changes. The world's statistics have shown an increasing trend in domestic milled rice production at its current peak of 462.75 million MT. This is closely followed by increasing rice consumption at 456.79 million MT (Wailes et al., 2012). Asia, the top producer of rice, is set for a 3% gain in production to 653 million MT in 2011–2012 and Africa is set for 1% gain to 25.5 MT. In the same period, the world rice production is also predicted to hit a new record of 721 million MT or a 3% increase (FAO, 2012). While rice production increases, there is also a foreseen increase in current global population to eight billion by 2020, thus it has been estimated that 25–40% more rice must be produced to meet the global demand for the crop and to cover up the impact of climate change such as drought, submergence and salinity (Datta, 2004). Other factors contributing to yield losses such as pests and diseases must also be given due attention to further reduce their impact on rice crop production.

The production of improved rice varieties that are grown in diverse agro-ecosystems are certainly subjected to biotic and abiotic stresses. Biotic stresses owing to pests and diseases had been estimated to cause yield losses by as much as 38–42% (Montesinos, 2003) with 14% of this had been attributed to pathogens (CABI, 2004). These figures are indicative of the fact that diseases are a major and continuous concern to crop production and are considered economically important in rice. Among the microorganisms that play an important role in affecting yield and seed quality of many crops, fungi are the most important and the largest group. Those that are seedborne in nature, numbering more than 200, are widely recognized due to the importance of seeds in international and local trade, and germplasm exchange.

Fungi are generally microscopic organisms with cell walls and true nuclei, but lack chlorophyll. Sexual or asexual reproductive structures are formed depending on the surrounding environmental conditions. Sexual fruiting structures, that develop through the fusion of unlike cells or gametes, or when two strains of opposite mating types fuse, form sexual spores. These sexual spores can be either ascospores or basidiospores depending on the class of fungi and fruiting structure. Ascospores are formed within a saclike zygote cell called ascus, which are contained in a fruiting body. Basidiospores are developed outside the club-like zygote cell called basidium.

Many fungi commonly produce asexual spores but produce sexual spores only under specific environmental conditions. The asexual spores called conidia are usually borne on conidiophores which are stem-like structures arising from terminal or lateral cells of a special hypha. Conidia are easily detached from the conidiophores and dispersed by wind, water and even insects. When the air is moist, they can attach onto the leaf surface tightly after landing (Hamer et al., 1988). When surrounding conditions are favorable for germination, the spore germinates by producing germ tube that eventually produce appresorium at the tip that help them adhere on the leaf surface. The appresorium produces infection peg that penetrates into plant tissues and invade the plant cells by growing in and between plant's cells until the pathogen overcome the host defense system. The fungus eventually proliferates within the host plant. In cases where both sexual and asexual structures are observed on the plant, the name applied to the causal pathogen is the sexual name or the teleomorph. If the teleomorph is known but rarely encountered or observed, then the asexual name or the anamorph is applied. Aside from spores, the fungus may also produce other structures to survive and initiate infection such as sclerotia. Sclerotia are compact masses of mycelia and are produced by a special group of fungi called mycelia sterilia.

Fungi can be seen on or in the plant at any growth stages and can be carried through the seed, soil, water, insects and the wind. The fungi that are seedborne are relatively difficult to control as their hyphae get established and become dormant in the seed. They can reduce seed viability and quality and cause infection on the succeeding crop.

Some fungi are also known in stored seeds affecting seed discoloration and reducing seed germinability (Neergaard, 1986; Mew and Gonzales, 2002; Butt et al., 2011). Seed in storage is also susceptible to insects and their infestations create entry points in the grain for the fungi (Pitt and Hocking, 1997). The movement of seeds can disseminate the fungi. The development of fungi in storage depends on conditions during storage and can result in quantitative and qualitative losses which often remain an unresolved problem. Storage fungi may also develop toxic metabolites or mycotoxins that are hazardous to animal and human health.

In this chapter, recent research efforts worldwide on genetic diversity of 10 fungal pathogens, considered to be of economic importance, causing rice blast, sheath blight, brown leaf spot, sheath rot, foot rot, narrow brown leaf spot, leaf scald, stem rot, false smut and head blight pathogens are discussed. However, there are other fungal pathogens infecting rice but they not elaborated here.

RICE BLAST PATHOGEN

Name of the disease: Rice Blast
Name of the pathogen: *Magnaporthe oryzae* Couch and Kohn (2002) (teleomorph)
Synonyms: *Dactylaria oryzae* (Cavara) Sawada (anamorph);
Magnaporthe grisea senso Yaegashi and Udagawa (1978);
Pyricularia oryzae Cavara

Rice blast is one of the most important diseases of rice. Its first occurrence was noticed in China as early as 1637 and the disease was known as rice fever disease. The disease is highly destructive in lowland rice in temperate and subtropical Asia, Latin America and Africa, as well as in rainfed lowland and upland rice in tropical Southeast Asia (Ou, 1985). Occurrence and severity of blast disease vary by year, location and even within a field depending on environmental conditions, and crop management practices (Kato, 2001). Historically rice blast epidemics in Asia and many other regions worldwide caused crop losses ranging from 10 to 50%. Typical endemic of panicle blast caused crop losses by 15 to 75% in rain-fed lowland in the humid tropics like in the Philippines during main crop seasons of 2008–2010 (further discussed below). Serious crop losses caused by neck blast include breakage of the panicle, incomplete grain filling, poor milling quality and chalky grains. Many investigators have considered blast to be a model disease for the study of genetics, epidemiology, molecular pathology of host parasite interactions and biology. Several disease management strategies have been studied intensively against rice blast, but often with limited success. An integration of host plant resistance and cultural management practices is always considered more practical to reduce losses in crop yield and produce cost-effective harvest.

Biology of the Pathogen

The fungus, *Pyricularia grisea* Sacc., was originally described in 1880 causing gray leaf spot on the grass *Digitaria sanguinalis*. The name *P. grisea* is applied to isolates from cereals and other grasses while *P. oryzae* was described in 1891 for isolates causing rice blast.

The sexual or teleomorphic stage of the rice blast pathogen, *Magnaporthe grisea*, has not been found in nature (Bonman, 1992a), but can be produced in the laboratory if isolates of opposite mating type are paired. As an ascomycete, it produces hyaline, fusiform (spindle-shaped with tapering ends) ascospores with three septa. The asci are unitunicate. This fungus is considered to be heterothallic with a bipolar mating system (mating controlled by two different alleles at a single locus) and additional genes controlling the sexual cycle of *P. oryzae*. Rossman et al. (1990) considered *P. oryzae* as a synonym of *P. grisea*. However, Couch and Kohn (2002) found that *M. grisea* contains two distinct monophyletic inter-sterile lineages, one associated with *Digitaria*, and the other associated with *O. sativa* and other grasses (*Eragrostris curvula, Eleusine, Eragrostris curvula, Eleusine coracana, Lolium perenne,* and *Setaria* species). These two lineages are differentiated on molecular, morphological and association of the host. They proposed the name *Magnaporthe oryzae* for the isolates of *Magnaporthe* from *O. sativa* and closely related isolates from other grasses, while other isolates from *Digitaria sanguinalis* (crabgrass) were distinct and were described as *Magnaporthe grisea.*

Physiological Races of the Pathogen

Rice blast pathogen has a large genetic diversity everywhere the rice crop is grown. The virulence of a blast pathogen population that was differentiated by using cultivars grown within a geographic region is recommended for all retested practical use (Bonman et al., 1986). Since the Near Isogenic Lines (NILs) with single resistance genes were developed in an *indica* genetic background as a differential set, seven races were reported from 46 Philippine isolates (Inukai et al., 1994). More complex pathogen population was revealed when the pathogenicity reactions were used with six *indica* and seven *japonica* NILs. Chen et al. (2001a) identified 48 pathotypes using the *indica* NILs, 82 pathotypes using the *japonica* NILs and a total of 344 pathotypes using both NILs from 792 single spore isolates from 13 major rice growing provinces of central and southern China. The large differences in the frequencies of the isolates producing compatible reactions on the NILs indicated the differences in frequencies of avirulence genes in the pathogen populations. These findings provided very useful information for formulating strategies for improving the blast resistance in rice breeding programs.

Disease Cycle

Blast infection starts when air-borne conidia land on the leaf surface and anchor to the cuticle of leaf with the help of spore-tip mucilage. Germination proceeds within 2–4 hours with the extension of germ tube which undergoes hooking, swelling at its tip and differentiation into appressorium. The formation of appresorium on the host surface marks the onset of the disease (Roy-Barman and Chattoo, 2005). After the appressorium accumulates melanin to achieve the needed mechanical stability, the fungal germ tube penetrates into the epidermal cell via the appressorium using its lytic enzymes (Chida and Sisler, 1987; Howard and Ferrari, 1989). The first infection hyphae inside the host cell can be seen 16–24 hours later. At first, the fungus only colonizes the cell it has penetrated, and after 2–3 days it spreads rapidly into the nearby cells. During its further development, the fungus spreads through the intercellular spaces to reach the stomatal chamber. From there, the melaninized conidiophores grow to the leaf surface. Subsequent sporulation of the fungus marks that it has completed its asexual cycle which take place within 4–7 days (Thieron et al., 1998).

Disease Development in the Field

In the rainfed lowland in Central and North western Luzon of the Philippines, leaf blast infection often starts in 10–14 days old seedlings established in dry seedbed during the rainy season. Its second cycle continues in about 34–44 days after transplanting. The disease is usually much more severe in plots that receive the highest amount of primary inoculum that may come from the infected seeds. In the second cycle, blast lesion produces more spores that are readily disseminated by wind to nearby densely seeded healthy seedlings. The likelihood of the third cycle to occur during the reproductive phase depends on the amount of air-borne conidia at the end of the vegetative growth stage. Spores produced near the end of the growing season may infect the collar of the flag leaf. This may lead to infection of the neck emerging from the infected collar on which the head will be supported to produce a condition called rotten neck or neck blast. Infection of the neck is generally considered to be the most injurious phase of the disease because it disrupts the entire panicle development. High relative humidity, short sunshine duration, and long duration of leaf wetness enhanced the sporulation and infection. Otherwise, non-flooded paddy applied with a high rate of nitrogen fertilizer enhanced the disease severity of panicle blast on susceptible cultivars such as NSIC Rc128 and Rc130. Under the same conditions, NSIC Rc160 was identified resistant.

Inoculum Source

Infected rice straw and stubbles are probably the most important sources of inoculum of the blast pathogen. The pathogen can survive from one season to the other on diseased crop residue and seed. Its primary infection starts where infected seed is sown densely in the seedbed. During continuous monitoring of the leaf blast on weeds commonly growing along the dikes surrounding paddies during the rainy seasons in the rained lowland in the Central Luzon, Philippines during 2008–2012, it was found that *Panicum repens* (Poaceae), *Eleucine indica* and *Echinochloa crus-galli* had the same leaf blast symptoms as that of rice plant. Also, the lesions contained similar conidia as those of *P. oryzae*. Although, there was no evidence that cross infection occurs between rice and weeds, the chances of spread are not unlikely.

Predisposing Factors

Rice blast infection in the tropics is favored during long periods of plant surface wetness, high relative humidity (90–96%), little or no wind at night (<2 m/s), and minimum temperatures between 22°C–24°C (Fig. 12.1). Extended leaf wetness (18–26%/6–10 hours), intermittent light rain (5–20 mm/week), low solar radiation (17–23 MJ/m²), and short sunshine duration (4–7 hour/day) favored disease development. Leaf wetness at 18–26°C for 8 to 12 hours is required for blast infection. Relative humidity above 90% favors the production and release of conidia with peak production between midnight from 3 to 8 days after the appearance of lesions. The pathogen

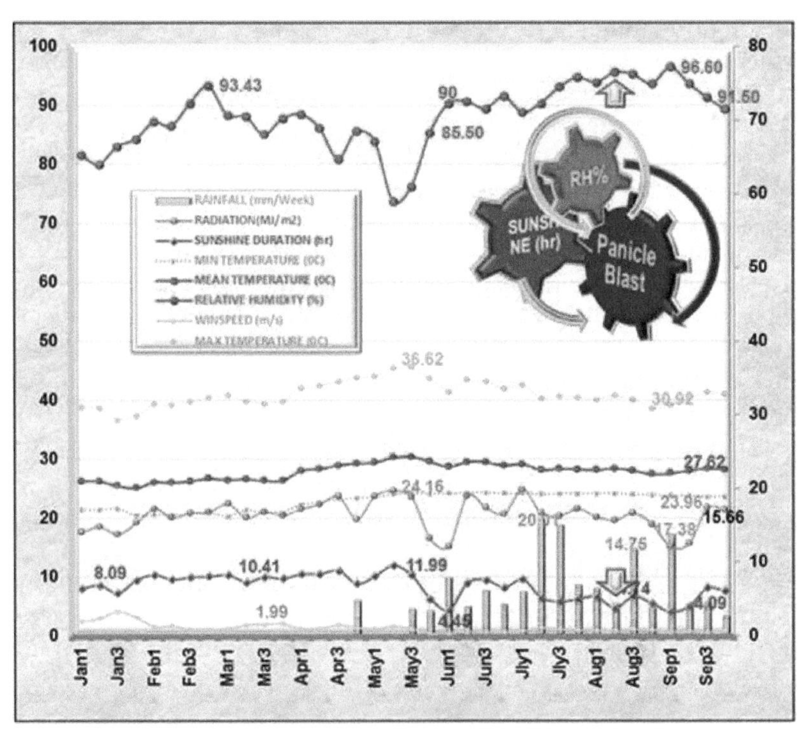

Figure 12.1. Prevalent weather parameters during July to September, 2008–10 were conducive to rice blast endemic in rainfed lowland rice locations, Central Luzon, Philippines.

Color image of this figure appears in the color plate section at the end of the book.

needs free moisture for pathogen penetration for infection to occur. The optimum temperature for sporulation and germination is 25–28°C, and 16–25°C for appressorium formation. Exposure of the diseased plants to around 32°C causes the blast lesions to expand rapidly and cease after eight days. Frequent periods of light rain (10–20 mm weekly) encourage infection more than heavy rain of short durations (30–70 mm weekly). High solar radiation, long sunshine duration (8–10 hour/day), RH <80%, frequent and high rainfall, and abundant water depth of 3–5 cm are conditions that are unfavorable for conidial production. However, the optimum day time temperatures of 25°C–27.8°C and night temperatures between 17°C–22.78°C for disease development in the semi-temperate rice ecosystem (TeBeest et al., 2007) were lower than those in the tropics.

In addition, the blast is favored by excessive nitrogen fertilization, aerobic soils and drought stress. High nitrogen rates and nitrate increase rice susceptibility to the disease. Extended soil aeration, and drought stress can lead to conversion of ammonium to nitrate and make rice more susceptible to blast (Scardaci et al., 1997).

Disease Symptoms

Airborne spores usually infect young leaves from seedling to the reproductive growth stage. The symptoms of rice blast include lesions that can be found on the leaves, collar, panicle node, stems, peduncles and panicles. The common name for the blast disease is usually associated with the affected part such as leaf blast, collar blast, neck or panicle blast or nodal blast. These symptoms are described below.

1. Leaf blast symptoms may vary according to the environmental conditions, age of the plant and the levels of host resistance (Fig. 12.2), sometimes resulting in death of young plants during the tillering stage. Lesions on susceptible cultivars initially appear as elliptical, gray-green and water-soaked with darker green margins, which can coalesce to become up to 2 cm long. Lesions often remain small in size (1–2 mm) and brown to dark brown on resistant cultivars.
2. Collar blast is the result of infection at the junction of the leaf blade and sheath resulting in a typical brown collar (Fig. 12.3). This infection can kill the entire leaf and may extend a few millimeters into and around the sheath. It can have a significant impact on yield when the flag leaf is affected.
3. Neck blast refers to the infected internode just below the panicle (Fig. 12.4). Panicle branches and glumes may also be infected. If infection occurs before the milk stage, the entire panicle may die leaving the grains unfilled resulting in whitehead-like damage caused by stemborer.

Figure 12.2. Leaf blast lesions initially appear as brown specks smaller than 1 mm in diameter without sporulation (a), turn roundish to elliptical about 1–3 mm with gray center surrounded by brown margins and/or yellow halo covered with conidia (b); and typical spindle-shaped lesions >3 mm in diameter with necrotic gray centers and reddish round margins which may coalescence (c).

Color image of this figure appears in the color plate section at the end of the book.

Figure 12.3. Collar blast symptoms. The infected leaf collar turned green to brownish which may expand to the leaf sheath and leaf base and cause collapsed or flabby leaves.

Color image of this figure appears in the color plate section at the end of the book.

Figure 12.4. Neck blast which causes death and brown internodes under the panicle and at one site of flag leaf. Under favorable weather, the fungus spreads onto the spikelets and produces white powder.

Color image of this figure appears in the color plate section at the end of the book.

Less severe infections may cause partial grain filling and poor milling quality. Panicle lesions are usually brown to dark brown.
Nodal blast infection exhibits brown or blackened necrotic nodes of culm and may occur before the plant approaches maturity. It can result in the complete death of the stem above the infection.
Symptoms of rice blast on seeds consist of brown spots and blotches (TeBeest et al., 2007). The fungal pathogen infects the florets resulting in unfilled spikelets, seeds with poor milling quality, and no seed production at all when pedicels become infected. Other parts of the panicle such as panicle branches and glumes may also be infected.

SHEATH BLIGHT PATHOGEN

Name of the disease: Sheath Blight
Name of the pathogen: *Rhizoctonia solani* Kühn (anamorph); *Thanathephorus cucumeris* (A.B. Frank) Donk (teleomorph)

Sheath Blight (ShB) is the most common disease, under intensified rice production systems worldwide, characterized by using high-yielding varieties, high crop density, and input of high rates of fertilizer, especially nitrogen (Ou, 1985; Teng et al., 1990; Savary et al., 2000a,b, 2006; Mew et al., 2004). Besides rice, the disease can occur in many other crops

(Gangopadhyay and Chakrabarti, 1982). Yield losses in tropical lowland rice in Asia were estimated to be about 5–10% (Savary et al., 2000a). Severe infestation attributed to high nitrogen application increases yield reduction up to of 20–30% (Savary et al., 2000a,b, 2006). In 2010, ShB was widespread across farmer community in Sto. Domingo, Nueva Ecija, Philippines, and the ShB caused crop losses in the susceptible variety NSIC Rc216 from 15 to 45% (Truong et al., unpubl.). When the disease started early from maximum tillering to early heading stages, grain filling was affected resulting in significant reduction in yield and grain quality.

Biology of the Pathogen

Rhizoctonia solani is the most widely recognized species of *Rhizoctonia.* The pathogen was originally described by Julius Kühn on potato in 1858. It is a species of Basidiomycetes that does not produce any asexual spores and only occasionally produces sexual spores (basidiospores). In nature, *R. solani* reproduces asexually and exists primarily as vegetative mycelium and sclerotia. Unlike many other Basidiomycetes, their basidiospores are not carried in a fleshy fruiting body. The number of nuclei (NCN) in the fungal mycelium and the structure of the septal spores have been studied extensively to differentiate *R. solani* from other *Rhizoctonia* species. Moore (1987) described the teleomorph, *Thanatephorus cucumeris* that contains more than two NCN and closed to the tips in young hyphae wider than 7 mm. This mycelium turns dark brown. The sclerotial bodies if present are irregular in shape, light to dark brown and undifferentiated into skin and medula. The *Rhizoctonia solani* AG1-1A causes most damage to the rice crop in locations where production is intensive and associated with several other crops and weed species (see below). The vegetative hyphae of *R. solani* are colorless when young and turn brown as they mature. The mycelium consists of hyphae divided into individual cells by a septum with a pore. The septal pore allows the mobility of cytoplasm, mitochondria and nuclei from cell to cell. The hyphae often branch out at approximately 90° angles and usually possess two or more nuclei per hyphal cell. The fungus survives in the soil as long-lived and hard sclerotial structure. Sclerotia are formed on the sheaths and leaves as early as a week after leaf and sheath lesions are seen, but are more typically observed on infected rice in the booting to heading stages. Sclerotia are initially white and slightly fuzzy, rapidly turning brown with an irregular shape (Fig. 12.5). Under natural conditions, sclerotia usually occur singly but may coalesce to form larger masses.

Figure 12.5. Typical symptom of sheath blight on the leaves (left), sclerotial bodies formed at the reproductive crop stage (center), and severe infection of panicles (right).

Color image of this figure appears in the color plate section at the end of the book.

Anastomosis Groups of *Rhizoctonia* Species

Rhizoctonia solani primarily attacks the under ground plant parts such as the seeds, hypocotyls and roots, but is also capable of infecting above ground plant parts (e.g., stem, sheath and leaves). Most rhizoctonia diseases are initiated by mycelium and sclerotia, but several important diseases of beans, sugar beet and tobacco are induced by basidiospore infection. The basidiospores can be dispersed to a long distance. The basidiospores germinate to produce hyphae that infect leaves during periods of high relative humidity and periods of extended wet weather. Under these conditions, basidiospores can often be observed on the base of stems near the soil surface or on the underside of leaves in the plant canopy.

Since *R. solani* and other *Rhizoctonia* spp. do not produce conidia and only rarely produce basidiospores, the classification of these fungi is often difficult. Prior to the 1960's, researchers relied mostly on differences in morphology and pathogenicity on various plant species to classify *Rhizoctonia*. Parmeter et al. (1969) introduced the concept of "hyphal anastomosis" to characterize and identify *Rhizoctonia*. The concept implies that isolates of *Rhizoctonia* that have the ability to recognize and fuse (anastomose) with each other are genetically related, whereas isolates of *Rhizoctonia* that do not have this ability are genetically unrelated.

Within an anastomosis AG, two types of hyphal interactions (C2 and C3) are most relevant for the study of population biology. The C2 reaction (also referred to as a killing reaction), represents a somatic incompatibility response between genetically distinct individuals. The C3 reaction (perfect fusion) between two isolates is indicative of genetic identity or near identity. Hyphal anastomosis and molecular methods are currently being used to further examine the taxonomy, ecology and pathology of *R. solani complex*. This aspect has been compiled and summarized by Genhua et al. (2012) as follows:

1. *Mutinucleated Rhizoctonia* spp.: Include 13 anastomosis groups, of which AG 1–4 is highly pathogenic to many host plants. Among them, AG-1 I A is capable of infecting the aerial parts of plants with sclerotial bodies which are the important sources of infection. It also induces different symptoms such as foliar blight, leaf blight, web-blight, head rot, bottom rot, and brown patch on rice, corn, barley, sorghum, potato, barnyard millet, common millet, soybean, peanut, lima bean, cabbage, leaf lettuce, *Stevia*, orchard grass, crimson clover, tall fescue, turfgrass, creeping bentgrass, perennial ryegrass, gentian, camphor, *Digitalis purpurea*, winter savory (*Satureja montana*), and ornamental foxglove (Li and Yan, 1990; Sneh et al., 1998; Fenille et al., 2002; Naito, 2004; Garibaldi et al., 2009, 2012).
   AG-2-2 III B induces damping-off, brown sheath blight, dry root rot, root rot, brown patch, large patch, black scurf, stem rot, stem blight, stem rot, collar rot, and crown brace rot on rice, soybean, corn, sugar beet, edible burdock (*Arctium lappa*), taro (*Colocasia esculenta*), *Dryopteris* spp., elephant foot, crocus, saffron (*Crocus sativus* Linn.), redtop, bentgrass, St. Augustine grass, turf, balloon flower (*Platycodon grandiflorum*), Christmas-bells (*Sandersonia aurantiaca*), *Hedera rhombea*, mat rash, *gladiolus*, ginger, and *Iris* Linn. (Sneh et al., 1998; Priyatmojo et al., 2001; Naito, 2004).
2. *Binucleated Rhizoctonia* spp.: Include 18 anastomosis groups. Among those, AG-Ba and AG-Bb cause the gray sclerotial disease and gray southern blight on rice, *Echinochloa crugalli* subsp. *submitica* var. *typica* and foxtail millet (Sneh et al., 1998; Sharon et al., 2008).

Genetic Diversity and Geographic Distribution of the Pathogen

Anastomosis group (AG)-1 I A of *R. solani* is a major pathogen causing sheath blight of rice. It infects many host plants. Recent protein and DNA-based studies support the separation of *R. solani* into genetically distinct groupings revealing a considerable genetic diversity within an anastomosis group. Frequent assessment of pathogen diversity is one of the most important criteria in designing disease management strategies. High genetic diversity of *R. solani* AG1-IA isolates collected from rice paddies was revealed using Inter Simple Sequence Repeats (ISSR) and enterobacterial repetitive consensus (ERIC) analysis (Khodayari et al., 2009). However, there was no correlation observed between genetic diversity, pathogenicity and geographical origins of isolates in this domain. In Venezuela, the fungus caused banded leaf and sheath blight on maize, which is considered an emerging disease problem where maize replaced traditional rice-cropping in adjacent fields (González-Vera et al., 2010). The effects of host specialization on gene flow between sympatric and allopatric rice, and maize-infecting fungal populations and an evidence of genetic recombination in the *R. solani* AG1 IA populations were detected. There was a symmetrical historical migration between the rice- and the maize-infecting populations at the location. Rice- and maize-derived isolates were able to infect both hosts but were more aggressive on their original hosts, consistent with host specialization (Bernardes de Assis et al., 2008; Gonzales-Vera et al., 2010). This finding indicates that a good understanding of increasing aggressiveness in *R. solani* AG1-IAI population structure over time and across geographical regions is a vital key to design an effective crop management system.

Disease Cycle

Rhizoctonia solani survives on infected rice straw, other crop residues, and in the soil as sclerotial bodies. When the field is irrigated, the sclerotia float out of the soil and are carried by water. The fungus is attracted to the chemical stimulants released by actively growing plant cells and decomposing plant residues. As the attraction process proceeds, the hyphae come in contact with the plant and cling on its external surface just above the water line. The fungus continues to grow and infects the plant by producing an infection cushion. The pathogen penetrates the plant and uses plant nutrients to continue its development. It usually infects the rice sheath in late tillering to early reproductive growth stages. During the infection process, it produces extracellular enzymes that degrade various components of plant cell walls including cellulose, cutin and pectin. The hyphae continue to proliferate and colonize dead tissues and form sclerotia. The developmental stage and the ability of fungal colonies to differentiate and form sclerotia are associated with an oxidative stress (Georgiou and Petropoulou, 2001). The fungus also survives as mycelium colonizing soil organic matter as a saprophyte. Sclerotia on dying leaves or rice stalk during the ripening stage of the rice crop, fall off during rice harvest and continue to survive in the field.

Sclerotia have a thick outer layer that allows them to float and survive in water. When new substrates become available, the sclerotia germinate and a new cycle of the fungus begins.

Inoculum Source

The pathogen is soilborne and waterborne. It produces sclerotial bodies on rice straw, stubbles and on some weeds. These sclerotial bodies are made of mycelial masses which are detachable and lightweight. They can be easily carried away through the paddy water. Mycelium present in crop residues and in seed is recognized as a source of primary inoculum. Weed hosts also serve as sources of inoculum.

Predisposing Factors

High relative humidity of about 96% under the canopy and a temperature range of 28°C–32°C favor infection and development of ShB. Increased cropping frequency and a wide host range also provide a continuous availability of host tissues for the pathogen and favor its survival across cropping seasons. Savary et al. (1995) showed that nitrogen levels affected the focal expansion of ShB owing to high leaf wetness under large and dense canopies. Higher nitrogen rate softens plant tissues, increases the density of canopies to retain moisture, and increases leaf-nitrogen content. The common recommended fertilizer rate in the Philippine rice production system is 90-40-40 NPK ha^{-1} and 120-60-60 kg ha^{-1} in wet and dry season, respectively (Palay Check System, 2007). In most cases, farmers apply either higher or lower rates than the recommended nitrogen rate without P and/ or K. Thus, analysis of disease incidence can be very useful considering the effect of nitrogen fertilizer supply and the crop density affected by seeding rates in direct seeded and transplanted rice. High crop density, either through increased sowing rates in direct-seeded rice or higher plant density in transplanted rice, has long been considered to be a major factor that favors the development of ShB. Canopy structure varying with methods of rice crop establishment could affect the spread of ShB (Willocquet et al., 2000). The apparent infection rate of ShB was lower in direct-seeded crops than in transplanted if 60–80 kg ha^{-1} of certified seeds was used. On the other hand, ShB could cause total crop failure due to continuous five cropping seasons within two years in Thailand and Vietnam (Mew et al., 2004b).

Disease Symptoms

Sheath blight symptoms consist of snake skin-like lesions that are usually observed on the leaf sheaths which may continue up to the leaf blades. The initial lesions are small, ellipsoid or ovoid and greenish-gray starting from the sheath near the water line in lowland fields. Lesions increase in size and may coalesce forming bigger lesions with irregular dark brown outlines and grayish-white centers. These alternating wavelike patterns of light tan and brown bands can extend up to the sheath and cover entire leaf surface including the flag leaf. Under favorable conditions, lesions may kill the whole leaf.

BROWN LEAF SPOT PATHOGEN

Common Name of the Disease: Brown Leaf Spot
Bipolaris oryzae (Breda de Haan) Shoemaker (Dela Paz et al., 2006) (anamorph)
Synonyms: *Drechslera oryzae* (Breda de Haan) Subramanian and P.C. Jain; *Helminthosporium oryzae* Breda de Haan; *Ophiobolus miyabeanus* Ito and Kuribayashi;
Cochliobolus miyabeanus (Ito and Kuribayashi) Drechsler ex Dastur (teleomorph)

Brown leaf spot is a common disease of rice in rainfed (Singh and Singh, 2000) and upland (Gupta and O'Toole, 1986) rice-growing areas of the world. It has been historically known as one of the most destructive rice diseases causing the great Bengal Famine of South Asia in 1942 (Padmanabhan, 1973). The disease causes 5% yield loss across all lowland rice production situations in South and Southeast Asia (Savary et al., 2000b). This, seed-borne pathogen can be carried from parental seed batches into the field crop adversely affecting the seedling vigor. Heavily infected seeds can cause 10–45% seedling mortality. In the poorly fertile soils, yield loss may reach upto 12–45% depending upon the disease severity levels.

Morphology of the Pathogen

The mycelium of the pathogen is hyaline to light olive. Conidia of *B. oryzae* isolates vary in size and number of cells per conidium. Conidia which are borne at the apical portion of the conidiophore are generally curved, boat or club-shaped to almost cylindrical, light brown to golden brown when mature, with 6 to 14 transverse septa or cross walls, 63–153 x 14–22 µm, and often with a minute, slightly protruding hilum (a dot at the point of attachment to a conidiophore). The unbranched conidiophores usually emerge from stomata singly or in groups of two or three, are brown, paler at the apex, cylindrical to clavate with 3–10 septa, and 20–60 x 5 µm (Ou, 1985). Mature conidia are brownish in color and germinate from one or both polar cells, which have thin walls. Less mature conidia are sub-hyaline and may produce germ tubes from intermediate segments. The conidiophore is multi-septate, about 600 µm long, and 4–8 µm wide produced singly or in groups of up to 17, straight to flexuous, sometimes geniculate, pale-brown to olive-brown, and lighter towards the apex (CABI, 2007).

The perithecia of *C. miyabeanus* are dark yellowish-brown, globose to elliptical, 360–780 µm in diameter, and each with a conical to cylindrical beak of 98–200 x 55–110 µm. The beak length is variable. Asci are straight to somewhat curved, 140–235 x 21–26 µm, clavate, and cylindrical clavate or broadly fusoid. The ascospores are 235–648 x 4–9 µm, hyaline to pale-olive green, filamentous, wider in the middle and attenuated towards the end, have 6–15 septa, surrounded by a mucilaginous sheath, and are coiled in a helix in the ascus (CABI, 2007).

Seed Transmission

The seed-borne transmission is a critical risk to rice productivity in the subsequent field crop. PCR technique-based on specific primers developed from ribosomal DNA (rDNA) regions coding for the rRNA subunits has proven as a complementary tool with the culture-based technique to detect such seed-borne pathogen population (Bruns et al., 1991; Hibbett, 1992; Brown et al., 1993; Bridge and Arora, 1998). Using this tool, seed samples taken recently from different geographic and rice agro-systems across Southeast Asia showed 3–6% infection. To assess the seed-borne transmission, 1–5% of the seed lots were evaluated for the next planting season. Molecular analysis showed that 22% of DNA fingerprints of some *B. oryzae* isolates obtained from the second generation of seedlings and seeds were identical to those from parental seed lots (Online:http://www.dfid.gov.uk/r4d/Project/291/Default. aspx). This indicates that the fungal spores survive on infected seeds. Recent phylogenetic studies using cultural characterization showed that *B. oryzae* isolates are highly diversified (Kamal and Mia, 2009; Kumar et al., 2011). Additionally, significant differences in aggressiveness have also been observed among isolates of the same fingerprint types. Some isolates in the pathogen lots represented 42 DNA fingerprints caused 21–85% germination failure, 3–14% post-emergence seedling death from 15 to 45 days after transplanting; and 19–23% yield loss (Kamal and Mia, 2009).

Disease Cycle

Development of seed-borne infection of *B. oryzae* starts from the dark hyphal cushions (Manandhar, 1999). The pathogen spreads from plant to plant in the field by airborne spores. The spores are airborne, allowing the pathogen to spread quickly. Unregulated seed exchange and poor seed quality and hygiene can facilitate the spread and carry-over of the disease within farms, villages, states and across countries. The fungus can also survive on infected rice straw and stubble and can cause brown spot on the subsequent crop. Its initial infection occurs on young seedlings. *Bipolaris oryzae* forms an appressorium, a specialized infection structure to infect rice (Ahn and Seok-Cheol, 2007). Biochemical processes controlled by the calcium/calmodulin signaling system seem to be involved in the induction of pre-penetration morphogenesis on rice. Ethylene glycol tetraacetic acid, calmodulin antagonists and calcineurin reportedly inhibited conidial germination and appressorium formation at the micromolar level. The calmodulin antagonist, W-7, also inhibited accumulation of mRNA of the calmodulin gene within germinating conidia and/or appressorium-forming germ tubes.

Inoculum Source

The fungus was also detected in switchgrass (*Panicum virgatum*) that may serve as an alternate host in the spread of the disease to rice (Waxman et al., 2011). The fungus can also survive on infected rice straw and stubble and cause brown leaf spot on the subsequent crop. Its initial infection occurs on young seedlings. Early identification of fungi based on morphological, physiological and disease characteristics hardly assessed the genetic variability in the seed-borne population and establish threshold fungal inoculum in the seed, soil, on debris, and adequately assess the field crop diseases because the fungus may be originally seed-borne and/or airborne in the pathosystem. Besides, rice seed lot usually carries a few pathogens at one time. In disease development and management, the importance of each individual pathogen or a combination of two or more pathogens carried by seeds for planting has not been addressed (Mew et al., 2004b). Recently developed molecular techniques are useful tools to look into the genetic population of pathogens, their ecosystem and seed transmission. This is a gateway to the disease epidemiology for optimization of disease management strategies with almost unlimited number of polymorphic loci to detect and direct assessment of genetic variation in a given fungal population (Bridge et al., 2004).

Predisposing Factors

Brown spot is favored by water stress; infection takes place at high relative humidity above 89% and is favored by leaf wetness (Ou, 1985). Long leaf wetness for 8–24 hours is required for infection to occur. The optimum temperature for infection ranging from 25 to 30°C for two months, unusually cloudy weather, and higher-than-normal temperatures and rainfall at the time of flowering and grain-filling stages predispose the crop for infection. Rice crop grown in nutrient-deficient and poorly drained soil are predisposed to brown spot infection.

The disease occurs in irrigated fields that are poorly drained and enriched with organic matter (Ou, 1985) and in acid soils in upland. Brown spot is considered as one of limiting factors in rice production of about 500,000 ha of acid sulfate soils in uplands in the Philippines (Sebastian et al., 2000). Zadoks (2002) considered the disease as the "poor rice farmer's disease". The disease also develops on plants affected by Akiochi (Ou, 1985; Moletti et al., 1996) and has been used as an indicator of this nutritional disorder. Akiochi is caused by excessive concentration of hydrogen sulfide in the soil and results in reduced nutrient uptake (Dobermann and Fairhurst, 2000). The disease is enhanced by low concentrations of silica, manganese and magnesium in soil (Ou, 1985). It commonly occurs in soil deficient with silica, potassium, manganese, magnesium, iron and calcium. Brown spots are also usually associated with low nitrogen and maturity of the plant becoming more susceptible with age particularly at flowering. Brown spot is also favored by reduced water supply, particularly when the rice crop is established by direct-seeding method (Savary et al., 2005). This may be because of rice plants in direct-seeded rice have a shallow root system (Castillo, 1962) and consequently become more sensitive to water stress.

Disease Symptoms

The disease can occur at all crop development stages. The pathogen infects the coleoptile, leaves, leaf sheath, panicle branches, glumes and spikelets. The disease causes seedling blight, with small, circular, yellow brown or brown lesions that may girdle the coleoptile and distort primary and secondary leaves (Webster and Gunnell, 1992). Initial symptoms appear as small circular to oval spots on the leaves and may vary in size, shape and color depending on the environmental conditions, age of the spots and susceptibility of the variety. Typical brown spot symptoms observed at the tillering stage and beyond are small and circular, dark brown to purple-brown with light brown to gray center and reddish brown border

(Fig. 12.6). Older spots may have bright yellow halos, which is due to a toxin produced by the pathogen (Vidyasekaran et al., 1986). Often the spots are 5 to 14 mm long (Webster and Gunnell, 1992) and can cause leaf wilting. Some spots become somewhat elongated with tiny, dark specks. On susceptible variety, lesions may coalesce causing dead leaves before maturity, resulting in lightweight or chalky grains. On resistant varieties, lesions are brown and pinhead-sized. Infected glumes and panicle branches have dark brown to black oval spots or discoloration on the entire surface under favorable conditions. The fungus may also infect the glumes, causing dark brown to black oval spots. Infection of florets leads to incomplete or disrupted grain filling and a reduction in grain quality. The pathogen can also penetrate grains, causing "pecky rice", a term used to describe spotting and discoloration of grains. Infection on the kernels significantly reduces grain yield, kernel weight and seed quality. Brown spot infection can adversely affect the grain yield and milling quality.

Figure 12.6. A traditional upland rice variety from Northeastern Central Luzon, Philippines found susceptible to brown leaf spot.

Color image of this figure appears in the color plate section at the end of the book.

SHEATH ROT PATHOGEN

Name of the disease: Sheath Rot
Name of the pathogen: *Sarocladium oryzae* (Sawada) W. Gams and D. Hawks.
Synonyms: *S. attenuatum* W. Gams & D. Hawks;
Acrocylindrium oryzae Sawada Hebert
Teleomorph: none reported

The disease is present in all rice growing countries worldwide (Ou, 1985). Recently, it has become more prevalent in rain-fed and irrigated lowland rice, particularly affecting female parental lines in intensive hybrid seed production. The highest losses recorded due to the disease ranged from 26 to 60%. The fungus is a weak pathogen and it is most often associated with the presence of stem borers and other forms of injury on the flag leaf sheath.

The syndrome is widespread in tropical Asia since the introduction of modern semi-dwarf and photoperiod-insensitive cultivars (Mew et al., 2004). *Sarocladium oryzae* is always associated with several fluorescent and non-fluorescent pseudomonad pathogens isolated from rotted sheaths and seeds with discoloration. These included *Burkholderia glumae*, *P. fuscovaginae*, and other nonpathogenic bacteria (Cottyn et al., 2001). In the process of intensifying rice production, rice seed appears to have become vulnerable to infection by many microorganisms, especially during the monsoon period. Collectively, seed discoloration and sheath rot usually occur during the rainy season owing to low solar radiation and high temperature and humidity.

Morphology of the Pathogen

The mycelium of the fungus is sparsely branched, septate and measures 1.5–2 μm in diameter. The conidiophores are branched once or twice, with 3–4 branches in a whorl and usually come out from mycelia that are slightly wider than the vegetative hyphae. The main hyphal axis is 15–22 x 2–2.5 μm with terminal branches tapering towards the tip and measuring 23–45 μm in length and 1.5 μm in width at the base. Conidia, which are simply produced consecutively, at the tip of the conidiophore are 4–9 x 1–2.5 μm, hyaline, smooth, single-celled and cylindrical.

Disease Cycle

The fungus invades the rice plant through stomata and wounds, and grows in between the cells of vascular bundles and mesophyl tissues (Ou, 1985). Genetic diversity study of *S. oryzae* isolates from the field experiments in central and northern Luzon, the Philippines, using RAPD primers with simple repetitive sequences derived from ITS detected 44 different DNA fingerprints from the parental seeds and 15% of fingerprints from seedlings of the second generation. All were identical to the collected parental seed (Holderness and Pearce, 1997). The results indicate that the technique is useful in tracking the fungal population structure changes through the seed transmission cycle from seed-to seed. Likewise, seed health testing showed the second generation seedlings had 3–6% incidence of *S. oryzae*. In North East and South India, pathogenicity test showed the genetic variability of *S. oryzae* isolates from different locations. The isolates were able to produce both cerulenin and helvolic acid responsible to induce the disease in higher percentage (Ayyadurai et al., 2005).

Inoculum Sources

The sheath rot fungus can survive as mycelium in the infected crop residues and on seeds. Some of its alternate hosts are maize, pearl millet, sorghum, and wild-rice-*O. rufipogon*, and grasses such as *Cyperus difformis, Echinochloa colonum, E. crus-galli, Eleusine indica, Hymenachne assamica, Leresia hexandra, Monochoria viginalis*, and *Panicum walense* (Balakrishnan and Nair, 1981, cited by Sreenivasaprasad and Johnson, 2001; Deka and Pookan, 1992). Pathogenicity studies confirmed that *S. oryzae* isolates can infect several species of grasses and sedges that are possible sources of inoculum for rice. *Sarocladium oryzae* has been reported as the causal organism of bamboo blight in Bangladesh such as *Bambusa balcoa, B. polymorpha, B. tulda, B. vulgaris* and *Meloccana baccifera* (Boa and Brady, 1987, cited by Sreenivasaprasad and Johnson, 2001). However, cross pathogenicity of rice isolates to bamboo was not conclusive. Further study on genetic variability on *S. oryzae* infected bamboo showed that four isolates had identical band pattern with rice isolates, except one band specific to bamboo isolate only (Bridge et al., 1997; Pearse et al., 2001). Thus far, PCR-based diagnosis has proven as a vital tool to shed light on the influence of seed-borne inoculum of *S. oryzae* on disease transmission, epidemic development and crop loss assessment.

Predisposing Factors

The disease is favored by high amounts of nitrogen, high relative humidity, dense crop growth and temperature from 20 to 28°C during heading to maturity stages of plant growth. The disease development retards panicle emergence. It is also more prevalent in the rainy season than in the dry season. Sheath rot is often associated with stem borer feeding which serves as entry points for this weak pathogen. The disease appears to be favored by high density of crop growth.

Disease Symptoms

Lesions start during the late booting stage as oblong or irregular oval spots, 0.5–1.5 cm long, with gray or light brown centers and a dark reddish-brown diffuse margin on the uppermost leaf sheath enclosing the young panicles (Fig. 12.7). The lesion is typically expressed as a reddish-brown discoloration of the flag leaf sheath. Symptoms are most severe on the uppermost leaf sheaths that enclose the young panicle at late boot stage. As the disease advances, lesions coalesce and cover almost the entire leaf sheath. Panicles remain within the sheath or may partially emerge.

Infection during the boot stage is most severe resulting in unfilled and discolored panicles. The panicles usually can not emerge completely; the young panicles remain inside the sheath. Panicles that have not emerged tend to rot, and florets turn red-brown to dark brown. Affected leaf sheaths have abundant whitish powdery mycelium. Grains from damaged panicles are reddish-brown to dark brown and may be sterile, shriveled, partially filled or unfilled.

Figure 12.7. Partially emerged panicles due to sheath rot infection showing rotten, red-brown to dark brown spikelets and unfilled, shriveled grains.

Color image of this figure appears in the color plate section at the end of the book.

FOOT ROT (BAKANAE) PATHOGEN

Name of the disease: Foot Rot, Bakanae
Name of the pathogen: *Fusarium fujikuroi* Nirenberg;
Fusarium moniliforme Sheld. (anamorph)
Synonyms: *F. heterosporum* Nees;
F. verticillioides (Sacc.) Nirenberg; *Lisea fujikuroi* Sawada
Gibberella fujikuroi (Sawada) Wollenworth;
G. moniliformis Wineland (teleomorph)

Bakanae means a "foolish disease of rice" in Japanese. It is also called the foot rot disease caused by *Fusarium moniliforme*. It was first described in Japan, and is now widely distributed in all rice growing countries in Asia and throughout the world. It is one of the diseases of rice that can cause yield losses from 20 to 50% in Asia (Ou, 1985). In the Philippines, the incidence of this seed-borne disease ranged from 1 to 70% in the var. PSB Rc82 (Gergon, E., unpubl. data). The disease occurs in both rainfed and irrigated lowland rice ecosystems and has also become prevalent in the temperate subtropical environment. It is caused by one or more *Fusarium* spp., forming a complex of pathogens and disease symptoms.

The disease is primarily seedborne and seed-transmitted. The fungus survives over adverse conditions in infected seeds and other diseased plant parts. The pathogen can also be isolated from seeds that appear healthy. Infected seeds after germination give rise to seedlings with bakanae symptoms. The fungus does cause considerable damage without producing visible symptoms on some rice varieties particularly PSB Rc82 and IR42. It produces gibberellins, plant hormones that cause elongation of the rice plants, resulting in bakanae disease (Sun and Snyder, 1981). Strains of *F. fujikuroi* (teleomorph *G. fujikuroi*) infect rice grown worldwide to produce gibberellins, moniliformin and beauvericin. Some rare strains of the fungus can produce fumonisin B1, B2, and B3 (Proctor et al., 2004; Glenn, 2007). Recently, bakanae is considered to be caused by a pathogen complex which is capable of producing mycotoxins. At least nine mating populations are now known within the *G. fujikuroi* species complex (Leslie, 1995; Klaasen and Nelson, 1996; Leslie and Klein, 1996; Nirenberg and O'Donnell, 1998; Britz et al., 1999; Samuels et al., 2001; Zeller et al., 2003). In Nepal, the predominant *Fusarium* species in seed coat include *Gibberella fujikuroi* complex consisting of mating type (MP) A (*F. verticillioides*), *G. fujikuroi* MP-C (*F. fujikuroi*), and *G. fujikuroi* MP-D (*Fusarium proliferatum*) (Desjardins et al., 2000). The widespread occurrence of MP-D suggests its significant role in the complex symptoms of bakanae disease of rice. Similarly, *G. fujikuroi* species complex also occur in Malaysia and Indonesia. Based on production of moniliformin, fumonisin, gibberellic acid and fusaric acid, and MPs within the *G. fujikuroi* species, Zainudin et al. (2008) reported several strains of the fungus from

76% of the isolates, namely MP-A (*G. moniliformis*), MP-B (*G. sacchari*), MP-C (*G. fujikuroi*) and MP-D (*G. intermedia*). The disease incidence and fumonisin contamination were highly prevalent in paddy seeds with 12.5% infection in Uganda (Taligoola et al., 2004), 20% infection in Pakistan (Bhalli et al., 2001), and 20% infection in Vietnam (Trung et al., 2001).

Knowledge of the genetic structure of the pathogen population might be useful to establish effective strategies for controlling the disease. Pathogenicity test of 30 *F. moniliforme* isolates collected within the province in Nueva Ecija, and PCR-based amplified rDNA-ITS target region showed two distinct types of symptoms and four fingerprints (Fig. 12.8, center) (Truong, unpubl.). In Iran, high genetic diversity among 41 isolates of *F. verticillioides* from different regions was detected using vegetative compatibility groups and RAPD (Bahmani et al., 2012). Phylogenetic relationship within the *G. fujikuroi* species complex was established using PCR product from 4 loci nuclear large subunit 28S, ITS region, mitochondrial small subunit (mtSSU) rDNA, and b-tubulin together with two protein-encoding nuclear genes. O'Donnell et al. (2000) also identified 10 new distinct species distributed in three areas: two within Asia and four within both Africa and South America. *F. verticillioides* isolates from paddy (*Oryza sativa* L.) were collected from different geographical regions in India and were capable of producing fumonisin (Maheshwar et al., 2009). The fumonisin producing ability of 11 out of 27 *F. moniliforme* isolates was also detected by PCR using vertf-1 and vertf-2 primers. This finding indicates that, *F. moniliforme* infection resulting in toxin production can be a serious problem and needs to be assessed as rice is the staple food for the people of Asia.

Figure 12.8. Foot rot or bakanae induces yellowish green leaves (left) and abnormally elongated tillers on infected PSB Rc82 rice variety (right). Four DNA pathogen profiles detected within the province surveyed for bakanae(insert).

Color image of this figure appears in the color plate section at the end of the book.

Moniliformin, a hydroxyl-cyclobutenedione, was initially characterized as a culture extract from *F. proliferatum* strain NRRL 6322, which was identified at that time as *F. moniliforme*, hence the name moniliformin (Burmeister et al., 1979). The metabolite is only known to be produced by species of *Fusarium*, and while moniliformin is known to be toxic to poultry based on laboratory toxicity studies (Burmeister et al., 1979; Rabie et al., 1982; Marasas et al., 1991; Leslie et al., 1996), it has not been associated with any acute or chronic disease outbreaks among animals.

Morphology of the Pathogen

The morphology, physiology, genetics and genomics of *Fusarium* species have been well documented in literature. Such data are critical for understanding these fungi and for managing their impact on the safety, value and yield of quality grain.

Asexual Structures

The fungus may produce dark blue, spherical sclerotia, about 80 x 100 µm in size. Its hyphae are branched and septate. It produces both micro- and macroconidia in culture. Macroconidia are long banana-shaped and multicellular borne on a basal cell bearing 2–3 apical phialides. Macroconidia have three to seven septa, slightly curved and ranging from 25–60 x 2.5–4 µm. The microconidia are fusiform to oval to spherical in shape and borne singly or in chains or in false heads laterally on conidiophores formed from the aerial hyphae measuring from 5–12 x 1.5–2.5 µm. Microconidia and macroconidia are important for wind and splash dispersal of the fungus. The conidia also are generally the propagules that result in infection of the host plants. Additionally, some species produce thick-walled resistant chlamydospores important for long-term survival.

Sexual Structures

The pathogen is a haploid, heterothallic ascomycete fungus. It produces ascospores sexually that are hyaline, elliptical, with one to sometimes three septa measuring 10–20 x 4–7 µm. The asci are cylindrical, piston-shaped, flattened above, and with dimensions of 90–102 x 7–9 µm. Each ascus contains four to six, seldom eight, ascospores. The perithecia are dark blue and measure 220–280 x 250–300 µm. They are spherical to oval and somewhat roughened outside.

Inoculum Sources

The pathogen predominantly spreads through infected seeds. Its microconidia and macroconidia are important for its wind and splash dispersal. These conidia are generally the propagules that result in infection of the host plants. Some species also produce thick-walled resistant chlamydospores which are important structures for their long-term survival. The fungus survives in infected seeds and other parts of the plants, and in the soil in infested crop residue as thick walled hyphae or macroconidia but for a shorter period. The fungus is also reported on maize, barley, sugarcane, sorghum and *Panicum miliaceum* (Ou, 1985).

Disease Cycle

Fusarium moniliforme is seedborne and primarily seed transmitted (Ou, 1985). The fungus may also be isolated from healthy looking seeds collected from an infested field. Seeds, therefore, provide one of the most efficient methods of pathogen dissemination at great distances and allow pathogen introduction into new areas. The seed-borne inoculum provides the initial site or focus for secondary infections. Its frequency of detection in the seed is a good estimate of initial inoculum in relation to disease development which ranges from 5 to 15% (Mew et al., 2004b). The fungus infects the plants through the roots or crowns and grows systemically in the plant but does not systemically infect the panicle. Seeds are infected at the flowering stage via air-borne ascospores and also from conidia that contaminate the seed during harvesting. Conidia production on infected tillers also coincides with the flowering and maturity of the crop, causing infection or contamination of the seeds. Infected crop residues from the previous season produce numerous conidia that subsequently infect neighboring plants when conditions are favorable. The conidia are disseminated easily by wind and water to cause new infections. The disease can reduce seed germination and the number of seedlings and mature plants.

Predisposing Factors

Sowing of seeds in infested soil often results in rapid progression of the disease and a high percentage of mortality. Higher application of nitrogen to the soil stimulates the development of disease. The disease is also favored by high temperatures (30–35°C).

Disease Symptoms

Symptoms of bakanae vary depending on the strain, inoculum levels and the presence of toxin. Surviving seedlings are usually elongated, taller, slender and paler green in color than normal plants either in the field or in seedbed (Fig. 12.8). Symptoms may also be stunting, chlorosis, root and crown rot, and eventual death of the plant depending on the pathogen strains. Stunted growth occurs usually in dry soil. Abnormal elongation or hypertrophic growth of the plant is induced by the hormone gibberellin. Infected plants are thin with yellowish green leaves, pale green flag leaves, and with a few tillers. The affected tillers usually die gradually at maximum tillering stage before reaching maturity. Infected older plants have few tillers, and produce adventitious roots at the lower nodes and culm. Leaf sheaths of infected plants may turn blue-black with the production of perithecia. Infected plants that survive until maturity do not produce panicles; if panicles are produced, the panicles have few partially filled sterile grains and most spikelets are empty. Panicles of healthy plants may be infected and turn pink, and the hulls or seeds may have a reddish color due to the growth and sporulation of the fungus. Infected seeds have white, fluffy mycelium often covering the entire seed. White powdery growth of conidiophores can also be seen on the lower portion of the culms.

NARROW BROWN LEAF SPOT

Name of the disease: Narrow Brown Leaf Spot
Name of the pathogen: *Cercospora janseana* (Racib.) O. Const. (anomorph)
Synonyms: *Cercospora oryzae* Miyake;
Sphaerulina oryzina K. Hara (teleomorph)

Narrow brown leaf spot (NBS) is considered a minor disease of rice that has been reported in almost all rice-growing countries in Asia, Latin America, Africa, the USA, Australia and Papua New Guinea. Severe leaf necrosis occurs on susceptible varieties as plants reach maturity. Infection causes reduction in the green lamina of leaves and may also cause premature death of leaves and sheaths, production of unfilled panicles, chalky grain and decreased milling recovery. The severe damage due to the disease has been reported to cause 40% yield loss in rice (Overwater, 1960). Losses due to the disease include reductions in yield, milling, and the cost of production.

Morphology of the Pathogen

The myelium of the pathogen is hyaline to light olive. Conidia on the host plant measure 12.9–47.2 x 3.9–6.3 µm and 10.6–72.9 x 3.3–6.4 in culture, cylindrical to clavate; 3–10 septate, hyaline or light olive, and air-borne at the apical portion of the conidiophore (Ou, 1985). The unbranched conidiophores are 20–60 x 5 µm and usually emerge from stomata in solitary or in groups of two or three. They are brown, paler at the apex, and cylindrical to clavate with 3–10 septate.

The teleomorph was described by Hara (1918). The globose, black perithecia are usually immersed in the epidermal tissues of the host plant. They are 60–100 µm in diameter with papiliform mouth with minute, rounded, blunt projection through which the spores escape. Asci are cylindrical or club-shaped, rounded at the top, stipitate, and with biseriate ascospores in two series or rows. Ascospores are 20–23 x 4–5 µm, spindle-shaped, straight or slightly curved, 3–10 septate, and hyaline.

Inoculum Source

The fungus persists on crop residues and on rice seeds. Spores are carried away by wind and splashing of rain over long distances. Under favorable conditions, the spores that have landed on leaf blades of susceptible varieties may initiate the infection. The fungus is also reported to infect *Panicum repens.*

Predisposing Factors

Narrow brown leaf spot is common in rainfed and upland rice fields. *Cercospora* development is favored by dense stands, high nitrogen rates and late planting. The disease is exceptionally severe in the second crop. Disease severity is affected by the susceptibility of varieties to one or more prevalent races of the pathogen and growth stage of the plant. Although rice plants are susceptible at all stages of growth, infection is usually more severe at maturity. This becomes more severe as the rice approaches maturity, causing premature ripening and yield reduction. It is reported that high levels of potash increase the disease. There were six physiological races reported in Arkansas and Texas (Sah and Rush, 1987) that can break the host resistance (Groth and Hollier, 2010).

Disease Symptoms

The common symptoms of NBS are short, linear, 2–10 x 1 mm, and light reddish brown lesions on leaf blades. The lesions also occur on leaf sheaths, pedicels, and glumes (Fig. 12.9). Premature leaf death occurs in severe cases. The symptoms may advance to the sheath of the flag leaf causing

Figure 12.9. Narrow brown leaf spot and leaf scald oftenly occur in poor soil with limited water supply in rain-fed lowland and upland in the humid tropics.

Color image of this figure appears in the color plate section at the end of the book.

large 2 cm-long necrotic cinnamon brown spots that typically encircle the uppermost internode below the panicle base, pedicels and glumes.

The lesions on the leaves and upper leaf sheaths of resistant varieties are narrower, shorter and darker brown than those observed on the susceptible varieties (Ou, 1985). On the lower sheaths, "net blotch" is distinguished by dark brown to yellow spots in between minor veins (Groth and Hollier, 2010). When disease pressure is severe, reddish brown discoloration on the sheath may occur. Abundant sporulation is observed on the lesions during extended wet periods. Net-like pattern of dark veins may appear on leaf sheaths. Head infections usually develop at the internodes with typical brown striations due to vein discoloration. On susceptible cultivars, wider and lighter brown lesions with gray necrotic centers may appear on the leaves (Groth and Hollier, 2010). Early maturing varieties tend to escape the major impact of the disease.

LEAF SCALD PATHOGEN

Common Name of the disease: Leaf Scald
Name of the pathogen: *Microdochium oryzae* (Hashioka and Yokogi) Samuels and I.C. Hallett
Synonyms: *Rhynchosporium oryzae* Hashioka and Yokogi;

Metasphaeria albescens Thümen;
Gerlachia oryzae (Hashioka and Yokogi) W. Gams (anamorph);
Monographella albescens (Thümen) Parkinson, Sivanesan and Booth (teleomorph)

Leaf scald had once reached an epidemic level in Japan (Matsuyama and Wakimoto, 1977). Humid tropical conditions favored the high disease severity resulting in crop loss up to 20–30% in India and Bangladesh and (Mia et al., 1986; Mondal et al., 1986). Leaf scald is known from all major rice growing regions of the world (Farr et al., 2008). The disease mainly appears on the leaves but it can also be observed on the panicles and seedlings. The leaf lesions can reduce the plant's nutrient absorption and photosynthetic area, which may result in decrease of tillering nodes. Infection on the panicle can cause sterility or abortion of developing kernels (Groth and Hollier, 2010). The pathogen is seedborne and can reduce grain quality.

The fungus usually invades host tissues from the tips and margins of the leaf through the stomata. The hyphae develop intercellularly through the mesophyll layer of the green leaf lamina ahead of the expanding necrotic zone.

Pathogenicity using detached leaf inoculation method as well as whole plant inoculation, using two isolates from the Philippines, supported the occurrence of pathogenic specialization in the pathogen which was reported by Bonman et al. (1990). Pathogenicity test that was done in West Africa by spraying seedlings using different isolates also showed a pronounced pathogenic diversity of the leaf scald pathogen.

Inoculum Source

The fungus survives on rice seeds, leaf tissues and in plant debris which are the primary sources of inoculum. However, the seeds are considered the major source of inoculum. The pathogen is transmitted from seed to coleoptile tips in symptomless seedlings at a rate of 2.18% (Guierrez et al., 2008). *Echinochloa crus-galli* (L.) Beauv., a weed, another host for *Rhynchosporium oryzae* (Singh and Gupta, 1980).

Predisposing Factors

Disease development usually occurs late in the season on mature leaves and is favored by wet weather, high nitrogen fertilization and close spacing. The fungus is a weak pathogen that develops faster if wounded leaf tissues are present.

Disease Symptoms

The symptoms of leaf scald may vary according to plant age, growth stage, variety and plant density. The disease usually occurs between maximum tillering and mature plant stages. Young lesions start at the leaf tips or from the edges of leaf blades forming an alternating zonate pattern of light-tan and dark reddish brown areas (Fig. 12.10). Lesions from the edges of leaf blades exhibit an indistinct, mottled pattern. As the lesions become old, zonation fades and lesions coalesce resulting in blighting of a larger part of the leaf blade. Eventually, affected leaves dry up, turn straw colored and appear scalded. Infected leaf tips near the midrib may split when there are strong winds. Seedling blight characterized by decaying coleoptile with red-brown discoloration and root rot, leaf sheath browning without necrosis, and head blight with flower deformation, sterility, and light-brown discoloration of the glumes are other symptoms (Fig. 12.10, right). Panicle infection causes a uniform light to dark, reddish-brown discoloration of the florets of the developing grain.

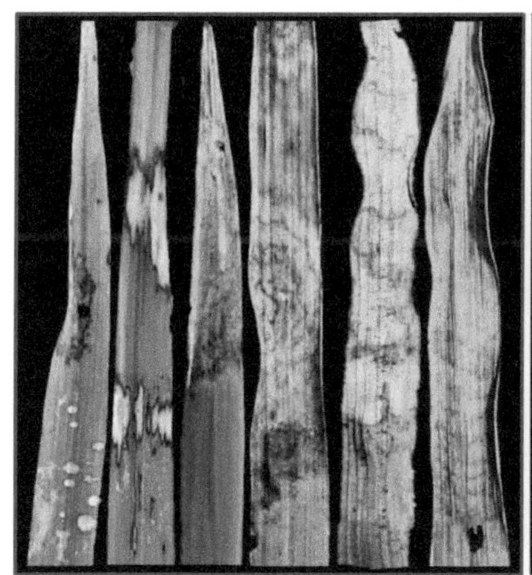

Figure 12.10. Leaf scald lesion initiates along leaf margin and turns gray with alternating narrow reddish-brown bands owing to intermittent rains and warm days in rain-fed and upland areas (left). Worsen microclimatic conditions can rot the spikelets (right).

Color image of this figure appears in the color plate section at the end of the book.

STEM ROT PATHOGEN

Common Name of disease: Stem Rot
Name of Causal organism: *Sclerotium oryzae*
Synonyms: *Sclerotium oryzae* Cattaneo (anamorph);

Nakataea sigmoidae (Cavara) K. Hara;
Magnaporthe salvinii (Cattaneo) R. Krause and Webster (teleomorph)

Most rice varieties are susceptible to stem rot. However, stem rot occurs only sporadically and is not considered to be a major rice disease in the Philippines and southern United States. The disease, however, becomes critical when the macro-nutrient elements are deficient in the soil and if stagnant water is present during the summer crop. Yield losses in rice owing to this disease ranges from 25 to 45%.

Predisposing Factors

The fungus initially infects the leaf sheath and invades into the culm often causing lodging. Stem rot development is favored by low potassium levels in the soil and rice following rice rotations.

Disease Symptoms

Sclerotium oryzae normally infects rice after tillering. The disease becomes severe usually during the reproductive stage. Stem rot symptoms initially appear as black angular lesions on leaf sheath near the water line on rice tillers (Fig. 12.11). The sheath later turns black. The culms develop dark-brown or black streaks causing the culm to collapse at maturity. Small round

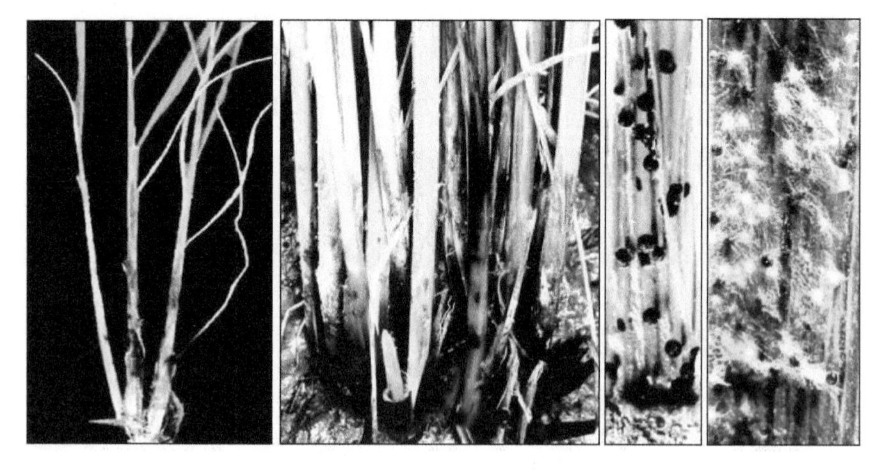

Figure 12.11. Initial infection of stem rot occurs above the water line and appears as black lesion at maximum tillering stage (left, first two photos). Small black sclerotial bodies are formed inside the plant tissues after the plant dried up (right two photos).

Color image of this figure appears in the color plate section at the end of the book.

black sclerotia are formed on dead tissues. They are incorporated into the soil with the crop debris and can survive for many years. The sclerotial structures which are light in weight float on the first flooded irrigation or rainfall water and come in contact with the rice plants. They tend to accumulate toward one side of the field due to wind.

RICE FALSE SMUT PATHOGEN

Name of the disease: Rice False Smut
Name of the pathogen: *Ustilaginoidea virens* (Cooke) Takahashi
Clavieps oryzae-sativae Hashioka (teleomorph)
Villosiclava virens Tanaka et al., 2008 (teleomorph)

Rice False Smut (RFS) caused by *Ustilaginoidea virens* has long been considered as a minor problem in rice production worldwide. RFS, once believed to be a sign of good harvest, is now-a-days considered an emerging disease in many rice-growing countries.

It is a sporadic disease in most rice growing areas in the Philippines during the rainy season. RFS infect inbreed variety NSIC Rc218 and hybrid variety NSIC Rc232, both do not show significant reduction in the yield. However, high incidence and severe damage of RFS in tropical and temperate environments such as in Egypt, India and the United States have increased, particularly in hybrid rice seed production systems, in China. RFS is truly a disease of rice which is gaining importance in hybrid rice production when spores are brought along with the pollen during the pollination stage. Parental lines of hybrid rice have similar genetic background (Virmani and Wan, 1988), which might have resulted in the selection of adapted virulent populations of the pathogen. Further, *U. virens* isolates in 14 counties in Sichuan, China using three rice hybrid varieties, confirmed significant differences in disease indices among the isolates, positive interaction between pathogen and varieties and variation in pathogen virulence associated with different host origin, and parental lines (Lu et al., 2009). In China, *U. virens* infected rice during the flowering stage, inhibited flower fertility, development of adjacent spikelets and decreased grain weight. The number of infected seed is positively correlated with yield loss. Spikelets in the lower part of the panicle are usually more severely infected than in the upper part (Hu, 1985). Damage owing to RFS is low (1–10%), but severe infections can reduce the grain yield by 50 to 75% (Li et al., 1986).

In 1997, false smut was considered to be an emerging concern in rice production and an increasingly significant pathogen of rice in Arkansas. RFS incidence ranges from 1 to 15% with at least 2–3 smuted balls per infected panicle (Rush et al., 2000). It is recently considered to be widespread within the state (TeBeest et al., 2011). RFS is reported as a new disease in Egypt and was first examined in the Nile Delta in 1997. The disease incidence and the number of infected grains became significantly higher in 2000 than in 2001. The disease usually affected a few grains (1–20 only) and occupied any part of the panicle (Atia, 2004). Yield losses caused by RFS ranged from 1–11%. The highest values in chaffiness occurred in samples collected from Abou-Hammad (29.33%). The percentage of chaffy grains increased with the disease severity and correlated with yield loss (Sharma and Joshi, 1975; Anand et al., 1985).

In India, RFS caused 44.37% yield loss in the severely infected cultivar coupled with decrease in grain weight with increasing number of smutted balls (Singh and Dube, 1978). A similar level of yield loss occurs due to smut balls with an increase in chaffiness and decreased grain weight (Hu, 1985). Loss in yield might also be due to increased sterility of spikelets adjacent to the false smut balls (Hashioka, 1971).

In addition to yield reduction, RFS also reduce seed quality. False smut balls formed on rice panicles produce a number of metabolites that are toxic to both plants and animals (Suwa, 1915). The fungus produces ustiloxin, which is toxic to domestic animals (Koiso et al., 1992). The ustiloxin structure was determined by a combination of X-ray crystallograph and amino acid analysis, and five ustiloxins A, B, C, D, and E were isolated (Koiso et al., 1994). Miyazaki et al. (2009) suggested that a routine monitoring of the contamination of ustiloxin A in forage rice silage can be done by high-performance liquid chromatograph.

Causal Pathogen

Rice false smut is caused by an ascomyceteous fungus, *U. virens* (teleomorph: *Claviceps oryzae-sativae* Hashioka). Recently, Tanaka et al. (2008) proposed that *U. virens* be named *Villosiclava virens* because the morphology and biology of the sexual form of the false smut causing fungus is closer to *Villosiclava* spp. than to *Claviceps* spp.

Ustilaginoidea virens can be cultured and produces conidia on the potato sucrose agar medium (Chen et al., 1994). The fungus produces chlamydospores on the smut balls which are borne laterally on minute sterigmata on radial hyphae. Chlamydospores when young are light colored and smooth, and become spherical to elliptical, warty, olivaceous and measuring 3–5 x 4–6 µm. They produce septate germ tubes that form conidiophores bearing ovoid and minute conidia at the tapering apex. Some of the green spore balls develop one to four sclerotia in the center, each one producing stalked stromata with globose swelling at the tip that contain perithecia around the periphery. A large number of asci are produced in each perthecium. Asci are cylindrical with a hemispherical apical appendage, 180–220 x 4 μm, and 8-spored. Ascospores are hyaline, filiform, unicellular and 120–180 μm.

Disease Cycle

The RFS was induced by inoculating the rice plant at booting stage (Ikegami, 1963). The initial inoculum is believed to be the sclerotia in the soil or ascospores, while airborne conidia lead to secondary infection. Furthermore, Schroud and TeBeest (2005) and Ditmore and TeBeest (2006) were able to infect the young seedlings by placing the spores onto the roots in a greenhouse and also under field conditions. However, knowledge of the disease cycle and epidemiology of *U. virens* remains incomplete (Lee and Gunnell, 1992) and that of seed transmission uncertain (Biswas, 2001a). The debate on the pathogen infection process has been going on for a long time because of the lack of experimental evidence.

Ditmore et al. (2007) and Zhou et al. (2008) used Nested PCR with primers aligned to specific ITS-rDNA from diverse geographical isolates and showed that 75% of rice seedlings grown from healthy seeds and planted in a field with a history of RFS were infected within three weeks after emergence. Similarly, rice seedlings that were grown from seeds treated with vacuum-infiltrated spores of *U. virens*, expressed signs or symptoms of infection such as chaffing and decreased panicle production, earlier in the growing season before smut balls were formed on panicles.

Recently, Ashizawa et al. (2012) using the green fluorescence protein-labeled pathogen strain, noted that when the conidia land on the outer spikelet surface, germinate and then invade the inner spikelet surface. Finally the hyphae cover the floral organs and eventually form the smut ball. The conidia that germinated on spikelet surfaces, grew extensively towards the apices of spikelets, and then colonized onto the inner surfaces of the lemma and palea. After heading, the floral organs are covered with hyphae. However, the pathogen does not penetrate the host cell walls directly and also does not form haustoria. The stigma and lodicules are also occasionally infected. In the smut balls, the ovary was not infected (Tang et al., 2013).

Inoculum Source

Rice cultivars play an important role in the degree of RFS infection (Singh et al., 1987; Biswas, 2001a). Furthermore, early maturing rice genotypes escape RFS infection, while the late maturing genotypes do not (Singh and Khan, 1989). Primary infection is believed to be initiated mainly by the ascospores produced from the sclerotia of the fungus while secondary infection is brought about by the chlamydospores. The chlamydospores can survive in the field for several months on infected rice straw and stubble (Webster and Gunnell, 1992). The fungus may survive on the previous crop and on alternate hosts, such as grasses, and can be sources of secondary inoculum. It can spread from plant to plant and in the field by air-borne spores. Shetty and Shetty (1985) and Atia (2004) found that the fungus can infect common rice weeds such as barnyard grass—*Echinochloa crus-galli*, cogon grass—*Imperata cylindrica*, and *Digitaria marginata*.

Further studies using real-time PCR assay, Zhou et al. (2008) and Ashizawa et al. (2010) demonstrated that PCR-based techniques are very useful in detecting the pathogen in symptomless rice seedlings. The technique can be deployed to optimize disease management strategies. These findings imply that RFS fungus is seed and soilborne and is transmitted through seeds. The results also confirmed the earlier observation that spores from smutted balls may germinate late in the growing season and infect rice flowers (Cartwright and Lee, 2006).

Predisposing Factors

Rice cultivars show different degree of infection with RFS. Long maturing cultivars are more susceptible to RFS infection (55% of smutted balls) than short maturing cultivars during the two successive growing seasons (Singh and Khan, 1989). The differences between tested rice cultivars to RFS might be attributed to host-pathogen interaction (Walker, 1975). Rice cultivars resistant to RFS are also reported by Anand et al. (1985), Singh et al. (1987), Singh and Khan (1989), Sugha et al. (1992), and Biswas (2001b).

High humidity (88 to 93% and leaf wetness from 18 to 22%) in the morning during the flowering stages enhances the pollination and grain development compensating for grain reduction by RFS. Percentage of infected tillers and number of smutted balls were lower in clay soil than in light soil with 47 and 14.33% incidence, respectively, and number of infected grains (6 and 5, respectively) during the two seasons (Atia, 2004). Thurston (1990) found that growing rice in flooded paddy reduces many plant pathogens. Rice plants directly seeded in the field showed higher RFS disease incidence and number of smutted balls than by the transplanted ones (Jiansheng et al. 1999). High disease severity was also noticed beside irrigation canals with relative humidity more than 90%, in the semi-water-deep cultivar, and as well as on the late maturing rice (Singh et al., 1987; Bhagat and Prasad, 1996; Biswas, 1999; Yashoda et al., 2000). Temperature ranging from 25 to 35°C favors the disease (Chen et al., 1994; Dodan and Singh, 1996; Yashoda et al., 2000). High disease incidence appears to be related to late sowing, late maturity of cultivars, and flowering stage that coincides with rainy days (Narinder and Singh, 1989; Ahonsi et al., 2000). High relative humidity and leaf wetness resulting from frequent rainy and cloudy days favored to false smut severity during heading stage (Cartwright et al., 1999). This report was however conflicting to the early finding on the effect of rainfall by Dodan and Singh (1996). Furthermore, higher input of fertilizer, susceptible cultivars and fungal virulence favor the development of false smut (Li et al., 1986; Fujita et al., 1989). High incidence of infected grains were recorded at the high application rate of N amount (190 kg N/ha). On the other hand, low disease incidence and smutted grains became very low at the half input of N (95 kg N/ha) over the two season trials. Excess rate of N causes a thinner wall in the cells of the host that can easily be penetrated by the pathogen (Patell et al., 1992; Cartwright et al., 1999).

Disease Symptoms

False smut is visible only after the panicle is exserted and when the rice crop is at the hard dough stage. The fungus produces clusters of sporangia. It can be composed of a single cell or can be multicellular that burst from the middle of glumes, forming velvety smut balls. During this period, the infected grains are transformed into large, velvety, green masses, which attain more than twice the diameter of the normal grains. The spore ball grows up to 1 cm or more in diameter enclosing the floral parts, covered with a membrane that bursts as it grows. The chlamydospores in the initial stages of development are orange yellow, turning later to dark green or almost black (Fig. 12.12). At this stage, the surface of the ball cracks. The outermost green layer of the ball consists of mature spores with mycelial fragments. The typical symptoms of RFS disease were described by Pandey (1982) and Verma and Singh (1988). In most cases, not all spikelets of a panicle are smutted, but some spikelets neighboring the smut balls are often unfilled (Webster and Gunnell, 1992).

HEAD BLIGHT PATHOGEN

Name of the disease: Head Blight
Name of pathogen: *Fusarium graminearum* Schwabe
Gibberella saubinetii (Mont.) Sacc., *Fusarium roseum* Link. (anamorph)
Gibberella zeae (Schwein.) Petch (teleomorph)

Fusarium graminearum causes head blight of small grains, including rice, wheat, and barley (Leslie and Summerell, 2006). This rice disease has since been reported in many countries, including Brazil, China, India, Japan, Nepal and Uganda (Ou, 1985; Desjardins et al., 2000). Head blight disease of rice is caused by the *F. graminearum* complex that includes 16

Figure 12.12. Orange-yellow smut balls are formed at early milking stage then turn dark green to almost black with cracked surface.

Color image of this figure appears in the color plate section at the end of the book.

phylogenetically distinct species worldwide (Sarver et al., 2011). Some isolates of *F. graminearum* can produce mycotoxins that can contaminate the grains and affect human beings and animals. The disease also reduces the germination capacity of infected seeds. However, not all wheat isolates of *F. graminearum* can produce trichothecenes (Goswami and Kistler, 2005). Likewise, Nepalese rice contained no detectable amount of trichothecenes (Desjardins et al., 2000).

Mycotoxin Production

Isolates of *F. graminearum* can produce mycotoxins nivalenol (NIV) and deoxynivalenol (DON) that are tightly linked to the TRI gene cluster for the synthesis of 8-ketotrichothecenes DON and NIV, and humans and farm animals consuming moldy cereals exhibited typical signs of trichothecene intoxication (Desjardins, 2006). Maize and wheat in North America and Europe commonly are contaminated with DON, while strains with NIV chemotypes are commonly recovered from cereal crops in Asia (Ichinoe et al., 1993; Kim et al., 1993). These chemotypes were detected by PCR using the Tri7 and Tri13 alleles in the trichothecene biosynthetic gene cluster (Lee et al., 2001, 2002). Using phylogenetic analysis, O'Donnell et al. (2000a,b) and Starkey et al. (2007) grouped *F. graminearum* into seven phylogenetic lineages and four additional lineages. The geographic distribution and agro-ecosystem may attribute to the lineage. Using UPGMA analysis of the AFLP banding patterns, Jungkwan et al. (2009) reported that major population of *F. graminearum* from rice in Korea are clustered with lineage 6 while major population in eastern Korea are clustered with lineage 7. Lineage 6 of *F. graminearum* may have a host preference for rice and that it may be fit for a rice agroecosystem than the other lineages present in Korea. Lineage 6 is predominant in warmer regions and lineage 7 dominates in cooler regions (Qu et al., 2008; Suga, 2008). The findings indicate that molecular diagnostic tools are vital to understand the agroecosystem of the pathogen for developing better disease management strategies.

Inoculum Source

The disease spreads predominantly through the infected seeds. Ascospores and conidia are released from infected seeds and can survive with crop residues. Symptoms of *Fusarium* head blight on diseased wild rice (*Zizania palustris*) are shrunken seeds with light tan to light brown sometimes with light pink mycelial growth (Nyvall et al., 1999). *F. graminearum* is found on whole seeds at all growth stages and also from shattered seeds. *F. anthophilum* and *F. subglutinans* are also frequently isolated at most growth stages, whereas *F. acuminatum*, *F. culmorum*, *F. solani*, and *F. semitectum* are infrequently isolated at one or more growth stages. However, *F. camptocera* is isolated only from shattered seeds. Furthermore, species of *Fusarium* are isolated in higher percentage during milking and dough stages. Survival of *Fusarium* spp. in diseased seed is noticed in whole dried seeds but the pathogen dies in seeds immersed in water. Survival is better in seed stored at 4°C. The strains that can infect the rice cultivar M201 by artificial inoculation under greenhouse conditions, show no trichothecenes from the infected rice florets (Goswami and Kistler, 2005).

Predisposing Factors

Infection is favored by high temperatures (25–30°C) and high relative humidiy (>85% and higher). In contrast, temperature (>31°C), humidity (<80%), leaf wetness (20–25°C), wind velocity (0 to 3 km/hr) under the canopy, and clay soil moisture content (15–20%) at ripening stage in the tropic can suppress the RFS. High nitrogen (>120 kg ha^{-1}) and the seeding rates (>60 kg ha^{-1}) coupled with flooded condition at the early ripening stage enhanced the RFS severity in the main cropping in the Philippines.

Disease Symptoms

The fungus forms white spots on the surface of glumes which later on become yellow, salmon or carmine. Infected grains are light, shrunken and brittle, and can have reddish appearance or brown spots (Fig. 12.13). Stem nodes are black and usually rot and later disintegrate. Stems wilt and break results into lodging of the plant.

Concluding Remarks

Cohesive relationship between molecular tools and the conventional techniques in establishing the genetic diversity and pathogenicity of nine major fungal pathogens causing important diseases of rice worldwide have been discussed here.

It is evident that the current trend in molecular technology has contributed to the development of cost-effective and ecologically sound rice disease management strategies, that helped to have positive long-term effect on rice productivity and consequently on food security, particularly for the third world countries. Thus, more research in this area will be further rewarding.

Figure 12.13. *Fusarium graminiarum* and *F. semitectum* are often found on the infected spikelets during hot and humid days in irrigated and rain-fed lowland rice.

Color image of this figure appears in the color plate section at the end of the book.

References

Ahonsi, M.O., Adeoti, A.A., Erinle, I.D., Alegbejo, T.A., Singh, B.N. and Sy., A.A. 2000. Effect of variety and sowing date on false smut incidence in upland rice in Edo State, Nigeria. IRRI Notes, 25: 14.

Ahn, H.P. and Seok-Cheol, S. 2007. Calcium/calmodulin-dependent signaling for prepenetration development in *Cochliobolus miyabeanus* infecting rice. J. Gen. Pl. Pathol., 73: 113–120

Anand, S.N., Kalhas, C.S. and Gupta, R.S. 1985. Incidence of false smut on some rice cultivars in Jammu (J & K). Res. Dev. Rept., 2: 67.

Ashizawa, T., Moriwak, J.J. and Hirayae, K. 2010. Quantification of the rice false smut pathogen *Ustilaginoidea virens* from soil in Japan using real-time PCR. Eur. J. Pl. Pathol., 128: 221–232.

Ashizawa, T., Takahashi, M., Arai, M. and Arie, T. 2012. Rice false smut pathogen, *Ustilaginoidea virens*, invades through small gap at the apex of a rice spikelet before heading. J. Gen. Pl. Pathol., 78: 255–259.

Atia, M.M.M. 2004. Rice false smut (*Ustilaginoidea virens*) in Egypt. J. Pl. Dis. Prot., 111: 71–82.

Ayyadurai, N., Kirubakaran, S.I., Srisha, S. and Sakthivel, N. 2005. Biological and molecular variability of *Sarocladium oryzae*, the sheath rot pathogen of rice (*Oryza sativa* L.). Curr. Microbiol., 50: 319–323.

Bahmani, Z., Nejad, R.F., Nourollahi, K., Fayazi, F. and Mahinpo, V. 2012. Investigation of *Fusarium verticillioides* on the basis of RAPD analysis, and vegetative compatibility in Iran. J. Pl. Pathol. Microbiol., 3: 147.

Balakrishnan, B. and Nair, M.C. 1981. Weed hosts of *Acrocylindrium oryzae* Saw. Sheath rot pathogen of rice. IRRI Newslet., 6: 13.

Bernardes de Assis, J., Peyer, P., Rush, M.C., Zala, M., McDonald, B.A. and Ceresini, P.C. 2008. Divergence between sympatric rice- and soybean-infecting populations of *Rhizoctonia solani* anastomosis group-1 IA. Phytopathol., 98: 1326–1333.

Bhagat, A. and Prasad, P.Y. 1996. Effect of irrigation on incidence of false smut of rice. J. Appl. Biol., 6: 131–132.

Bhalli, J.A., Aurangazeb, M. and Ilyar, M.B. 2001. Chemical control of bakanae disease of rice caused by *Fusarium verticillioides*. J. Biol. Sci., 1: 483–484.

Biswas, A. 1999. Occurrence of false smut and kernel smut disease in shallow water rice selections in West Bengal, India. Environ. Ecol., 17: 1035–1036.

Biswas, A. 2001a. False smut disease of rice: a review. Environ. Biol., 19: 67–83.

Biswas, A. 2001b. Field reaction of hybrid rice varieties to false smut (FS) and kernel smut (KS) disease in West Bengal, India. Environ. Ecol., 19: 229–230.

Bonman, J.M., Vergel De Dios, T.L. and Khin, M.M. 1986. Physiologic specialization of *Pyricularia oryzae* in the Philippines. Pl. Dis., 70: 767–769.

Bonman, J.M. 1992a. Rice blast. pp. 14–18. *In*: Webster, R.K. and Gunnel, P.S. (eds.). Compendium of Rice Diseases. APS Press, St. Paul, Minnesota. USA.

Bridge, P.D., Pearce D.A., Rutherford, M.A. and Rivero, A. 1997. VNTR-derived oligonucleotides as PCR primers for population studies in filamentous fungi. Letters Appl. Microbiol., 24: 426–430.

Bridge, P.D. and Arora, D.K. 1998. Interpretation of PCR method for species definition. pp. 64–83. *In*: Bridge P.D., Arora D.K., Elander R.P. and Reddy C.A. (eds.). Application of PCR in Mycology. Wallingford: CAB International.

Bridge, P.D., Singh, T. and Arora, D.K. 2004. The application of molecular markers in the epidemiology of plant pathogenic fungi. pp. 57–68. *In*: Arora, D.K. (ed.). Fungal Biotechnology in Agricultural, Food, and Environmental Applications. Mycology Series, ISBN 0-8247-4770-4.

Britz, H., Coutinho, T.A., Wingfield, M.J., Marasas, W.F.O., Gordon, T.R. and Leslie, J.F. 1999. *Fusarium subglutinans* f. sp. *pini* represents a distinct mating population in the *Gibberella fujikuroi* species complex. Appl. Environ. Microbiol., 65: 1198–1201.

Brown, A.E., Muthumeenakshi, S., Sreenivasaprasad, S., Mills, P.R. and Swinburne, T.R. 1993. A PCR primer-specific to Cylindro-carpon heteronema for detection of the pathogen in apple wood. FEMS Microbiol. Lett., 108: 117–120.

Bruns, T.D., White, T.J. and Taylor, J.W. 1991. Fungal molecular systematics. Ann. Rev. Ecol. Systemat., 22: 525–564.

Burmeister, H.R., Ciegler, A. and Vesonder, R.F. 1979. Moniliformin, a metabolite of *Fusarium moniliforme* NRRL 6322: purification and toxicity. Appl. Environ. Microbiol., 37: 11–13.

Butt, A.R., Yaseen, S.I. and Javaid, A. 2011. Seed-borne mycoflora of stored rice grains and its chemical control. J. Animal & Plant Sci., 21: 193–196.

CAB International. 2004. Crop Protection Compendium, 2004 Edition. Wallingford, UK: CAB International. Online: http://www.cabicompendium.org/cpc.

CAB International. 2007. Cochliobolus miyabeanus. Crop Protection Compendium. Wallingford, UK: CAB International. Online: http://www.cabicompendium.org/ cpc.

Cartwright, R. and Lee, F. 2006. Management of Rice Diseases. Rice Production Handbook. Online: http://www.uaex.edu/Other_Areas/publications/PDF/MP192/chapter10. pd.

Cartwright, R.D., Lee, F.N., Ross, W.J., Wann, S.R., Overton, R. and Pasons, C.E. 1999. Evaluation of rice germplasm for reaction to kernel smut, false smut, stem rot and black sheath rot of rice. Arkansas Agric. Experiment Station, Univ. Arkansas, Fayetteville, USA. Series-Arkansas Agric. Exp. Stat., 486: 157–168.

Castillo, P.S. 1962. A comparative study of directly-seeded and transplanted crops of rice. Philipp. Agric., 45: 560–566.

Chen, Z.Y., Yin, S. and Chen, Y.L. 1994. Biological characteristics of *Ustilaginodea virens* and the screening *in vitro* of the causal pathogen of false smut of rice and fungicides. Chinese Rice Res. Newsllet., 2(1): 4–6.

Chen, H.L., Chen, B.T., Zhang, D.P., Xie, Y.F. and Zhang, Q. 2001a. Pathotypes of *Pyricularia grisea* in rice fields of central and southern China. Pl. Dis., 85: 843–850.

Chida, T. and Sisler, H.D. 1987. Restoration of appressorial penetration ability by melanin precursors in *Pyricularia oryzae* treated with antipenetrants and in melanin-deficient mutants. J. Pesticide Sci., 12: 49–55.

Cottyn, B., Regalado, E., Lanoot, B., De Cleene, M., Mew, T.W. and Swings, J. 2001. Bacterial populations associated with rice seed in the tropical environment. Phytopathol., 91: 282–292.

Couch, B.C. and Kohn, L.M. 2002. A multilocus gene genealogy concordant with host preference indicates segregation of a new species, *Magnaporthe oryzae* from *M. grisea*. Mycol., 94: 683–693.

Datta, S.K. 2004. Rice Biotechnology: A need for developing countries. AgBio. Forum, 7: 31–35.

Deka, A.K. and Phoodan, A.K. 1992. Some weeds hosts of *Sarocladium oryzae* in Assam, India. IRRI News Lett. Rice Res. Notes, 17: 25.

Desjardins, A.E., Manandhar, H.K., Plattner, R.D., Manandhar, G.G., Poling, S.M. and Maragos, C.M. 2000. *Fusarium* species from Nepalese rice and production of mycotoxins and gibberellic acid by selected species. Appl. Environ. Microbiol., 66: 1020–1025.

Desjardins, A.E. 2006. Fusarium Mycotoxins: Chemistry, Genetics, and Biology. APS Press, St. Paul, Minnesota, USA.

Ditmore, M. and TeBeest, D.O. 2006. Detection of seed-borne *Ustilaginoidea virens* by nested-PCR. *In*: Norman, R.J., Meullenet, J.F. and Moldenhauer, K.A.K. (eds.). B.R. Wells Rice Research Series. University of Arkansas Agricultural Experiment Station Research Series, 540: 121–125.

Ditmore, M., Moore, J.W. and TeBeest, D.O. 2007. Infection of plants of selected rice cultivars by the false smut fungus, *Ustilaginoidea virens*, in Arkansas. *In:* Norman, R.J., Meullenet,

J.F. and Moldenhauer, K.A.K. (eds.). B.R. Wells Rice Research Studies. University of Arkansas Agricultural Experiment Station Research Series, 550: 132–138.

Dobermann, A. and Fairhurst, T. 2000. Rice: nutrient disorders and nutrient management. Potash and Phosphate Institute, Singapore, and International Rice Research Institute, Los Baños, Philippines, 191 p.

Dodan, D.S. and Singh, R. 1996. False smut of rice: present status. Agric. Res., 17: 227–240.

Farr, D.F., Rossman, A.Y., Palm, M.E. and McCray, E.B. 2008. Fungal databases: Systematic mycology and microbiology. Online: http://www.nt.ars-grin.gov/fungaldatabases/.

Fenille, R.C., Luizde, S.N. and Kuramae, E.E. 2002. Characterization of *Rhizoctonia solani* associated with soybean in Brazil. Eur. J. Pl. Pathol., 108: 783–792.

Food and Agriculture Organization (FAO). 2012. Global rice production set to hit record in 2011–2012. Online: http://business.inquirer.net/42733/global-rice-production-set-to-hit-record-in-2011-2012%E2%80%94fao.

Fujita, Y., Sonoda, R. and Yaegashi, H. 1989. Inoculations with conidiospores of false smut fungus to rice panicels at the booting stage. Ann. Phytopathol. Soc. Japan, 55: 629–634.

Garibaldi, A., Gilardi, G., Bertetti, D. and Gullino, M.L. 2009. First report of leaf blight on foxglove (*Digitalis purpurea*) caused by *Rhizoctonia solani* AG-1-IA in Italy. Pl. Dis., 93: 318.

Genhua, Y. and Chengyun, L. 2012. General description of *Rhizoctonia* species complex. pp. 41–52. *In*: Cumagun, C.J. (ed.). Plant Pathology. Online: http://www.intechopen.com/books/plant-pathology/general-description-of-rhizocotonia-species-complex.

Georgiou, C.D. and Petropoulou, K.P. 2001. Effect of the antioxidant ascorbic acid on sclerotial differentiation in *Rhizoctonia solani*. Pl. Pathol., 50: 594–600.

Glenn, A.E. 2007. Mycotoxigenic *Fusarium* species in animal feed. Animal Feed Sci. Technol., 137: 213–240.

Gonzales-Vera, A.D.J., Bernardes-de-Assis, M., Zala, B.A., McDonald, F., Correa-Victoria, E.J., Graterol-Matute, E.J. and Ceresini, P.C. 2010. Divergence between sympatric rice- and maize-infecting populations of *Rhizoctonia solani* AG-1 IA from Latin America. Phytopathol., 100: 172–182.

Goswami, R.S. and Kistler, H.C. 2005. Pathogenicity and in plant a mycotoxin accumulation among members of the *Fusarium graminearum* species complex on wheat and rice. Phytopathol., 95: 1397–1404.

Groth, D. and Hollier, C. 2010. Leaf scald of rice. Louisiana Plant Pathology Disease identification and management series. Online: http:// www. Isuagcenter.com., November 21, 2012.

Guierrez, S.A., Reis, E.M. and Carmona, M.A. 2008. Detection and transmission of *Microdochium oryzae* from rice seed in Argentina. Australian Pl. Dis. Notes, 3: 75–77.

Gupta, P.C. and O'Toole, J.C. 1986. Upland rice: a global perspective. International Rice Research Institute, Los Baños, Philippines. 360 p.

Hamer, J.E., Howard, R.J., Chumley, F.G. and Valent, B. 1988. A mechanism for a surface attachment in spores of plant pathogenic fungus. Science, 239: 288–290.

Hara, K. 1918. Diseases of a rice plant. Tokyo, Japan.

Hashioka, Y. 1971. Rice disease in the world. VIII. Disease due to *hypocreales ascomycetes* (fungal disease-5). Riso, 20: 235–258.

Hibbett, D.S. 1992. Ribosomal RNA and fungal systematics. Trans. Mycol. Soc. Japan., 33: 533–556.

Holderness, M. and Pearce, D.A. 1997. Assessing the role of seed as a vector of rice pathogens. Proceedings, International Seed Testing Association Pest and Disease Committee, Mycology Working Group, Ottawa, Canada. October 1997.

Howard, R.J. and Ferrari, M.A. 1989. Role of melanin in appressorium function. Exp. Mycol., 13: 403–418.

Hu, D.J. 1985. Damage of false smut to rice and effect of the spores of *Ustilaginodea virens* on germination of rice seed. Zhejiang Agric. Sci., 4: 164–167.

Ichinoe, M., Kurata, H., Sugiura, Y. and Ueno, Y. 1993. Chemotaxonomy of *Gibberella zeae* with special reference to production of trichothecenes and zearalenone. Appl. Environ. Microbiol., 46: 1346–1369.

Ikegami, H. 1963. Occurrence and development of sclerotia of the rice false smut fungus. Res. Bull. Fac. Agric., Gifu Univ., No. 20.

Inukai, T., Nelson, R.J., Zigler, R.S., Sarkarung, S., Takamurel, I. and Kinoshita, T. 1994. Differentiation of pathogenic races of rice blast fungus by using near-isogenic lines with Indica genetic background. J. Fac. Agric. Hokkaido Univ., 66: 27–35.

Jiansheng, C., Yuelan, G., Baoli, L., Liqing, G., Songhan, Z. and Chang, A.H. 1999. Study on occurrence tendency and controlling strategy of pests and diseases in direct-sowing paddy fields in Shanghai. Acta Agric. Shanghai, 15: 72–75.

Jungkwan, L., Hun, K., Yun, S.H., Leslie, J.F. and Lee, Y.W. 2009. Genetic Diversity and Fitness of *Fusarium graminearum* Populations from Rice in Korea. Appl. Environ. Microbiol., 75: 3289–3295.

Kamal, M.M. and Mia, M.A. 2009. Diversity and pathogenicity of the rice brown spot pathogen, *Bipolaris oryzae* (Breda de Haan) Shoemker in Bangladesh assessed by genetic fingerprint analysis. Bangladesh J. Bot., 38: 119–125.

Kato, H. 2001. Rice blast disease. Pesticide Outlook, 12: 23–25.

Khodayari, M., Safaie, N. and Shamsbakhsh, M. 2009. Genetic diversity of Iranian AG1-IA isolates of *Rhizoctonia solani*, the cause of rice sheath blight, using morphological and molecular markers. J. Phytopathol., 157: 708–714.

Kim, J.C., Kang, H.J., Lee, D.H., Lee, Y.W. and Yoshizawa, T. 1993. Natural occurrence of Fusarium mycotoxins (trichothecenes and zearalenone) in barley and maize in Korea. Appl. Environ. Microbiol., 59: 3798–3802.

Koiso, Y., Natori, M., Iwasaki, S., Sato, S., Sonoda, R., Fujita, Y., Yaegashi, H. and Sato, Z. 1992. Ustiloxin: a phytotoxin and a mycotoxin from false smut balls on rice panicles. Tetrahedron Lett., 33: 4157–4160.

Koiso, Y., Li, Y., Iwasaki, S., Hanaoka, K. and Kobayashi, T. 1994. Ustiloxins, antimitotic cyclic peptides from false smut balls on rice panicles caused by *Ustilaginoidea virens*. J. Antibiot. (Tokyo), 47: 765–773.

Kumar, P., Anshu, V. and Kumar, S. 2011. Morpho-pathological and molecular characterization of *Bipolaris oryzae* in Rice (*Oryzae sativa*). J. Phytopathol., 1: 51–56.

Lee, F.N. and Gunnell, P.S. 1992. False smut. *In:* Webster, R.K. and Gunnell, P.S. (eds.). Compendium of Rice Diseases. APS, St. Paul, Minn. pp. 28.

Lee, T., Han, Y.K., Kim, K.H., Yun, S.H. and Lee, Y.W. 2002. Tri13 and Tri7 determine deoxynivalenol- and nivalenol-producing chemotypes of *Gibberella zeae.* Appl. Environ. Microbiol., 68: 2148–2154.

Lee, T., Oh, D.W., Kim, H.S., Lee, J., Kim,Y.H., Yun, S.H. and Lee, Y.W. 2001. Identification of deoxynivalenol- and nivalenol-producing chemotypes of *Gibberella zeae* by using PCR. Appl. Environ. Microbiol., 67: 2966–2972.

Leslie, J.F. 1995. *Gibberella fujikuroi*: available populations and variable traits. Can. J. Bot., 73: S282–S291.

Leslie, J.F. and Klein, K.K. 1996. Female fertility and mating type effects on effective population size and evolution in filamentous fungi. Genetics, 144: 557–567.

Leslie, J.F., Marasas, W.F.O., Shephard, G.S., Sydenham, E.W., Stockenstrom, S. and Thiel, P.G. 1996. Duckling toxicity and the production of fumonisin and moniliformin by isolates in the A and E mating populations of *Gibberella fujikuroi* (*Fusarium moniliforme*). Appl. Environ. Microbiol., 62: 1182–1187.

Leslie, J.F. and Summerell, B.A. 2006. The Fusarium Laboratory Manual. Blackwell Publishing, Ames, Iowa, USA.

Li, Y.G., Kang, B.J., Zhang, B.D., Zeng, H.Z., Xie, K.X., Lan, Y.T., Ma, H. and Li, T.F. 1986. A preliminary study on rice false smut. Guangdong Agric. Sci. I, 4: 45–47.

Li, H.R. and Yan, S.Q. 1990. Studies on the strains of pathogens of sheath blight of rice in the east and south of Sichuan Province. Acta Mycol. Sinica, 9: 41–49.

Lu, D.H, Yang, X.Q., Mao, J.H., Ye, H.L., Wang, P., Chen, Y.P., He, Z.Q. and Chen, F. 2009. Characterising the pathogenicity diversity of *Ustilaginoidea virens* in hybrid rice in China. J. Pl. Path., 91: 443–451.

Maheshwar, P.K.S., Moharram, A. and Janardhana, G.R. 2009. Detection of fumonisin producing *Fusarium verticillioides* in paddy (*Oryza sativa* L.) using polymerase chain reaction (PCR). Brazilian J. Microbiol., 40: 10.

Manandhar, J.B. 1999. Hyphal cushion formation around rice embryos by *Cochiobolus miyabeanus*. Phytopathol. Mediterranea, 38: 89–94.

Marasas, W.F.O., Thiel, P.G., Sydenham, E.W., Rabie, C.J., Lubben, A. and Nelson, P.E. 1991. Toxicity and moniliformin production by four recently described species of *Fusarium* and two uncertain taxa. Mycopathol., 113: 191–197.

Matsuyama, N. and Wakimoto, S. 1977. A comparison of the esterase and catalase zymograms of *Fusarium* species with special reference to the classification of a causal fungus of *Fusarium* leaf spot of rice. Ann. Phytopathol. Soc. Japan, 43: 463–470.

Mew, T.W. and Gonzales, P. 2002. A Handbook of Rice Seedborne Fungi. International Rice Research Institute, Los Banos, Philippines, and Science Publishers, Inc., Enfield, NH, USA. 83 p.

Mew, T.W., Leung, H., Savary, S., Vera Cruz, C.M. and Leach, J.E. 2004. Looking Ahead in Rice Disease Research and Management. Crit. Rev. Pl. Sci., 23: 103–127.

Mia, M.A.T., Safeeulla, K.M. and Shetty, H.S. 1986. Seed-borne nature of *Gerlachia oryzae*, the incitant of leaf scald of rice in Karnataka. Indian Phytopathol., 39: 92–93.

Miyazaki, S., Matsumoto, Y., Uchihara, T. and Morimoto, K. 2009. High-performance liquid chromatographic determination of ustiloxin A in forage rice silage. J. Vet. Med. Sci., 71: 239–241.

Moletti, M., Giudici, M.L. and Villa, B. 1996. Rice Akiochi-brown spot disease in Italy: agronomic and chemical control. Inf. Fitopatol., 46: 41–46.

Mondal, A.S., Ahmed, H.U. and Miah, S.A. 1986. Effect of nitrogen on the development leaf scald disease of rice. Bangladesh J. Bot., 15: 213–215.

Montesinos, E. 2003. Development, registration and commercialization of microbial pesticides for plant protection. Intl. Microbiol., 6: 245–252.

Moore, R.T. 1987. The genera of *Rhizoctonia*-like fungi: *Asorhizoctonia, Ceratorhiza* gen. nov., *Epulorhiza* gen. nov., *Moniliopsis* and *Rhizoctonia*. Mycotaxon, 29: 91–99.

Naito, S. 2004. *Rhizoctonia* Diseases: Taxonomy and population biology. Proceeding of the International Seminar on Biological Control of Soilborne Plant Diseases. Japan-Argentina Joint Study, Buenos Aires, Argentina, p. 18–31.

Narinder, S. and Singh, M.S. 1989. Effect of different levels of nitrogen and dates of transplanting on the incidence of false smut of paddy in Punjab. Indian J. Ecol., 14(1): 164–167.

Neergaard, P. 1986. Seed Pathology. S. Chand and Company Ltd., New Delhi.

Nirenberg, H.I. and O'Donnell, K. 1998. New *Fusarium* species and combinations within the *Gibberella fujikuroi* species complex. Mycologia, 90: 434–458.

Nyvall, R.F., Percich, J.A. and Mirocha, C.J. 1999. *Fusarium* head blight of cultivated and natural wild rice (*Zizania palustris*) in Minnesota caused by *Fusarium graminearum* and associated *Fusarium* spp. Pl. Dis., 83: 159–164.

O'Donnell, K., Nirenberg, H.I., Aoki, T. and Cigelnik, E. 2000a. A Multigene phylogeny of the *Gibberella fujikuroi* species complex: Detection of additional phylogenetically distinct species. Mycoscience, 41: 61–78.

O'Donnell, K., Kistler, H.C., Tacke, B.K. and Casper, H.H. 2000b. Gene genealogies reveal global phylogeographic structure and reproductive isolation among lineages of *Fusarium graminearum*, the fungus causing wheat scab. Proc. Natl. Acad. Sci. USA, 97: 7905–7910.

Ou, S.H. 1985. Rice diseases. 2nd Ed. Commonwealth Mycological Institute, Kew, Surrey, England, 380 p.

Overwater, C. 1960. The ten-year old Bernhart polder, 1950–1960. Surinaam Landbouw, 8: 159–218.

Padmanabhan, S.Y. 1973. The great Bengal famine. Annu. Rev. Phytopathol., 11: 11–26.

Palay Check System for Irrigated Lowland Rice. 2007. PhilRice and FAO. 91 pp.

Pandey, B.P. 1982. A text book of plant pathology, pathogen and plant disease. S. Chand and Company Ltd., Ram Nagar, New Delhi-110055, p. 452.

Parmeter, J.R.J., Sherwood, R.T. and Platt, W.D. 1969. Anastomosis grouping among isolates of *Thanatephorus cucumeris*. Phytopathol., 59: 1270–1278.

Patell, K.V., Vala, D.G., Mehta, B.P. and Patell, T.C. 1992. Effect of nitrogen doses on incidence of false smut of rice. Indian J. Mycol. Pl. Pathol., 22: 260–262.

Pearse, D.A., Bridge, P.D. and Hawksworth, D.L. 2001. Species concept in *Sarocladium*, causal agent of sheath rot in rice and bamboo blight. pp. 285–292. *In*: Sreenivasaprasad, S. and Johnson, R. (eds.). Major Fungal Disease of Rice. Recent Advances. Kluwer Academic Publisher, 17, 3300 AA Doedrecht, the Netherlands.

Pitt, J.I. and Hocking, A.D. 1997. Fungi and food spoilage. Blackie Academic & Professional, London, 593 pp.

Priyatmojo, A., Escopalao, V.E., Tangonan, N.G., Pascual, C.B., Suga, H., Kageyama, K. and Hyakumachi, M. 2001. Characterization of a new subgroup of *Rhizoctonia solani* anastomosis group 1 (AG-1 ID), causal agent of a necrotic leaf spot on coffee. Phytopathol., 91: 1054–1061.

Proctor, R.H., Plattner, R.D., Brown, D.W., Seo, J.A. and Lee, Y.W. 2004. Discontinuous distribution of fumonisin biosynthetic genes in the *Gibberella fujikuroi* species complex. Mycol. Res., 108: 815–822.

Qu, B., Li H.P., Zhang, J.B., Huang, T., Carter, J., Liao, Y.C. and Nicholson, P. 2008. Comparison of genetic diversity and pathogenicity of *Fusarium* head blight pathogens from china and Europe by SSCP and seedling assays on wheat. Plant Pathol., 57: 642–651.

Rabie, C.J., Marasas, W.F.O., Thiel, P.G., Lubben, A. and Vleggaar, R. 1982. Moniliformin production and toxicity of different *Fusarium* species from Southern Africa. Appl. Environ. Microbiol., 43: 517–521.

Rossman, A.Y., Howard, R.J. and Valent, B. 1990. *Pyricularia grisea*, the correct name for the rice blast disease fungus. Mycologia, 82: 509–512.

Roy-Barman, S. and Chattoo, B.B. 2005. Rice blast fungus sequenced. Curr. Sci., 89: 930–931.

Rush, M.C., Shahjahan, A.K.M., Jones, J.P. and Groth, D.E. 2000. Outbreak of false smut of rice in Louisiana. Pl. Dis., 84: 100.

Samuels, G.J., Nirenberg, H.I. and Seifert, K.A. 2001. Perithecial species of Fusarium. pp. 1–14. *In*: Summerell, B.A., Leslie, J.F., Backhouse, D., Bryden, W.L. and Burgess, L.W. (eds.). *Fusarium*: Paul E. Nelson Memorial Symposium. APS Press, St. Paul, MN.

Sarver, B.A.J., Ward, T.J., Gale, L.R., Broz, K., Kistler, H.C., Aoki, T., Nicholson, P., Carter, J. and O'Donnell, K. 2011. Novel *Fusarium* head blight pathogens from Nepal and Louisiana revealed by multilocus genealogical concordance. Fungal Gen. Biol., 48: 1096–1107.

Savary, S., Castilla, N., Elazegui, F.A., McLaren, C.G., Ynalvez, M.A. and Teng, P.S. 1995. Direct and indirect effects of nitrogen supply and disease source structure on rice sheath blight spread. Phytopathol., 85: 959–965.

Savary, S., Castilla, N.P., Elazegui, F.A., Teng, P.S., Du, P.V., Tang, Q., Huang, S., Lin, X., Singh, H.M. and Srivastava, R.K. 2000a. Rice pest constraints in tropical Asia: characterization of injury profiles in relation to production situations. Pl. Dis., 84: 341–356.

Savary, S., Willocquet, L., Elazegui, F.A., Castilla, N. and Teng, P.S. 2000b. Rice pest constraints in tropical Asia: quantification of yield losses due to rice pests in a range of production situations. Pl. Dis., 84: 357–369.

Savary, S., Castilla, N., Elazegui, F.A. and Teng, P.S. 2005. Multiple effects of two drivers of agricultural change, labor shortage and water scarcity, on rice pest profiles in tropical Asia. Field Crops Res., 91: 263–271.

Savary, S., Teng, P.S., Willocquet, L. and Nutter, F.W., Jr. 2006. Quantification and modeling of crop losses: a review of purposes. Annu. Rev. Phytopathol., 44: 89–112.

Schroud, P. and TeBeest, D.O. 2005. Germination and infection of rice roots by spores of *Ustilaginoidea virens*. *In*: Norman, R.J., Meullenet, J.F., Moldenhauer, K.A. and Wells,

B.R. (eds.). Rice Research Series. Univ. Arkansas Agri. Exper. Station Res. Series, 540: 143–151.

Sebastian, L.S., Alviola, P.A. and Francisco, S.R. 2000. Bridging the rice yield gap in the Philippines. pp.112–134. *In*: Papademetriou, M.K., Dent, F.J. and Herath, E.M. (eds.). Bridging the Rice Yield Gap in the Asia-Pacific Region. Food and Agriculture Organization of the United Nations Regional office for Asia and the Pacific, Bangkok, Thailand.

Sharma, N.D. and Joshi, R. 1975. Effect of different nutrient media on the growth and sporulation of *Ustilaginodea virens* (Cooke) Takahashi. Curr. Sci., 44: 352–354.

Sharon, M., Kuninaga, S., Hyakumachi, M., Naito, S. and Sneh, B. 2008. Classification of *Rhizoctonia* spp. using rDNA-ITS sequence analysis supports the genetic basis of the classical anastomosis grouping. Mycosci., 49: 93–114.

Shetty, S.A. and Shetty, H.S. 1985. An alternative host for *Ustilaginodea virens* (Cooke) Takahashi. IRRI Newslett., 10: 11.

Singh, R.A. and Dube, K.S. 1978. Assessment of loss in seven rice cultivars due to false smut. Indian Phytopath., 31: 186–188.

Singh, S.A. and Gupta, P.K.S. 1980. *Echinochloa crus-galli* (L). Beauv.: a new host for *Rhynchosporium oryzae* Hashioka and Yokogi. Intl. Rice Res. Newslett., 5: 17.

Singh, G.P., Singh, R.N. and Singh, A. 1987. Status of false smut (FS) of rice in eastern Uttar Pradesh, India. IRRI Newslett., 12: 28.

Singh, R.A. and. Khan, A.T. 1989. Field resistance to false smut and narrow brown leaf spot in Eastern Uttar Pradesh, India. IRRI Newslett., 14: 16–17.

Singh, V.P. and Singh, R.K. 2000. Rainfed rice: a sourcebook of best practices and strategies in eastern India. International Rice Research Institute, Los Baños, Philippines, 292 pp.

Sneh, B., Burpee, L. and Ogoshi, A.A. 1998. Identification of *Rhizoctonia* species. APS, St. Paul, Minnesota.

Sreenivasaprasad, S. and Johnson, R. 2001. Major fungal disease of rice, recent advances. Kluwer Academic Publishers, 17- 3300 AA Doedrecht, the Netherlands.

Starkey, D.E., Ward, T.J., Aoki, T., Gale, L.R., Kistler, H.C., Geiser, D.M., Suga, H., Tóth, B., Varga, J. and O'Donnell, K. 2007. Global molecular surveillance reveals novel *Fusarium* head blight species and trichothecene toxin diversity. Fungal Genet. Biol., 44: 1191–1204.

Suga, H., Karugia, G.W., Ward, T., Gale, L.R., Tomimura, K., Nakajima, T., Miyasaka, A., Koizumi, S., Kageyama, K. and Hyakumachi, M. 2008. Molecular characterization of the *Fusarium graminearum* species complex in Japan. Phytopathol., 98: 159–166.

Sugha, S., Sharma, O.P. and Kaushik, R.P. 1992. Performance of rice genotypes against rice false smut pathogen under rainfed conditions. Pl. Dis. Res., 8: 76–77.

Sun, S.K. and Snyder, W.C. 1981. The bakanae disease of the rice plant. pp. 104–113. *In*: Nelson, P.E., Toussoun, T.A. and Cook, R.J. (eds.). Fusarium: Diseases, Biology and Taxonomy. The Pennsylvania State University Press, University Park, PA.

Suwa, M. 1915. Experimental *Ustilaginoidea* toxicosis. Igaku Chuo Zasshi, 13: 661–686. (in Japanese).

Taligoola, H.K., Ismail, M.A. and Chebonm, S.K. 2004. Mycobiota associated with rice grains marketed in Uganda. J. Biol. Sci., 4: 271–278.

Tang, Y.X., Jin, J., Hu, D.W., Yong, M.L. and Xu, Y. 2013. Elucidation of the infection process of *Ustilaginoidea virens* (teleomorph: *Villosiclava virens*) in rice spikelets. Pl. Pathol., 62: 1–8.

TeBeest, D.O., Guerber, C. and Ditmore, M. 2007. Rice blast. The Plant Health Instructor. DOI: 10.1094/PHI-I-2007-0313-07.

TeBeest, D.O., Jecmen, A.C. and Ditmore, M. 2011. Infection of rice by the false smut fungus, *Ustilaginoidea virens*. *In*: Norman, R.J. and Moldenhauer, K.A.K. (eds.). B. R. Wells Rice Research Studies 2010, Univ. Arkansas Agric. Experi. Station Res. Series Fayetteville, 591: 70–81.

Teng, P.S., Torries, C.Q., Nuque, F.L. and Calvero, S.B. 1990. Current knowledge on crop losses in tropical rice. pp. 39–54. *In*: IRRI, Crop Loss assessment in Rice. International Rice Research Institute, Los Banos

Thieron, M., Pontzen, R. and Kurahashi, Y. 1998. Carpropamid: a rice fungicide with two modes of action. Pflanzenschutz-Nachrichten Bayer 51, 3: 257–278.

Thurston, H.D. 1990. Plant diseases management practices of traditional farmers. Pl. Dis., 74: 96–102.

Trung, T., Bailly, J.D., Querin, A., Le bras, P. and Guerre, P. 2001. Fungal contamination of rice from South Vietnam, mycotoxinogenesis of selected strains and residues in rice. Rev. Méd. Vét. Res., 152(7): 555–560.

Verma, R.K. and Singh, R.A. 1988. Variations in *Claviceps oryzae-sativae* the incitant of false smut of rice. Indian Phytopathol., 41: 48–50.

Vidhyasekaran, P.E., Borromeo, S. and Mew, T.W. 1986. Host-specific toxin production by *Helminthosporium oryzae*. Phytopathol., 76: 261–266.

Wailes, E.J. and Chavez, E.C. 2012. World Rice Outlook International Rice Baseline with Deterministic and Stochastic Projections, 2012–2021. Department of Agricultural Economics and Agribusiness Division of Agriculture, Room 217, Agriculture Building, University of Arkansas Fayetteville, AR 72701. Online: http://ageconsearch.umn.edu/bitstream/123203/2/March%202012%20World%20Rice%20Outlook.

Walker, J.C. 1975. Plant pathology. 3rd ed. MacGrawHill Book Company, Inc, New York.

Waxman, K.D., Bergstrom, G.C., Waxman, K.D. and Bergstrom, G.C. 2011. First report of a leaf spot caused by *Bipolaris oryzae* on switchgrass in New York. Pl. Dis., 95: 1192–1192.

Webster, R.K. and Gunnell, P.S. 1992. Compendium of rice diseases. APS, St. Paul, Minnesota, USA, 62 p.

Willocquet, L., Fernandez, L. and Savary, S. 2000. Effect of various crop establishment methods practiced by Asian farmers on epidemics of rice sheath blight caused by *Rhizoctonia solani*. Pl. Pathol., 49: 346–354.

Xu, J.L., Xue, Q.Z., Luo, L.J. and Li, Z.K. 2002. Preliminary report on quantitative trait loci mapping of false smut resistance using near-isogenic introgression lines in rice. Acta Agric. Zhejiangensis, 14: 14–19.

Yashoda, H., Anahosur, K.K., Kulkarni, S., Yashoda, H. and Anahosur, K.H. 2000. Influence of weather parameters of the incidence of false smut of rice. Adv. Agric. Res. India, 14: 161–165.

Zadoks, J.C. 2002. Fifty years of crop protection, 1950–2000. Neth. J. Agric. Sci., 2002: 181–193.

Zainudin, N.A.I.M., Razak, A.A. and Salleh, B. 2008. Secondary metabolite profiles and mating populations of *Fusarium* species in section Liseola associated with bakanae disease of rice. Malaysian J. Microbiol., 4: 6–13.

Zeller, K.A., Summerell, B.A., Bullock, S. and Leslie, J.F. 2003. *Gibberella konza* (*Fusarium konzum*) sp. nov. from prairie grasses, a new species in the *Gibberella fujikuroi* species complex. Mycologia, 95: 943–954.

Zhou, Y.L., Pan, Y.J., Zie, X.W., Zhu, L.H., Wang, S. and Li, Z.K. 2008. Genetic diversity of rice false smut fungus, *Ustilaginoidea virens* and its pronounced differentiation in populations in north China. J. Phytopathol., 156: 559–564.

CHAPTER 13

Fungi and Fungi-like Organisms Associated with Brassicas, with Special Reference to Those Causing Decomposition of its Debris

Jalpa P. Tewari

ABSTRACT

Brassica spp. plants, including their living and dead parts (debris or residue), are colonized by pathogenic and saprophytic fungi, respectively. The living plant surfaces also have many phylloplane fungi. The pathogenic fungi are broadly classifiable into biotrophic and necrotrophic types. The biotrophs, *Albugo candida, Plasmodiophora brassicae* and *Urocystis brassicae,* induce galls in the affected plant parts whereas *Erysiphe polygoni* and *Hyaloperonospora brassicae* do not. The fungal necrotrophs on Brassicas are a larger group that includes pathogens such as *Alternaria* spp., *Leptosphaeria maculans, Sclerotinia sclerotiorum* and others. Biomass from the vegetative and reproductive (except seeds) phases of plants after completion of the life cycle and senescence eventually transforms into plant debris. Degradation of debris, which is the perennation (overwintering/oversummering)

Department of Agricultural, Food, and Nutritional Science, University of Alberta, Edmonton, Alberta, Canada, T6J 3C9 and Lung, Allergy, Sleep Centers of America, 224 W. Exchange Street, Ste. 380, Akron, Ohio 44302, USA.
Email: jalpaptewari@gmail.com

habitat for many fungal pathogens, is important from the disease-mitigation standpoint. It is also significant from agricultural agronomic perspectives. None of the biotrophic or necrotrophic fungi cause any noticeable degradation of living plant parts or debris except for *S. sclerotiorum* which causes shredding of some fibrous elements from the infected stems. Little work has been done on saprophytic fungi degrading *Brassica* debris in the field. The bird's nest fungus, *Cyathus olla*, is common in fields in Alberta, Canada and causes appreciable breakdown of canola (*Brassica napus*) debris. Three forms of this fungus are found in fields in Alberta. They include *C. olla* f. *olla*, *C. olla* f. *anglicus* and *C. olla* f. *Brodii*. The last variant was described earlier as a new form from Alberta. *Cyathus olla* secretes oxalic acid, sequesters calcium and produces crystals of calcium oxalate. It also produces the enzymes laccase, manganese peroxidase, aryl-alcohol oxidase and polygalacturonase. Cellulose and hemicellulose are also degraded in solid-state fermentation of canola substrate (ground root and basal stem). Debris of canola is quite recalcitrant to degradation in the field. However, when infested with *C. olla* in the field, it becomes soft, macerated and shows ample hyphae and hyphal cords emanating from it. This fungus has the potential of being formulated as a microbial inoculant for managing the debris of canola in the field.

Introduction

Many *Brassica* spp. are agriculturally important crop plants and are sources of a number of common commodities such as oil and other forms of human food, animal feed and others. Hence, fungi which affect them as pathogenic and saprophytic factors are of great significance.

The Brassicas have a number of growth and senescence phases and all of them are subject to colonization (infection or infestation, as appropriate) by fungi. In addition, the Brassica plants are a special habitat as they contain sulfur-containing compounds, the glucosinolates (Fahay et al., 2001; Mikkelsen et al., 2002). The glucosinolates are present in 16 families of dicotyledonous plants, including Brassicaceae, and they along with their breakdown products have wide-ranging biological properties including being fungicidal (Fahay et al., 2001; Mikkelsen et al., 2002). The Brassicas also have a sulfur-containing dynamic defense response system consisting of phytoalexins (Browne et al., 1991; Pedras and Yaya, 2010). The vegetative phase of Brassicas, like other plants, provides the developmental, physiological/biochemical, nutritional, and mechanical framework for the reproductive phase. Also, the vegetative and a part of the reproductive phases after completion of the life cycle and senescence eventually transform into plant debris or residue. Hence, the plant debris is an inevitable consequence of the cycle of plant growth and reproduction. The basal parts or stumps of plants attached to the soil after the crop is harvested are sometimes called stubble. However, due to the inherent nature of plant structural and growth processes, the debris becomes a reservoir of many elements comprising various inorganic and organic components of the plant. The soil will become deficient in nutrients if these constituents are not returned back to it. The plant debris also affects a number of other parameters including soil structure, soil hydrology and plant pathogen carry-over from one growing season to the next. Hence, it is important that the crop debris be broken down within a reasonably short span of time. A number of environmental, biological (fungi, bacteria and microfauna), and mechanical management (human interventions by farm machinery or manually in some countries) factors interact in this process. Some constituent materials of plant debris (such as lignin) are recalcitrant to degradation by microbes. Hence, the plant debris needs to be in the soil environment exposed to lignin-degrading microbes under favorable conditions often for extended periods of time.

Efficient crop agronomy requires prudent management of debris. These practices affect the integrity and microbiology of debris, and are briefly discussed here. Little is known of fungi associated with the debris of *Brassica* spp. except for the epidemiology of disease-causing pathogens. This information relates to agricultural ecosystems and none to natural ecosystems where a number of wild *Brassica* spp. grow. This chapter presents a general review of the various fungi and fungi-like organisms associated with *Brassica* spp. as pathogens and phylloplane residents, and their possible role in the decomposition of debris. Finally, the chapter focuses on decomposition of Brassica debris by fungi, especially by *Cyathus olla* (Batch) ex Pers., which is present in agricultural soils in Alberta, Canada where oleiferous Brassicas are one of the principal crops. This fungus may have possible application in agricultural management of debris as a microbial inoculant.

Management of *Brassica* Debris

Various aspects of combining canola (specially bred cultivars of *B. juncea* (L.) Cosson, *B. napus* L. and *B. rapa* L.) and management of debris emanating therefrom, especially from the Canadian perspective, are given in Blenis et al. (1999) and Anonymous (2011). The crop can be combine-swathed (Fig. 13.1) and threshed later for uniform maturity. During this process, the swath, if moist, can become colonized by the fungal pathogen, *Sclerotinia sclerotiorum* (Lib.) de Bary. This can cause some breakdown of tissues (see below). On the other hand, the crop can be directly combined and the resulting straw and chaff broadcast widely and evenly. It is important to watch the stage of crop maturity at harvest as canola siliquae have a tendency to shatter and release seed, if over mature, resulting in loss of yield.

Siliquae infected with *Alternaria* spp. (see below) are particularly prone to shattering (Tewari and Mithen, 1999). Further infection with *Alternaria* spp. can also take place if the swaths are moist. However, tissues infected with *Alternaria* spp. are firm and relatively intact. Hence, *Alternaria* spp. do not cause any noticeable breakdown of infected tissues. In countries such as Canada, harvesting is done using machinery whereas in countries such as India it is often done manually (Fig. 13.2).

Figure 13.1. A machine-swathed field of canola (*Brassica napus*) in central Alberta, Canada. Note rows of crop swath and standing stubble. **Figure 13.2.** Manually swathed field of Indian mustard (*B. juncea*) in Pantnager area, U.P., India. Note the bundles of crop and standing stubble.

After harvesting, the field is prepared for the next cropping season which, based on good crop rotation practices, should not include the same crop (Anonymous, 2011). Crop rotation reduces carry-over of plant pathogens to the following growing season.

Fungi and Fungi-like Organisms Parasitic on *Brassica* spp.

Many fungi and fungi-like organisms are known to be parasitic on *Brassica* spp., both cultivated and wild (Mundkur, 1938; Vaartnou and Tewari, 1972a,b; Rai et al., 1974; Kolte, 1985; Tewari, 1985, 1991; Paul and Rawlinson, 1992; Saharan and Verma, 1992; Williams and Saha, 1993; Verma and Saharan, 1994; Tewari and Mithen, 1999; Kharbanda et al., 2001; Rimmer et al., 2007; Choi et al., 2009; Kaur et al., 2011). Broadly speaking these pathogens can be classified into the categories biotrophs and necrotrophs. Examples of the biotrophs include *Albugo candida* (Pers.) Kuntze, *Eysiphe polygoni* DC, *Hyaloperonospora brassicae* (Gäum.) Göker, Voglmayr, Riethm., Weiss & Oberw., *Plasmodiophora brassicae* Wor., and *Urocystis brassicae* Mundkur. All these pathogens, except *E. polygoni* and *H. brassicae*, cause noticeable hypertrophy and hyperplasia but none of the biotrophs cause any evident maceration of tissues which may aid in breakdown of the affected plant parts. *Plasmodiophora brassicae* and *U. brassicae* infection produces galls which are full of spores. Once mature, the spores are released through breakdown of the galls. *Albugo candida* infection results in the production of the white-rust stage producing asexual sporangia. Also, malformations of stems called stagheads are full of oospores which are sexual propagules of the pathogen (Figs. 13.3, 13.4). *Hyaloperonospora brassicae* mostly causes infections of leaves and stems (generally stagheads produced as a result of complex infection with *A. candida*, see above) and produces conidia and oospores. The vesicular-arbuscular mycorrhizal fungi (*Glomus* spp.) are also rarely reported in a biotrophic mutually beneficial symbiotic relationship with Brassica roots although most studies have concluded that the crucifers are non-mycorrhizal (Nelson and Achar, 2000; Vierheilig et al., 2000). The necrotrophs constitute a much larger group of fungi and include pathogens of Brassicas such as *Alternaria alternata* (Fr.) Keissler, *A. brassicae* (Berk.) Sacc., *A. brassicicola* (Schw.) Wiltshire, *A. japonica* Yoshii, *Fusarium* spp., *Leptosphaeria maculans* (Desm.) Ces. and de Not. [anamorph *Phoma lingam* (Tode ex Fr.) Desm.], *Pyrenopeziza brassicae* Sutton and Rawlinson, *Pythium* spp., *Rhizoctonia solani* Kühn, *S. sclerotiorum*, *Sclerotium rolfsii* Sacc., *Verticillium dahliae* Kleb., and *V. longisporum* Karapapa, Bainbridge and Heale. A more complete list of pathogens is available in the references given above. Most necrotrophs also do not cause any appreciable breakdown of affected tissues. However, *S. sclerotiorum* and *S. rolfsii* produce appreciable quantities of oxalic acid, pectolytic, cellulolytic and some other cell wall-

Figure 13.3. Malformation of a lateral branch of canola (*Brassica rapa*) into a staghead by *Albugo candida*. Also note the white-rust stage on the surface of the staghead. **Figure 13.4.** Malformations of the terminal branches of canola (*B. rapa*) into stagheads by *A. candida*. Both photographs were taken in central Alberta, Canada.

degrading enzymes (Lumsden, 1976; Punja, 1985; Punja and Damiani, 1996; Li et al., 2004a,b), and cause a limited maceration of the infected tissues. Oxalic acid is a strong chelator of divalent cations including calcium (Figs. 13.5, 13.6), lowers the pH, and facilitates the activity of cell wall degrading enzymes (Godoy et al., 1990; Franceschi and Loewus, 1995). In fact, as a result of oxalic acid production, one of the symptoms of *Brassica napus* stems infected with *S. sclerotiorum* is shredding of fibrous tissues. The fibrous elements left behind are lignified elements since ligninolytic

Figure 13.5. Scanning electron micrograph of the surface of a stem-lesion of canola (*Brassica napus*) caused by *Sclerotinia sclerotiorum* in central Alberta, Canada, showing bipyramidal crystals of calcium oxalate.

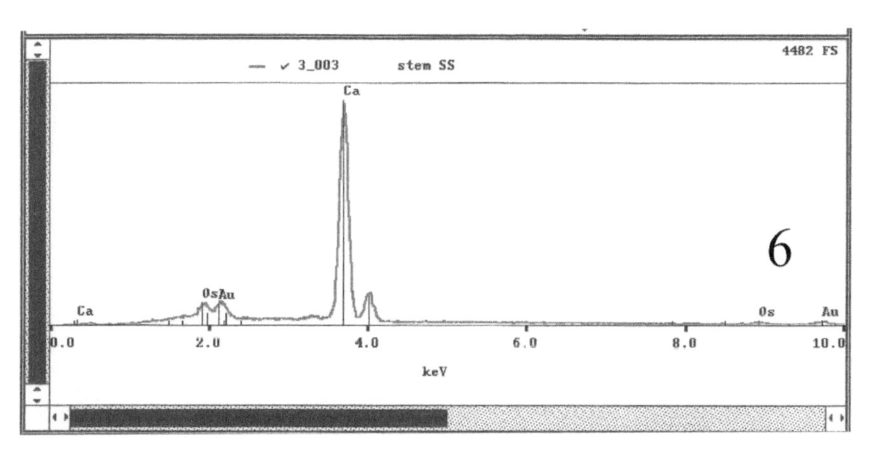

Figure 13.6. Energy-dispersive X-ray microanalysis spectrum of calcium oxalate crystals shown in Fig. 13.5. Note the emission lines for calcium. The emission lines from osmium and gold have originated from specimen fixation and coating of the specimen for scanning electron microscopy, respectively.

enzymes are not produced by *S. sclerotiorum* (Peltier et al., 2004, 2009). *Sclerotium rolfsii* is similar in activity. In the *S. sclerotiorum* pathosystem, oxalic acid also suppresses oxidative burst of the host tissue (Cessna et al., 2000) and deregulates guard cells during infection (Guimarães and Stotz, 2004). *Alternaria* spp. (Figs. 13.7–13.9) and *L. maculans* (Figs. 13.10, 13.11) cause infection of leaves, stems and siliquae. *Leptosphaeria maculans* sometimes also causes infection of the roots. Both these pathogens do not cause any appreciable maceration of tissues. The other pathogens listed above also colonize stems and some other plant parts but do not reveal any appreciable breakdown of tissues. Some pathogens of Brassicas also infect seeds. Additionally, some of them become seed borne and become a source of disease occurrence in the next season.

Similar to other plants, the spectrum of fungi infecting/infesting growth and senescent stages (including plant debris) of Brassicas are not quite the same. This is due to various pathological factors regulating fungal infection in the growing plant and biochemical differences between the growing plant and the debris. While the biotrophs can survive (overwinter/oversummer, as appropriate) in association with the debris, they can grow only in association with a living host. The necrotrophs are facultative parasites to varying degrees and can survive and even grow/sporulate in association with the host debris.

Figure 13.7. Blackspot lesions on the leaves of *Brassica* sp. caused by *Alternaria brassicae* in Delhi area, India. **Figure 13.8.** Blackspot lesions on the siliquae of *Brassica juncea* caused by *A. brassicae* in Delhi area, India.

Figure 13.9. Scanning electron micrograph of a siliqua of canola (*Brassica napus*) from chaff showing groups of conidiophores of *Alternaria brassicae*. Each conidiophore bears a large apically beaked conidium.

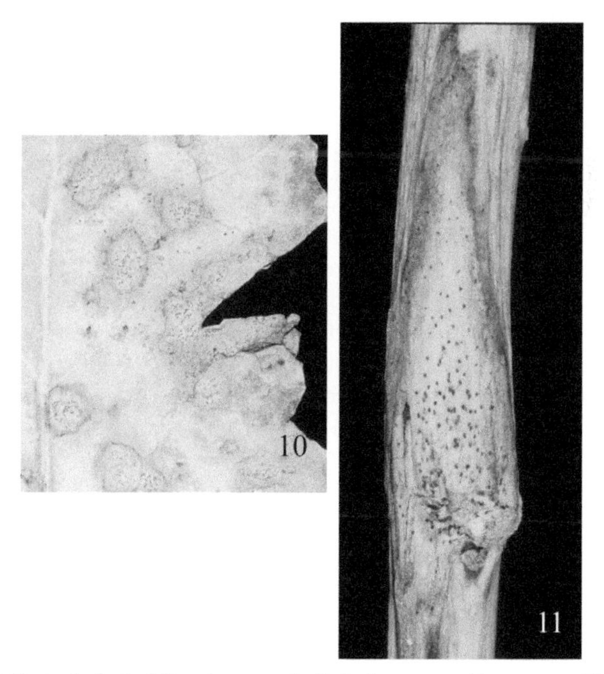

Figure 13.10. Part of a leaf of *Brassica napus* in Paderborn area, Germany, with foliar lesions caused by *Leptosphaeria maculans* (anamorph *Phoma lingam*) showing pycnidia. **Figure 13.11.** Stem of canola (*B. napus*) in central Alberta, Canada, with a canker caused by *Leptosphaeria maculans* (anamorph *Phoma lingam*) showing pycnidia.

Fungi Present on the Phylloplane of *Brassica* spp.

A large number of fungi have been reported from the phylloplane of *B. napus* and *B. rapa* (Vaartnou and Tewari, 1973; Vaartnou et al., 1974; Tsuneda et al., 1976; Tsuneda and Skoropad, 1978, 1980; Singh and Rai, 1980). These fungi are of three kinds including being saprophytic, parasitic and mycoparasitic. The saprophytic fungi vary in types and numbers among the different growth stages/periods of the host. The parasitic fungi are pathogens such as *Alternaria* spp. and *Fusarium* spp. The fungus, *Verticillium* state of *Nectria inventa,* isolated from the phylloplane proved to be a mycoparasite of *Alternaria brassicae* and a number of other phylloplane fungi (Tsuneda et al., 1976; Tsuneda and Skoropad, 1980).

Fungi Present in Association with Brassica Debris

A large number of fungal pathogens of Brassicas are present in the debris which serves as a habitat for them to perennate. Much agricultural management work has been done to reduce the length of their perennation (Kharbanda and Tewari, 1996). This is aimed to alleviate the diseases caused by them such as that by *Leptosphaeria maculans* which continues to produce primary inoculum (ascospores) from Brassica debris for an extended period (Petri, 1995). Some pathogens such as *Alternaria japonica* (Vaartnou and Tewari, 1972b) and *A. brassicae* (Tsuneda and Skorapad, 1977) produce chlamydospores and microsclerotia as resting structures, which can potentially extend the periods of their survival in the field. However, as mentioned earlier, none of them cause any appreciable destruction of the host tissues. Cultural management practices (crop rotation, conventional tillage and others) have been developed to accelerate the breakdown of debris; thereby destroying the habitat of pathogen. However, the current management practices advocate the use of conservation tillage (direct seeding, no-till or zero-till and minimum or reduced tillage), especially in the drier areas. This reduces soil erosion, improves hydrology and reduces operating costs (Anonymous, 2011). Using this management strategy, the crop debris either sits on or is close to the soil surface (Fig. 13.12). This keeps the debris drier and extends the time for its breakdown. As an example, in Saskatchewan, Alberta, Canada which has a semi-arid environment, the canola stubble can remain inoculum-producing for 5–7 years (Petri, 1995). Hence, conservation tillage is at cross-purposes with speedy destruction of the pathogen habitat but is considered prudent with overall crop agronomic practices.

Figure 13.12. A field of oats (*Avena sativa*) in south-eastern Alberta, Canada which is a relatively dry area. The oats were direct-seeded in a standing-stubble field of canola (*Brassica napus*).

While much research has been done with Brassica debris-borne disease-causing fungi, so far there has been only little focus on saprophytic fungi which may accelerate debris decomposition. This research is outlined below.

Saprophytic Fungi Present in Association with Brassica Debris

One saprophyte, *Gliocladium roseum*, isolated from Brassica debris was found to be a mycoparasite of *Alternaria brassicae* (Boyko and Tewari, 1993).

A wood decay bird's nest fungus, *Cyathus* sp. (later identified as *C. olla*), was found in the fields in Alberta, British Columbia, and Manitoba, Canada mostly in association with canola debris (Tewari and Briggs, 1995). In the field, the fungus grew profusely on canola debris but only sparsely on wheat and barley debris (Figs. 13.13, 13.14). It seemed to be tolerant to intensive agricultural management and appeared in a field year-after-year. The colonized debris was soft and macerated and showed conspicuous hyphal growth radiating from the debris pieces as hyphal cords (Figs. 13.15, 13.16). These characteristics indicated that the fungus accelerates decomposition of canola debris in the field and may have potential in management of debris-borne fungal diseases such as the blackleg and blackspot caused by *Leptosphaeria maculans* and *Alternaria brassicae*, respectively. The hyphae, hyphal cords and basidiocarps of the saprophyte were adorned with numerous crystals rich in calcium as studied by light microscopy

Figure 13.13. A field in central Alberta, Canada, where canola (*Brassica napus*) was grown in the previous season. Note the basidiocarps of *Cyathus olla* and pieces of the debris of canola. Note a clump of the mycelial cords of *C. olla* at about the center of the photograph. **Figure 13.14.** Basidiocarps of *C. olla* attached to debris of canola (*B. napus*) and a cereal (one piece in about the middle of the group). Most basidiocarps contain peridiola. Note a Canadian dime (diameter 18.03 mm) for an estimate of magnification which is about the same in Fig. 13.13 also.

(Fig. 13.17) and Scanning Electron Microscopy (SEM) in conjunction with energy-dispersive X-ray microanalysis (Figs. 13.18–13.21). Further characterization of the crystals produced by *C. olla* (growing on Brassica debris in the field) and *C. striatus* (Huds.) ex Pers. (growing on wood chips used as mulch) was done using SEM, FT/IR spectroscopy and ^{13}C-NMR spectroscopy (Tewari et al., 1997). Many pieces of canola debris colonized with *C. olla* revealed evidence of structural damage by SEM whereas wood chips colonized with *C. striatus* presented only minimal damage. Also, the crystal morphologies were appreciably different in the two species. The FT/IR spectra of the crystals revealed absorbance bands similar to those

Figures 13.15, 13.16. Pieces of degraded debris of canola (*Brassica napus*) infested with *Cyathus olla* in a field in central Alberta. Note the basidiocarps and the mycelial cords. The coin in Fig. 13.15, a Canadian dime (diameter 18.03 mm), is for an estimate of magnification which is about three times of this in Fig. 13.16.

of a control sample of calcium oxalate. The ^{13}C-NMR spectrum revealed a characteristic carbonyl carbon resonance for oxalate and this along with a strong peak for calcium present in X-ray microanalysis spectrum proved that the crystals were those of calcium oxalate.

Hyphae of many species of bird's nest fungi were examined for the presence of crystals on V8 juice and Difco potato-dextrose agars, and crystals were found to be present only in some species (Shinners and Tewari, 1997). In some species, crystal production was dependent on the medium of growth. Where present, the crystals were classified as styloid, raphide or bipyramidal types. Energy-dispersive X-ray microanalyses confirmed that all crystals were rich in calcium and X-ray diffraction spectra revealed

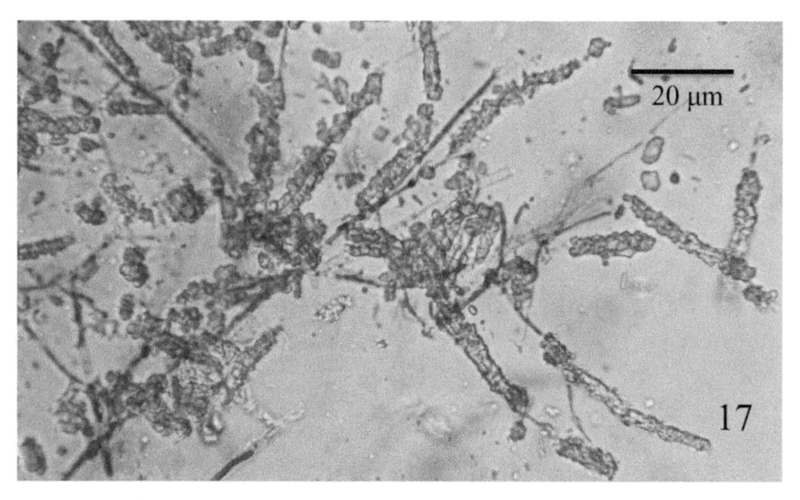

Figure 13.17. Light micrograph of the submerged hyphae of *Cyathus olla* grown in V8 juice agar and mounted in lactophenol cotton blue. Note the crystals of calcium oxalate covering the hyphae.

Figure 13.18. Debris of canola (*Brassica napus*) infested with *Cyathus olla* in the field in central Alberta, Canada. Note the clay granules (see the Fig. 13.20), hyphae, and hyphal cords on the surface of the debris.

that both dihydrate (weddellite) and monohydrate (whewellite) forms of calcium oxalate were present (Shinners and Tewari, 1997). Calcium oxalate production in fungi is a good correlate of cell wall-degrading enzyme activity. Hence, it would be of interest to compare these parameters in both bird's nest fungi with and without crystals on the hyphae.

Figure 13.19. Scanning electron micrograph of the hyphae and hyphal cords of *Cyathus olla* seen in Fig. 13.18. Note the various types of calcium oxalate crystals (see Fig. 13.21).

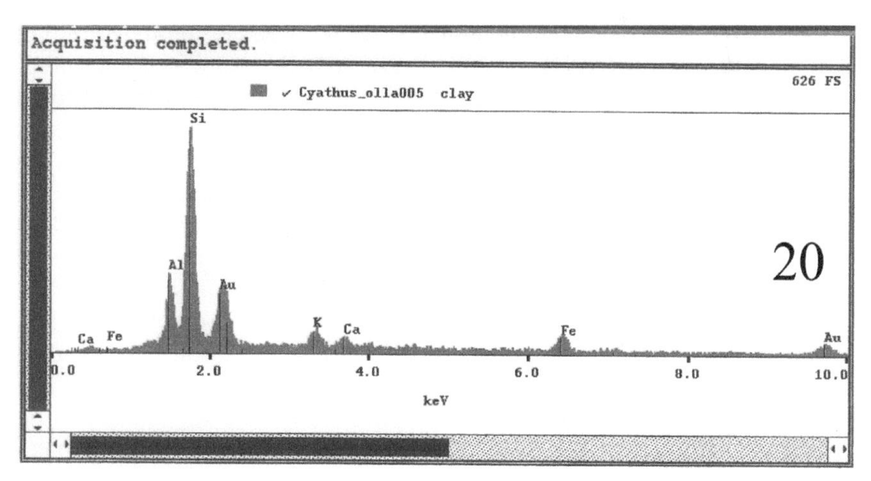

Figure 13.20. Energy-dispersive X-ray microanalysis spectrum of a clay granule seen in Fig. 13.18. Note the emission lines for silicon, calcium, and potassium which are common constituents of clay. The emission lines from gold have originated from coating of the specimen for scanning electron microscopy.

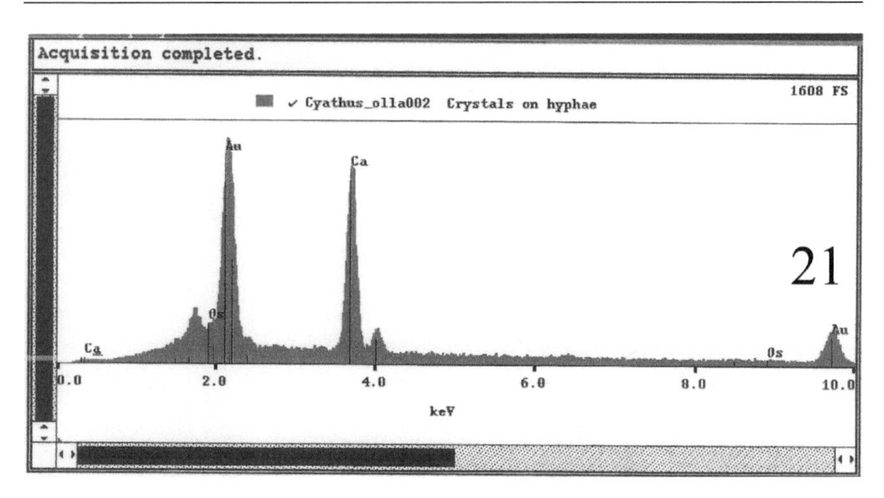

Figure 13.21. Energy-dispersive X-ray microanalysis spectrum of calcium oxalate crystals on the hyphal structures seen in Fig. 13.19. Note the emission lines for calcium. The emission lines from osmium and gold have originated from specimen fixation and coating of the specimen for scanning electron microscopy, respectively.

Brodie (1952, 1975, 1978) noticed that basidiocarps of the species *Cyathus olla* are variable in size, color and form, and recognized three forms under this species (*C. olla* f. *olla, C. olla* f. *anglicus,* and *C. olla* f. *lanatus*). Basidiocarps found in the fields in Alberta, Canada attributable to *C. olla* were also found to be variable. Polymerase Chain Reaction (PCR)-based Random Amplified Polymorphic DNA (RAPD) analysis distinguished among a large number of accessions and UPGMA analysis generated a dendrogram consisting of three distantly branched groups (Shinners and Tewari, 1998). Considered together, the morphological and RAPD data supported the presence of *C. olla* f. *olla,* C. *olla* f. *anglicus,* and a third new plicate peridium basidiocarp form which was named as *C. olla* f. *brodiensis* Shinners and Tewari (Fig. 22). Both *C. olla* f. *olla* and C. *olla* f. *anglicus* had non-plicate peridium basidiocarps and the latter had relatively larger basidiocarps (Shinners and Tewari, 1998).

The new form epithet *brodiensis,* formulated in the honor of late Dr. H.J. Brodie (Shinners and Tewari, 1998), was not appropriate, and the new form is hereby renamed as *C. olla* f. *Brodii* Shinners and Tewari.

Shinners et al. (2002) reported on the plant cell wall decomposing enzymatic activities of 42 *C. olla* f. *olla, C. olla* f. *anglicus* and *C. olla* f. *Brodii* isolates collected mostly from Alberta, Canada. All isolates, except two, degraded a model lignin substrate, the polymeric dye Poly R-478, during plate assays indicating ligninolytic activities by them. *Phanerochaete chrysosporium* Burdsall, used as a positive lignin-degrading control, also degraded the polymeric dye, revealing ligninolytic activity. Further

Figure 13.22. Wood chips used as mulch in central Alberta, Canada showing the basidiocarps of *Cyathus olla* f. *Brodii*. Note the plicate peridium of basidiocarps. Note a Canadian penny (diameter 19.1 mm) for an estimate of magnification.

biochemical studies on an isolate *C. olla* f. *anglicus* isolated from canola roots detected laccase and manganese peroxidase in both 1- and 4-week incubations during solid state fermentations of canola roots while aryl-alcohol oxidase was detected only after four week incubation. Assays also detected polygalacturonase but not cellulases. Peláez et al. (1995) reported aryl-alcohol oxidase activity in *C. olla* but not the other two enzymes.

The basal stems and roots of canola plants are woody and are especially resistant to breakdown in the field. Standing stubble of these plant parts from five different *Brassica napus* canola cultivars were collected and their fiber content was studied by the Goering Van Soest method (Goering and Van Soest, 1970; Shinners-Carnelley and Tewari, 2000). The cultivar Cyclone had higher Neutral Detergent Fiber (NDF) than the other four cultivars (range 70.1–77.7%). Cyclone also had the highest cellulose and hemicellulose contents. It had 14.5% lignin content which was superseded only by the cultivar Westar (15.5%).

Decomposition of the ground substrate of the cultivar Cyclone was studied with solid-state fermentation for 45 days at 25°C by 15 accessions of *C. olla* (five each of *C. olla* f. *olla*, *C. olla* f. *anglicus* and *C. olla* f. *Brodii*; Shinners and Tewari, 2000; Shinners-Carnelley and Tewari, 2000). *Cyathus olla* f. *Brodii* caused the greatest reduction in dry weight and the substrate inoculated with it had 60.6 and 75.3% of the original lignin and hemicellulose remaining. The NDF was increased and cellulose was degraded. Overall, *C. olla* f. *Brodii* was most effective at reducing the dry weight, lignin and hemicellulose of the ground substrate.

The bird's nest fungi are white-wood rotters and all, except *C. stercoreus* (Schwein) de Toni, are strictly saprophytic. Besides being saprophytic, *C. stercoreus* is also reported to cause disease in turf grass in the form of fairy rings (Mercier et al., 1999; Buzi, 2004). *Cyathus olla* has been reported as a saprophyte a number of times from agricultural ecosystems (see Baird et al., 1993; Tewari et al., 1997; Shinners and Tewari, 1998). An unnamed *Cyathus* sp. is also reported from rape fields in the UK (Maksymiak and Hall, 2000).

Cyathus olla is capable of decomposing lignin (Shinners et al., 2002) as some other *Cyathus* spp. (Abbott and Wicklow, 1984; Maksymiak and Hall, 2000). Canola stubble, especially from the basal stem and root regions, is woody and lignified, and is hence particularly recalcitrant to decomposition (Blenis and Chow, 2005; Shinners-Carnelley and Tewari, 2000). The fungal pathogen, *Leptosphaeria maculans*, causal agent of the blackleg disease, overwinters on canola stubble and produces primary inocula during the following growing seasons. The disease is less severe in fields where *Cyathus* sp. is present (Maksymiak and Hall, 2000). Maksymiak and Hall (2002) reported reduced production of primary inoculum of *L. maculans* as a result of the application of *C. striatus*. Many studies by the author and members of his group described in this chapter have shown profuse growth of *Cyathus olla* in the colonized debris and soil and maceration of infested canola debris. Solid-state fermentation studies using ground canola residue have revealed reduction in dry weight of the substrate as well as of lignin, hemicellulose and cellulose. Also, production of the enzymes laccase, manganese peroxidase, aryl-alcohol oxidase and polygalacturonase enzymes by *C. olla* was shown. Additionally, morphological and PCR-based RAPD analyses distinguished among a large number of accessions and UPGMA analysis generated a dendrogram consisting of three distantly branched groups. Based on these considerations, a new form known as *C. olla* f. *Brodii* (formerly named *C. olla* f. *brodiensis*) was recognized and was most effective in the stubble decomposition parameters discussed above. *Cyathus olla* is capable of degrading the overwintering/oversummering habitat (i.e., debris) of *L. maculans* and other such pathogens, and has the potential of being developed as microbial inoculant for management of such debris-borne pathogens. Blenis and Chow (2005) found little potential of developing the wood decay fungi, including *C. olla*, as biological control agents as they were unable to colonize and decompose nonsterile canola stubble. However, ample infestation and decomposition of canola stubble by *C. olla* is observed in the field (Tewari and Briggs, 1995; Tewari et al., 1997), indicating colonization of nonsterile stubble. Hence, further work is required to investigate the conditions under which this occurs so that suitable inoculant formulation and application strategies can be developed.

As mentioned earlier that *C. olla* shows profuse hyphal growth in association with colonized canola debris and production of mycelial cords which emanate from it (Tewari and Brigs, 1995). No doubt the metabolic vigor originating from stubble-decomposition and antifungal- and antibacterial antibiotic-producing activities of this fungus (Allbutt et al., 1971; Liu and Zhang, 2004) may be contributing to these capabilities. Antibiotic-production may be an important attribute that allows uninhibited growth of this fungus on canola stubble in the field.

Calcium biomineralization serves a number of functions including being part of the natural calcium cycle. Hyphae, hyphal cords and basidiocarps of *C. olla* are heavily encrusted with crystals of calcium oxalate. This phenomenon is common in fungi especially Basidiomycetes, Ascomycetes and Mucorales (Arnott, 1995; Oyarbide et al., 2001). The mechanism of calcium oxalate crystal formation is not well understood but super-saturation must be a driving force for crystallization (Kavanagh, 1995). Oxalic acid production and sequestration of calcium from the cell wall leads to calcium oxalate crystal production. This also acts in concert with the activities of cell wall degrading enzymes and consequent breakdown of the cell wall. The result is decomposition of canola stubble infested with *C. olla*.

Acknowledgements

Help received from Prof. J.K. Misra and Prof. Tamás Papp for confirming the name f. *Brodii* is gratefully acknowledged. Steven G. DeVries (Figs. 13.5, 13.6) and Bryn Jonzon (Figs. 13.18–13.21) helped prepare some of the figures used in this chapter. Both were undergraduate students of the author at that time.

References

Abbott, T.P. and Wicklow, D.T. 1984. Degradation of lignin by *Cyathus* species. Appl. Environ. Microbiol., 47: 585–587.

Allbutt, A.D., Ayer, W.A., Brodie, H.J., Johri, B.N. and Taube, H. 1971. Cyathin, a new antibiotic complex produced by *Cyathus helenae*. Can. J. Microbiol., 17: 1401–1407.

Anonymous. 2011. Canola Grower's Manual. Canola Council of Canada, Winnipeg, MN, Canada.

Arnott, H.J. 1995. Calcium oxalate in fungi. pp. 73–111. *In*: Khan, S.R. (ed.). Calcium Oxalate in Biological Systems. CRC Press, Boca Raton, FL, USA.

Baird, R.E., Summer, D.R., Mullinix, B.G., Dowler, C.C., Phatak, S.C., Johnson, A.W., Chalfant, R.B., Gay, J.D., Chandler, L.D. and Baker, S.H. 1993. Occurrence of fleshy fungi from agricultural fields. Mycopathologia, 122: 29–34.

Blenis, P.V. and Chow, P.S. 2005. Evaluating fungi from wood and canola for their ability to decompose canola stubble. Can. J. Plant Pathol., 27: 259–267.

Blenis, P.V., Chow, P.S. and Stringam, G.R. 1999. Effects of burial, stem portion and cultivar on the decomposition of canola straw. Can. J. Plant Sci., 79: 97–100.

Boyko, L.A. and Tewari, J.P. 1993. Interactions of *Myrothecium verrucaria* and *Gliocladium roseum* with *Alternaria brassicae*. Ann. Meeting Plant Path. Soc. of Alberta, Vegreville, AB, Nov. 1–3.

Brodie, H.J. 1952. Infertility between two distinct forms of *Cyathus olla*. Mycologia, 44: 413–423.

Brodie, H.J. 1975. The Bird's Nest Fungi. University of Toronto, Press, Toronto, Ont., Canada.

Brodie, H.J. 1978. A hitherto un-named form of *Cyathus olla* (Nidulariaceae). Bot. Notiser, 131: 31–34.

Browne, L.M., Conn, K.L., Ayer, W.A. and Tewari, J.P. 1991. The camalexins: new phyoalexins produced in the leaves of *Camelina sativa* (Cruciferae). Tetrahedron, 47: 3909–3914.

Buzi, A. 2004. Investigation of the detrimental effects of *Cyathus stercoreus* on turf grass. Informatore Fitopatologico, 54: 47–51.

Cessna, S.G., Sears, V.E., Dickman, M.B. and Low, P.S., 2000. Oxalic acid, a pathogenicity factor for *Sclerotinia sclerotiorum*, suppresses the oxidative burst of the host plant. Plant Cell, 12: 2191–2200.

Choi, Y.-J., Shin, H.-D. and Thines, M. 2009. The host range of *Albugo candida* extends from Brassicaceae through Cleomaceae to Capparaceae. Mycol. Progress, 8: 329–335.

Fahey, J.W., Zalcmann, A.T. and Talalay, P. 2001. The chemical diversity and distribution of glucosinolates and isothiocyanates among plants. Phytochemistry, 56: 5–51.

Franceschi, V.R. and Loewus, F.A. 1995. Oxalate biosynthesis and function in plants and fungi. pp. 113–130. *In*: Khan, S.R. (ed.). Calcium Oxalate in Biological Systems. CRC Press, Boca Raton, FL, USA.

Godoy, G., Steadman, J.R., Dickman, M.B. and Dam, R. 1990. Use of mutants to demonstrate the role of oxalic acid in pathogenicity of *Sclerotinia sclerotiorum* on *Phaseolus vulgaris*. Physiol. Mol. Plant Pathol., 37: 179–191.

Goering, H.K. and Van Soest, P.J. 1970. Forage fiber analyses (apparatus, reagents, procedures, and some applications). *In*: Agriculture Handbook (n. 379). Agriculture Service, USDA, Washington, D.C., USA.

Guimarães, R.L. and Stotz, H.U. 2004. Oxalate production by *Sclerotinia sclerotiorum* deregulates guard cells during infection. Plant Physiol., 136: 3703–3711.

Kaur, P., Sivasithamparam, K. and Barbetti, M.J. 2011. Host range and phylogenetic relationships of *Albugo candida* from cruciferous hosts in Western Australia, with special reference to *Brassica juncea*. Plant Dis., 95: 712–718.

Kavanagh, J.P. 1995. Calcium oxalate crystallization *in vitro*. pp. 1–21. *In*: Khan, S.R. (ed.). Calcium Oxalate in Biological Systems. CRC Press, Boca Raton, FL, USA.

Kharbanda, P.D., Fitt, B.D.L., Lange, R.M., West, J.S., Lamey, A.H. and Phillips, D.V., Primary collators (last update 3/8/01). Common Names of Plant Diseases. Diseases of Rapeseed = Canola (*B. napus* L. and *Brassica rapa* L. (= *B. campestris* L.)). APSNet. American Phytopathological Society, Saint Paul, MN, USA. http://www.apsnet.org/publications/commonnames/Pages/Rapeseed.aspx.

Kharbanda, P.D. and Tewari, J.P. 1996. Integrated management of canola diseases using cultural methods. Can. J. Plant Pathol., 18: 168–175.

Kolte, S.J. 1985. Diseases of annual edible oilseed crops. *In*: Rapeseed, Mustard and Sesame Diseases. CRC Press, Boca Raton, FL, USA.

Li, R., Rimmer, R., Buchwaldt, L., Sharpe, A.G., Seguin-Schwartz, G., Coutu, C. and Hegedus, D.D. 2004a. Interaction of *Sclerotinia sclerotiorum* with a resistant *Brassica napus* cultivar:

expressed sequence tag analysis identifies genes associated with fungal pathogenesis. Fungal Genet. Biol., 41: 735–753.

Li, R., Rimmer, R., Buchwaldt, L., Sharpe, A.G., Seguin-Schwartz, G. and Hegedus, D.D. 2004b. Interaction of *Sclerotinia sclerotiorum* with *Brassica napus*: cloning and characterization of endo- and exo-polygalacturonases expressed during saprophytic and parasitic modes. Fungal Genet. Biol., 41: 754–765.

Liu, Y.-J. and Zhang, K.-Q. 2004. Antimicrobial activities of selected *Cyathus* species. Mycopathologia, 157: 185–189.

Lumsden, R.D. 1976. Pectolytic enzymes of *Sclerotinia sclerotiorum* and their localization in infected bean. Can. J. Bot., 54: 2630–2641.

Maksymiak, M.S. and Hall, A.M. 2000. Biological control of *Leptosphaeria maculans* (anamorph *Phoma lingam*) causal agent of blackleg/canker on oil seed rape by *Cyathus striatus*, a bird's nest fungus. *In*: BCPC Conference: Pests and diseases 2000, 1: 507–510.

Maksymiak, M.S. and Hall, A.M. 2002. Biocontrol of canker on oilseed rape by reduction and inhibition of initial inoculum. *In*: Proc. BCPC Conference: Pests and diseases, 2002, 1, 2: 769–772.

Mikkelsen, M.D., Petersen, B.L., Olsen, C.E. and Halkier, B.A. 2002. Biosynthesis and metabolic engineering of glucosinolates. Amino Acids, 22: 279–295.

Mercier, J., Carson, T.D. and White, D.B. 1999. Fairy rings in turf associated with the bird's nest fungus *Cyathus stercoreus*. Plant Dis., 83: 781.

Mundkur, B.B. 1938. Host range and identity of the smut causing root galls in the genus *Brassica*. Phytopathology, 28: 134–142.

Nelson, R. and Achar, P.N. 2000. First report of vesicular-arbuscular mycorrhizal fungi on *Brassica* and its effect on the downy mildew disease. APS Pacific Div. Meeting, June 18–20, Victoria, B.C., Canada, Abs (Joint with Can. Phytopathol. Soc.).

Oyarbide, F., Osterrieth, M.L. and Cabello, M. 2001. *Trichoderma koningii* as a biomineralizing fungus agent of calcium oxalate crystals in typical Argiudolls of the Los Padres Lake natural reserve (Buenos Aires, Argentina). Microbiol. Res., 156: 113–119.

Paul, V.H. and Rawlinson, C.J. 1992. Diseases and Pests of Rape. Verlag Th. Mann., Gelsenkirchen-Buer, Germany.

Pedras, M.S.C. and Yaya, E.E. 2010. Phytoalexins from Brassicaceae: News from the front. Phytochemistry, 71: 1191–1197.

Peláez, F., Martínez, M.J. and Martínez, A.T. 1995. Screening of 68 species of basidiomycetes for enzymes involved in lignin degradation. Mycol. Res., 99: 37–42.

Peltier, A.J., Hatfield, R.D. and Grau, C.R. 2004. *Sclerotinia sclerotiorum* does not detectably modify or metabolize lignin in mature soybean stems. North Central Division Meeting Abstracts. (Online). Available at http://www.apsnet.org/meetings/div/nc04absaps Publication no. P-2005-0031-NCA.

Peltier, A.J., Hatfield, R.D. and Grau, C.R. 2009. Soybean stem lignin concentration relates to resistance to *Sclerotinia sclerotiorum*. Plant Dis., 93: 149–154.

Petri, G.A. 1995. Long-term survival and sporulation of *Leptosphaeria maculans* (blackleg) on naturally-infected rapeseed/canola stubble in Saskatchewan. Can. Plant Dis. Surv., 75: 23–34.

Punja, Z.K. 1985. The biology, ecology, and control of *Sclerotium rolfsii*. Ann. Rev. Phytopathol., 23: 97–127.

Punja, Z.K. and Damiani, A. 1996. Comparative growth, morphology, and physiology of three *Sclerotium* species. Mycologia, 88: 694–706.

Rai, J.N., Tewari, J.P., Singh, R.P. and Saxena, V.C. 1974. Fungal diseases of Indian crucifers. Beihefte zur Nova Hedwigia, 47: 447–86.

Rimmer, S.R., Shattuck, V.I. and Buchwaldt, L. (eds.). 2007. Compendium of Brassica Disease. American Phytopathological Society, Saint Paul, MN, USA.

Saharan, G.S. and Verma, P.R. 1992. White rusts: A review of economically important species. IDRC-MR315e. International Development Research Centre, Ottawa, ON, Canada.

Shinners-Carnelly, T.C., Szpacenko, A., Tewari, J.P. and Palcic, M.M. 2002. Enzymatic activity of *Cyathus olla* during solid state fermentation of canola roots. Phytoprotection, 83: 31–40.
Shinners, T.C. and Tewari, J.P. 1997. Diversity in crystal production by some bird's nest fungi (Nidulariaceae) in culture. Can. J. Chem., 75: 850–856.
Shinners, T.C. and Tewari, J.P. 1998. Mophological and RAPD analyses of *Cyathus olla* from crop residue. Mycologia, 90: 980–989.
Shinners-Carnelly, T.C. and Tewari, J.P. 2000. Decomposition of canola stubble by solid state fermentation with *Cyathus olla*. Phytoprotection, 81: 87–94.
Shinners, T.C. and Tewari, J.P. 2000. Accelerated stubble decomposition using *Cyathus olla*. A strategy for control of stubble-borne diseases of canola. IOBC/WPRS Bull., 23: 77–81.
Singh, D.B. and Rai, B. 1980. Studies on the leaf surface mycoflora of mustard (*Brassica campestris* L. cv. YS-42). Bull. Torrey Bot. Club, 107: 447–452.
Tewari, J.P. 1985. Diseases of canola caused by fungi in the Canadian prairies. Agri. For. Bull., The University of Alberta, 8: 13–20.
Tewari, J.P. 1991. Structural and biochemical bases of the blackspot of crucifers. pp. 325–330. *In*: S.K. Malhotra (ed.). Advances in Structural Biology, Vol. 1. Jai Press Inc. Greenwich, Conn.
Tewari, J.P. and Briggs, K.G. 1995. Field infestation of canola stubble by a bird's-nest fungus. Can. J. Plant Pathol., 17: 291.
Tewari, J.P. and Mithen, R.F. 1999. Diseases. pp. 375–411. *In*: Gomez-Campo, C. (ed.). Biology of Brassica Coenospecies. Elsevier Science B.V., Amsterdam, The Netherlands.
Tewari, J.P., Shinners, T.C. and Briggs, K.G. 1997. Production of calcium oxalate crystals by two species of *Cyathus* in culture and infested plant debris. Z. Naturforsch., 52c: 421–425.
Tsuneda, A. and Skoropad, W.P. 1977. Formation of microsclerotia and chlamydospores from conidia of *Alternaria brassicae*. Can. J. Bot., 55: 1276–1281.
Tsuneda, A. and Skoropad, W.P. 1978. Phylloplane fungal flora of rapeseed. Trans. Brit. Mycol. Soc., 70: 329–333.
Tsuneda, A. and Skoropad, W.P. 1980. Interactions between *Nectria inventa*, a destructive mycoparasite, and fourteen fungi associated with rapeseed. Trans. Brit. Mycol. Soc., 74: 501–507.
Tsuneda, A., Skoropad, W.P. and Tewari, J.P. 1976. Mode of parasitism of *Alternaria brassicae* by *Nectria inventa*. Phytopathology, 66: 1056–1064.
Vaartnou, H. and Tewari, I. 1972a. *Alternaria alternata* parasitic on rape in Alberta. Plant Dis. Rep., 56: 676–677.
Vaartnou, H. and Tewari, I. 1972b. *Alternaria* on Polish-type rape in Alberta. Plant Dis. Rep., 56: 633–635.
Vaartnou, H. and Tewari, I. 1973. A strain of *Chaetomium olivaceum*. Mycopath. et Mycol. Applic., 51: 239–241.
Vaartnou, H., Tewari, I. and Horricks, J. 1974. Fungi associated with diseases on Polish-type rape in Alberta. Mycopath. et Mycol. Applic., 52: 255–260.
Verma, G.S. and Saharan, G.S. 1994. Monograph of *Alternaria* Diseases of Crucifers. Saskatoon Research Centre Technical Bulletin 1994-6E, Agric. Agri-Food Canada, Saskatoon, Sask., Canada.
Vierheilig, H., Bennett, R., Kiddle, G., Kaldorf, M. and Ludwig-Müller, J. 2000. Differences in glucosinolate patterns and arbuscular mycorrhizal status of glucosinolate-containing plant species. New Phytol., 146: 343–352.
Williams, P.H. and Saha, L.R. 1993. Diseases of Mustard (*Brassica juncea* (L.) Czernj. & J.M. Coulter var. *crispifolia* L.H. Bailey and *B. nigra* (L.) W. Koch). APSNet. American Phytopathological Society, Saint Paul, MN, USA. http://www.apsnet.org/publications/commonnames/Pages/Mustard.aspx.

CHAPTER 14

Fungi Associated with Soybean [*Glycine max* (L.) Merrill] Diseases

S.K. Sharma[a] and S.K. Srivastava[b,*]

ABSTRACT

Soybean [*Glycine max* (L.) Merrill], a native of eastern Asia, is a promising leguminous crop which is receiving wide acceptance due to its high protein and oil content. Its cultivation is increasing each year by becoming an important source of a variety of low cost nutritious food items. Therefore, the crop has received attention of a large number of researchers of different disciplines including plant pathology in order to have a higher and stable yield as well as quality produce.

Soybean hosts a variety of pathogens including fungi. Almost all parts of a soybean plant are susceptible to diseases and more than 100 pathogens have so far been recorded on this crop. Some of them are known to cause economic damage. This chapter presents a review of the work done so far on some important diseases viz., Collar rot (*Sclerotium rolfsii*), Charcoal rot (*Macrophomina phaseolina*), Anthracnose (*Colletotrichum truncatum*), Rust (*Phakopsora pachyrizi*), Myrothecium leaf spot and blight (*Myrothecium roridum*), Frog-eye leaf spot (*Cercospora sojina*) and Purple seed stain (*Cercospora kikuchii*). The aspects that have been covered include the distribution, symptoms, causal organisms, taxonomy, morphology, biology, ecology, epidemiology and yield losses caused by the diseases.

Directorate of Soybean Research, Khandwa Road, Indore–452 001, Madhya Pradesh, India.

[a] Email: sharmask1759@rediff.com

[b] Email: dsrdirector@gmail.com

* Corresponding author

Introduction

The Soybean *Glycine max* (L.) Merrill, a native of eastern Asia, is a promising leguminous crop. It was domesticated by farmers in the eastern half of northern China during the Shang dynasty. For several thousand years, people in eastern Asia have used soybeans for food and animal feed. Today, soybeans are grown to some extent in most parts of the world and are a primary source of protein and oil.

Soybean is a known host/substrate of a variety of pathogens including fungi, bacteria, viruses, nematodes and phytoplasma. So far, 29 fungal pathogens, six bacterial pathogens, 18 viruses, six nematodes and three phytoplasma diseases have been recorded on this crop (Sinclair and Dhingra, 1975; Sinclair and Shurtleff, 1975; Tisselli et al., 1980; Verma et al., 1988; Sinclair and Backman, 1989; Srivastava and Agarwal, 1989; Sharma, 1990; Srivastava, 2001). Out of them, only 35 pathogens are known to cause economic damage (Sinclair, 1983).

The first record of the soybean disease dates back to 1882, when Frank included a brief description of root gall symptoms due to *Meloidogyne* sp. in soybean plants grown in a green-house in Berlin (Frank, 1882). The first description of a fungal disease affecting aerial parts came about two decades later when Massalongo (1900) reported *Phyllosticta sojaecola* as a new species affecting soybean leaves. Similarly, Butler and Bisby (1931) listed the fungal diseases recorded in India. Morse and his co-workers for the first time described the diseases of soybean common in USA in the form of a farmer's bulletin (Morse and Carter, 1939; Morse et al., 1949).

In the North-eastern India, the occurrence of soybean diseases had remained unexplored until the 70s. Saikia (1986) listed the earliest report of a soybean disease from the northeastern region of India. However, during the next few years, several diseases affecting the crop were reported by the ICAR Research Complex for North Eastern Hill Region (NEH), Shillong, for example, important diseases caused by the fungi: *Phakospora pachyrhizi* (Maiti et al., 1981), *Setospheria glycines* (Srivastava et al., 1982), *Aristostoma camarographiodis* (Maiti et al., 1983) and *Colletotrichum capsici* (Sharma et al., 1991). Also, Maiti et al. (1983), Verma et al. (1988), and Sharma et al. (1989) have presented good accounts of the diseases affecting the crop in the NEH Region of India.

In the present chapter, major fungal diseases of soybean reported in India have been described and discussed with regards to their distribution of causal organism, symptoms and diagnosis, taxonomy and biology of causal fungus, etiology and epidemiology of the diseases.

Collar Rot

Collar rot of soybean is caused by *Sclerotium rolfsii* Sacc. [teleomorph: *Athelia rolfsii* (Curzi) Tu and Kimbrough]. It is a devastating soil-borne fungus with a wide host range (Aycock, 1966; Punja, 1988). The disease becomes serious in most of the soybean growing areas. Under heavy soil moisture, some times more than 60–70% crop has been found damaged due to this disease. Borkar (1992) reported that the collar rot disease caused by *S. rolfsii* severely reduced the nodulation by *Rhizobium* even during the early stage of wilting, thus affecting the yield considerably.

Distribution

S. rolfsii is thought to have caused serious crop losses over many centuries and the first unmistakable report of the fungus dates back to 1892 with Peter Henry Rolfs' discovery of the organism in association with tomato blight in Florida (Aycock, 1966). Silfs' report in the late 19th century and over 2000 publications on the pathogen support its worldwide distribution, particularly in the tropical and subtropical regions (Weber, 1931; Aycock, 1966).

The occurrence of *S. rolfsii* on soybean was recorded for the first time in 1926 in the United States by Haskell (1926). The fungus isolated from tomato and pepper was found to be pathogenic on soybean in the Philippines upon wound inoculation (Atienza, 1927). The soybean strain of *S. rolfsii* in the Philippines was similar to that of cotton strain of the fungus (Celina, 1936). The disease was also reported from Java (Indonesia) causing substantial losses to soybean under humid conditions (Goot and Muller, 1932). Thompson (1928) reported the occurrence of *S. rolfsii* in Malaya for the first time in 1927. In Trinidad, Briton-Jones and Baker (1934) noticed the fungus on soybean. Nobile et al. (1935) recorded the occurrence of *S. rolfsii* on soybeans for the first time in New South Wales (Australia).

There are several reports from the USA on the occurrence of *S. rolfsii* on soybean (Koehler, 1931; Pinckard, 1942; Atkinson, 1943, 1944; Allington, 1944; Bain, 1944). The occurrence of *S. rolfsii* on soybean was recorded in Mississippi (Kilpatrick, 1955; Johnson and Kilpatrick, 1953), in Louisiana (Person, 1944), in Alabama (Stone and Seal, 1944), and from Tingo Maria zone of the Peruvian Montana (Crandall and Dieguez, 1948).

Survey of soybean nurseries in Alabama, Georgia, Louisiana, Mississippi and South Carolina made during 1944–1946 showed the presence of a disease caused by *S. rolfsii* (Weimer, 1947). The occurrence of *S. rolfsii* on soybean was also reported from Virginia (Fenne, 1949), Sarawak (Turner, 1964) and Indiana (Laviolatta et al., 1973).

In the Asian continent, the occurrence of *S. rolfsii* on soybean was reported from several researchers of India (Agarwal and Kotasthane, 1971), Japan (Kurata, 1960), the Philippines (Mejia, 1954), Singapore (Turner, 1964), Taiwan (Chandrasrikul, 1962), Abkhazia (former USSR) (Vardaniya, 1971). The occurrence of *S. rolfsii* on soybean was also reported from African countries, in Mozambique (De Carvalho and Mends, 1958). Lists of its occurrence on soybean in the Salishbury area of Southern Rhodesia was published by Hopkins (1950) and Whiteside (1960). De Guerpel (1942) has described the occurrence of *S. rolfsii* on soybean from France.

From South America, reports are available for the occurrence of *S. rolfsii* on soybean from Venezuela (Muller, 1941), La Molina (Peru) (Bazan De Segura, 1947), Malaya (Thompson and Johnston, 1953), Fiji (Morwood, 1956), Guatemala (Muller, 1950), and from Colombia (Patino, 1967).

Symptom

S. rolfsii, the causal fungus of the disease, primarily attacks the stem parts of soybean. But can cause infection on any part of the plant, including roots, pods, petioles, leaves and flowers under favorable environmental conditions. The first signs of the infection, though usually undetectable, are dark-brown lesions on the stem at or just beneath the soil level. However, the first visible symptoms are, progressive yellowing and wilting of the leaves (Fig. 14.1a). Following this, the fungus produces abundant white, fluffy mycelium on infected tissues and on the soil (Fig. 14.1b). The fungus produces oxalic acid and other toxins causing degradation and death of the host tissues. Characteristic disease symptom in seedlings occurs in the form of their damping-off. The collar region starts decaying. Plants become light yellow, which later start drooping off and wilting (Fig. 14.1d). The white mycelial growth later on produces small, mustard like, white to cream colored sclerotia gradually becoming pink to light brown resembling mustard seeds at maturity (Fig. 14.1c, e). The fungus occasionally produces basidiospores at the margins of the lesions under humid conditions, though this is not common. Leaf petioles may rot around the stem base causing wilting and yellowing of the lower leaves. The pathogen often girdles the stem at the soil line and white string-like mycelium can grow out from the stem base onto the soil surface. Small spherical sclerotia, around 1–2 mm in diameter, first white then brown, develop soon afterwards on the soil and affected plant parts (Fig. 14.1c, e). Patches of infected plants often appear. Seedlings are very susceptible and die quicker than the older plants that have formed woody tissues.

Figure 14.1. Soybean plants infected with collar rot disease caused by *Sclerotium rolfsii.*

a. Affected plants showing yellowing, drying of leaves and wilting symptoms
b. Mycelial growth of on the lower stem of the affected plants near soil surface
c. Initiation of sclerotia formation on the mycelim of *Sclerotium rolfsii* over the affected stem of soybean
d. Root and collar region of plant infected with *Sclerotium rolfsii* exhibiting the growth of fungal mycelium and drying of other dead host tissues
e. Pink colored, mustard seed like sclerotia produced on the infected stem of soybean

Taxonomy

The species was first described in 1911 by an Italian mycologist, Pier Andrea Saccardo, based on specimens sent to him by Peter Henry Rolfs who considered the unnamed fungus to be the cause of tomato blight in

Florida. The specimens sent to Saccardo were sterile, consisting of hyphae and sclerotia. He placed the species in the old form genus *Sclerotium*, naming it *Sclerotium rolfsii*. It is, however, not a species of *Sclerotium* in the strict sense.

In 1932, Mario Curzi discovered that the teleomorph spore-bearing stage was a corticied fungus and accordingly placed the species in the form genus *Corticium*. The teleomorph of *Sclerotium rolfsii* Curzi was first placed in *Corticium centrifungum* (Lev.) Bres. Later on *Corticium rolfsii* was proposed to be the basidial state of *S. rolfsii* by Curzi (1931). *Pellicularia rolfsii* (Curzi) E. West has also been described as perfect stage of the fungus. Talbot (1973) suggested that the basidial stage of *S. rolfsii* belonged to the genus *Athelia* and in 1978 *Corticium rolfsii* was transferred to the genus *Athelia* as Athelia rolfsii (Curzi) Tu and Kimbr. (Tu and Kimbrough, 1978).

Morphology

The mycelium is white in early stages of development and becomes tan as it matures and finally forms strands of brown pigmented hyphae. Hyphae measure 6–9 µm wide; have clamp connections and numerous dikaryotic cells. The sclerotia resemble mustard seeds, are round to irregular in shape, and are generally tan on the outside.

Colonies of *S. rolfsii* are readily distinguished on plant material or on artificial media by their gross morphological characteristics. Rapidly growing, silky white hyphae tend to aggregate into rhizomorphic cords (Aycock, 1966). In culture, the whole area of a Petri plate is rapidly covered with mycelium, including the aerial hyphae which may also cover the lid of the plate. Both in culture and in plant tissue, a fan-shaped mycelial expansion may be observed growing outward and branching acutely (Takahashi, 1927).

Sclerotia may be spherical or irregular in shape and at maturity resemble the mustard seed (Taubenhaus, 1919; Barnett and Hunter, 1972; Mahmood et al., 1976; Boonthong and Sommort, 1985; Naidu, 2000; Mohan et al., 2000). Sclerotial size was reported to vary from 0.1 to 3.0 mm (Om Prakash and Singh, 1976; Anahosur, 2001).

At least two types of hyphae are produced (Aycock, 1966). Coarse, straight, large cells (2–9 x 150–250 µm) have two clamp connections at each septation, but may exhibit branching in place of one of the clamps. Branching is common in the slender hyphae (1.5–2.5 µm in diam.) which tend to grow irregularly and lack clamp connections. Slender hyphae are often observed penetrating the substrate (Aycock, 1966).

Sclerotia begin to develop after seven days of mycelial growth (Weber, 1931; Backman and Brenneman, 1984). Townsend and Willetts (1954) recognized four zones in the mature sclerotium: i) thick skin, considered to be the dried exudates giving a smooth appearance to the sclerotia, ii) rind of thickened cells, iii) cortex of thin walled cells, and iv) medulla containing filamentous hyphae. Sclerotia, formed on a host, tend to have a smooth texture, whereas those produced in culture may be pitted or folded (Paola, 1933).

Biology and Ecology

S. rolfsii prefers hot weather conditions (28–30°C). It occurs in tropical and sub-tropical areas of the world where the temperature and relative humidity are high for most of the time.

Mostly *S. rolfsii* diseases have been reported on dicotyledonous hosts, but several monocotyledonous species are also infected (Aycock, 1966; Mordue, 1974). Humid weather is favorable for sclerotial germination and mycelia growth. The fungus growing on plant surface kills the plant tissue by producing cell wall degrading enzymes and high amount of oxalic acid. In the process, the cell walls and middle lamellae are destroyed which facilitate infection. The pathogen survives in the soil for several years in plant residues in the upper 10 cm. The fungus starts growing on the senescent leaves which drop to the soil, and from there invades the base of the plant. The fungus spreads through the mycelium growing on the soil surface, or by mycelial fragments and sclerotia that are carried away by the surface water, or by mechanical means. The sexual stage of *S. rolfsii* is rare in nature and its role in the life cycle of the fungus is unknown. Genetic exchange in mycelia of *S. rolfsii* isolates is largely thought to be limited to mycelial compatibility (Nalim et al., 1995).

S. rolfsii is able to survive within a wide range of environmental conditions. Growth is possible over a broad pH range, though best at acidic conditions. The optimum pH range for mycelial growth is 3.0 to 5.0, and sclerotial germination occurs between 2.0 to 5.0. Germination is inhibited at a pH above 7.0. Maximum mycelial growth occurs between 25 and 35°C with little or none at 10 or 40°C. Mycelium is killed at 0°C, but sclerotia can survive at temperatures as low as –10°C. High moisture is required for optimal growth of the fungus. Sclerotia fail to germinate when the relative humidity is much below saturation.

Epidemiology

Collar rot of soybean is more common in sandy than in the clay soils. The pathogen perpetuates in the soil. Sclerotia produced during crop season remain viable in the soil and serve as a primary source of inoculum for initiating the disease in the successive crop. The sclerotia germinate when suitable moisture and temperature conditions are available and cause infection in young germinating soybean seedlings. Secondary infection is limited due to restricted movement of sclerotia and mycelium. Borkar (1992), while studying the influence of weather on the incidence of *Sclerotium* root rot, reported that 100% death of young plants occurred after a dry spell for three continuous days with a soil temperature between 29°C and 30°C followed by a shower of light rain.

High temperatures and moist conditions are associated with germination of sclerotia (Punja, 1985). High soil moisture, dense planting and frequent irrigation promote infection (Paola, 1933; Aycock, 1966; Clark and Moyer, 1988).

Many proteins are also known to be released from cell walls of infected soybean cv "*Hawkeye* 63" hypocotyls incubated with culture filtrate obtained from 4-week old cultures growing on an autoclaved mixture 1:1 (w/v) of fresh soybean hypocotyls and water. Incubation in 10 ml of culture filtrate at pH 4 in 0.1 M acetate buffer for 30 minutes at 30°C released many bound cell wall proteins. The mixtures of cell wall proteins were then subjected to either disk electrophoresis using isoelectric focusing in polyacrylamide gels in pH 3–10 gradients. The gels were photopolymerized with riboflavin rather than sulfate-polymerized. Tests of peroxidase and esterase activities were made. Best separation into discrete bands of peroxidase and esterase activities occurred using isoelectric focusing techniques. Fungal culture filtrate contained no detectable peroxidase, but the filtrate did have esterase activity (Curtis and Barnett, 1972).

When the fungus *S. rolfsii* is cultured by the use of sclerotia as inoculum repeatedly on medium containing 10^{-4} $HgCl_2$, its growth becomes inferior to the original strain in the first generation, but in the fourth generation it begins to grow better than the latter and becomes so resistant as to able to grow on medium containing 10^{-4} mol. $HgCl_2$. Also in the eight generation the fungus grows to the same degree as in the fourth generation, but such acquired resistance is instantly lost by return to a non-mercuric chloride medium. This strain is also resistant to $HgNO_3$, $CuSO_4$, and $CuCl_2$, and its growth is better than the original strain. Its pathogenic intensity on soybean is almost the same as the original culture (Kawase, 1955).

Enzymatic Activity

Up to 24% of the peroxide of purified cell walls of soybean hypocotyls was released by incubation of cell walls with hydrolytic enzymes secreted by the fungus *S. rolfsii* (Barnett, 1974). This estimation is based on comparison of peroxidase activity recorded in the medium with peroxides activity in unincubated cell walls. The peroxides release reaction occurs at 0°C. The peroxides release reaction occurs almost equally fast in the pH range of 3.5–8.0. The release of peroxides from cell walls cannot be attributed solely to arabanase, polygalacturonase or cellulase in the culture filtrate; although on Sephadex G-75 chromatography these activities overlap the peroxides releasing activity. Culture filtrate released less than 5% of the hydroxyl proline proteins of the cell walls.

Yield Loss

Collar rot disease caused by *S. rolfsii* has become a serious disease in most of the soybean growing areas, under heavy soil moisture. Sometimes more than 60–70% of the crop has been found damaged due to this disease. Borkar (1992) reported that the disease severely reduced the nodulation by rhizobium even during the early stage of wilting, thus affecting the yield considerably.

Charcoal Rot

Soybean is also one of the important hosts of the fungus *Macrophomina phaseolina* (Tassi) Goid. The fungus causes charcoal rot, dry rot and seedling blight disease (Reichert and Hellinger, 1947; Su et al., 2001) in soybean. The fungus is primarily soil inhabiting but is also seedborne in many crops including soybean. It survives in the soil mainly as microsclerotia, which germinate repeatedly during the crop growing season.

Distribution

The occurrence of charcoal rot disease of soybean caused by *Macrophomina phaseolina* was reported for the first time from Bermuda, in the year 1927 (Waterson, 1939). Thereafter, its occurrence was recorded on soybean in the southern Rhodesia during 1930 (Hopkins, 1931), whereas its occurrence on infected soybean in the field test in India was reported in 1936 (Likhite, 1936). The disease is now common in the states of Madhya Pradesh, Maharastra, Rajasthan and Delhi (Agarwal, 1973; Agarwal et al., 1973; Gangopadhyay et al., 1973; Gupta and Chauhan, 2005).

Charcoal rot disease is currently wide spread and is known to occur on soybean in North and South America, Australia, Asia, Europe and the African continents (McGee, 1991; Vishunavat, 2003). In the USA, it usually occurs in its states—Missouri, Mississippi, Alabama, Illinois and Indiana (Moore, 1984; Wyllie, 1988).

Symptoms

The disease symptoms vary, depending on the growth stages at which the soybean plant is infected. Usually, charcoal rot develops later in the season but it can cause seedling disease as well, since the roots of the plant can be infected at any time during the soybean-growing season.

On seedlings: Seedlings can be infected in those exceptional years when soils are dry and soil temperature is continuously above 35°C for 2–3 weeks. After emergence, symptoms can be visible on cotyledonary leaves as brown to dark spots. Sometimes, the margins of the cotyledonary leaves become brown to black, and such leaves fall at an early stage. From the unifoliate leaf stage onwards, the symptoms appear on emerging hypocotyls of infected seedlings as circular to oblong, reddish brown lesions that may turn dark brown to black after several days. These lesions may extend up into the stem.

On the adult plant: The pathogen causes lesions on the roots, stem, pods and seeds. From the ground level upward, brown to gray superficial lesions infrequently appear on the stem. Microsclerotia are formed in the vascular tissue and in the pith, giving a grayish black appearance to the subepidermal tissues of the stem. Such discoloration is first visible at nodes as profuse small, black, randomly distributed specks. The pathogen also produces ashy stem blight symptoms (Dhar and Sarbhoy, 1987). A reddish-brown discoloration of the vascular elements of roots and lower stem preceeds the premature yellowing as the fungus spreads up the stem during the season. The fungus enters the internal tissues and kills the vascular tissues resulting in the yellowing of leaves and untimely death of the plants (Fig. 14.2a, b). Irregular, big gray to black colored lesions are formed on the stem surfaces of affected plants producing the characteristic charcoal rot disease symptom. Large number of dark black colored sclerotia and flask shaped pycnidia are produced on these lesions which can be seen readily when the epidermal tissue of the lower stems and roots are peeled off from the affected parts (Fig. 14.2c). The ostioles of these pycnidia protrude out from the plant surface. When infection occurs on mature and dry pods they are covered with locally or widely distributed black bodies (microsclerotia).

Biology and Ecology

The sclerotial stage of the causal fungus is described as *Rhizoctonia bataticola* (Taub) Butler, while the pycnidia producing stage is referred to as *Macrophomina phaseolina* (Tassai) Goid. The perfect stage of the fungus has been described as *Orbilia obscura*by Ghosh et al. (1964). Individual

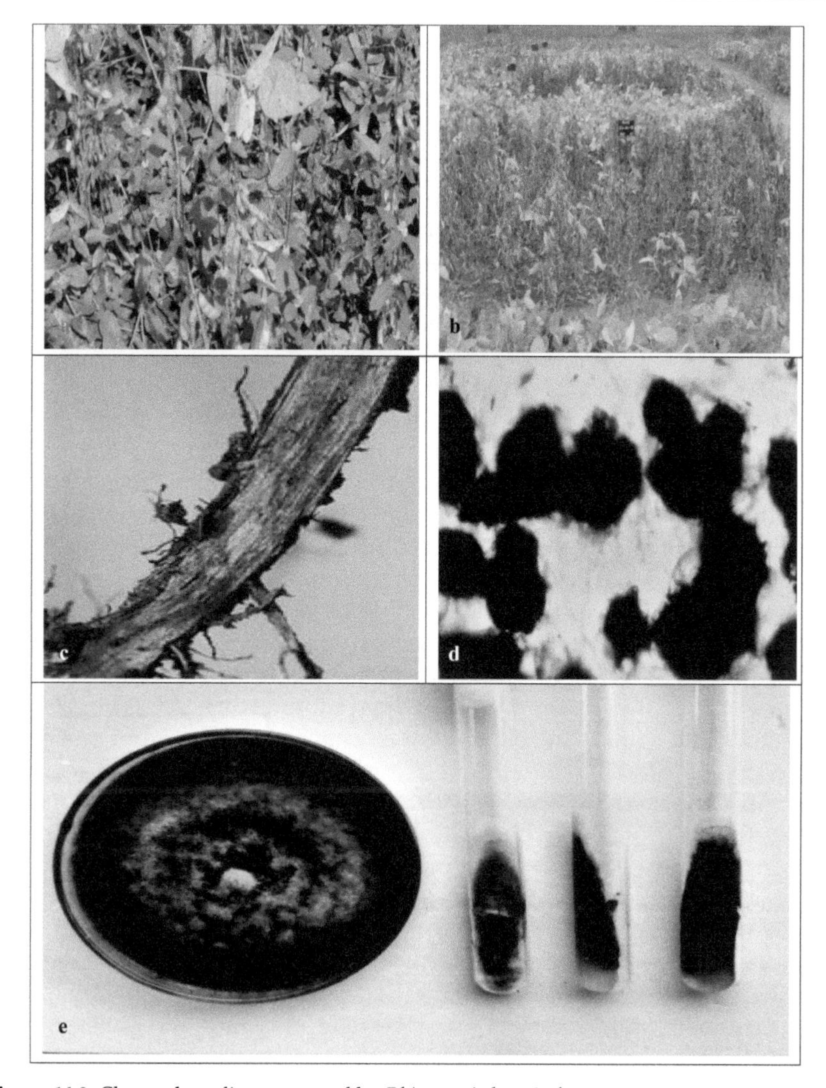

Figure 14.2. Charcoal rot disease caused by *Rhizoctonia bataticola*.

a. Crop affected with charcoal rot disease showing the yellowing, drooping, wilting and drying of affected plants
b. A susceptible cultivar showing complete drying and mortality of the plants infected with *R. bataticola*
c. Longitudinally splitted root of the plants infected with *R. bataticola* showing black discoloration, peeling of the epidermal layer and large number of pin-head size, black colored sclerotia of the pathogen
d. Black colored sclerotia of various shapes and sizes produced in the culture of *Rhizoctonia bataticola*
e. Mycelial growth and sclerotia of *Rhizoctonia bataticola* produced in culture medium on Petri-plates and also on slants

microsclerotia are made up of 50–200 or more individual cells united by a septal pore in each cell and can germinate repeatedly during the crop-growing season. The microsclerotia maybe of variable sizes (50–150 µm) depending on the available nutrients in the substrate (Fig. 14.2d, e) on which the propagules are produced (Short and Wyllie, 1978). The fungus sometimes also produces pycnidia in host crops (Mihail and Taylor, 1995), but their importance in the epidemiology of the fungus depends on the host species involved as well as the fungal isolate (Ahmed and Ahmed, 1969). Pycnidia are black, flask shaped with well developed ostiole.The pycnidiospores are hyaline, ellipsoid to obovoid, and measure (16-)20–24(-32) x (6-)7–9(-11) µm. Variability in the morphology, biology and pathogenicity has been reported amongst the isolates/strains isolated from various hosts and locations (Srivastava and Dhawan, 1980).

Growth of the fungus in soil is fueled by nutrients stored in the microsclerotia and continues till soil nutrient levels are insufficient for fungal competitors. Because of this characteristic, the pathogen competes well with other soil pathogens, when soil nutrient level is low and temperature is above 30°C. Microsclerotia germinate from a few cells at a time on the surface of or in close proximity to roots. Roots exudates induce germination of microsclerotia and thus help in infection of host roots. On the root surface, numerous germ tubes from microsclerotia are formed which give rise to appressoria on the anticlinal walls of the epidermal cells. These appressoria penetrate the epidermal cell wall by mechanical pressure and enzymatic digestion or through wounds and natural openings (Hartman et al., 1999). Ammon et al. (1975) observed that within three days of inoculation, appressoria are produced by the tips of primary hyphae on the root surface. During the initial stages of pathogenesis, the mycelium penetrates the root epidermis of mostly 7–42 days old soybean plants and is primarily restricted to the intercellular spaces of the cortex of the primary roots. Consequently, adjacent cells collapse and heavily infected plantlets may die. At flowering, the fungal hyphae grow intercellularly through the xylem of lateral roots, primary roots, and the lower part of the stem, and form microsclerotia that plug the vessels (Wyllie, 1988; Mayek-Perez et al., 2002), thus disrupting the host cell functions. At this stage of infection, the infected plants show necrotic lesions on stem branches and peduncles.

In soybean, formation of microsclerotia is triggered by flowering and pod setting (Wyllie and Calvert, 1969) and maybe indicative of initiation of death of the host (Short and Wyllie, 1978). Disease results into poor seed setting, reduced seed size, and finally to yield loss. Generally, disease manifestation and yield losses are more at higher temperatures, soil moisture stress and under light soil situations (Srivastava and Dhawan, 1979). After plant death, colonization by mycelia and formation of microsclerotia in host tissue continues until tissues have dried. After decay of roots and the plant debris, microsclerotia are released into the soil and the disease cycle continues.

Yield Loss

The estimated yield loss due to charcoal rot in soybean in top 10 soybean producing countries during 1994 was 1.234 million metric tones (Wrether et al., 1997). In other reports, average annual losses were estimated to be 5% in the Missouri state (USA) with some growers experiencing 30–50% loss (Wyllie, 1988). Charcoal rot was responsible for greater losses in soybean in comparison to other diseases known from central Mississippi and Alabama to central Illinois and Indiana (Moore, 1984). In India, epiphytotics occur in areas where temperature ranges from 36–40°C during the cropping season. During the 1997 season, charcoal rot caused substantial losses to plant stands and yield in the Guna district of the state of Madhya Pradesh (M.P.) (Gupta and Chauhan, 2005). And depending on the temperature and soil moisture status, the disease had been inflicting serious losses to the crop in certain localized pockets of Bhopal, Sehore, Hoshangabad and Harda districts of M.P. (Fig. 14.2b) and Amravati, Akola districts of Maharashtra.

Anthracnose

Soybean anthracnose caused by *Colletotrichum truncatum* (Schw.) Andrus and W.D. Moore is one of the most important seed-borne fungal pathogen of soybean (Sinclair and Backmam, 1989). This is the most common species associated with this disease, but several other species of *Colletotrichum* have also been identified to be involved. Anthracnose caused by *Colletotrichum capsici* was also reported on soybean as a new host from India (Sharma et al., 1992), the fungus produces crowded, black acervuli on infected tissues. These turn into dark bodies looking much like "pin cushions" on the plant surface (Fig. 14.3d, e). Acervuli produce many spores (conidia) through which, the pathogen spreads. The fungus overwinters in infected crop debries and has been shown to be seed-borne.

Distribution

The first report of anthracnose leaf spot caused by *C. dematium* f. *truncatum* was from Korea in the year 1917 (Nakata and Takimoto, 1934). Subsequently, Hara (1930) gave the descriptive account and Hemmi (1920) provided the full description and the morphology of *C. glycines* (*C. dematium* f. *truncatum*) on soybean from Japan.

Figure 14.3. Anthracnose disease of caused by *Colletotrichum* sp.

a. Crop affected with anthracnose disease exhibiting fire like appearance of dried affected plants
b. Leaves infected with *Colletotrichum* sp. showing the veinal necrosis symptoms of anthracnose on undersurface
c. Leaves infected with *Colletotrichum* sp. showing the burning symptoms of anthracnose
d. Plants infected with anthracnose showing the drying of stem and branches and small black colored spore producing fruiting bodies on infected tissues
e. Anthracnose symptoms on pods showing dot-like black colored acervulli of the *Colletotrichum* sp. arranged in concentric rings
f. Spore-producing acervulli of *Colletotrichum* sp. with long pointed needle like setae

Anthracnose (*C. dematium* f. *truncatum*) was reported to cause a black spotting on the pods of soybean in North Borneo (Johnston, 1960) and leaf drop and seed rot under prevailing wet weather condition in Java (Goot and Muller, 1932). Ling (1948) reported *Colletotrichum* as a parasitic fungus on soybean from Szechwan, China. Han (1959a) noticed the presence of *Colletotrichum glycines* (= *C. dematium* f. *truncatum*) on soybean in Taiwan.

Wolf and Lehman (1924), from North Carolina, reported that the disease is seed-borne and that *Colletotrichum glycineum* (= *C. dematium* f. *truncatum*) is distinct from *Glomerella cingulata*.

Colletotrichum glycines on soybean has been reported world over and was documented in Delaware (Manns and Adams, 1929), Jamaica (Martyn, 1942), Canada (Conners and Savile, 1943; Creelman, 1965), Florida (Rhoads, 1944), Georgia (Weimer, 1947), Iowa (Crall, 1951), China (Ling, 1948) and Taiwan (Han, 1959a), North Carolina (Johnston, 1960), former Soviet Union (Nelen, 1968; Skripka et al., 1986), Iran (Zad, 1979), Hungary (Ersek, 1979), Cameroon (Bernaux, 1979), Senagal (Girard, 1979), Yugoslavia (Jasnic, 1983), Bangladesh (Mridha et al., 1984; Fakir and Mridha, 1985), Indonesia (Kobayashi and Zenno, 1984), Nepal (Manandhar and Sinclair, 1982), Malaysia (Nik and Lim, 1984), the Philippines (Kobayashi and Zenno, 1984), Pakistan (Mirza and Rehman, 1985), Mozambique (Vernetti et al., 1985), UAE (Burhan, 1986), Italy (Calzolari et al., 1987), Ghana (Twumasi et al., 1989), South Africa (Koch et al., 1989), Gabon (Ndzoumba et al., 1990), Slovakia (Ondrej, 1994), Nigeria (Amusa, 1994), Argentina (Pioli et al., 1997), India (Poharkar and Raut, 1997), Bulgaria (Kaiser et al., 1998), Turkey (Eken and Demirci, 2000) and Australia (Parbery and Lee, 1972).

The existence of fungus *C. truncatum* which causes the pod blight of soybean was reported for the first time from India by Nene and Srivastava (1971), and subsequently its occurrence was recorded from the other parts of the country by several workers viz., Singh et al. (1973) and Saxena and Sinha (1978). It was also reported by Singh and Shukla (1987) from Uttar Pradesh; Saikia and Phukan (1983) from Andhra Pradesh; Bhardwaj and Thakur (1991) from Himachal Pradesh; Banu et al. (1990) from Karnataka; Nicholson and Sinclair (1973); Singh (1993) from Madhya Pradesh; Rao et al. (1989) from Maharashtra; and Singh and Srivastava (1989) from Rajasthan.

Symptoms

Infected plant parts like leaves, stems and pods may not show any symptoms in the early stages. At an advanced stage of disease, black fruiting bodies (acervuli) with minute black spines (setae) are visible even with the necked eyes and are the diagnostic (Figs. 14.3d, e).

Pre-mature defoliation may occur throughout the canopy when cankers girdle the leaf petiole leading to drying of the plants (Fig. 14.3a). Infected plants may be shorter than the healthy ones. Morgan and Johnson (1964) gave a descriptive account of the cankers that girdled petioles and caused leaf blades to shed, leaving only the shriveled petioles attached. The petiole lesions resemble target spot but occur earlier. Leaf rolling is also noticed.

Melhus (1942) reported that the infected plants get stunted at apical region; tip of the stem gets curled and brown upper leaves show small necrotic lesions (Figs. 14.3b, c). Sometimes, the leaves exhibit veinal necrosis on the under surface (Fig. 14.3b) leading to yellowing and drying of foliage (Fig. 14.3a). Rhoads (1944), and Parbery and Lee (1972) also observed poor seed germination and cankers on cotyledons due to *C. dematium.*

Infection at an early stage may cause no seed formation, or if seeds develop, they may be smaller and fewer in number. Infected seeds are shriveled or moldy and can develop brown irregularly stained, gray areas with black specks or may not show any symptoms. The fungus is confined at first to the seed coat. Such seeds may die during germination or if they germinate, they may produce infected seedlings. Infected seeds may also lead to pre- and post-emergence damping-off. Cotyledons may have the dark-brown, sunken lesions which may gradually extend up to epicotyls and radical and may become water-soaked.

Seed-borne inoculum of *C. truncatum* is responsible for three types of infection on soybean: (i) pre-emergence killing (ii) seedling blight and (iii) symptoms-less establishment of internal mycelium. At the host maturity, under proper environmental conditions, the fungus may fruit abundantly on the stem and pods (Figs. 14.3d, e, f) (Tiffany, 1951).

Taxonomy

The genus *Colletotrichum* was first described as *Vermicularia* in 1790 (Sutton, 1992; Hyde et al., 2009). Thereafter, the genus *Colletotrichum*, characterized by hyaline, straight or fulcate conidia and setose acervuli, was established by Corda (1831). Later, Arx von (1957) studied the taxonomy of this genus carefully and reduced the number of described taxa from several 100 to 11 accepted species. Later on, however, the number of accepted species of *Colletotrichum* were increased to 39 (Sutton, 1992), 60 (Kirk et al., 2008) and 66 (Hyde et al., 2009) by different workers. But, according to Nguyen (2010), the taxonomic position of some species is still unclear. Several species of *Collecttrichum*, i.e., *C. gleosporiodes, C. acutatum, C. graminicola* and *C. dematium* are broadly defined and considered to be species complexes or group species. The currently defined species boundaries are vague and relationships within some of these species complexes are not well-resolved (Sutton, 1992; Cannon et al., 2000). However, DNA sequencing of fungal genomes can be of immense value in assisting species identification (Sreenivasaprasad et al., 1996; Farr et al., 2006; Cannon et al., 2008; Cai et al., 2009; Hyde et al., 2009) and, therefore should be used to complement the morphological data.

Biology and Ecology

Transmission of *Colletotrichum truncatum* is achieved through the infected soybean seeds. The pathogen was recorded from anthracnose lesions on seedlings grown in sterilized soil from naturally infected seeds (Khare and Chacko, 1983) and also from inoculated seeds grown in sterilized sand under greenhouse conditions (Dhingra et al., 1978; Roy, 1982). The pathogen has also been detected in the cotyledons and cortex of the stem that subsequently advanced longitudinally to the pod and cotyledons of the developing seeds (Tiffany, 1951). Symptomless-infected seedlings developing in unsterilized soil in the greenhouse from seeds inoculated with a conidial suspension of *C. truncatum* are also reported.

The perfect stage of the fungus, *Glomerella dematium* (Hori) Lehman & Wolf, occurs on decaying stems left in fields. The fungus also survives on soybean stem residues (Lehman and Wolf, 1926). The ascogenous stage of *C. truncatum* was found on the diseased stems of soybean and also in culture (Lehman and Wolf, 1926). Isolates of *C. truncatum* from soybean seeds produced microsclerotia in culture and in soybean tissues (Khan and Sinclair, 1992). The black colored, erumpent acervuli formed on the infected host surface of stems, branches and pods (Figs. 14.3d, e) have black needle shaped pointed setae of 3–8 X 60–300 µm protruding out from the host surface (Fig. 14.3f). Single-celled, hyaline, orbiculate, 3.4 X 18–30 µm conidia are produced inside these acervuli.

The studies on the effect of environmental conditions on the development of anthracnose revealed that maximum disease incidence is noticed in the second fortnight of September and the first fortnight of October when the average temperature, relative humidity and rainfall were 28.4°C, 76% and 92.5 mm, respectively, whereas minimum disease incidence has been recorded in the second fortnight of July when higher temperature coupled with lower relative humidity and rainfall prevailed (Singh et al., 2001).

The role of light, temperature and relative humidity on the germination of *C. truncatum* and soybean pod infection studied under laboratory conditions in India (Kaushal et al., 1998) indicated that the optimum temperature for conidial germination and germ tube elongation was 20°C, and for soybean pod infection was 25°C. Three hours of light followed by nine hours of dark were the best for spore germination and germ tube elongation. Twelve hours of light followed by 12 hours darkness were most suitable for pod infection. Pod infection and the development of acervuli took longer time in continuous light. High (100%) relative humidity was required for pod infection and the development of acervuli.

Disease cycle initiates with the infection from the mycelium of the pathogen in infected seeds or debris. The fungus after germination produces numerous small, deep seated lesions on cotyledons, keeps on multiplying and moving into the developing seedlings to cause post-emergence damping off of seedlings under humid conditions. Alternatevely, mycelium may also establish itself in the infected seedlings without the development of symptom until plants begin to mature. Conidia produced in acervuli on infected plant parts under favorable conditions may initiate secondary infection by producing appressoria after germination. It takes about 60–65 hours for the symptoms to develop (Kuo et al., 1999).

Pathogenicity

Seed viability and vigor are the two most important characteristics of seed quality for propagation of soybean. Severe seed infection by *C. truncatum* may be able to inflict considerable damage to seeds after harvest, consequently posing a serious problem to the economy of this crop in the world trade. Begum et al. (2008) observed that under the glass house conditions, *C. truncatum* was highly pathogenic at seed and seedling stages. The fungus reduced seed germination significantly (46.4%) as compared to the control. The highest frequency of pre-emergence damping-off (48.0%) was observed in *C. truncatum* inoculated seeds compared to 3.0% in control seeds. In case of post-emergence damping-off, 28.5% of seedlings die within 14 days after sowing. In these experiments, the infected seeds and seedlings gave positive re-isolation of *C. truncatum* on PDA plate thus confirming its pathogenicity. Seedling survivability was also reduced by 75.8% in the inoculated seed treatment (Begum et al., 2008).

Begum et al. (2008) reported that the incidence of *C. truncatum* infection was much more prevalent in seed coat followed by cotyledons and embryonic axes without any external symptoms during incubation period. Thus, the appearance of the disease on seedlings could be attributed to the latent infection of *C. truncatum* into soybean seed coats (Sinclair, 1991). Germination was reduced (29.2%) by *C. truncatum* in soybean seeds as compared to the uninoculated seeds (control). Elias (2006) reported that the low viability refers to low germination which is a well known indicator of seed deterioration. However, Hepperley et al. (1983) and Srichuwong (1992) noticed that the *C. truncatum* decreased the seed viability and consequently reduced germination because of its infection in seed tissues and thus damaging the seed coat, cotyledons and embryonic axes.

C. truncatum significantly reduced seed germination by 46.4%, viability by 26.8%, causing 48.0% pre-emergence damping-off and 28.5% post-emergence damping-off within 14 days of sowing (Begum et al., 2008). Significantly higher electrolyte leakage was found in inoculated seeds than those of uninoculated seeds, which may probably be responsible for low seed vigor of inoculated soybean seeds.

Yield Loss

The disease is wide spread in temperate soybean production zones but causes minor losses (Athow, 1987; Tiffany, 1951). However, losses are more severe in warm, humid regions. In Alabama, yields were reduced by an average of 19.4% for three cultivars compared to plots in which the disease was controlled by fungicide applications (Backman et al., 1982). Yield losses of 30% were attributed to anthracnose in Nigeria (Rheenen, 1975) and a survey in two states in Brazil detected the disease in 57% of the fields (Lehman et al., 1976). Anthracnose of maturing plants causes serious losses, particularly during the rainy period when shaded lower branches and leaves are killed (Fig. 14.3a). Yield losses up to 50% in Thailand and 100% in India have been reported due to anthracnose (Sinclair and Backman, 1989; Ploper and Backman, 1992; Manandhar and Hartman, 1999).

In 2006, yield losses of 2.54 million tons and 1.18 lakh tons were estimated as being caused by anthracnose in the top eight soybean producing countries of the world and in India alone, respectively (Wrather et al., 2010).

Rust

Soybean rust, caused by *Phakopsora pachyrhizi*, is the most serious disease of economic importance in the Orient. It is known to occur in continents like Asia, Europe, Africa, Australia and America. Of late, this has become an important disease of soybean world wide. The disease was ealier thought to be of minor importance and was known to occur in the low hills of U.P., West Bengal and North east regions of India, but now it is of great significance, because the losses in yield range from 20–100%. Furthermore, after 1993, the rust scenario took a serious turn by recurring every year and occupying larger areas of soybean cultivation. Hence, it has became imperative to know more about the disease and its eco-friendly integrated management.

Distribution

Soybean rust disease was first recorded in Japan in the year 1902 by Nakanishiki who identified the fungus as *Uredo sojae* (Yang, 1977). Later, Hennings (1903) confirmed the fungus as *Uredo sojae* on leaves of the wild grown soybean (*Glycine soja*) collected from Tosca province in Japan by

Yoshinaga. However, the rust fungus in the Western Hemisphere on *Vigna* was reported in 1891, which was referred to as *Uromyces vignae* (cited by Sinclair and Backman, 1989).

From India, a report of soybean rust came as early as in 1906 from Poona on a host, which was wrongly identified as *Glycine hispida* (Sydow et al., 1906). But, Butler (1953) clarified that the host was actually a species of *Mucuna* and, therefore, the report of rust on *Glycine* sp. was incorrect. Sathe (1972) re-examined the above specimen and concluded that the rust pathogen on *Mucuna* sp. was actually *Uromyces mucunae* and *U. sojae*. But soybean rust in India was reported from Madras by Ramakrishnan (1951) who named it *Uromyces sojae*.

The name *Phakopsora sojae* was first given by Sawada as early as in 1933, which was subsequently renamed as *P. pachyrhizi* by Hiratsuka (1935, 1936). The first authenticated report of rust in India was available from Pantnagar in 1970 (Sarbhoy et al., 1972). Later on, this rust was seen at low hills of U.P. and Kalyani in Bengal (Singh and Thapliyal, 1977). Till 1974, the rust remained restricted in and around Pantnagar and subsequently disappeared from India. However, in 1980, after a lapse of almost five years, the rust reappeared at the high altitudes of Meghalaya and later on at the plains of Assam in the Northeast region of India (Maiti et al., 1981; Sharma, 1990). Till 1993, the rust remained confined to certain regions but afterwards it spread to other soybean growing areas of the country and now it is widely observed in various states of India like M.P., Maharashtra, Karnataka, Rajasthan, Andhra Pradesh, Tamil Nadu, Kerala and Himachal Pradesh.

Symptoms

Soybean rust symptoms are similar for both, *Pakopsora pachyrhizi* and *P. meibomiae* infections. Symptoms begin on the lower leaves of the plant as small lesions that increase in size and change from gray to tan or reddish brown on the undersides of the leaves (Fig. 14.4b). Although, lesions are most common on leaves but may occur on petioles, stems and pods as well. Thus, soybean rust produces two types of lesions; tan and reddish brown. Tan lesions, when mature, consist of small pustules (uredinia) surrounded by slightly discolored necrotic areas with masses of tan spores on the lower surface. Reddish brown lesions have larger reddish brown necrotic areas with a limited number of pustules (uredinia) on the lower leaf surface (Fig. 14.4c). Once the pod set begins, infection can spread rapidly to the middle and upper leaves of the plant. The pustules or urediosori formed on the stem and petioles are usually larger than those formed on leaves. The urediosori are light brown to brown in color having a diameter of 100–200 nm.

Environmental conditions have impact on the incidence and severity of soybean rust. Prolonged leaf wetness combined with temperature between

Figure 14.4. Rust disease caused by *Phakopsora pachyrhizii.*

a. Crop affected with rust disease exhibiting drying of affected plants
b. Pustules of the rust fungus on the under surface of the leaves
c. and d. Enlarged view of the rust pustules on leaves

15°C and 30°C and humidity of 75–80% is required for spore germination and infection. Under these conditions pustules appear within 5–10 days and spores are produced in 10–21 days. High level of infection in soybean field results in a distinct yellowing and browning of field and commonly, premature senescence of plant leaves (Fig. 14.4a).

Causal Organism

The causal fungus was first reported as *Uromyces vignae* in 1891 by Bresadola in the Western Hemisphere (cited by Sinclair and Backman, 1989) and as *U. sojae* by Hennings (1903) in the Eastern Hemisphere, obviously because of the close resemblance of its uredospores with those of *Uromyces*. In 1906, Sydow et al. (1906) recorded the occurrence of telial stage. Later on, Sydow and Sydow (1914) recorded the pathogen on *Pachyrizus erosus* (L.) Urban and named it as *P. pachyprhizi*. However, the fungus continued to be referred by different names even after Arthur (1917) transferred *Uromyces vignae* of Bresadola to *Phakopsora vignae*. Sawada (1928) described soybean rust pathogen as *P. sojae*. The controversy seems to have ended after Hiratsuka (1932) adapted the name *P. pachyrhizi* for soybean rust fungus and treated *P. vignae* and *P. sojae* as synonyms. The fungus produces hyaline or yellow to brown colored, round to elliptical 15–25 X 15–34 nm urediospores inside the urediosori. They are surrounded by clavate, hyaline or straw colored paraphyses bending inwardly, which remain united at the base forming a dome-shaped structure around urediosori. The surface of the urediospores may be minutely echinulated and may change their color from dirty white, yellow, pink or brown depending upon the age of the host plant, relative humidity, light and temperature of the atmosphere.

Another species, *Phakopsora meibomiae* was found to be the causal organism of soybean rust in south of North America, Carribean and South America down to Argentina. However, *P. pachyrhizi* is more aggressive than *P. meibomiae*.

Epidemiology

The epidemiology and disease cycle of *P. pachyrhizi* was studied by several workers. The maximum infection was found at 20–25°C with 10–12 hours dew period (Kitani and Inoue, 1960; Hsu and Wu, 1968). Marchetti et al. (1975) and Kochman (1979) also studied the effect of temperature on disease development and survival of the uredospores. They found that the maximum disease development occurred under a temperature regime of 17–27°C and the germination was significantly reduced at the temperature range of 28.5–42.5°C for eight hours. Earlier, Ilag (1977) reported normal germination between 10–20°C, however, the germ tube remained shorter between 5–15°C and attained normal length between 15–30°C. The germ tube attained the maximum length under 15 hours of darkness and nine hours of light at 20°C. Kumar and Verma (1985) reported that the uredospores stored in polythene bags at 18–25°C gradually lost their germinability in 15 days. However, when the spores were suspended in 1000 ppm sucrose solution, significant increase in germination percentage was observed. Melching et al. (1979) compared the four rust isolates and found that all the four cultures produced lesions in seven days and secondary uredospores in nine days. However, the Indian cultures produced more lesions per unit leaf area. The mean lesion areas on upper and lower leaf surfaces were similar in Indian, Taiwanese and Indonesian cultures but were smaller in the Australian culture. The number of uredia per lesion was also less in Australian culture. However, the uredospores of all the cultures were similar in length and width as well as in germination potential. The virulence and aggressiveness of the four rust isolates on soybean were also compared by Bromfield et al. (1980).

Sharma (1990) studied the effect of host age on the incidence and severity of rust and indicated that generally the disease appeared in the first half of September, irrespective of the host age. However, its appearance was delayed on the August sown crop, which indicated that climatic factors like poor rainfall might also determine its time of appearance of the disease. However, by early October, the disease became much more severe on both May- and June-sown crops.

The host pathogen relationship was studied by Bonde et al. (1976) and they found that the uredospores started germinating within two hours after inoculation in a dew chamber at 20°C. Appressoria developed within two hours and the earliest penetration was observed in seven hours. The penetration occurred directly through the cuticle with the help of a trans -epidermal vesicle, which subsequently emerged from the invaded cell and formed intercellular mycelium in the mesophyll. Mc Lean and Byth (1981) observed that the number of uredospores germinated on the leaf surface, number of appressoria formed, and the number of host cell penetrated, differed among susceptible and resistant soybean cultivars. Bonde and Brown (1980) examined isolates from Australia, India, the Philippines, Taiwan and Puerto Rico on the cv. *Wayne*. They observed that the isolates were indistinguishable in their pre- and post-penetration colonization phase and morphology of uredia. The only difference that could be observed was in the germpore appearance, which may be due to thinner germpore plugs. Morphological differences between the anamorphic and teleomorphic stages of the rust fungus were based on layering of telia and the wall thickness of teliospores, which were studied by Ono et al. (1992). However, the molecular differentiation of the *P. pachyrhizi* and *P. meibomiae* was reported for the first time by Frederick et al. (2002), which was based on four and two sets of polymerase chain reaction, respectively.

Reports of the telial stage have contributed to a better understanding of the taxonomy of *Phakopsora pachyrhizi* but their role in the life cycle of the rust fungus and the epidemiology of the disease is still unknown, though attempts to induce their germination have succeeded (Koch and Hope, 1987). They could create appropriate conditions under which the teliospore could germinate and form basidia and basidiospores in the laboratory. Although, the inducive treatments closely resembled the environmental conditions in the field, which may suggest that the teliospores might be giving rise to basidiospores in nature also, yet, information on the infectivity of basidiospores and the host(s), they can infect, is still lacking. Without this information, the life cycle of *Phakopsora pachyrhizi* will continue to remain incompletely understood.

Host Range

Phakopsora pachyrhizi has an extremely wide host range since it infects a large number of dicotyledonous plants for example, the common bean (*Phaseolus vulgaris*) and wild soybean (*Glycine usseriensis*), both in the field and in the laboratory (Yang, 1977). Keogh (1974) has made a rather thorough study of the host plants indigenous to Australia. A list of the hosts of *P. pachyrhizi* Syd. is also available (Kitani and Inoue, 1960; Lin, 1966; Keogh,1974). However, the nine isolates were separated into six pathogenic groups differing mainly in their reaction types with and without sporulation or no infection on *Vigna unguiculata*, *P. vulgaris*, and *Pachyrhizus erosus*. Burdon and Speer (1984) established a set of differential *Glycine* hosts for the identification of *P. pachyrhizi*. Legume-species that have been identified as host of *P. pachyrhizi* in Thailand are: *Canavaria gladiata*, *Carjinus* sp., *Centrosema pubescens*, *Pachyrhizus rerosus*, *Vigna mungo*, *Dolichos lablab*, *Phaseolus aureus*, *P. lathyiodes*, *Pueraria thunbergiana*, *Pisum sativum*, and *Vigna simensis* (Poonpolgul and Surin, 1980).

Yield Losses

Soybean rust caused by *Phakopsora pachyrhizi* is recognized as the most destructive disease in both the Eastern and Western Hemisphere (Sinclair, 1977). In the Eastern Hemisphere, it is a major constraint to soybean cultivation, affecting the crop both in the tropics and sub-tropics (Yang, 1977). Though, reports of losses on account of rust range from 10 to 90%, under congenial conditions for the occurrence of disease (Gupta, 2004). Sharma and Gupta (2006) compiled the yield loss data of soybean rust caused by *P. pachyrhizi*.

Myrothecium Leaf Spot and Blight

The Myrothecium leaf spot and blight disease has become quite serious in almost all soybean growing regions of India (Srivastava, 2003) and some countries of Asia. Under favorable conditions of high relative humidity and temperature, it causes severe damage to the foliage entailing substantial yield losses.

Distribution

Bain (1944) reported the occurrence of *Myrothecium roridum* as an incitant of the leaf spot disease of cowpea and soybean from USA. Later on *Myrothecium* sp. was reported to cause the leaf spot on soybean (McIntosh, 1951). The disease has been reported from Illinois, USA (Schiller et al., 1978), Brazil (Almerida et al., 1980), Pakistan (Ali et al., 1995), Korea (Yum and Park, 1990), Malasiya (Nik, 1983). In India, the disease was first reported by Munjal (1960) and later described by Saksena and Tripathi (1971). The fungus was reported as intercepts on the seeds imported from USA (Lambat et al., 1969). The disease was reported from Jabalpur (Madhya Pradesh) by Lakshminarayana and Joshi (1978). Gradually, the disease became prevalent in soybean growing areas of Madhya Pradesh (Srivastava and Agrawal, 1989; Singh and Srivastava, 1994; Singh and Pandey, 1998; Singh et al., 1994; Bhale, 2002; Srivastava, 2003).

Symptoms

The symptoms of the disease have been reported almost from all parts of the plant. The first symptom appears on the cotyledonary leaves as light brown, rounded or irregular spots with reddish brown margins. Sometimes these spots increase in size covering the major portion of the cotyledons even causing their death or of the seedling itself.

The main symptoms are observed on the leaves as oval to circular light brown spots with violet margins finally becoming brown with reddish, dark brown margins (Fig. 14.5a) (Lakshminarayana and Joshi, 1978; Srivastava et al., 1995; Singh and Pandey, 1998). After some time these spots increase in size, coalesce with each other becoming irregular in shape with reddish brown margins. Some times concentric rings are also observed on these spots. Lower leaves of the plants get affected first and later on the upper leaves also exhibit identical symptoms. Under favorable conditions of continuous rains, high humidity and temperature, these spots spread rapidly covering major portions of the leaves resulting in the yellowing and ultimate drying of the foliage. After some time, minute, pin-head sized, cream or pink colored sporodochia of the fungus are produced on the suface of the spots. Usually, there is a single solitary sporodochium in the center of the spots, but some times, more than one sporodochium, arranged in a circle on these spots are also observed (Fig. 14.5b, d). With the lapse of time, these sporodochia turn black in color. There is a great degree of variation in the color, size and shape of the spots as well as in the number and arrangement of sporodochia on them.

Besides the leaves, fungus also affects the stems, branches, flowers and pods of soybean, producing similar types of spots but of different shapes and sizes (Fig. 14.5e). In case of severe infection, the affected plants exhibit drying symptoms before pod formation causing considerable reduction in the yield (Srivastava, 2003; Srivastava and Khan, 1997).

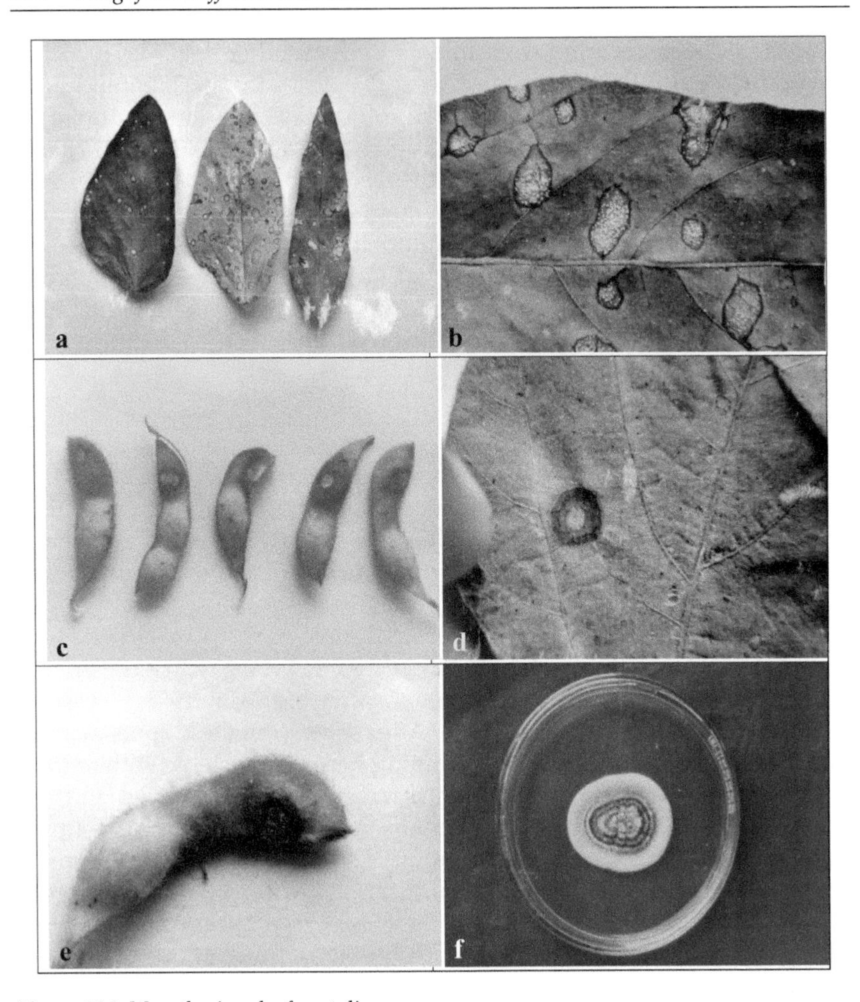

Figure 14.5. Myrothecium leaf spot disease.

a. Light brown colored spots of various shapes on leaves caused by *Myrothecium roridum*
b. Brown colored spots surrounded by reddish brown margins having large number of cream colored erumpent sporodochia of the causal fungus *Myrothecium roridum*
c. Brown colored spots of various shapes formed on the pods as a result of infection with *Myrothecium roridum*
d. Enlarged view of a spot caused by *Myrothecium roridum*
e. Dark brown colored sunken spot on the pod showing cream to black colored sporodochia of causal fungus *Myrothecium roridum*
f. Mycelial growth of colony of causal fungus *Myrothecium roridum* on PDA showing black colored sporodochia of the fungus arranged in concentric rings

Causal Organism

The causal fungus, *Myrothecium roridum* Tode ex Fries, is a polyphagus pathogen of cosmopolitan occurrence affecting a large number of plants. The genus *Myrothecium* is classified under the family Tuberculariaceae of the subdivision Deuteromycotina (Ainsworth, 1973). It was established by Tode (1790) and was adopted by Fries (1829) with four species, three of which he distinguished according to the nature of the disks (sporodochia) and the shape of the spores. Thus, *M. roridum* has swollen disks and cylindrical spores; *M. verrucaris* (Aib and Schwein) Ditmar has flattened disks and ovate spores; *M. inundatum* Tode has flat disks and globose spores. The fourth species, *M. scybalorum* (Schum.) Fr., is known only from the description of Schumacher (1803). However, Tode's generic description was amended by Link (1809) but definite description of the fungus *Myrothecium roridum* Tode ex. Fries was recognized for the first time by Saccardo (1886). Later on, Preston (1943) carried out a detailed study of the genus *Myrothecium* Tode and provided an elaborate account. He described *M. roridum* Tode ex Fries, *M gramineum* Lib., *M. jollymannii* Preston and *M. striatisporum* Preston as species of the genus *Myrothecium*.

The causal fungus, *M. roridum* is relatively slow growing, producing a white colony on PDA containing black dot-like sporodochia arranged in concentric rings (Fig. 14.1f). Sometimes, the sporodochia may be scattered all over the surface of the colony. Sporodochia are sessile, 0.1–2.0 (average 1.05) mm in diameter, about 0.05–0.1 mm deep, discoid or irregular in shape, initially light green in color turning to black with white raised margins. Conidiophores produced inside sporodochia are hyaline, erect, branched and septate, main axis tapering, 3–4 celled, cells about 10 X 1.5 µm, branches one or two celled, arising singly or in pairs or whorls either directly from the terminal cells of the parent axis or immediately below the septum of an intermediate cell, each branch terminating in a whorl of phialides. Phialides are slender, clavate, straight, hyaline, usually arranged in whorls of 3–7 at the apex of the main axis and its branches. Condia are cylindrical, slightly tapering with rounded ends, single celled, hyaline, changing to olive green to black when in mass, 5.0–11.5 X 1.5–2.5 (average 8.25 X 2.0) µm in size.

The soybean leaf diffusate have been reported to contain a phytoalexin, "Glyceolin". The biosynthesis of the glyceolin in leaves is supposed to be elicited by the mycelial wall extract of *M. roridum* (Purkayastha and Ghosh, 1983). The fungus is known to produce three toxins. Myrothecin, trichothecin and roridin are known to have been isolated from its culture filtrate (Nespik et al., 1961). A nicrotin-like metabolite have also been reported from its culture filtrate (Pawar and Thirumalachar, 1966).

Epidemiology and Disease Cycle

The disease intensity is favored by high relative humidity and temperature (Srivastava and Singh, 1973; Kuhne and Leupold, 1985). Higher levels of NPK nutrition is reported to favor the disease severity (Srivastava and Khan, 1995; Srivastava et al., 1988, 1991). The age and concentration of inocula have a direct bearing on the intensity of disease (Srivastava and Khan, 1996). The disease severity is usually high under higher plant population levels (Srivastava et al., 1995). The age of the host plant at the time of infection is also an important factor determining the intensity of disease (Srivastava and Khan, 1994).

The causal fungus survives both as sporodochia and mycelium in seeds, crop residues, and soil or on collateral hosts. Infected seeds result into seed and seedling rots, and cotyledonary spots (Singh and Pandey, 1998; Srivastava and Agrawal, 1989) which serve as sources of inoculum and cause fresh infections on trifloliate leaves. The sporodochia and conidia produced on the leaf surface are disseminated through air, rain water and rain splash causing secondary infections and increasing the severity of disease. The fungus travels to stems, branches, petioles and ultimately infects the pods and seeds.

Frog-eye Leaf Spot

Frog-eye leaf spot disease of soybean, caused by *Cercospora sojina* Hara, is world wide in distribution and causes substantial yield losses. Frog-eye leaf spot, as the name indicates, is primarily a foliar disease of soybean. Frog-eye leaf spot is most likely to become a problem if infected seed is planted or if the disease occurred in the previous year's soybean crop and the land is not rotated. Extended periods of wet weather during the growing season also favor disease development.

Distribution

Frog-eye leaf spot disease occurs in almost all soybean growing countries of the world and is one of the important foliar diseases of soybean. Hara (1915, 1930), for the first time, described the frog-eye spot caused by *Cercospora sojina* as a new leaf spot disease of soybean from Japan. Similarly, Adams (1933), Lehman (1928, 1934, 1942), and Sun (1958) recorded this disease from Australia, USA and China. Mann's and Adams (1933), Kornfield (1935), Tai (1936), Takasugi (1936) and Miura (1930) delineated the occurrence of the frog-eye leaf spot on soybean leaf, stem and pods from the USA, Europe, China and India, respectively. Muller and Chupp (1942) collected leaves infected with *C. sojina* from Venezuela. Reports of this disease are also available from Canada (Conners and Savile, 1944; Koch and Hildebrand, 1944). Fenne (1942, 1949) reported the disease from Virginia (USA). Thirumalachar and Chupp (1948) recorded from India the occurrence of *Cercospora sojina* on soybean that survived at an altitude of 3,000 ft. and 30 inches average rainfall. Mehta et al. (1950) observed this disease during November and December throughout Uttar Pradesh and also in the rest of India. Muller and Chupp (1950) and Muller (1950) found the disease at Chimaltenago, Guatemala. Weimer (1947) and Crall (1952) reported the occurrence of soybean disease caused by *C. sojina, C. kikuchii* and *Phakospora pachyrhizi* from Iowa, and some other states of the USA. Han (1959) surveyed the diseases of soybean in Taiwan and found that frog-eye leaf spot does occur there and in varying degrees of severity. The disease has also been known to be prevalent in Primorsk, Pacific coastal region of Russia (Mikhalenko, 1965; Ovchinnikova, 1968). Frog-eye leaf spot on soybean was reported as a new disease in Zambia (Javaid and Ashraf, 1978). Its occurrence (*C. sojina*) from all over China has also been noted (Feng and Wang, 2004).

During disease surveys in the five states of the North eastern region of India viz., Arunachal Pradesh, Meghalaya, Manipur, Mizoram and Nagaland, the occurrence of this disease was found with 22% yield loss (Gupta, 2003, 2004). Other Indian states where it has been recorded are: Himachal Pradesh, Madhya Pradesh, Karnataka and Uttaranchal.

Symptoms

Frog-eye leaf spot is primarily a foliar disease. However, infection also occurs on stems, pods and seeds. Minute, discrete spots, that are reddish-brown and circular to sub-circular or angular, 0.5–5 mm in diameter, first appear on the upper surface of the leaves (Figs. 14.6b, c). As the lesions enlarge and age, the central area becomes olive-gray or ash-gray and is surrounded by a narrow, dark reddish-brown border (Figs. 14.6b, c). There is no chlorotic zone surrounding the lesions. On the lower leaf surface, the spots are darker brown or gray. Conidiophores are dark, borne singly or in very dense fascicles in the center of each lesion, mostly on the underside of the leaf. Older spots become very thin, often paper-white, and translucent and sometimes due to removal of the dead tissues, shot holes are produced on the leaf lamina (Figs. 14.6b, c). Several spots may coalesce to form larger, irregular spots. When the lesions are numerous, leaves wither and fall prematurely.

Infection on stem is conspicuous but not very common and appears later in the season. Young lesions on stems are deep-red with narrow, dark-brown to black margins (Fig. 14.6a). The central area of the lesion is slightly sunken. As the lesions enlarge, the center becomes brown to light

Figure 14.6. Frog-eye leaf spot disease caused by *Cercospora sojina.*

a. Purple brown discoloration of apical region of the seedlings infected with *Cercospora sojina*

b and c. Characteristic symptoms of the frog-eye leaf spot disease seen as light brown colored spots surrounded by reddish brown margins and shot holes formed due to tearing away of affected tissues

d. Pods infected with *Cercospora sojina* exhibiting dark brown colored lesions on pods

gray, usually with a narrow, dark-brown border. Infected seeds develop conspicuous light to dark gray or brown areas varying from minute specks to large blotches that can cover the entire seed coat. Usually, there is some cracking or flaking of the seed coat.

Lehman (1934) recorded that the fungus was frequently observed on stems, pods (Fig. 14.6d) and seeds of soybean. In stems, the fungus is chiefly confined to the cortex and the injury to phloem and cambium is usually due to diffusion of the toxic substance from the necrotic cortex. In pods, the mycelium penetrates through the pod wall, entering the thin white membranes lining the pod and closely infecting the seeds.

Causal Organism

The causal organism of the frog-eye leaf spot disease was first diagnosed by Hara (1915), who described the fungus as a new species of *Cercospora* and named it as *Cercospora sojina*. Lehman (1928) first published a detailed report of the disease in the USA. However, he identified the fungus as *C. daizu* (= *C. sojina*). The fungus was primarily confined to leaves, rarely occurred on stems but never observed on pods. It caused minute reddish-brown spots on the upper surface of the leaves and greatly reduced the seed yield. The fungus was found perpetuating on diseased leaves, stems and seeds. Hara (1930) and Miura (1930) have given the descriptive accounts of the soybean pathogens including *C. sojina*. However, Chupp (1954) gave a complete mycological description of *C. sojina* occurring on *Glycine max* in a monograph on the genus *Cercospora*.

Detailed cultural and physiological investigations on the frog-eye leaf spot was done by Kurata (1960). *Cercospora sojina* has been found to be a highly variable fungus. Athow et al. (1962) reported a new physiological race of *C. sojina* from soybean cultivars *Clark* and *Wabash* other than *race* 1 and designated it as *race* 2. Soon, *race* 3 and 4 were found in North Carolina (Rose, 1968) and *race* 5 in Georgia (Phillips and Boema, 1981), which became a threat to soybean production in North eastern USA. Yorinori (1981) studied nine new isolates, identified seven new races, and designated them. He suggested some differential cultivars with *Bienville, Blackhawk, Bragg, Capital, Comet, Davis, Dorman, Flambeau, Hampton, Hood, Lee, Mandarin (Ottawa), Patoka, Tanner* and *Wabash*. Earlier, Vasudeva (1963) described the morphology of the causal fungus *C. sojina* and distinguished it from *C. cruenta, C. flagellifera, C. kikuchii,* and *C. glycines* in having cylindrical and wide conidia. Sinclair and Backman (1989) described that the conidia of *C. sojina* are 0–10 septate, hyaline and elongated to fusiform, tapering towards the tip when they are young. The conidia of *C. sojina* are much shorter (24–108 X 3–9 µm) than those of *C. kikuchii*. Generally, 1–3 but sometimes upto 11 conidia are formed on a single conidiophore. Conidiophores arise singly or in fascicles of 2–25 from a thin brown stroma and measure 52–120 X 4–6 µm. The fungus has isolates, which sporulate abundantly on V-8 juice agar and lima bean agar, but other isolates sporulate poorly on most media.

Pure culture of *C. sojina* was obtained on PDA (Sharma, 1990). However, it failed to sporulate on a variety of culture media including seed coat agar, soybean cotyledon and embryo agar, soybean leaf, carrot leaf and turnip leaf decoction agar. But, it sporulated on PDA after 10 months of incubation under ambient temperature (21°C to 28°C).

Seed Borne Nature

C. sojina is seed borne. Infected seeds exhibit light to dark gray or brown discolorations as specks to large blotches and in some cases cover the entire seed surface. Alternate light to brown bands develop on the seed coat. Sometimes, cracks are also formed on seed coat. Histopathology of infected soybean seeds exhibited the mycelium in aggregates in the hilar region, seed coat layers, in the space between seed coat and embryo. The cotyledons were not infected (Singh and Sinclair, 1985). The seeds get infected when in pods. The hyphae penetrate through the pores and cracks in the seed coats as well as through hilar tracheids. Plant to seed transmission of *C. sojina* is at the beginning of pod formation and at the pod filling stage (Lin et al., 1991). Production of cercosporin by *C. sojina* facilitates seed colonization by the fungus as it inhibits other seed borne fungi (Velicheti and Sinclair, 1992). That the pathogen is seed borne, has been described by Khare and his co-workers (Khare et al., 2000). They have also provided in detail the method of detection and isolation of the fungus from the seeds along with its management.

Disease Cycle

C. sojina is seed borne and can survive in the seed for upto three years. The fungus perpetuates as mycelium on infected plant debris for one year. These sources are responsible for the recurrence of the disease. Conidia, formed after initial infection, spread to short distances by air currents. Warm and humid weather results in heavy sporulation, and secondary spread of the disease takes place through long distance dissemination of conidia from one field to another.

Epidemiology

The lesions of frog-eye leaf spot appear, both on the upper and lower leaf surfaces; and when the spots become numerous, the leaf withers and falls prematurely causing great losses in grain yield. Lehman (1928), however, found lesions only on the leaves and stems. He reported that the disease was capable of causing great reduction in seed yields, with maximum injuries to late maturing varieties. He also found that the fungus overwinters on diseased leaves, stems and seeds. However, later Lehman (1934) observed the fungus on pods also. In pods the myecelium penetrates through the pod wall and closely infests the seeds. In stems, the myecelium remains chiefly confined to the cortex and the injuries to phloem and cambium were due to diffusion of toxic substances produced by the fungus in the cortex. He also found that the fungus remains superficially borne on the seeds. Lehman and Poole (1929) indicated that *C. sojina* might survive on moist diseased leaves until the next planting season. Ovchinnikova (1968) studied epidemiology of the disease in the Pacific coastal region of Russia. He noticed that the disease was more harmful in early stages of growth, during which the infected leaves fall and the plants became weak. The pathogen attacked all local cultivars and caused 8-fold reduction in seed yields. He also established that the over wintered leaves harbored conidia of *C. sojina*, which acted as the main source of inoculum. Inoculation experiments have also shown that the host lesions become visible after 9–12 days, and the first conidia were produced in these lesions within 24–48 hours (Sinclair and Backman, 1989). The conidia were disseminated by air currents and rain and caused secondary infection. Sufficient moisture was necessary for the infection and the leaves maturing during dry period often escaped infection. In the Jiamusi region of Heilongjiang province of China, the relationship between epidemiology of soybean frog-eye leaf spot and metorological conditions was studied by Huang et al. (1998). They found positive linear correlations between the yield loss and disease severity on leaves, pods, and seeds of resistant and susceptible cultivars.

Purple Seed Stain

Purple seed stain disease caused by *Cercospora kikuchii* (Matsomoto and Tomoyasu) M.W. Gardner, is one of the major diseases of soybean, and can be found wherever soybeans are grown. The fungus causes leaf blight and purple discoloration of the seed coat and adversely affects the germination and quality of the seeds. The disease is known by several colorful symptoms such as purple blotch, purple speck, and purple spot. Isolates of the fungus differ in the level of seed infection they produce and in their ability to infect foliage. *Cercospora kikuchii* overwinters as mycelium in the infected seeds and plant debris. The fungus is present in the seed coat and grows through the tissues to the cotyledons and into the stem. In case of mild seed infection, the seed coat may shed before infection can occur. The young plant may be killed at an early age or the fungus may become established but produce no symptoms. *Cercospora kikuchii* sporulates on the surface of debris and tissues may become colonized from infected seed during warm humid weather. Conidia are wind blown or rainsplashed. Secondary infections usually begin at flowering. Seeds become infected when the fungus invades the pod and grows through the adaxial vein to the hilum and eventually to the seedcoat. High rainfall and cool climate favor the pathogen to survive and rebuild for causing disease during the succeeding season.

Distribution

Suzuki first recorded the purple seed stain disease in the Orient in the year 1921 but he could not attribute it to any fungal organism and thought it to be caused by climatic factors. Matsumoto and Tomoyasu (1925) recorded the occurrence of the purple seed stain in the Western Hemisphere.

In the Eastern Hemisphere, Yoshii (1927) and Hara (1930) noticed the occurrence of *C. kikuchii* in Japan and Miura (1930) in Manchuria, China. Later, Takasugi (1936) and Nakata and Asyuama (1941) also recorded the incidence of *C. kikuchii* as well as *C. sojina* in China. Occurrence of this disease was recorded in Korea by Nakata and Takimoto (1934); in Borneo by Johnston (1960); and in New Guinea by Johnston (1961). In Taiwan, Ling (1948) and Liu (1948), while describing the assorted parasitic fungi on soybean, recorded the occurrence of purple seed stain (*C. kikuchii*) disease. Han (1959a,b) and Sawada (1959) also reported the occurrence of *C. kikuchii* on soybean and the former considered the purple seed stain (*C. kikuchii*) as a destructive disease of soybean in Taiwan. The first report of the outbreak of purple seed stain in India was published by Nene and Srivastava (1971). In the same year, Agarwal and Joshi (1971) reported the wide spread occurrence of the disease in Tarai area in Nainital district of U.P. (now in Uttaranchal) and they studied the effect of the disease on seed quality and germination. Reports of the occurrence of disease are also available from the North-eastern hills of India (Maiti et al., 1983; Vishwadhar and Chaudhary, 1982; Verma et al., 1988; Sharma, 1990, 2003; Sharma et al., 1988, 1989, 1993, 2000). Sharma et al. (2000) reported that the disease is prevalent in all the six North-eastern states of India in varying magnitudes and at different altitudes and agro-climatic conditions, right from the plains of Assam to the high hills of Meghalaya.

Besides Asia, the purple seed stain disease has also been recorded from America (Gardner, 1924, 1927, 1928; Petty, 1943; McNew, 1948; Haenseler, 1946, 1947; Johnson and Kilpatrick, 1953; Kilpatrick, 1955; Patino, 1967), Africa (Riley, 1960) and Europe (Lusin, 1960; Nagata, 1962). Do Amaral (1951) reported this disease from Sao Paulo in Brazil. The disease has also been recorded from Venezuela which was possibly introduced in the country with imported seeds (Diaz, 1966). Diaz Polanco and Casanova (1966) described the symptoms and other details of the disease. Litzenberger and Stevenson (1957) recorded this disease affecting soybean on the Pacific side of Nicaragua. In Europe, the disease has been reported from Yugoslavia (Lusin, 1960; Nagata, 1962). In Africa the disease was noticed in Tanzania by Riley (1960).

Symptom

The purple seed stain disease is also known as purple spot, purple blotch, purple speck or lavender spot. Symptoms of purple seed stain are quite conspicuous and are easily distinguished on the seeds. The seed discoloration varies from pink or a pale purple to dark purple and ranges from minute specks to large, irregular blotches, which may cover the entire surface of the seed coat (Fig. 14.7a). This pathogen also causes leaf blight,

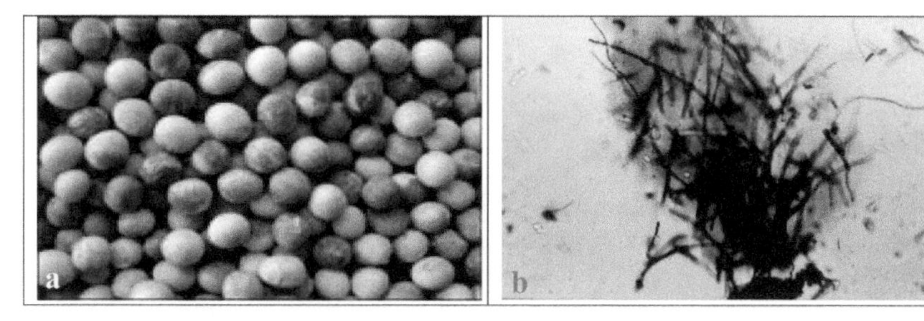

Figure 14.7. Purple seed stain disease caused by *Cercospora kikuchi.*

a. Purple discoloration of seeds affected with *Cercospora kikuchi*
b. Conidiophores and conidia of *Cercospora* sp. growing on seeds

which is characterized by reddish-brown to purple, angular to irregular lesions, which expand irregularly. On the lower side of the leaf, irregularly shaped reddish-brown to light purple blotches appear which cause necrosis. Sometimes their upper leaves appear to be light purple, leathery and dark. Blighting of younger upper leaves over large areas is a striking symptom of the disease. Veinal necrosis, defoliation along with early senescence of the plant is also very common.

Causal organism

The causal organism of purple seed stain disease of soybean was first identified in Japan as *Cercosporina* sp. by Kikuchi (1924) and Tomoyasu (1924). Matsumoto and Tomoyasu (1925) gave the complete diagnosis of the causal fungus and erected the new species *Cercosporina kikuchii* after Kikuchi who first encountered the fungus on soybean seeds in 1922. Diseased soybean seeds kept in damp Petri dishes support profuse sporulation of *Cercosporina* (= *Cercospora*) *kikuchii* at 15–20°C. Conidia of the fungus were also observed on the inner surface of infected pods by Matsumoto (1928). While studying the cultural characterstics, he observed that the fungus did not sporulate in culture media. However, the fungus sporulated on living seeds in laboratory at 18–27°C.

In a study undertaken by Sharma (1990), pure culture of *C. kikuchii* obtained on PDA from infected seeds, failed to sporulate under different incubation conditions. And also, the fungus did not sporulate on a variety of other culture media tried viz., soybean seed-coat agar, soybean cotyledon and embryo agar, soybean leaf decoction agar, carrot leaf decoction agar and turnip leaf decoction agar. White dense and uniform mycelial colonies on PDA slants turn grayish with age at the center. Hyphae from fresh culture are hyaline and septate turning pale brown with age. Conidia, from the host are, slender, straight to variously curved, truncate to subtruncate at the base, measuring 2.69–5.38 x 5.38–113.68 µm with 2.22 µm septa. Condiophores from the host are yellowish to dark brown and hyaline at the tip (Fig. 14.7b).

The growth of *C. kikuchii* was good on soybean seed coat medium and PDA (Crane and Crittenden, 1967). The fungus grows well on malt extract agar and PDA but without conidial formation, no growth was observed on water agar (Chen et al., 1979). Conidia in good numbers are formed on V-8 juice agar and carrot leaf decoction agar. In V-8 juice and potato dextrose broth, the fungus shows good growth.

Optimum temperature for the hyphal growth and spore germination is 20–30°C and optimum temperatures for disease development and sporulation on leaves, pods and seeds are 20 and 27°C, respectively. Germ tubes from conidia penetrate the leaf and pod through wounds or stomata, but the hyphae directly penetrate the cuticle of seeds (Fujita, 1990). Rate of spore germination, elongation of germ tubes, formation of infection structures and penetration are important steps in causing infection (Fujita and Suzuki, 1988).

Seed Borne Nature

In the infected seeds *C. kikuchii* mycelia are found to be equally abundant in the hourglass and parenchyma cell layers of the seed coat with a high concentration in the hilum region but sparse in the palisade layer (Ilyas et al., 1975). Occasionally the mycelium is found in cotyledonary and embryobic tissues. Sasaki (1982) observed first visible symptoms of the disease on cotyledons 10 to 15 days after sowing. The infection spreads to young stems, unifoliate leaves, trifoliate leaves, branches, petioles and finally to pods and seeds. Secondary infection of new tissues occurs simultaneously by conidia are produced on infected tissues also.

Chen et al. (1979) observed that hyphae found on the seed coat and the crevices of purple stained seeds grew into the cotyledons through seed coat pores during germination. Hyphae were also observed invading the ovary.

Soybean seeds infected with *C. kikuchii* exhibited degradation of seed coat proteins but not cotyledonary proteins (Velichati and Sinclair, 1992). Lepoxygenase is degraded only in seed coats infected with *C. kikuchii*. The infection increases free fatty acid content and protein and reduces oil content (Pathan et al., 1989). According to Agarwal and Joshi (1971), protein and oil contents were the same in healthy and diseased seeds.

Toxin Production

Cerospora kikuchii produces red, photoactivated, phytotoxic polyketide toxin cercosporin, which is a crucial pathogenicity factor in the development of leaf and pod blight. The gene LE6 is essential for cercosporin production behavior of *C. kikuchii*. Transcription of LE6 is regulated by light. The soybean genotype capable of producing strong LE6 inhibiting compound possesses resistance to the pathogen. Hence it is necessary to understand the genetic and physiological regulation of cercosporin for the development of soybean lines with durable resistance to *C. kikuchii* (Upchurch, 1995). Velicheti and Sinclair (1994) studied cercosporin from *C. kikuchii* in plant tissues with reference to pathogenesis. It facilitates soybean seed coat tissues colonization by the pathogen. A correlation has been established between cercosporin production and the extent of purple stain in the seed due to *C. kikuchii* colonization. Based on colony coloration on potato-dextrose-agar medium, cercosporin has been divided into three groups. Ckcp isolates produce purple coloration, Ckcy isolates have a yellow background due to lipids and Ckcc isolates produce red coloration. Most of the isolates were of Ckcp group with purple coloration. Production of cercosporin by the three isolates in seed coats was in the order of Ckcc>Ckcp>Ckcy. All the isolates on seed inoculation caused purple stain on the seed coat.

Almeida et al. (2004) noticed a strong correlation between cercosporin content and virulence. Genetic differentiation among and within populations was observed based on 86 RADP loci. No relationship could be identified between isolates and geographic origin or cercosporin. They observed that population of *C. kikuchii* are pathogenically, genotypically and geographically variable.

Epidemiology

The infection of *C. kikuchii* is influenced by temperature, length of the period of pod wetness and pod stage. The optimum temperature for infection is 25°C. There is no infection at 15 and 35°C. Disease incidence increases with increasing pod wetness period upto 30 hours with no disease at less than 24 hours (Schuh, 1992). No infection occurs at pre- or post-bloom stages (Schuh, 1992). Immature seeds harvested early had more purple stain (Ortiz et al., 1988).

Delaying soybean maturity by the promotion of growth, delaying harvest or wet condition caused an increase in the incidence of purple seed stain.

The disease is seed borne and the fungus is found on the inner surface of the seed coat. Gardner (1928) noted that the infection of the seed occurs mostly through hilum. Matsumoto (1928) found the presence of conidia of *C. kikuchii* on the inner surface of infected pods as well as cotyledons of seeds. He could get profuse sporulation of *C. kikuchii* when diseased seeds were incubated in damp Petri dishes at 15–20°C. Kilpatrick (1956) in his study on survival of *C. kikuchii* on soybean stem found that during the first 18 months, the fungus survived and sporulated profusely after which a gradual decline in sporulation was noticed with time. However, after 42 months the fungus could be isolated from debris. He concluded that *C. kikuchii* overwintered on stem lesions besides the seeds.

Jones (1958) isolated a sporulating strain by transferring spores from soybean to PDA in Petri-plates, it formed spores within 5–6 days. However, Jones (1968) also found that the *C. kikuchii* survived on infected soybean debris. However, it ceased to sporulate within 2 month, if the debris were buried under the soil, but it continued to sportulate abundantly if the infected debris was left on the surface of the soil. He, therefore, suggested the removal of the soybean crop debris from the fields so as to eliminate this source of inoculum.

The purple pigment was isolated by Kuyama and Tamura (1957) from cultured mycelia of *C. kikuchii*. The pigment cercosporin was purified and found to be a polyhydroxy derivative of a polycyclic quinon having an extended quinon system. Studies have, however, also shown that *C. kikuchii* is often present in seeds with no discoloration and apparently looking healthy. Crane and Crittondon (1962) attempted to determine the cause of variation in the incidence of purple stain in the same cultivars. They concluded that delay in normal maturity of plants caused a marked increase in the number of purple stain seeds. Gardner (1927) reported purple discoloration of seeds to be due to development of mycelium of *C. kikuchii* in the seed.

Laviolette and Athow (1972) have determined the optimum time of infection by *C. kikuchii*. In an artificial inoculation experiment they found that one or two inoculations during full flowering period gave the maximum infection in up to 90% seeds. The purple seed stain disease not only affects the quality of the seeds, but also the crop stands. However, the disease did not affect maturity, lodging, plant height and seed yield.

The role of *C. kikuchii* in soybean seed health is very poorly understood. Studies on germination and field emergence in relation to discolored seeds are very contradictory. Several reports conclude that germination is reduced up to 30% in a laboratory test. While some believe that discoloration is either a cosmetic seed quality problem, primarily related to marketability or it may produce weak seedlings. Surface discoloration over nearly 50% could delay germination.

Sharma (1990) studied the primary and secondary sources of infection of purple seed stain disease at Barapani (Meghalaya) conditions of India. He recorded that the plants grown in a net-house exhibited almost no disease incidence both in case of plants grown from treated and untreated healthy seeds. Mild disease was recorded on seeds, obtained from plants grown from treated and untreated diseased seeds, kept inside the net-house. Seeds from plants grown out-door exhibited higher incidence of the disease; in case of healthy and diseased as well as treated and untreated seeds. The maximum numbers of infected seeds were obtained from plants grown from untreated diseased seeds (Sharma, 2003). He found that the seed stored in a seed-bin at Barapani contained viable inoculum to initiate the primary infection of the disease. However, it would be interesting to ascertain whether secondary spread of the disease is caused by the conidia produced on infected parts of the same plant or from another source or both.

Acknowledgement

Authors are deeply indebted to the authorities of the Indian Council of Agricultural Research, New Delhi, for providing the facilities and encouragement for preparing this compilation. We express our sincere gratitude and thanks to the authors and publishers whose publications have been consulted and used in preparing this manuscript. We are very thankful to the editors for their guidance, suggestions and help, and the publishers of the book for this chapter.

References

Adams, J.F. 1933. Reports of the plant pathologist for 1932. Dalaware State Board Agr. Quart. Bull., 23: 3–16.

Agarwal, D.K. 1973. Pathological and physiological studies of some root infecting fungi. Ph.D. Thesis. Agra University, Agra, India.

Agarwal, D.K., Gangopadhyay, S. and Sarbhoy, A.K. 1973. Effect of temperature on charcoal rot disease of soybean. Indian Phytopathology, 26(3): 587–598.

Agarwal, S.C. and Kotasthane, S.R. 1971. Resistance in some soybean varieties against *Sclerotium rolfsii* Sacc. Indian Phytopathology, 24: 401–403.

Agarwal, V.K. and Joshi, A.R. 1971. A preliminary note on the purple stain disease of Soybean. Indian Phytopathology, 24: 811–814.

Ahmed, N. and Ahmed, Q.A. 1969. Physiologic specialization in *Macrophomina Phaseoli* (Maubl.) Ashby, causing stem rot of jute, *Corchorus* species. Mycopathologia, 39(2): 129–138.

Ainsworth, G.C. 1973. Introduction and key to higher taxa. *In*: Ainsworth, G.C., Sparrow, F.W. and Sussman, A.S. (eds.). The Fungi: An Advanced Treatise IV B 1-7. Academic Press, New York.

Allington, W.B. 1944. Soybean disease investigations at the US. Regional Soybean Laboratory. Soybean Dig., 4(11): 60–65.

Almeida, A.M.R., Kaster, M. and Albuquerque, F.C. 1980. Occurrence of *Myrothecium roridum* on soybeanin Piaui State (of Brazil). Fitopathologia Brasileria, 5: 129–133.

Almeida, A.M.R., Marin, S.R.R., Binneck, Piuga, F.F., Satori, F., Costamilan, L.m., Texeria, M.R.O and Lopes. 2004. Pathogenicity, moleculant analysis, and cerosporin content of Brazilian isolates of *Cerospora kikuchii*. Proc. VII th world Soybean Research Conf., Brazil, 29th Feb. 5th March 2004. 69–70.

Ali, M.I., Dogar, M.A. and Ahmad, R. 1995. Seed borne diseases of soybean and their chemical control. Pakistan J. Phytopath., 7(2): 160–162.

Ammon, V., Wyllie, T.D. and Brown, M.F. 1975. Investigation of the infection process of the *Macrophomina phaseoline* on the surface of soybean roots using scanning electron microscopy. Mycopathologia, 55: 77–81.

Amusa, N.A. 1994. Production, partial purification and bioassay of toxic metabolites of three plant pathogenic species of *Colletotrichum* in Nigeria. Mycopathologia, 128(3): 161–166.

Anahosur, K.H. 2001. Integrated management of potato *Sclerotium* wilt caused by *Sclerotium rolfsii.* Indian Phytopathology, 54: 158–166.

Arthur, J.C. 1917. Relationship of the genus Keuhneola. Bull. Torrey Bot. Club., 44: 501–517.

Arx von, J.A. 1957. Die Arten der Gathung *Colletotrichum* Cda. Phytopathologische Zeitschrift, 29: 414–468.

Athow, K.L. 1987. Fungal diseases. pp. 687–727. *In*: Wilcox J.R. (ed.). Soybeans: Improvement, Production and Uses, 2nd edition, Monogr. 16. Madison, USA: American Society of Agronomy.

Athow, K.L., Probst, A.H., Kurtznan, C.P. and Lauiolette, F.A. 1962. A newly identified physiological race of *Cercospora sojina* on soybean. Phytopathology, 52: 712–714.

Atienza, M. 1927. *Sclerotium* disease of tomato and pepper. Phillipp. Agr., 15: 579–588.

Atkinson, R.E. 1943. Soybean diseases in the Carolina. Plant Dis. Reptr., 27: 603–604.

Aycock, R. 1966. Stem rot and other diseases caused by *Sclerotium rolfsii*. NC Agric. Exp. Stn. Tech. Bull., 174. 202 pp.

Backman, P.A. and Brenneman, T.B. 1984. Comendium of Peanut Diseases. Amer. Phytopath. Soc., St. Paul, Minnesota.

Backman, P.A., Willams, J.C. and Craw Ford, M.A. 1982. Yield losses in soybean from anthracnose caused by *Colletotrichum truncatum*. Plant Disease, 66(11): 1032–1034.

Bain, D.C. 1944. Soybean diseases in Mississippi and Louisiana. Plant Dis. Reptr., 28: 834.

Banu, I.S.K.F., Shivanna, M.B. and Shetty, H.S. 1990. Seed-borne nature and transmission of *Colletotrichum truncatum* in chilli. Advances in Plant Sciences, 3(2): 200–206.

Barnett, N.M. 1974. Release of peroxidase from soybean hypocotyls cell walls by *Sclerotium rolfsii* culture filtrates. Canad. J. Bot., 52: 265–271.

Barnett, H.H. and Hunter, B.B. 1972. Illustrated genera of imperfect fungi. Burgess Publishing Company, Minnesota.

Bazan De Segura, C. 1947. Some fungi of Peru. Lima, Peru, Estac-Expt. Agr. La Molina. Bol., 33, 28 pp.

Begum, M.M., Sariah, M., Puteh, A.B. and Zainal Abidin. 2008. Pathogenicity of *Colletotrichum truncatum* and its Influence on soybean seed quality. International Journal of Agriculture & Biology, 10: 393–398.

Bernaux, P. 1979. Identification of some soybean diseases in Cameroon. Agronomie Tropicale, 34(43): 301–304.

Bhale, M.S. 2002. Seed health management in soybean. In Asian Congress of Mycology and Plant pathology. University of Mysore, Mysore, Karnataka, India, October 1–4, 2002.

Bhardwaj, C.L. and Thakur, D.R. 1991. Efficacy and economics of fungicide spray schedules for control of leaf spots and pod blight in urdbean. Indian Phytopathology, 44(4): 470–475.

Bonde, M.R. and Brown, M.F. 1980. Morphological comparison of isolates of *Phakopsora pachyrhizi* from different areas of the world. Canadian Jr. of Microbiology, 26(12): 1443–1449.

Bonde, M.R., Melching, J.S. and Bromfield, K.R. 1976. Histology of the suscept pathogen relationship between Glycine max and phakospora pachyrhizi, the cause of soybean rust. Phytopathology, 66: 1290–1293.

Boonthong, A. and Sommort, T. 1985. Southern blight of peanut (*Arachis hypogeal* L.) caused by *Sclerotium rolfsii*. Proceeding of the fourth Thailand National Groundnut Research Meeting for 1984. pp. 203–209.

Borkar, S.G. 1992. Influence of weather on *Sclerotium* root rot and wilt in soybean. Indian Mycol. Pl. Pathol., 22(2): 193–194.

Briton-Jones, H.R. and Baker, R.E.D. 1934. Notes on some other fungus diseases in Trinidad. Trop. Agr. (Trinidad), 11: 67–68.
Bromfield, K.R., Melching, J.S. and Kingsolver, C.H. 1980. Virulence and aggressiveness of *Phakospora pachyrhizi* isolates causing soybean rust. Phytopathology, 70: 17–21.
Burdon, J.J. and Speer, S.S. 1984. A set of differential *Glycine* hosts for the identification of races of *Phakopsora pachyrhizi* Syd. Euphytica., 33: 891–896.
Burhan, M.J. 1986. Outbreaks and new records. United Arab Emirates, anthracnose of alfalfa. FAO Plant Protection Bulletin, 34(2): 111.
Butler, E.J. 1953. Fungi of India (revised by Vasudeva, R.S.). IARI, New Delhi.
Butler, E.J. and Bisby, G.R. 1931. The fungi of India. Imp. Counc. Agr. Res. India Sci. Mono, 1: 44 pp.
Cai, L., Hyde, K., Taylor, P., Weir, B., Waller, J., Abang, M., Zhang, J., Yang, Y., Phoulivong, S. and Liu, Z. 2009. A polyphasic approach for studying *Colletotrichum*. Fungal Diversity, 39: 183–204.
Calzolari, A., Cavanni, P. and Ponti, I. 1987. Bacterial and fungus diseases of soybean. Informatore Fitopatologica, 37(3): 7–12.
Cannon, P.F., Bridge, P.D. and Monte, E. 2000. Linking the past, present and future of *Colletotrichum* systematic. pp. 1–20. *In*: Prusky, Freeman, S. and Dickman, M.B. (eds. 1). *Colletotrichum*. Host Specificity, Pathology and Host-Pathogen Interaction. APS Press, St. Paul, Minnesota.
Cannon, P.F., Buddie, A.G. and Bridge, P.D. 2008. The typification of *colletrotrichum gloeosporioides*. Mycotaxon, 104: 189–204.
Celina, M.S. 1936. Disease of cotton in the Phillippines: I. *Sclerotium* stem rot with rot with notes on other diseases. Philipp. Agr., 25: 302–320.
Chandrasrikul, A. 1962. A preliminary host list of plant diseases in Thailand. Thailand Dept. Agr. Tech. Bull., 6. 23 pp.
Chen, M.D., Lyda, S.D. and Halliweli, R.S. 1979. Infection of soybean with conidia of *Cercospora kikuchii*. Mycologia, 71: 1158–1165.
Chupp, C. 1954. A monograph of the fungus genus *Cercospora* Ithaca N. Y. 667 pp.
Clark, C.A. and Moyer, J.W. 1988. Compendium of Sweet Potato Diseases. Amer. Phytopath. Soc., St. Paul, Minnesota.
Conners, I.L. and Savile, B.O. 1943. Twenty-second annual report of the Canadian plant disease survey, 1942. 110 pp.
Conners, I.L. and Savile, D.B.O. 1944. Twenty third annual report of the Canadian Plant Disease Survey, 1943: 122.
Corda, A.C. 1831. Die Pilze Deutschlands. pp. 1–144. *In*: Strum, J. (ed.). dungen nach der Natur mit Beschreibungen 3.
Crall, J.M. 1951. Soybean diseases in Iowa in 1950. Plant Disease Reporter, 35: 320–321.
Crall, J.M. 1952. Soybean diseases in Iowa in 1951. Plant Disease Reporter, 36: 302.
Crandall, B.S. and Dieguez, J. 1948. A check list of diseases of economic plant in the Tingo Maria zone of the Peruvian Montana. Plant Dis. Reptr., 32: 20–27.
Crane, J.L. and Crittenden, H.W. 1962. Effect of deflowering and tobacco ring virus on the occurrence of *Cerospora kikuchii* on soybean. Phytopathology, 52: 1217.
Crane, J.L. and Crittenden, H.W. 1967. Growth of *Cercospora kikuchii* on various media. Disease Reporter, 51: 112–114.
Creelman, D.W. 1965. Summary of the prevalence of plant diseases in Canada 1964. Plant Disease Survey, 45: 37–83.
Curtis, C.R. and Barnett, N.M. 1972. Polycrylamid disc-gel electrophoresis and isoelectric focusing of proteins released from soybean hypocotyls cell walls by cultura fíltrate of *Sclerotium rolfsii*. Phytopathology, 62: 688.
Curzi, M. 1931. Studi Sulo. *Sclerotium rolfsii* Bollettino della station di Catalogia Vegetable di Roma N S, 11: 306–373.
De Carvalho, T. and Mends, O. 1958. Diseases of plants in Mozambique. Lourence Marques. 84 pp.

De Guerpel, H. 1942. The pests and diseases of soybean. Bot. Appl. And Agr. Trop., 17: 195–201.
Dhar, V. and Sarbhoy, A.K. 1987. Ashy stem blight of soybean in India. Current Science, 56(22): 1182–1183.
Dhingra, O.D., Sediyama, C., Carraro, I.M. and Reis, M.S. 1978. Behavior of four soybean cultivars to seed-infecting fungi in delayed harvest. Fitopathol. Brasil., 3: 277–282.
Diaz, P.C. 1966. *Cercospora Kikuchii* on soybean, a new pathogen in Venezuela. Agr. Trop., 16: 213–221.
Diaz, Polanco C. and Casanova, J.R. 1966. *Cercospora kikuchii* on soybean, a new pathogen in Venezuela. 6th Agron. Cof. Maracaibo, March 1966: Rpt. Vol. III. Caracas. Venezuela.
Do Amaral, J.E. 1951. Principal diseases of cultivated plants in the State of Sao Paulo and their control. Biologico, 17: 179–188.
Eken, C. and Demirci, E. 2000. First report of *Colletotrichum truncatum* on alfalfa in Turkey. Plant Disease, 84(1): 100.
Elias, S. 2006. Seed quality testing. pp. 561–599. *In*: Basra, A.S. (ed.). Handbook of Seed Science and Technology. Food Product Press; An Imprint of the Hawoth Press, New York.
Ersek, T. 1979. Occurrence of charcoal rot and anthracnose of soybean in Hungary. Acta Phytopathologica Academiae Scientiorum Hungaricae, 14(1/2): 17–21.
Fakir, G.A. and Mridha, U. 1985. Dia-back, a new disease of lady's finger (*Hibiscus esculentus* L.) in Bangladesh. Bangladesh Journal of Plant Pathology, 1(1): 25–28.
Farr, D.F., Aime, M.C., Rossman, A.Y. and Palm, M.E. 2006. Species of Colletotrichum on Agavaceae. Mycological Research, 110: 1395–1408.
Feng, F. and Wang, H.G. 2004. Soybean diseases report in China. Proc. VII World Soybean Research Cant Brazil. 29th Feb.–5th March 2004. pp. 331–334.
Fenne, S.B. 1942. Two new records for the frog eye leaf spot: Virginia. Plant Disease Reporter, 26: 383.
Fenne, S.B. 1949. Alfalfa and soybean diseases in Virginia, 1948. Plant Dis. Reptr., 33: 90–91.
Frank, A.B. 1882. Gallen der *Angaillula radicicola* Greff and *Soja Hispida*, Medicago sativa, Lactuca sativa, and Pirus communnis. Verb. Bot. Prov. Brandenburg., 23(2): 54–55.
Frederick, R.D., Snyder, C.L., Peterson, G.L. and Bonde, M.K. 2002. Polymerase chain reaction assays for the detection and discrimination of the soybean rust pathogen *Phakopsora pachyrhizi* and *P. meibomiae*. Phytopathology, 92(2): 217–227.2
Fries, E.M. 1829. Systema Mycologicum III., 1: 216–218.
Fujita, Y. 1990. Ecology and control of purple seed stain of soybean caused by *Cercospora kikuchii*. Bulletin of the Tohoku National Agricultural Experiment Station, 81: 51–109.
Fujita, Y. and Suzuki, H. 1988. Histological study on the resistance in soybean (*Glycine max*) and a wild soybean (*G. soja*) to purple seed stain caused by *Cercospora Kikuchii*. Annals of the Phytopathological Society of Japan, 54: 151–157.
Gangopadhyay, S., Agarwal, D.K., Sarbhoy, A. and Wadhi, S.R. 1973. Charcoal rot disease of soybean in India. Indian Phytopathology, 26(4): 730–732.
Gardner, M.W. 1924. Indiana plant disease, (1921). Proc. Indiana Acad. Sci., 33: 163–201.
Gardner, M.W. 1927. Indiana plant diseases, 1925. Proceedings Indiana Academy of Sciences, 36: 231–247.
Gardner, M.W. 1928. Indiana plant diseases (1927). Proc. Indiana Acad. Sci., 38: 143–157.
Ghosh, T., Mukherji, N. and Basat. M. 1964. On the occurrence of a new species of Orbilia Fr. Jute Bull., 27: 134–141.
Girard, J.C. 1979. Detection of two diseases on soybean seeds from Senegal. Agronomie Tropicale, 34(3): 305–307.
Goot, P.V.D. and Muller, H.R.A. 1932. Pests and diseases of the soybean crop in Java. Preliminary report Landhuw Tijdschr. *Vereen. Landb.* Nederl.-Indie., 7: 683–704
Gupta, G.K. 2003. Integrated diseases management of soybean diseases. Training mannual No. DRS/02/2003, Dept. pf Plant Breeding and Genetics, J.N. Krishi Vishwa Vidhyalaya. Jabalpur, M.P., 114–125.

Gupta, G.K. 2004. Soybean diseases and their management. pp. 145–168. *In*: Singh, N.B., Chauhan, G.S. Vyas, A.K. and Joshi, O.P. (eds.). Soybean Production and Improvement in India. National Research Centre for Soybean, Indore, M.P., India.

Gupta, G.K. and Chauhan, G.S. 2005. Symptoms, Identification and Management of Soybean Diseases. Technical Bulletin no. 10. National Research for Soybean. Indore, India. p 92.

Haenseler, C.M. 1946. Soybean diseases in New Jersey, New Jersey Agr. Expt. Sta. Plant Disease Notes, 23: 17–20.

Haenseler, C.M. 1947. Pathologist describes four principal diseases of fluid soybeans in Jersey. New Jersey Agr., 29(3): 4.

Han, Y.S. 1959. Studies on purple spot of soybean. J. Agr. And For., 8: 1–32.

Hara, K. 1930. Pathologia agriculturis Plantarum. Agr. Country, 12: 18.

Han, Y.S. 1959a. Soybean diseases in Taiwan. Agr. Assoc. China (Taiwan) Jr., 26: 31–38.

Han, Y.S. 1959b. Studies on purple spot of soybean. Jr. Agr. And For, 1–32.

Hara, K. 1915. Spot disease of soybean. Agr. Country, 9: 28.

Hartman, G.L., Sinclair, J.B. and Rupe, J.C. 1999. Compendium of soybean diseases. The American Phytopathological Society (Edition IV), APS Press. p 100.

Haskell, R.J. 1926. Diseases of Cereal and Forage crops in the United States in 1925. Plant Dis. Reptr. Suppl., 48: 301–381.

Hemmi, T. 1920. Beitrage zur kenntnis der morphologie und physiologie der *Japanischen Gloesporien*. J. Coll. Agr. Hokkaido Imp. Univ., 9: 1–159.

Hepperly, P.R., Mignucci, J.S., Sinclair, J.B. and Mendoza, J.B. 1983. Soybean anthracnose and its seed assay in Puerto Rico. Seed Science Technology, 11: 371–380.

Hennings, P. 1903. Some new Japanese Uredinales IV (in German) Hedwigia. Suppl., 42: 107–108.

Hiratsuka, N. 1932. Notes on soybean rust. Trans. Biol. Soc. Tottori, 1: 8–11.

Hiratsuka, N. 1935. Phakospora of Japan I. Bot. Mag. (Tokyo), 49: 781–788.

Hiratsuaka, N. 1936. Uredinales Collected in Kushu (Japan) III. J. Japan Bot., 12: 265–272.

Hopkins, J.C.F. 1931. Plant pathology in Southern Rhodesia during the year 1930. Rhodesia Agr. J., 28: 384–389.

Hopkins, J.C.F. 1950. A descriptive list of plant diseases in Southern Rhodesia and list of bacteria and fungi. Southern Rhodesia Dept. Agr. Mem. 2 (rev. 2nd ed.). 106 pp.

Hsu, C.M. and Wu, L.C. 1968. Study on soybean rust. Sci. Agric. (Taiwan), 16: 186–188.

Huang Chunyan, Ma Shumei, Zhu Chuanyin, Zhang Zengmin, Guo Mei, Li Baoying and Tang Lixin. 1998. Study on forecast for epidemical tendency and yield loss of soybean frogeye leaf spot. Soybean Science, 17: 48–52

Hyde, K., Cai, L., Cannon, P., Crouch, J., Crous, P., Damm, U., Goodwin, P., Chen, H., Johnston, P. and Jones, E. 2009. Colletotrichum names in current use. Fungal Diversity, 39: 147–183.

Ilag, L.L. 1977. Studies on the biology of the soybean rust fungus in the Phillipines. Rust of Soybean: The problem and research needs. INTSOY Series No., 12: 16–17.

Ilyas, M.B., Ellis, M.A. and Sinclair, J.B. 1975. Evaluation of soil fungicides for control of charcoal rot of soybeans. Plant Disease Reporter, 59(4): 360–364.

Jasnic, S. 1983. *Colletotrichum dematium* (Pers. Ex Fr.) Grove var. *truncate* (Schw.) Arx., the pathogen of anthracnose of soybean in Yugoslavia. Zastira Bilja, 34(3): 381–389.

Javaid, I. and Ashraf, M. 1978. Some observations on soybean diseases in Zambia and occurrence of *Pyrenochaeta glycines* on certain varieties. Plant Diseases Reporter, 62(1): 46–47.

Johnson, H.W. and Kilpatrick, R.A. 1953. Soybean diseases in Mississippi in 1951–1952. Plant Dis. Reptr. 37: 154–155.

Johnston, A. 1960. A preliminary plant disease survey in North Borneo. Plant Prol. And Prot. Div., Food and Agr. Org. Rome, 43 p.

Jones, J.P. 1958. Isolation of sporulation strain of *Cercospora kikuchii*, by selective sub-culturing. Phytopathology, 48: 287–288.

Jones, J.P. 1968. Survival of *Cercospora kikuchii* on soybean stems in the field. Plants Disease Reporter, 52: 931–934.

Kaiser, W.J., Mihov, M., Muehlbauer, F.J. and Hannan, R.M. 1998. First report of anthracnose of lentil incited by *Colletotrichum truncatum* in Bulgaria. Plant Disease, 82(1): 128.

Kaushal, R.P., Kumar, Anil, Tyagi, P.D. and Kumar, A. 1998. Role of light, temperature and relative humidity on germination of *Colletotrichum truncatum* and soybean pod infection under laboratory conditions. Journal of Mycology and Plant Pathology, 28(1): 1–4.

Kawase, Y. 1955. On theresistance of *Pellicularia rolfsii* causing sclerotial blight of soybean to mercuric chloride. Osaka Pref. Univ. Bull. Ser. B., 5: 167–174.

Keogh, R.C. 1974. *Phakospora pachyrhizi* Syd. The causal agent of soybean rust. Australian Plant Path. Soc. News Ltr., 3: 5.

Khan, M. and Sinclair, J.B. 1992. Factors affecting seed infection and transmission of *Colletotrichum dematium f.* sp. *Truncate* in soybean. Seed Science Technol., 11: 853–858.

Khare, M.N. and Chacko, S. 1983. Factors affecting seed infection and transmission of *Colletotrichum dematium f.* sp. *Truncate* in soybean. Seed Science Technology, 11: 853–858.

Khare, M.N., Bhale, M.S. and Kumar, Kumud. 2000. Seed borne pathogens causing diseases in soybean, their detection, diagnosis and management. pp. 77–94. *In*: Narain, U., Kumar, K. and Srivastava, M. (eds.). Advances in Plant Disease Mangement. Creative Offset Press, New Delhi.

Kikuchi, R. 1924. On a disease of the soybean caused by *Cercosporin.* Utsunomiya Agr. Coll. Bull., 1: 1–19.

Kilpatrick, R.A. 1955. Soybean diseases in the delta are of Mississipi in 1954. Plant Disease Reporter, 39: 578–579.

Kilpatrick, R.A. 1956. Longesvity of *Cercospora kikuchii* on soybean stems. Phytopathology, 46: 58.

Kirk, P.M., Cannon, P.F., Minter, D.W. and Stalpers, J.A. 2008. The Dictionary of the Fungi, 10th edition. CAB Bioscience, UK.

Kitani, K. and Inoue, Y. 1960. Studies on the soybean rust and its control measures. (Part 1). Studies on the soybean rust. Shikoku Agr. Expt. Sta. (Zentsuji, Japan). Bull., 5: 319–342.

Kobayashi, T. and Zenno, Y. 1984. Anthracnose of legume tree seedlings in the Phillippines and Indonesia. Journal of the Japanese Forestry Society, 66(3): 113–116.

Koch, E. and Hoppe, H.H. 1987. Germination of the teliospores of Phakospores pachyrhizi. Soybean Rust News Letr., 8: 5–8.

Koch, L.W. and Hilderbrand, A.A. 1944. Soybean diseases in South Western Ontaria in 1943. Canada P1. Dis. Surv Annual Report, 23: 29–32.

Koch, S.H., Baxter, A.P. and Knox-Davies, P.S. 1989. Identity and pathogenicity of *Colletotrichum* species from *Medicago sativa* in South Africa. Phytophylactica, 21(1): 69–78.

Kochman, J.K. 1979. The effect of temperature on development of soybean rust (*Phakospora pachyrhizi*). Aust. J. Agric. Res., 30: 273–277.

Koehler, B. 1931. Diseases of soybean in Illinois. Proc. Amer. Soybean, Assoc., 3: 60–64.

Kornfield, A. 1935. Schadigungen and Krankheiten der Olbohne (Soja), Soweit sie bisher in Europe bekannt geworden sind. Zeitscher Pflanenkr., 45: 577–613.

Kuhne, H. and Leupold, R. 1985. Myrothecium likes high humidity. Planzenschutz., 85: 1364–1366.

Kumar, S. and Verma, R.N. 1985. Soybean rust in NEH hills of India: Further observations. Soybean Rust News Letter, 7: 17–19.

Kuo, K.C., Lee, C.Y., Chiang, S.S. and Zheng, Y.S. 1999. Causal agent, pathogenesis and fungicide screening of lima bean anthracnose. Plant Protection Bulletin, Taipei, 41(4): 265–275.

Kurata, H. 1960. Studies on fungal diseases of soybean in Japan. Natl. Inst. Agr. Sci. (Tokyo) Bull. Ser. C., 12: 1–154.

Kuyama, S. and Tamura, T. 1957. *Cercosporin:* A pigment of *Cercospora Kikuchii* et Tomoyasu. I. Cultivation of the fungus, isolation and purification of pigment. II. Physical and chemical properties of *Cercosporin* and its derivatives. Amer. Chem. Soc. J., 79: 5725–5729.

Lambat, A.K., Raychoudhary, S.P., Lele, V.C. and Nath, R. 1969. Fungi intercepted on imported soybean seed. Indian Phytopath., 22: 327–330.
Laviolette, F.A. and Athow, K.L. 1972. *Cercospora kikuchii* infection of soybean as affected by stage of plant development. Phytopathology, 62: 771.
Laviolatte, F.A., Abney, T.S., Wilcox, J.R. and Athow, K.L. 1973. Indiana soybean disease and crop condition survey. Indiana Agr. Expt. Sta. Bull. SB 22, September 1973, 8 pp.
Lakshminarayana, C.S. and Joshi, L.K. 1978. Myrothecium leaf spot of soybean in India. Plant Dis. Reptr., 62: 231–234.
Lehman, S.C. 1928. Frog eye leaf spot of soybean caused by *Cercospora daizu* Miura. J. Agr. Res., 36: 811–833.
Lehman, S.G. 1934. Frog eye (*Cercospora daizu* Miura) on stems, pods, and seeds of soybean. and the relation of these infections torecurrence of the disease. J. Agr. Res., 48: 131–147.
Lehman, S.C. 1942. Notes on plant diseases in North Carolina in 1941 Soybean. Plant Disease Reporter, 26: 111.
Lehman, S.G. and Wolf, F.A. 1926. Soy-bean anthracnose. J. Agric. Res., 33: 381–390.
Lehman, S.C. and Poole, R.F. 1929. Research in Botany North Carolina Agr. Expt. Sta. Annual Report, 51: 59–67.
Lehman, P.S., Machado, C.C. and Tarrago, M.T. 1976. Frequency and severity of soybean diseases in the States of Rio Grande do Sul and Santa Catarina. Fitopatol. Brasil., 1: 183–193.
Likhite, V.N. 1936. Host range of the Gujrat cotton root rot. Proc. Assoc. Econ. Biol. Coimbatore., 3: 18–20.
Lin, S. 1966. Studies on physiologic races of soybean rust fungus, *Phakopsora pachyrhizi* Syd. Taiwan. Agr. Res., 51: 24–28.
Lin, P.L., Li, Y.P., Liu, J., Wu, B.Z. and Li, Y. 1991. Study on time of fungicide application for controlling soybean frog eye spot (*Cercospora sojina* Hara). Soybean Science, 10: 135–138.
Ling, L. 1948. Host index of the parasitic fungi of Szechwan, China. Plant Disease Reporter Suppl., 173: 1–38.
Link, H.F. 1809. Observations in ordines plantarum naturals. Dis. Mag. Gesnaturf Freunds, Lokula., 3: 23.
Litzenberger, S.C. and Stevenson, J.A. 1957. A preliminary list of Nicaraguan plant diseases. Plant Disease Reporter Suppl., 243: 1–19.
Liu, S.T. 1948. Seed brone diseases of soybean. Bot. Bull. Acad. Sinica, 2: 69–80.
Lusin, V. 1960. *Cercospora kikuchii*—Soybean disease. Savremena Poljoprivreda, 8: 601–604.
Mahmood, M., Mahmood, A., Gupta, S.K. and Kumar, S. 1976. Studies on root rot disease of groundnut caused by *Sclerotium rolfsii*. Proceeding of Bihar Academy of Agricultural Science, 13: 157–158.
Maiti, S., Vishwadhar and Verma, R.N. 1981. Rust of Soybean in India—A reappraisal. Soybean Rust Newsletter, 4(1): 14–16.
Maiti, S., Kumar, S., Verma, R.N. and Vishwadhar. 1983. Current status of soybean diseases in North East India. Soybean Rust New sletter, 6(1): 14–21.
Manandhar, J.B. and Sinclair, J.B. 1982. Occurrence of soybean diseases and their importance in Nepal. FAO Plant Protection Bulletin, 30(1): 13–16.
Manandhar, J.B. and Hartman, G.L. 1999. Anthracnose. pp. 13–14. *In*: Hartman, G.L., Sinclair, J.B. and Rupe, J.C. (eds.). Compendium of Soybean Diseases, 4th edition. St. Paul, APS Press, USA.
Manns, T.F. and Adams, J.F. 1929. Department of plant pathology. Delware Agr. Expt. Sta. Bull., 162: 53–37.
Manns, T.F. and Adams, J.F. 1934. Department of Plant Pathology. Delaware Agr. Expt. Sta. Bull., 188: 3646.
Marchetti, M.A., Uecker, F.A. and Bromfield, K.R. 1975. Uredial development of *Phakospora pachrhizi* in soybeans. Phytopathology, 65: 822–823.
Martyn, E.B. 1942. Diseases of plants in Jamaica. Jamaica Dept. Sci. and Agr. Bull., 32. 34 pp.

Massalongo, C. 1900. De nonnullis speciebus novis micromycetum agri veronensis. Atti. R. Inst. Veneto Sci. Letl. Ed. Arti., 59: 683–690.

Matsumoto, T. 1928. Observations on spore formation in the fungi *Cercosporakikuchii.* Annals of Phytopathological Society Japan, 2: 65–69.

Matsumoto, T. and Tomoyasu, R. 1925. Studies on purple speck of soybean seed. Annals of Phytopathological Society Japan, 1: 1–14.

Mayek-Pérez, N., Garcia-Espinosa, R., López-Castañeda, C., Acosta-Gallegos, J.A. and Simpson, J. 2002. Water relations, histopathology, and growth of common bean (*Phaseolus vulgaris* L.) during pathogenesis of *Marcophomina phaseolina* under drought stress. Physiological and Molecular Plant Pathology, 60(4): 185–195.

Mc Gee, D.C. 1991. Soybean Diseases. A Reference Source for Seed Technologist. APS Press, St. Paul, Minnesota. p 151.

McIntosh, A.E.S. 1951. Annual Report of the Department of Agriculture, Malaya for the year 1949. 87 pp.

Mc Lean, R.J. and Byth, D.E. 1981. Histological Studies of the pre-penetration development and penetration of soybeans by rust *Phakospora pachyrhizi* Syd. Aust. J. Agr. Res., 32: 435–443.

Mc New, G.L. 1948. Study of soybean diseases and their control. Iowa Agr. Expt. Sta. Rpt. On Agr. Res. For the yea ending June 30, 1948. pp. 188–189.

Mehta, P.R., Garg, D.N. and Mathur, S.C. 1950. Important diseases of food crops, their distribution in India and Uttar Pradesh (India). Dept. Agr. Tech. Bull., 2: 13.

Mejia, A.S. 1954. *Sclerotium* wilt of supa (*Sindora supa* Merr.). Phillipp. J. For., 9: 119–132.

Melching, J.S., Bromfield, K.R. and Kingsolver, C.H. 1979. Infection, colonization and uderospore production on *Wayne* Soybean by four cultures of *Phakospora pachyrhizi*, the cause of soybean rust. Phytopathology, 69(12): 1262–1265.

Melhus, I.E. 1942. Soybean diseases in Iowa in 1942. Plant Disease Reporter, 26: 431–432.

Mihail, J.D. and Taylor, S.J. 1995. Interpreting variability among isolates of *Macrophomina phaseolina* in pathogenicity, pycnidium production, and chlorate utilization. Canadian Journal of Botany, 73(10): 1596–1603.

Mikhalenko, A. 1965. Disease of legumes in the Primorsk region. Zashch. Rast. Vredit. Bolez., 10: 4143.

Mirza, M.S. and Rehman, A. 1985. Outbreaks and new records. Pakistan. *Colletotrichum dematium* var. *truncatum* on soybean. FAO Plant Protection Bulletin, 33(1): 44.

Miura, M. 1930. Diseases of the main agricultural crops in Manchuria. Manchuria Railway Co. Agr. Expt. Sta. Bull. 11 (Rev. ed.) 56 p.

Mohan, L., Paranidharan, V. and Prema, S. 2000. New diseases of timla fig (*Ficus auriculata*) in India. Indian Phytopathology, 53: 496.

Moore, W.F. 1984. Stem canker and charcoal rot. pp. 68–72. *In*: Proceedings Annual Soybean Research Conference 14, American Seed Trade Association, Washington, D.C.

Mordue, J.E.M. 1974. *Corticium rolfsii*, CMI Descriptions of Pathogenic Fungi and Bacteria. No. 410, 2 p.

Morgan, F.L. and Johnson, H.W. 1964. Leaf symptoms of soybean anthracnose. Phytopathology, 54: 625.

Morse, W.J. and Carter, J.L. 1939. Soybean culture and varieties: US Dept. Agr. Farmer Bull. 1520.

Morse, W.J., Carter, J.L. and Williams, L.F. 1949. Soybean: culture and varieties. US Dept. Agr. Res. Serv. Farmers Bull. 1520(rev.) 38.

Morwood, R.B. 1956. A preliminary list of plant diseases in Fiji. Fiji Dept. Agr., Agr. J., 27: 51–54.

Mridha, A.U., Fakir, G.A. and Miam, M.A.W. 1984. A record of seedling diseases from Rasulpur forest nursery, Modupur. Bano Biggyan Patrika, 13(1/2): 45–50.

Muller, A.S. 1941. Survey of diseases of cultivated plants in Venezuela, 1937–1941. Soc. Venezolana de Cienc. Natl., Bol., 7: 99–113.

Muller, A.S. 1950. A preliminary survey of plant diseases in Guatemala. Plant Disease Reporter, 34: 161–164.

Muller, A.S. and Chupp, C. 1942. Soc. Venezolana de Cience, Nail. Bol., 8: 35–59.

Muller, A.S. and Chupp, C. 1950. *Cercospora* in Guatemala. Cieba, 1: 171–178.

Munjal, R.L. 1960. A commonly occurring leaf spot disease caused by *Myrothecium roridum*. Indian Phytopatah., 13: 150–155.

Nagata, T. 1962. Report to the Govt. of Yugoslavia on improvement of soybean cultivation. FAO Expanded Tech. Asstt. Prog. Fao Rept 1465: 22 p.

Naidu, Harinath. 2000. Crossandra—a new host record for Sclerotium rolfsii. Indian Phytopathology, 53: 496–497.

Nakata, K. and Takimoto, K. 1934. A list of crop diseases in Korea. Agr. Expt. Sta. Govt. Central Chosen Res. Rpt., 15: 1–146.

Nakata, K. and Asyuama, H. 1941. Report on diseases on main agricultural crops. Manchuria —Bereau Indus. Manchuria Rpt., 32: 1–166.

Nalim, F.A., Start, J.L., Woodward, K.G., Sengar, S. and Keller, N.P. 1995. Mycelial compatibility groups in Taxas Peanut field populations of *Sclerotium rolfsii*. Phytopathology, 85: 1507–1512.

Ndzoumba, B., Conca, G.P. and Prta Puglia, A. 1990. Observations on the mycoflora of seeds produced in Gabom. FAO Plant Protection Bulletin, 38(4): 203–212.

Nelen, E.S. 1968. Fungus diseases of agricultural plants new for Soviet Union. Mikol. i. Fitopatol., 2: 128–133.

Nene, Y.L. and Srivastava, S.S.L. 1971. Report on the newly recorded purple stain of soybean seed caused by *Cercospora kikuchii* and pod blight caused by *Colletotrichum damatium* var. *truncate*. Plant Protection Bull. FAO, 19(3): 66–67.

Nespick, A., Koear, M. and Siewinski, A. 1961. Antibiotic properties of mycelium and metabolite of *Myrothecium roridum* and their pathogenic potential. Trans. British Mycol. Soc., 61: 347–354.

Nguyen Phuong Thi Hang. 2010. *Colletotrichum* spp. Associated with anthracnose disease on coffee in Vietnnam and on some other major tropical crops. Doctoral Thesis summated to Swedish University of Agricultural Sciences, Alnarp. Available at http: / / pub.epsilon. slu. se/2269 /1/nguyen.p-100421.pdf.

Nicholson, J.F. and Sinclair, J.B. 1973. Effect of planting date, storage conditions and seedborne fungi on soybean seed quality. Plant Dis. Reptr., 57: 770–774.

Nik, W.Z. 1983. Seed borne fungi of soybean and mungbean and their pathogenic potential. Malaysian Applied Biology, 12: 21–28.

Nik, W.Z. and Lim, T.K. 1984. Occurrence and site of infection of *Colletotrichum dematium* f. sp. *Truncatumin* naturraly infected soybean seeds. Journal of Plant Protection in the Tropics, 1(2): 87–91.

Noble, R.J., Hynes, H.J., Mc-Cleery, F.C. and Birmingham, W.A. 1935. Plant diseases recorded in New South Wales, Dept. Agr. New South Wales Sci. Bull. 46.

Om Prakash and Singh, U.N. 1976. Basal root of mango seedlings caused by *Sclerotium rolfsii*. Indian Journal of Mycology and Plant Pathology., 6: 75.

Ondrej, M. 1994. Evaluation of the gene pod of soybean (*Glycine max* Merr.) for resistance to diseases (*Phoma, Colletotrichum, Rhizoctonia, Fusarium*). Rocenka Geneticks zdroje rastlin, 24–28.

Ono, Y., Buritica, P. and Hennen, J.F. 1992. Delimination of *Phakopsora pachyrhizi* and *Cerotelium* and their species on Leguminosae. Mycol. Res., 96: 825–850.

Ortiz, C., de Cianzio, S.R. and Hepperly, P.R. 1988. Fungi and insect damage to soybean harvested at immature stages in tropical environments. Journal of Agriculture of the University of Puerto Rico, 72: 73–79.

Ovchinnikova, A.M. 1968. *Cercospora* on soybean in Primorsk (Pecific Coastal) region, *Zashch Rast*. Mosk., 13: 27–28.

Paola, M.A. 1933. A *Sclerotium* seed rot and seedling stem rot of mango. Philippine Journal of Science, 52: 232–261.

Parbery, D.G. and Lee, C.K. 1972. Anthracnose of soybean. Australian Plant Pathological Society, Newslatter, 1: 10–11.

Pathan, M.A., Sinclair, J.B. and Mc Clary, R.D. 1989. Effects of *Cercospora kikuchii* on soybean seed germination and quality. Plant Disease, 73: 720–723.

Patino, H.C. 1967. Diseases of oleaginous annuals in Colombia. Agr. Trop., 23: 532–539.

Pawar, V.H. and Thirumalachar, M.J. 1966. Necrotin as a wilt toxin (anti fungal and antibiotic) from Myrothecium roridum. Hindustan Anatibiotic Bull., 8: 126–128.

Person, L.H. 1944. List of plant diseases observed during survey in Mississippi and Louisiana, August to November, 1943. Plant Dis. Reptr. Suppl., 148: 280–283.

Petty, A.M. 1943. Soybean disease incidence in Maryland in 1942 and 1943. Plant Disease Repts., 27: 347–349.

Pinckard, J.A. 1942. Diseases of soybeans and peanuts in Mississippi. Plant Dis. Reptr., 26: 472–473.

Phillips, D.V. and Boerma, H.R. 1981. *Cercospora sojina* race 5: A threat to soybean in the South Eastern United states. Phytopathology, 71: 334–336.

Pioli, R.N., Benavidez, R. and Morandi, E.N. 1997. Preliminary studies on pathogen incidence in Fresh consumption of soybean seeds. Fitopthologia, 32: 116–120.

Ploper, L.D. and Backman, P.A. 1992. Nature and management of fungal diseases affecting soybean stems, pods and seeds. pp. 174–184. *In*: Copping, L.G., Green, M.B. and Ress, R.T. (eds.). Pest Management in Soybean. SCI: Elsevier Applied Science, London and New York.

Poharkar, V.C. and Raut, J.G. 1997. Fungi associated with soybean seed in Vidarbh. PKV Research Jr., 21: 99–100.

Poonpolgul, S. and Surin, P. 1980. Study on host range of soybean rust fungus in Thailand. Soybean Rust Newsletter, AVRDC., 3: 30–31.

Preston, N.C. 1943. Observations on the genus Myrotheicum Tode, I. The three classic species. Trans. Brit. Myciol. Soc., 26: 58–168.

Punja, Z.K. 1985. The biology, ecology and control of *Sclerotium rolfsii*. Annual Review of Phytopathology, 23: 97–127.

Punja, Z.K. 1988. *Sclerotium (Athelia) rolfsii* a pathogen of many plant species. pp. 523–534. *In*: Sidhu, G.S. (ed.). Advances in Plant Pathology. Vol. 6, Genetic of Plant Pathogenic Fungi. Academic Press, London.

Purkayastha, R.P. and Ghosh, S. 1983. Elicitation and inhibition of phytoalexin biosynthesis in myrothecium infected soybean leaves. Indian J. Experimental Biol., 21: 216–218.

Ramakrishnan, T.S. 1951. Additions to fungi of Madras XI. Pre. Indian Acad. Sci. Sect. B., 34: 157–164.

Rao, V.G., Pande, Alaka and Patwardhan, P.G. 1989. Three new records of fungal diseases of economic plants in Maharashtra State. Biovigyanam, 15(1): 51–53.

Reichert, I. and Hellinger, E. 1947. On the Occurrence, morphology and parasitism of *Sclerotium bataticola*. Palestine Journal of Botany Rehovot Series, 6: 107–147.

Rheenen, H.A. 1975. Soybean in the northern states of Nigernia. pp. 158–159. *In*: Luse, R.A. and Rachie, K.O. (eds.). Proceedings of IITA Collaborators meeting on Grain Legume Imporvement Ibadan, Nigeria, IITA.

Rhoads, A.S. 1944. Summary of observations on plant diseases in Florida during the emergency plant disease prevention project surveys, July 25 to December 31, 1943. Plant Disease Reportor (supp.), 148: 262–276.

Riley, E.A. 1960 . A revised list of plant diseases in Tanganyika Territory. Commonw. Mycol. Inst., Mycol. Papers, 75: 42 pp.

Rose, J.P. 1968. Effect of single and double infection of soybean mosaic and bean mottle viruses on soybean yield and seed characteristics. Plant Disease Reporter, 52: 344–348.

Roy, K.W. 1982. Seedling disease caused in soybean by species of *Colletotrichum* and *Glomerella*. Phytopathology, 72(8): 1093–1096.

Saccardo, P.A. 1886. Sylloge Fungorum., 4: 750.

Saikia, V.N. and Phukan, A.K. 1983. Occurrence of seedling blight of soybean in Assam. Journal of Research, Assam Agricultural University, 4(2): 171–172.
Sarbhoy, A., Thapiyal, P.N. and Payak, M.M. 1972. *Phakospora pachyrhizi* on soybean in India. Science and Culture, 38(4): 198.
Saksena, H.K. and Tripathi, R.C. 1971. Myrothecium leaf spot of soybean in India. Indian J. Mycol. And Pl. Pathol., 1: 75–76.
Sasaki, Y. 1982. The progress of disease caused by *Cercospora kikuchii* Matsumoto et Tomoyasu on soybean. Bull. Of the Hiroshima Prefectural Agr. Expt. Sta., 45: 43–52.
Sathe, A.V. 1972. Identity and nomenclature of soybean rust from India. Curr. Sci., 41(7): 264–265.
Sawada, K. 1928. Descriptive catalogue of the Formation Fungi. Part 4 Dept. Agr. Res. Inst. Formosa. Rpt. 33 pp. 1–123.
Sawada, K. 1959. Descriptive catalogue of Taiwan (Formosan) fungi. Part XI. Natl. Taiwan Univ. (Taipei) Coll. Agr. Spec. Bull., 8: 268 pp.
Saxena, R.M. and Sinha, S. 1978. Seed-borne infection on *Vigna radiate* (L.) Wilczek var. *radiate* in Uttar Pradesh—new records. Science and Culture, 44(8): 377–379.
Schiller, C.T., Hepperly, P.R. and Sinclair, J.B. 1978. Pathogenicity of *Myrothecium roridum* from Illinois on soybean. Pl. Dis. Reptr., 62: 882–885.
Schuh, W. 1992. Effect of pod development stage, temperature, and pod wetness duration on the incidence of purple seed stain of soybeans. Phytopathology, 82: 446–451.
Schumacher, C.F. 1803. Enumeratis planataru in partibus. Saellandiae septentrionalea et orientales. II, 418, cited from Fries, E.M., Systema Mycologicum. 218.
Sharma, S.K. 1990. Phytopathological studies on important fungal diseases of soybean of NEH India with special reference to soybean rust and purple seed stain. Ph.D thesis, Jiwaji University, Gwalior, M.P., India 168 pp.
Sharma, S.K. 2003. Field study on purple seed stain (*Cercospora kikuchii*) disease of soybean at mid altitude of Meghalaya. Crop Research, 26(3): 512–514.
Sharma, S.K. and Gupta, G.K. 2006. Current status of Soybean rust (*Phakospora Pachyrhizi*). A review. Agricultural Review, 27(2): 91–102.
Sharma, S.K., Verma, R.N. and Chauhan, S. 1988. Studies on the field reaction of soybean cultivar against purple seed stain disease (*Cercospora kikuchii*). Biological Bull. of India, 10(3): 117–121.
Sharma, S.K., Verma, R.N. and Chauhan, S. 1989. Seed borne microbes of purple stain infected seeds of soybean cultivars. Indian Jr. of Hill Farming, 2(2): 109–111.
Sharma, S.K., Verma, R.N. and Chauhan, S. 1991. A new host Record of *Colletotrichum capsici* on soybean from India. Inter. Jr. of Tropical Plant Diseases, 10(2): 277.
Sharma, S.K., Verma, R.N. and Chauhan, S. 1992. Soybean—A new host record for *Colletotrichum capsici* from India. International Journal of Tropical Plant Diseases, 10(2): 277.
Sharma, S.K., Verma, R.N. and Chauhan, S. 1993. Sources of resistance to Cercospora kikuchii incitant of purple stain disease of soybean. In current trends in life sciences, Vol. 29. Recent trend in plant disease control. pp. 415–427. Pub. By Today and Tomorrow's printer & Publishers. New Delhi.
Sharma, S.K., Verma, R.N. and Chauhan, S. 2000. Distribution and magnitude of purple seed stain disease of soybean in North–East India. Journal of Hill Research, 13(2): 105–107.
Short, G.E. and Wyllie, T.D. 1978. Inoculum potential of *Macrophomina phaseolina*. Phytopathology, 68(5): 742–746.
Sinclair, J.B. 1977. Infections soybean diseases of World importance. PANS, 23(1): 49–57.
Sinclair, J.B. 1983. Fungicide sprays for control of internally seedborne fungi. Seed Science and Technology, 11.
Sinclair, J.B. 1991 Latent infection of soybean plants and seeds by fungi. Plant Disease, 75: 220–224.
Sinclair, J.B. and Dhingra, O.D. 1975. An Annolated Bibliography of Soybean Diseases 1882–1974. Urbana, USA. INTSOY, Univ. Illinois, 280 pp.

Sinclair, J.B. and Shurtleff, M.C. 1975. Compendium of Soybean Diseases. American Phytopathological Soc., 68 pp.

Sinclair, J.B. and Backman, P.A. 1989. Compendium of Soybean Diseases. 3rd edition. St. Paul, MN, USA: American Phytopathological Society.

Singh, D.P. 1993. Relative susceptibility of soybean cultivars to pod blight caused by *Colletotrichum truncatum* (Schw.). Agricultural Science Digest Karnal, 13(2): 90–92.

Singh, B.B. and Thapliyal, P.N. 1977. Breeding for resistance to soybean rust in India. Rust of Soybean: The problem and research needs. INTSOY Series No., 12: 62–65.

Singh, Tribhuwan and Sinclair, J.B. 1985. Histopathology of *Cercospora sojina* in soybean seeds. Phytopathology, 75(2): 185–189.

Singh, R.R. and Shukla, P. 1987. Amino acid changes in black Gramm leaves infected with *Colletotrichum truncatum*. Indian Phytopathology, 40(2): 241–242.

Singh, D.S. and Pandey, K.K. 1998. Diseases of soybean and their management. pp. 253–270. *In*: Thind, T.S. (ed.). Diseasers of field Crops and their management. National Agricultural Technology Information Centre, Ludhiana, India.

Singh, D.S., Shukla, A.K. and Vasuniya, S.S. 1994. Varietal screening of soybean against Myrothecium leaf spot. Indian Phytopatahol., 47: 340–341.

Singh, S.K. and Srivastava, H.P. 1989. Some new fungal diseases of moth bean. Indian Phytopathology, 42(1): 164–167.

Singh, S.N. and Srivastava, S.K. 1994. Screening of soybean varieties against leaf spot disease caused by *Myrothecium roridum*. Indian Journal of Mycology & Plant Pathology, 24(3): 222.

Singh, O.V., Agarwal, V.K. and Nene, Y.L. 1973. Seed health studies in soybean raised in the Nainital Tarai. Indian Phytopathol., 26: 260–267.

Singh, R., Singh, S.B. and Singh, P.N. 2001. Effect of environmental conditions on development of anthracnose of soybean. Annals of Plant Protection Sciences, 9(1): 146–147.

Skripka, O.V., Sizova, T.P. and Babevo, E.N. 1986. Fungi on soybean seeds in Abkhazia Mikologiya. Fitopatologia, 20(4): 306–308.

Sreenivasaprasad, S., Mills, P.R., Meehan, B.M. and Brown, A.E. 1996. Phylogency and systematic of 18 Colletotrichum species based on ribosomal DNA Spacer sequences. Genome, 39: 499–512.

Srichuwong, S. 1992. Seed-borne infection of *Colletotrichum truncatum* (Schw.) andrus and moore in soybean seeds and its control. Ph.D. Thesis. Universiti Pertanian Malasia, Serdang, Selangor.

Srivastava, L.S., Vishwadhar, Sahambi, H.S. and Choudhary, D.N. 1982. A new leaf spot of soybean caused by *Setoshpaeria glycines*. Ind. Jr. Mycol. And Plant Pathology.

Srivastava, M.P. and Singh, A. 1973. Myrothecium disease of cotton. Haryana Agric. Univ. Jour. Res., 3: 221–223.

Srivastava, S.K. 2001. Soybean Disease Management in India. pp. 114–118. *In*: Bhatnagar, P.S. (ed.). Proceedings of India Soy Forum. 2001- Harnessing the Soy potential for health and wealth The Soybean Processor Association of India. (SOPA), Indore.

Srivastava, S.K. 2003. Myrothecium leaf spot, pod blight and seed decay of Soybean. pp. 83–87. *In*: Bhale, M.S., Khare, D., Sharma, N.D. and Verma, R.K. (eds.). On The Spot Diagnosis And Identification Of Major Seed Borne Diseases Of Soybean. Department of Plant Breeding and Genetics, J.N. Krishi Vishwa Vidyalaya, Jabalpur.

Srivastava, S.K. and Dhawan, S. 1979. Epidemiology of *Macrophomina* stem and root rot of *Brassica juncea* (L.) Czern & Coss in northern India. Proceedings Indian National Science Academy B., 45: 617–622.

Srivastava, S.K. and Dhawan, S. 1980. Relative pathogenicity of different isolates of *Macrophomina phaseolina* on *Brassica juncea*. Indian Journal of Botany, 3(2): 172–175.

Srivastava, S.K., Thakur, M.P. and Singh, P.P. 1988. Severity of *Myrothecium roridum* Tode ex Fries on soybean cv. JS 72-44 in relation to varying doses of nitrogen, phosphorus and potash. Indian Jour. Mycology & Plant Pathology, 18: 76.

Srivastava, S.K. and Agarwal, S.C. 1989. Rog Niyantran. pp. 133–167. *In*: Singh, O.P. and Srivastava, S.K. (eds.). Soybean Agro Botanical Publishers, Bikaner.

Srivastava, S.K., Thakur, M.P. and Singh, P.P. 1991. Effect of nitrogen, phosphorus and potassium on myrothecium leaf spot (*Myrothecium roridum*) of soybean (*Glycine max*). Indian Jour. Agric. Sciences., 61(5): 341–342.

Srivastava, S.K., Singh, P.P. and Khan, S.U. 1995. Effect of plant population and varieties on the myrothecium leaf spot (*Myrothecium roridum*) of soybean (*Glycine max*). Indian Phytopathology, 48(3): 360–362.

Srivastava, S.K. and Khan, S.U. 1994. Impact of host age at infection time on the severity of *Myrothecium* leaf spot disease of soybean. Indian Phytopathology, 47(2): 190–191.

Srivastava, S.K. and Khan, S.U. 1995. Effect of NPK fertilizer levels on the severity of Myrothecium leaf spot of soybean. Indian Jour. Of Mycology & Plant Pathology, 25(3): 322–323.

Srivastava, S.K. and Khan, S.U. 1996. Influence of age and density of inoculum on the severity of myrothecium leaf spot of soybean. Indian Jour. Of Mycology & Plant Pathology, 26(1): 1–3.

Srivastava, S.K. and Khan, S.U. 1997. Estimation of yield losses in soybean [*Glycine max* (L.) Merrill] due to myrothecium leaf spot disease in relation to crop age at the time of inoculation in India. International Jour. of Pest Management, UK, 43(2): 105–107.

Stone, G.M. and Seal, J.L. 1944. Plant diseases observed in Alabama in 1943. Plant Dis. Reptr. Suppl., 148: 276–280.

Su, G., Suh, S.O., Schneider, R.W. and Russin, J.S. 2001. Host specialization in the charcoal rot fungus, *Macrophomina phaseolina*. Phytopathology, 91(2): 120–126.

Sun, S.D. 1958. Soybean. 248 pp. Moscow.

Sutton, B.C. 1992. The genus *Glomerella* and its anamorph *Colletotrichum*. pp. 1–26. *In*: Bailey, J.A. and Jeger, M.J. (eds.). *Colletotrichum*: Biology, Pathology and Control CAB International: Wallingford.

Sydow, H. and Sydow, P. 1914. A contribution to knowledge of the parasitic fungi of the Island of Formosa. Ann. Mycol., 12: 105.

Sydow, H., Sydow, P. and Butler, E.J. 1906. Fungi India Orientalis. I. Ann. Mycol., 4: 424–445.

Tai, F.L. 1936. Notes on Chinese fungi. VII. Chinese Bot. Soc. Bull., 2: 45–66.

Takahashi, T. 1927. A Sclerotium disease of larkspur. Phytopathology, 17: 239–245.

Takasugi, H. 1936. Division of plant pathology and entomology. Control. Agr. Expt. Sta. Manchurian R.R., 1933, pp. 583–739.

Talbot, P.H.B. 1973. Aphyllophorale. pp. 327–344. *In*: Ainsworth, G.C., Sparrow, F.K. and Susman, A.S. (eds.). The Fungi—An Advanced Treatise. Vol. IV B Academic Press, New York.

Taubenhaus, J.J. 1919. Recent studies on *Sclerotium rolfsii*. Journal of Agricultural Research, 10: 127–138.

Thirumalachar, M.J. and Chupp, C. 1948. Notes on some *Cercospora* in India. Mycologia, 40: 352–362.

Thompson, A. 1928. Report of the mycologist. In Annual report for 1927 of heads of divisions of the department of agriculture, Federated Malaya States and Straits Settlements. Malaya Agr. J., 16: 161–168.

Thompson, A. and Johnston, A. 1953. A host list of plant diseases in Malaya. Commonweaslth Mycol. Inst., Mycol. Papers 52. 38 pp.

Tiffany, L.H. 1951. Delayed sporulation of *Colletotrichum* on soybean. Phytopathology, 41: 975–985.

Tisselli, O., Sinclair, J.B. and Hymowitx, T. 1980. Sources of resistance to selected fungal, bacterial, viral and nematode diseases of soybeans. INTSOY Series No., 18. Univ. of Illinois at Urbana 134 pp.

Tode, H.J. 1790. Fungi Mecklenburgensis select, 1: 25.

Tomoyasu, R. 1924. The causal fungus of purple seed of soybean. Jr. Plant. Prot. (Tokyo), 11: 310–315.

Townsend, B.B. and Willetts, H.J. 1954. The development of sclerotia of certain fungi. Ann. Bot., 21: 153–166.

Tu, C.C. and Kimbrough, J.W. 1978. Systematics and phylogeny of fungi in Rhizoctonia complex. Botanical Gaz., 139: 454–466.

Turner, G.J. 1964. New records of plant diseases in Sarawak for the year 1962. Gardens' Bull. (Singapore), 20: 369–376.

Twumasi, J.K., Hossain, A.M. and Fenteng, P.K. 1989. Occurrence of some cowpea diseases and the evaluation of resistance of selected cowpea lines to brown blotch disease in the forest ecology of Ghana. Tropical Grain Legume Bulletin No., 36: 7–9.

Upchruch, R.G. 1995. Genectic regulation of cercosporin production in *Cercospora kikuchii.* Journal of the American Oil Chemistry Society, 72: 1435–1438.

Vardaniya, I.Y. 1971. Diseases of soybean in Abkhazia. Zashch East. Mosk., 16: 40–41.

Vasudeva, R.S. 1963. Indian *Cercospore.* New Delhi, 245 pp.

Velicheti, R.K. and Sinclair, J.B. 1992. Reaction of seed borne soybean fungal pathogens to *Cercosporin.* Seed Science & Technology, 20: 149–154.

Velicheti, R.K. and Sinclair, J.B. 1994. Production of the cercosporin and colonization of soybean seed coats by *Cercospora kikuchii.* Plant Disease, 78: 342–346.

Verma, R.N., Kumar, S., Chandra, S., Sharma, S.K. and Chauhan, S. 1988. Diseases of soybean in North-Eastern states of India. Nat. Symp. On Insect Pests and Diseases of Soybean, 30–31. (Abstr.).

Vernetti, F.J., Tomm, G.O. and Casela, C.R. 1985. Occurrence of *Mycosphaerella phaseolorum, Phyllosticta glyconeum* and *Colletotrichum truncatum* in Mozambique. Fitopatologia Brasileira, 10(1): 163–165.

Vishunavat, K. 2003. Charcoal rot of soybean. pp. 55–59. *In*: Bhale, M.S., Khare, D., Sharma, N.D. and Varma, R.K. (eds.). On the Spot Diagnosis and Identification of Major Seed-borne Diseases of Soybean. Training Mannual No. DRS/02/2003, Department of Plant Breeding & Genetics, JN Krishi Vishwa Vidyalaya, Jabalpur, India, p. 125.

Vishwadhar and Chaudhary, D.N. 1982. Field reaction of some soybean varieties to purple stain disease. Indian J. Pl. Pathology, 12(3): 312–313.

Waterson, J.M. 1939. Annoteted list of diseases of cultivated plants in Bermuda. Bermuda Dept. Agr. Bull., 18.38 pp.

Weber, G.F. 1931. Blight of carrots caused by *Sclerotium rolfsii,* with geographic distribution and host range of the fungus. Phytopathology, 21: 1129–1140.

Weimer, J.L. 1947. Disease survey of soybean nurseries in the South. Plant Dis. Reptr. Suppl., 168: 27–53.

Whiteside, J.O. 1960. Diseases of legume crops in Southern Rhodesia. Rhodesia and Nyasaland Dept. Res. and Spec. Sera., Proc. Ann. Conf. off., 4: 52–57.

Wolf, F.A. and Lehman, S.G. 1924. Report of division of plant pathology. North Carolina Agr. Ext. Sta. Ann. Rpt., 47 pp. 83–85.

Wrether, J.A., Anderson, T.R., Arysad, D.M., Gai, J., Ploper, L.D., Porta-Puglia, A., Ram, H.H. and Yorimori, J.T. 1997. Soybean disease loss estimates for the top ten soybean producing countries in 1994. Plant Disease, 81: 107–110.

Wrater, A., Shannon, G., Balardin, R., Carregal, L., Escobar, R., Gupta, G.K., Ma, Z., Morel, W., Ploper, D. and Tenuta, A. 2010. Diseases effects on soybean yield in the top eight soybean-producing countries in 2006. Online. Plant Health Progress. Doi 10.1094/PHP-2010-70125-01P.S.

Wyllie, T. D. 1988. Charcoal rot of soybean—Current status. pp. 106–113. *In*: Wyllie, T.D. and Scott, D.H. (eds.). Soybean Disease of the North Central Region. St. Paul. MH, USA, APS Press.

Wyllie, T.D. and Calvert, O.H. 1969. Effect of flower removal and pod set on formation of sclerotia and infection of *Glycine max* by *Microphomina phascolina.* Phytopathology, 59: 1243–1245.

Yang, C.Y. 1977. Soybean rust in the Eastern Hemisphere. Rust of Soybean: The problem and research needs. INTSOY Series No., 12: 22–33.

Yoshii, H. 1927. Crop diseases in 1926. Ann. Agr. Expt. Sta. Chosan, 7: 21–34.

Yorinori, J.T. 1981. *Cercospora sojina*. Pathogenicity, new races and seed transmission in soybeans. Dissertation Abstracts International B., 42(2): 448B–449B.

Yum, K.J. and Park, E.W. 1990. Soybean leaf diseases caused by Myrothecium roridum. Korean Jour. Pl. Pathol., 6: 313.

Zad, J. 1979. Mycoflora of soybean seeds. Iranian Journal of Plant Pathology, 15(1/4): 42–47; 32–33.

CHAPTER 15

Ecology of Endemic Dimorphic Pathogenic Fungi

Harish C. Gugnani

ABSTRACT

This chapter presents an update of the present state of our knowledge on the ecology of endemic dimorphic pathogenic fungi—*Histoplasma capsulatum* var. *capsulatum*, *H. capsulatum* var. *duboisii*, *Blastomyces dermatitidis*, *Paracoccidioides brasiliensis*, *Coccidioides immitis*, *C. posadasii*, principal agents of chromoblastomycosis, viz. *Fonsecaea pedrosoi*, *Phialophora verrucosa*, and *Cladophialophora carrionii*, and *Penicillium marneffei*. These fungi have a specific niche in nature. The microclimatic conditions for survival and multiplication of these endemic fungi in soil are described and discussed in detail in relation to epidemiology of human infections caused by them. Association of these pathogenic fungi with animals, viz. birds, rodents, canines, felines and armadillos, and their dispersal in the environment in relation to human infections are reviewed. Preventive measures for the control of human infections caused by systemic dimorphic pathogenic fungi are also described.

Introduction

The endemic dimorphic pathogenic fungi have two distinct morphological forms. These fungi include *Histoplasma capsulatum*, *Blastomyces dermatitidis*, *Paracoccidioides brasiliensis*, *Coccidioides immitis*, *C. posadasii*, *Sporothrix schenckii*, agents of chromoblastomycosis (mainly *Fonsecaea pedrosoi*, *Phialophora verrucosa*, *Cladophialophora carrionii*) and *Penicillium marneffei*

Saint James School of Medicine, Kralendjik, Bonaire (Dutch Caribbean), West Indies.
Email: harish.gugnani@gmail.com, harishgugnani@yahoo.com

(Kwon-Chung and Bennett, 1992; de Hoog et al., 2000; Brandt and Warnock, 2007; DiSalvo, 2009). In culture and in the environment, they grow as filaments with characteristic spores; while in the infected tissue they grow mostly as yeasts, and sometimes as other forms, viz. spherules containing endospores or septate bodies. The yeast form can also be produced *in vitro* by cultivating the fungi at 37°C on a rich medium, e.g., brain-heart infusion agar or blood agar supplemented with 0.1% cysteine. Except for *S. schenckii* and agents of chromoblastomycosis, human infections due to these fungi occur initially in the lungs by inhalation of infectious spores or hyphal elements from an environmental source; later the infection disseminates to other internal organs. Infections due to *S. schenckii* and agents of chromoblastomycosis are initiated by traumatic implantation of the causative fungus from the environment into the skin/subcutaneous tissue (Kwon-Chung and Bennett, 1992), the fungi invading the dermis, subcutaneous tissue, muscle and facia. Dimorphic fungi are identified in the laboratory by several morphological and sometimes also by biochemical characteristics. The morphological characters include the colonial characters, appearance and arrangement of their asexual spores, which may be large (macroconidia) or small (microconidia, arthroconidia) and the tissue form, which may be yeast, sometimes spherules or septate bodies. *In vitro* conversion to yeast from and exo-antigen tests are required where applicable, for confirming the identity of the isolate. Salient clinical symptoms of human diseases caused by dimorphic fungi, the diagnostic characters of the mycelial and tissue forms of the causative agents are given in Table 15.1. The microscopic features of the culture and tissue forms of the dimorphic fungi are illustrated in Figures 15.1–15.15. Ecology of these fungi and the mechanism of spread of their infectious spores to humans are of great importance in understanding the epidemiology of their infections, and that has implications in public health (Ajello and Hay, 1998; Brandt and Warnock, 2007). This chapter attempts to present a concise update of our knowledge of the ecology of different species of dimorphic pathogenic fungi.

Ecology of *Histoplasma capsulatum*

Histoplasma capsulatum is a soil saprophyte that produces mycelium bearing microconidia (infective spores), and macroconidia in culture at room temperature and in the environment, and narrow budding yeast cells in the infected tissue. Inhalation of airborne microconidia is generally the route of infection. *H. capsulatum* has two varieties, viz. var. *capsulatum* and var. *duboisii*. *H. capsulatum* var. *capsulatum* is the etiological agent of histoplasmosis (also called classical, American histoplasmosis or histoplasmosis capsulati) endemic in North America, and parts of Asia, Africa and South America. *H. capsulatum* var. *duboisii* (producing much

Table 15.1. Endemic dimorphic pathogenic fungi, diseases, caused by them and their diagnostic characters.

Species of fungus	Disease caused	Symptoms	Diagnostic characters
Histoplasma capsulatum var. *capsulatum*	Histoplasmosis	Pulmonary infection, asymptomatic to progressive, often disseminating to hilar and mediastinal lymph nodes, liver, spleen and bones	Delicate mycelium, gray to gray-brown producing large, rounded, single-celled spiny (tuberculate) macroconidia, 8–14µm in diameter, and small oval micro-conidia (Fig. 15.1). 2–4 µm at room temperature (RT). Small narrowly budding yeast cells (2.5–4 µm) in the macrophages and outside in the infected tissue, and *in vitro* at 37°C (Fig. 15.2)
Histoplasma capsulatum var. *duboisii*	African histoplasmosis	Papules and nodules on the skin with a characteristic hyperpigmented halo around them, ulcerative lesions on skin and subcutaneous abscesses may occur. Skull, ribs, vertebrae, humerus femur, tibia, wrist are frequently involved.	Mycelial form as in *H. capsulatum* var. *capsulatum* at RT. Large oval (8–15 µm) narrowly budding yeast cells in giant cells (Fig. 15.3) and macrophages in the infected tissue, and *in vitro* at 37°C
Blastomyces dermatitidis	Blastomycosis	Chronic pneumonia. Disseminating most commonly to the skin, bone, and, in males, prostate	Gray-white to light-buff delicate mycelium forming ovoid to pyriform, one-celled, smooth-walled conidia 2–10 µm (Fig. 15.4) at RT. Large yeast cells, 6–10 µm that have characteristically a single bud attached to the parent cell by a broad base in the infected tissue (Fig. 15.5), and *in vitro* at 37°C
Paracoccidioides brasiliensis	Paracoccidioidomycosis	Cough, dyspnea, malaise, fever and weight loss are common symptoms. Muco-cutaneous, oral laryngeal and pharyngeal lesions occur frequently in patients with acute pulmonary infection	Pyriform microconidia, chlamydospores and arthroconidia formed on hyphae (Fig. 15.6). Multiple budding cells in the infected tissue, and *in vitro* at 37°C (Fig. 15.7)

Coccidioidoides immitis and *C. posadasii*	Coccidioidomycosis	Pulmonary infection with fever, chills and cough, occasionally with pleuritic pain. May heal or progress to disseminated form with very high rate of mortality	Mycelial form produces single celled, rectangular to barrel-shaped arthroconida, 2.5–4 x 3.5 μm, separated by disjunctor cells (Fig. 15.8), Spherules containing endospores in the infected tissue (Fig. 15.9)
Sporothrix schenckii	Sprorotrichosis	Nodules (at the site of introduction) that ulcerate and spread along lymphatics	Ovoid or elongated, 3–6 x 2–3 μm, one-celled conidia formed in petaloid clusters on tiny denticles by sympodial proliferation at the apex of the conidiophore (Fig. 15.10). Oval, elongate or cigar shaped budding yeast cells in the infected tissue, and *in vitro* at 37°C (Fig. 15.11)
Fonsecaea pedrosoi, Phialophora verrucosa, Cladiophialophora carionii	Chromoblastomycosis	Warty nodules that progress to have "cauliflower-like" topography (at the site of introduction)	Dark-colored mycelium with characteristic sporulation in each species at RT. (Fig. 15.12), and brown sclerotic cells dividing by fission (septate bodies) in the infected tissue (Fig. 15.13)
Penicillium marneffei	Penicilliosis marneffei	Fever, weight loss, hepatomegaly and characteristic umblicated, necrotic papules on the skin	Brush like conidiophores with long curving chains of conidia at room temperature (Fig. 15.14); fission yeast in the infected tissue, and *in vitro* at 37°C (Fig. 15.15)

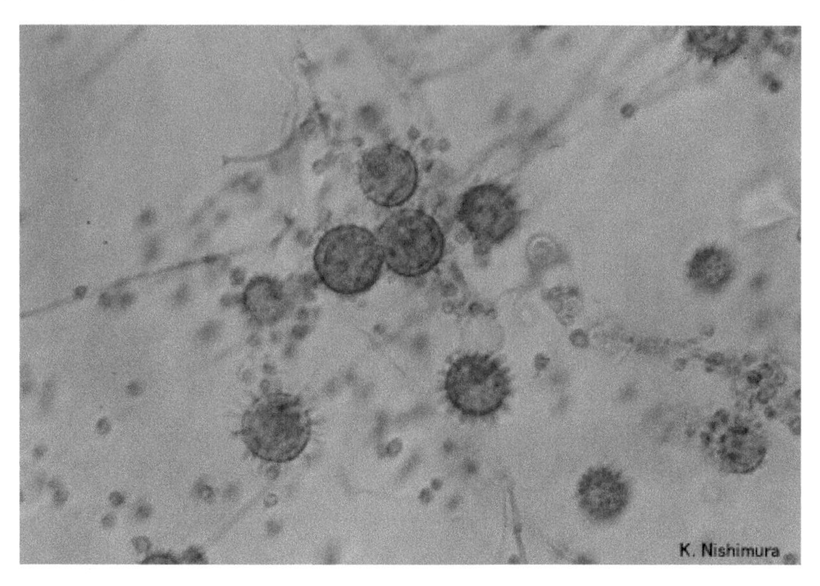

Figure 15.1. Tubercualte macroconidia and pyriform microconidia of *H. capsulatum* var. *capsulatum*, x 425.

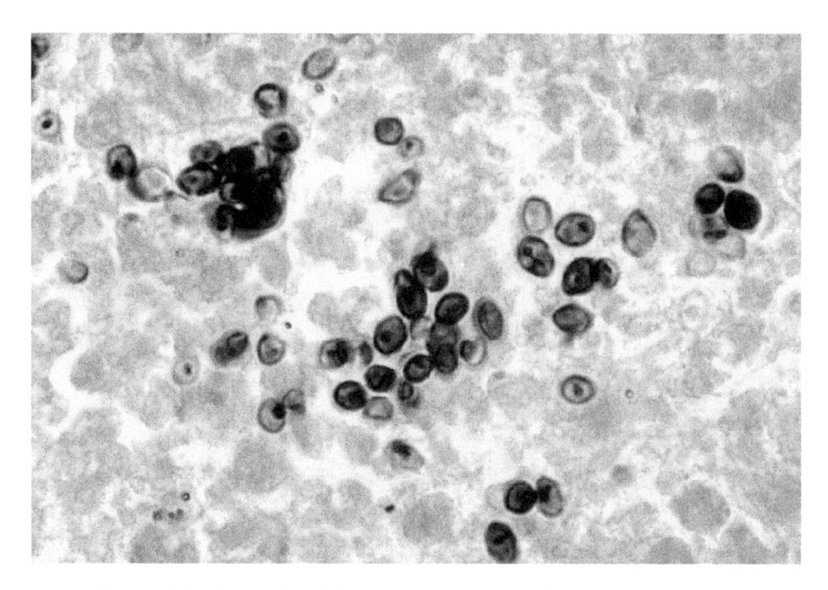

Figure 15.2. Yeast cells of *H. capsulatum* var. *capsulatum* in tissue, x 825.

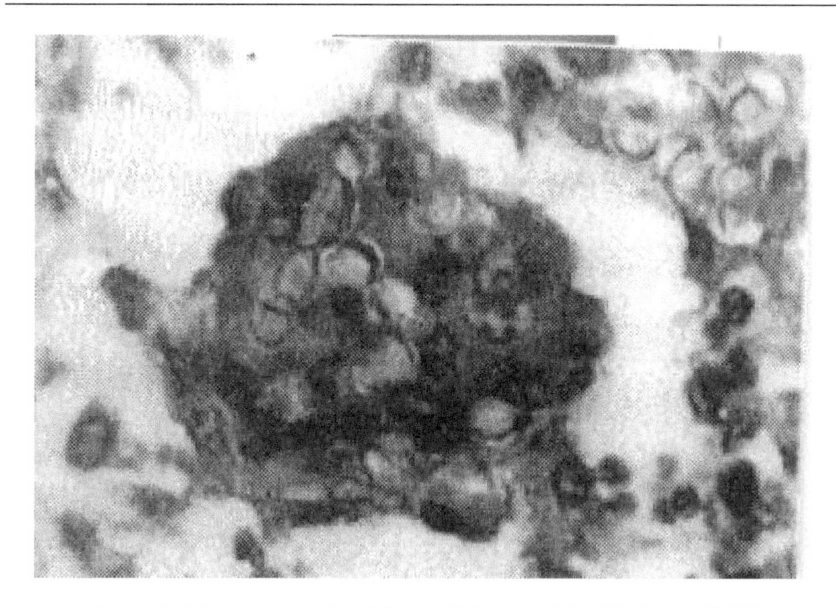

Figutr 15.3. Large yeast cells of *H. capsulatum* var. *duboisii* in tissue, X 475.

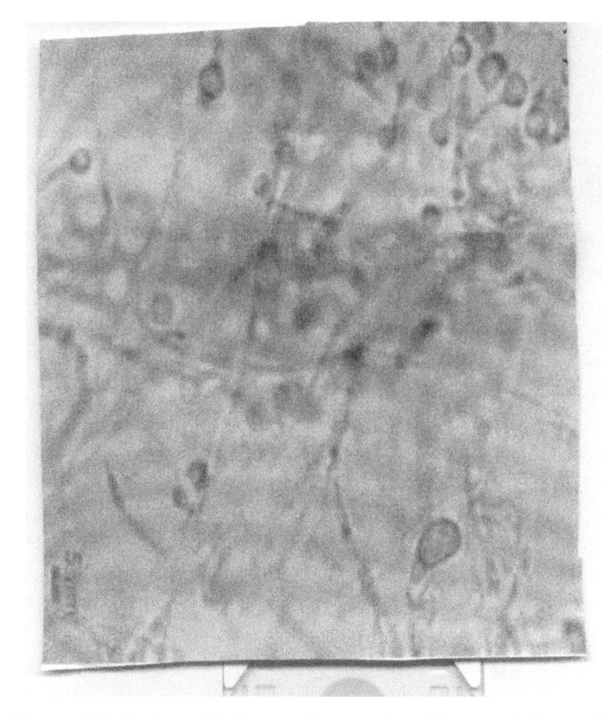

Figure 15.4. *B. dermatitidis;* mycelium and pyriform conidia, x 425.

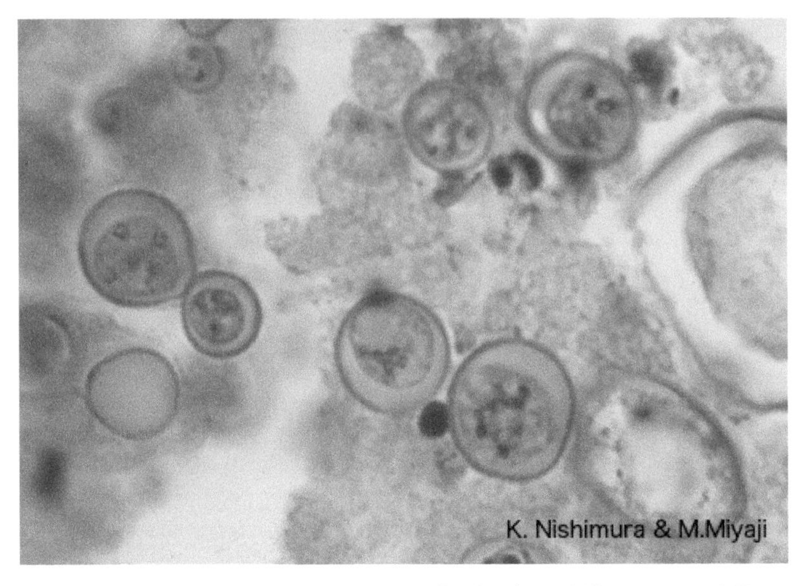

Figure 15.5. Broad-based budding yeast cells of *B. dermatitidis* in tissue, x 875.

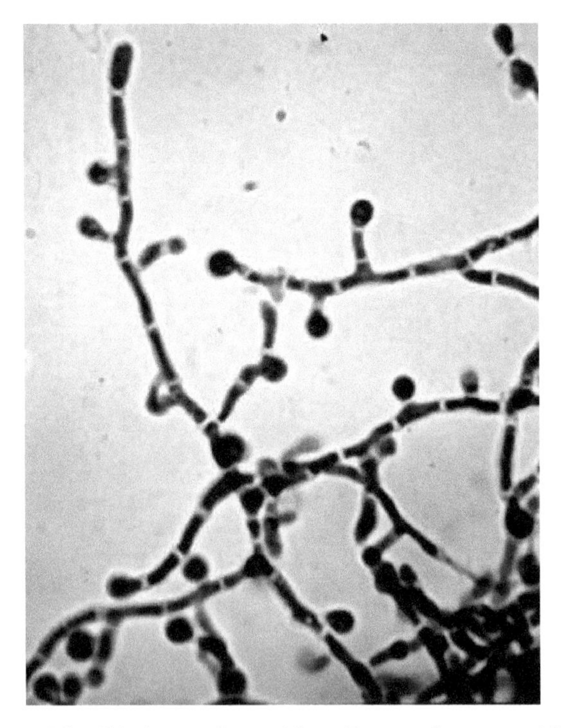

Figure 15.6. *P. brasileinsis*; mycelium with pyriform to elongate conidia, x 475.

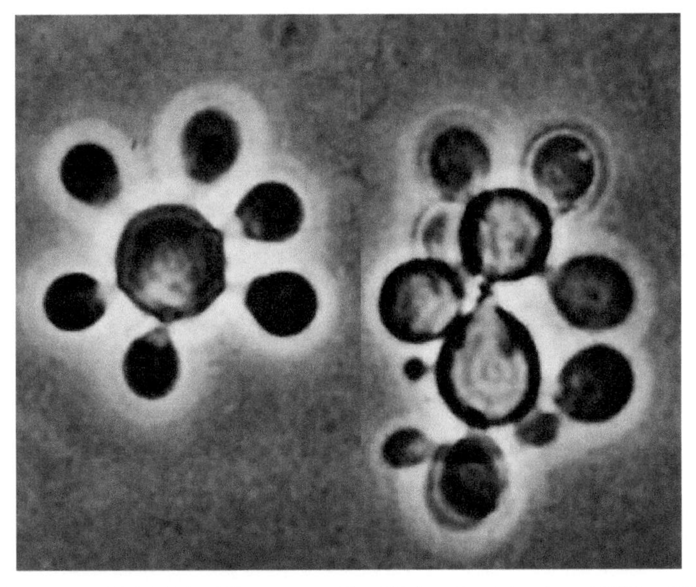

Figure 15.7. Multiple budding yeast cells of *P. brasiiensis in vitro* at 37°C, x 575.

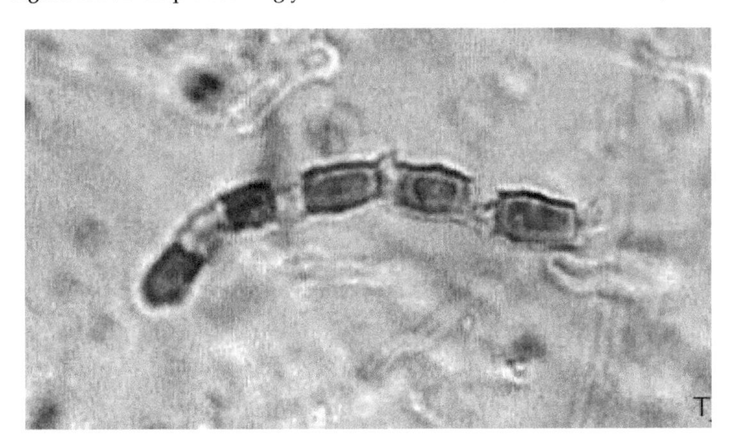

Figure 15.8. *C. immiti* and *C. posadasii* arthroconida separated by disjunctor cells, x 875.

larger yeast cells in the infected tissue) causes histoplasmosis duboisii (also called African histoplasmosis) is endemic in Africa (Ajello, 1983; Gugnani, 2000). Classical histoplasmosis primarily affects the lungs often disseminating to other internal organs, while histoplasmosis doboisii is characterized by the presence of granulomatous and ulcerative lesions in the skin, subcutaneous tissues and bones. Lungs and other internal organs are rarely involved (Ajello, 1983; Kwon-Chung and Bennett, 1992; Gugnani and Mouto-Okafor, 1998; Brandt and Warnock, 2007).

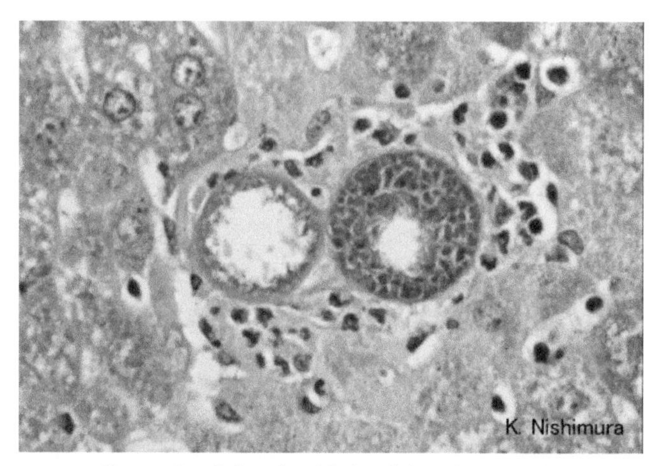

Figure 15.9. Spherule of *C. immiitis* in tissue, x 525.

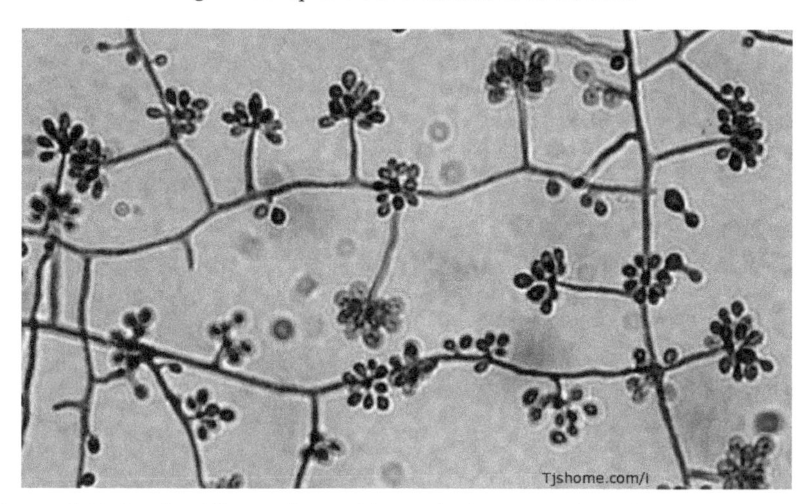

Figure 15.10. *S. schenckii*; mycelium with conidia borne on short denticles, in petaloid arrangment on conidiophores, x 575.

H. capsulatum var. *capsulatum* is a soil-based fungus that has been isolated from many regions of the world and is most often associated with river valleys; the most highly endemic region is the Ohio and Mississippi River valleys. The factors favoring the growth of this fungus in soil are a mean temperature of 22° to 29°C, an annual precipitation of 35 to 50 inches, and a relative humidity of 67–87%. Such conditions are typically found in the temperate zone between latitudes 45° North to 30° South (Kwon-Chung and Bennett, 1992; Tewari et al., 1998). According to Centers for Disease Control and Prevention (CDC) estimates, approximately 50 million people had been infected in North America by 2010, most asymptomatically. The

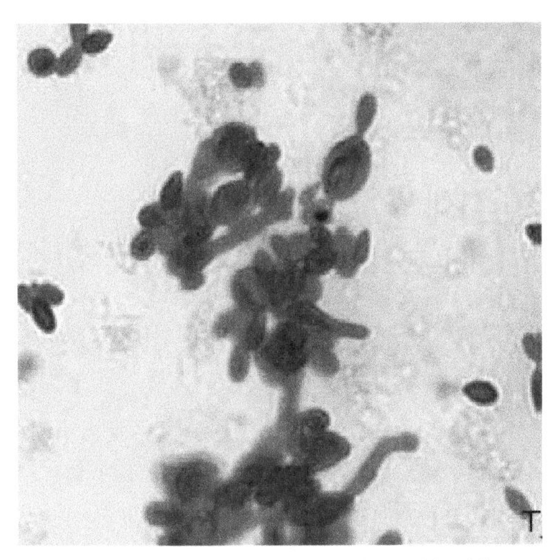

Figure 15.11. *S. schenckii*; oval, elongate or cigar shaped budding yeast cells, x 575.

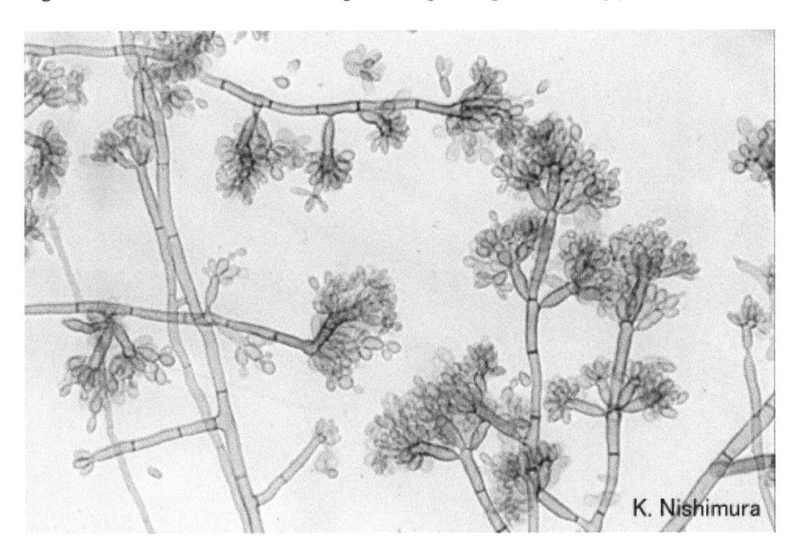

Figure 15.12. *F. pedrosoi*; conidiophores bearing clavate to ellipsoidal conidia in short chains.

organism is typically found within 20 cm of the soil surface, and it prefers soil that is moist, acidic, and has high nitrogen content. When suitable microclimatic conditions are present and soils are enriched with feces of birds in the chicken houses, or with bat guano in roosting sites of bats, the environment is favorable for the growth of *H. capsulatum* var. *capsulatum*. Avian habitats appear to be especially suitable for the proliferation of *H. capsulatum* (CDC 24/7, 2013a). In such areas, infectious particles can

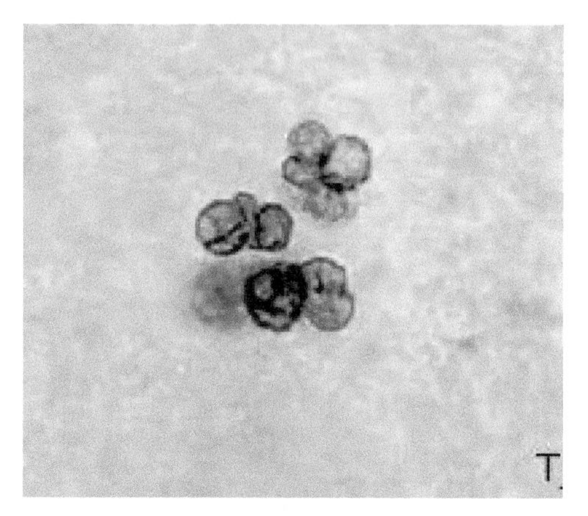

Figure 15.13. Septate bodies in wet mount from a lesion in a case of chromoblastomycosis, x 475.

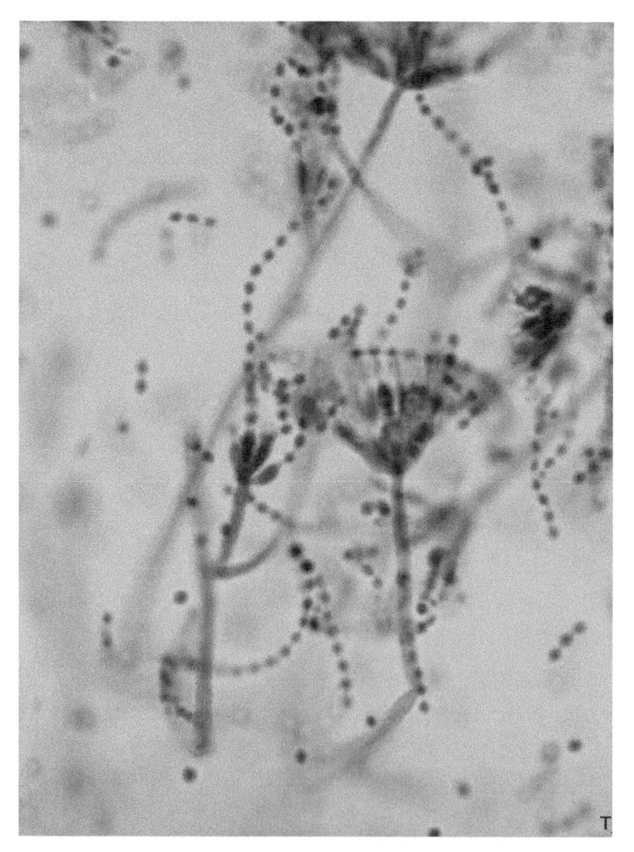

Figure 15.14. *P. marneffei*; biverticillate and monverticillate penicill bearing curved chains of conidia, x 425.

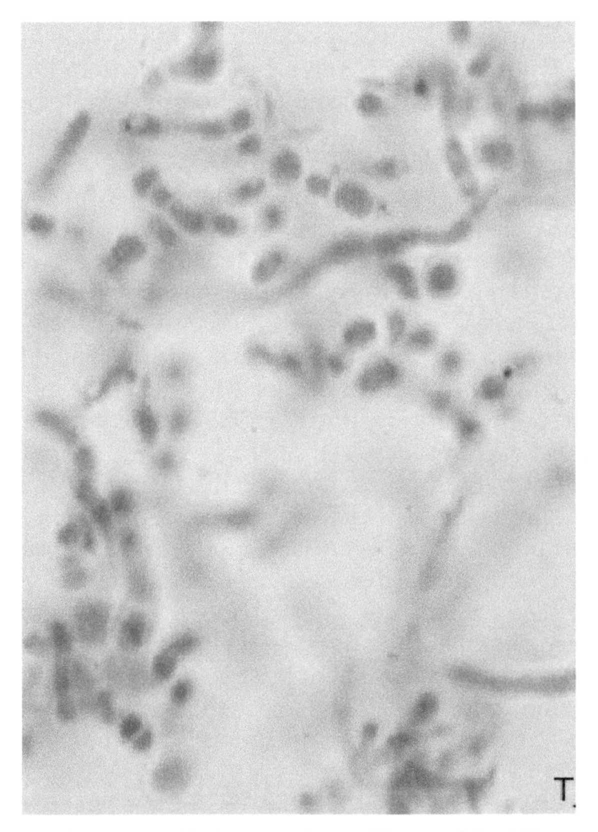

Figure 15.15. Fission yeast form of *P. marneffei*, x 425.

exceed 10^5/g of soil. Natural infection has never been demonstrated in birds presumably because of their high body temperature. Persons spreading chicken manure around their garden often develop pulmonary infection a few weeks later. Roosting sites of other birds also promote the growth of the fungus. Outbreaks of histoplasmosis have been associated with construction and renovation activities that disrupt contaminated soil. Several outbreaks have been reported from South Carolina where workers used bulldozers to clean the canebrakes which served as blackbird roosts. All workers and bystanders, who were exposed, contracted histoplasmosis (DiSalvo and Johnson, 1979). Contaminated soil can be potentially infectious for years.

Cases of histoplasmosis have also been described in animals including cats and dogs from some countries (Kwon-Chung and Bennett, 1992).

Chiroptera are considered to play a significant role in the epidemiology of histoplasmosis. In contrast to birds, bats may become infected, and transmit the infection through their droppings. Recovery of *H. capsulatum* var. *capsulatum* from various organs of free flying bats has been well documented, with most studies relying on cultures from the liver, spleen and lung. Demonstration of tissue infection by histologic techniques correlates poorly with the positive culture for *H. capsulatum*, and no cellular response to tissue infection has been noted (Dierks, 1965). Talyor et al. (1999) isolated *H. capsulatum* from the gut, lung, liver, and spleen of 17 of the 208 captured bats representing six species of six different genera in a Mexican cave environment. All were adult bats: six males (one *Pteronotus parnellii*, two *Natalus stramineus*, two *Artibeus hirsutus*, and one *Leptonycteris nivalis*) and 11 females (one *Myotis californicus*, one *Mormoops megalophylla*, eight *A. hirsutus*, and one *L. nivalis*). Dias et al. (2011) isolated *H. capsulatum* var. *capsulatum* from 87 (3·6%) of 2427 bats in Brazil. The infected bats, all insectivorous species were identified as *Molossus molossus*-74, *Nyctinomops macrotis*-10, *Tadarida brasiliensis*-1, *M. rufus*-1 and *Eumops glaucinus*-1. The presence of bats in caves, attics, ceilings and roofs is important epidemiologically as they can increase the chance of human acquisition of pathogens, including *H. capsulatum* var. *capsulatum*. Histoplasmosis occurs as a significant occupational disease in bat caves in Mexico, when workers harvest the guano for fertilizer. Fresh guano is less likely to contain any infectious microconidia of the fungus (Taylor et al., 1999). Experimental observations have suggested that at least in some bat species, controlled chronic infection occurs thus allowing the bat to excrete viable cells of *H. capsulatum* var. *capsulatum* over a long period of time (McMurray and Greer, 1979).

Information on the natural habitats of *H. capsulatum* var. *capsulatum* in India has been lacking, though histoplasmosis has been sporadically reported from several parts of the country (Gopalakrishnan et al., 2012). There is only one report of natural occurrence *H. capsulatum* var. *capsulatum* from India. It concerns isolation of the fungus from one of the three samples of soil admixed with bat guano collected from an abandoned room of a 350-years old palatial building infested with insectivorous bats of the species, *Scotophilus heathi* in Serampur near Kolkata in West Bengal (Sanyal and Thammaya, 1975). Attempts to isolate *H. capsulatum* var. *capsulatum* from any of the 630 wild animals belonging to 11 species and 10 genera of small mammals examined in India were negative (Gugnani, 1972); the fungus also could not be recovered from any of the 40 bats examined (Gugnani, unpubl. data).

Information on the ecology of *H. capsulatum* var. *duboisii* is scanty. A few reports of var. *duboisii* infections of humans associated with chicken coops and caves infested with bats had suggested that the var. *duboisii* has the same ecological niche as the var. *capsulatum*. Al-Doory and Kalter (1967) reported isolation of *H. capsulatum* var. *duboisii* from pooled soil samples collected in Kenya but it was a misidentification as the isolate was later identified to be

H. capsulatum var. *capsulatum* (Ajello, 1983). The environmental occurrence of *H. capsulatum* var. *duboisiii* was unequivocally demonstrated with the discovery of a natural focus of this fungus in a bat cave in a rural area, viz. Ogbunike, in eastern Nigeria, by Gugnani et al. (1994). Eight of the 45 samples of soil admixed with bat guano were positive for the fungus. Also, out of the 35 bats of two species, viz. *Nycteris hispida* (The hairy slit-faced bat with long ears) and *Tadarida pumila* (The little free-tailed bat) examined from the cave, one of them (*N. hispida*) yielded *H. capsulatum* var. *duboisii* from its intestinal contents. Identity of the isolates was confirmed by exoantigen tests, mating with tester strains of *H. capsulatum*, and by *in vitro* conversion to large yeast form (Gugnani et al., 1994). Further, an investigation of skin and serum reactivity among humans to histoplasmin in the vicinity of the natural focus (bat cave) of *H. capsulatum* var. *duboisii* found a very high prevalence (35%) of skin sensitivity among the cave guides, traders and farmers (Muotoe-Okafor et al., 1996). Also, 17 (9.4%) of 181 young adults, including farmers, palm oil workers resident in the vicinity of cave demonstrated precipitating antibodies to histoplasmin in their sera. These findings established the natural reservoir of *H. capsulatum* var. *duboisii*, and the role of a bat habitat in the epidemiology of human infections due to this fungus (Muotoe-Okafor et al., 1996). Spontaneous natural infections (including an outbreak) due to *H. capsulatum* var. *duboisii* have been reported in baboons (*Cynocephalus babuin, Papio cynocephalus papio*). All infected baboons were detected in France, and the USA and they had originated from countries in West Africa (Butler et al., 1988; Butler and Hubbard, 1991).

Ecology of *Blastomyces dermatitidis*

Blastomyces dermatitidis, the etiological agent of blastomycosis is a dimorphic fungus occurring as mycelial form with pyriform conidia in culture, and broad-based budding yeast cells in the infected tissue (Brandt and Warncok, 2007). The fungus possibly occurs as a saprobe in soil and decaying woody material (Baumgardner et al., 1999). Isolation of *B. dermatitidis* from the environment is most likely when the sample contains soil rich in organic material such as animal feces, plant fragments, insect remains, dust and has a pH less than 6.0 (Klein et al., 1987). Blastomycosis is endemic in North America with highest incidence of infection occurring in the eastern part of the USA. Most reported cases outside of the United States are from Canada (Ontario and Manitoba) and the African continent. Cases have also been reported from Mexico, South America, Middle East and India (DiSalvo, 2009). Infection commonly occurs in people collecting firewood, tearing down old buildings or engaged in other outdoor activities which disturb the soil (Kaufman, 2011). It is difficult to isolate *B. dermatitidis* from soil. With only a few exceptions, attempts to recover *B. dermatitidis* from soil in the areas endemic for the disease have been unsuccessful (DiSalvo, 2009). The fungus thrives in wet environments, such as riverbanks, lakes and swamps, where damp soil lacking direct sunlight fosters growth of the fungus. Changing climatic factors such as rainfall, temperature and humidity, as well as soil pH and organic content, are believed to play a role in the environmental occurrence of *B. dermatitidis* (Klein et al., 1986). Exposure to soil, the most identified risk factor for infection during outbreaks, may be increased during excavation and construction activities as well as recreational activities along waterways (Klein et al., 1986, 1987). Although *B. dermatitidis* grows more abundantly in moist acidic soils, the spores aerosolize less readily in moist soil than in dry soil conditions if the soil is disturbed (Klein et al., 1987).

Contrary to the frequent recovery of *Histoplasma capsulatum* var. *capsulatum* from a wide variety of bat species in the American continent (Dierks, 1965; Taylor et al., 1999), isolation of *B. dermatitidis* from the visceral organs of bats is extremely rare. In 1967, Tesh and Schneidau (1967) had suggested the possibility of bats serving as carriers or hosts of *B. dermatitidis* on the basis of experimental infection of bats of the species, *Tadarida brasiliensis* (the free-tailed bat) with this fungus. Evidence for the association of bats with *B. dermatitidis* was provided by the reports of the recovery of this pathogen, from the visceral organs of bats (*Rhinopoma hardwickei hardwickei*) in India by Khan et al. (1982) and Randhawa et al. (1985). A very recent critical appraisal of the reports on blastomycosis from India reveals that only two human and one canine case can be accepted as autochthonous and valid (Randhawa et al., 2013).

Dogs that are frequently exposed to environments where *B. dermatitidis* exists are at increased risk of developing blastomycosis. Dogs are infected by inhaling the infectious particles and are good indicators of the existence of *B. dermatitidis* in the nearby environment. Dogs that are at greatest risk of developing clinically apparent blastomycosis are 2–4 year old intact male large-breed dogs living in endemic areas (Legendre, 2006). However, there seems to be no correlation between the existence of dogs with blastomycosis and emergence of human cases in a defined environmental area. Also, there is no evidence of animal-to-human transmission of the disease.

Ecology of *Paracoccidiodes brasiliensis*

Paracoccidioides brasiliensis, a thermally dimorphic fungus is the etiological agent of paracoccidioidomycosis (PCM), the most important systemic mycosis in Latin America (Franco et al., 1994). When grown at room temperature, *P. brasiliensis* is a mold composed of thin septate hyphae and conidia (infective form). At the human body temperature (37°C) and in the infected host, the fungus exists as yeast reproducing by multiple budding. Naturally acquired *P. brasiliensis* infection in animals has been reported only in armadillos (Baggali et al., 2003). *P. brasiliensis* has also been recovered from the feces of bats, *Artibeus lituratus* (Grosse and Tramsitt, 1966). Ecology of PCM is not completely understood. There are only a few reports on the recovery of *P. brasiliensis* from soil samples and that generally could not be repeated from the same location. *P. brasiliensis* has been isolated from soil, only once from Argentina (Negroni, 1966), twice from the same farm in Venezuela (Albornoz, 1971), and once from a coffee growing area in Brazil (Silva-Vergara, 1998). This implies that soil is most probably not the actual permanent habitat of the pathogen. A very limited number of isolations of *P. brasiliensis* from soil samples despite many attempts, usually not repeated in samples taken from the same localities, make it difficult to accept soil as the pathogen's permanent habitat. Probably *P. brasiliensis* resides in the soil only transiently and for a variable time (Franco et al., 2000; Conti-Dias, 2007).

With the help of a molecular method (nested PCR), Terçarioli et al. (2007) were able to detect *P. brasiliensis* in samples of the sandy and clayey soils collected from armadillo burrows. The authors observed that the fungus can develop both in clayey and sandy soils when the humidity is high. The demonstration of a high incidence of *P. brasiliensis* infection in the nine-banded armadillo, *Dasypus novemcinctus*, a wild mammal that typically has evolved in South America and lives burrowed in soil (Baggali et al., 2003; Baggali and Barrozo, 2005), has opened new perspectives for understanding the pathogen's ecology and evolution. From *D. novemcinctus* and also from the naked-tailed armadillo, *Cabassous centralis*, *P. brasiliensis* can be isolated regularly. Current epidemiological evidence indicates that *D. novemcinctus* is a key factor in the epidemiology of PCM. In addition to being a natural host and possible reservoir, its geographical distribution in Latin America clearly overlaps with the endemic areas of PCM (Baggali and Barrozo, 2005). The systematic recovery of *P. brasiliensis* from armadillo tissues has demonstrated the importance of this animal in PCM endemic areas, helping to locate hot spots of fungal occurrence in some environments. It has also suggested valuable insights about its evolutionary aspects (Bagagli et al., 2008). Very recently, Arentes et al. (2013) demonstrated that the aerosol sampling associated with molecular detection through Nested PCR proved the best method for discovering *P. brasiliensis* in the environment.

Ecology of *Coccidioides* spp.

Coccidioides immitis and *C. posadasii*, etiological agents of coccidioidomycosis are thermally dimorphic fungi found in soil particularly in warm and dry areas with low rainfall (<600 mm annual rainfall), high summer temperatures and latitudes located between 40° North and 40° South (Ajello,

1971; de Hoog et al., 2000; Galgiani, 2010; Colombo et al., 2011). The two species are morphologically identical but genetically and epidemiologically distinct (Kleiman, 2008). *C. immitis/posadasii* are found 5–30 cm below the surface of alkaline soils, and are also isolated from rodent burrows in desert-like areas of Southwest United States. *C. immitis* is geographically limited to California's San Joaquin valley region, whereas *C. posadasii* is found in the desert Southwest of the United States, Mexico and South America. The two species appear to co-exist in the desert southwest and Mexico (Sifuentes-Osorino et al., 2012).

Coccidioidomycosis results from inhaling spores (arthroconidia) of *C. immitis*/*C. posadasii*. According to CDC, at least 30–60% of people who live in endemic areas are exposed to the fungus at some point during their lives. In most people, the infection goes away on its own. Each year the fungus infects more than 150,000 people, many of whom are sick without knowing the cause or have mild infections that are not detected (Galgiani, 2010). There is a relationship between temperature and precipitation, and outbreaks of coccidioidomycosis. Incidence of the disease varies seasonally as well as annually due to changing climatic conditions. However, the specific environmental conditions that may produce an outbreak of coccidioidomycosis are not well understood in space and time (Kolivaras et al., 2001). The ecology of the pathogen, *Coccidioides*, remains obscure and there is limited knowledge of the environmental antecedents of disease outbreaks (Barker et al., 2012). Using BALB/c mice as a biosensor and molecular techniques, 8.9% of soils analyzed from the Tucson area (Pima County, Arizona) were found to contain *C. posadasii* (Barker et al., 2012). *C. immiitis* has also been recovered from armadillos (Eulalio et al., 2001). Reported point source outbreaks are infrequent and have followed diverse soil-disrupting activities or events, such as archeological sites (Perera and Stone, 2002; Peterson et al., 2004) or anthropological digs and construction work (Cordeiro et al., 2006; Galgiani, 2010), earthquakes (Schneider et al., 1997) dust storms (Williams et al., 1979; Kwok et al., 2009), and armadillo hunting (Lima et al., 2012). The repeated isolation of *C. immitis* from the soil has been from small, well defined areas (Green et al., 2000). Animals, although not the major natural reservoir, cannot be ignored as possible factors in the ecology of *C. immitis*.

Ecology of *Sporothrix shenckii*

Sporothrix shenckii, a dimorphic fungus is the agent of sporotrichosis, a chronic fungal disease usually limited to cutaneous, lymphocutaneous and subcutaneous tissues. *S. schenckii* occurs as a filamentous mold with conidia in petaloid arrangement in culture and in the environment, and as elongate, oval or cigar shaped budding yeast cells in the infected tissue. *S. schenckii* is a cosmopolitan fungus, isolated from soil and decaying plant materials. *Sphagnum* moss and rose bushes are well known sources of this organism (Roberts, 1975; Dixon and Salkin, 1991). The primary habitats for *S. schenckii* are the soil and plants. Several factors such as temperature, humidity and necessary nutrients influence the survival of *S. schenckii* in nature. The most common way of the infection is the skin trauma by a prick from a thorn or a piece from a variety of barks, trees and bushes. *S. schenckii* has been isolated from humid straw, prairie hay, desiccated mushrooms, potting soil, armadillo burrows and from stored *Sphagnum* moss, but not from any moss directly from bogs nor from living plants (Zhang and Andrews, 1993).

Occupations that predispose persons to infection include: farmers, gardeners, florists, tree planters, forestry workers, horticulturists, workers of the greenhouses, orchids growers, outdoor laborers, masons, veterinarians and laboratory workers (microbiologists). People involved in activities and exposure to contaminated soils, wood, *sphagnum* moss, wheat, straw, prairie hay, corn husks and strubble, barks, thorny plants and shrubs, reeds, rose bushes, cat's skin ulcers, rodent bites and punctures caused by insects are also more prone to infection (Bravo, 2012). Several domestic animals and rodents are carriers of this fungus (Bravo, 2012). In the spring of 1988, the largest documented US-outbreak of cutaneous sporotrichosis occurred, with 84 culture-confirmed cases among persons from 15 states, who were exposed to Wisconsin-grown *Sphagnum*-moss used in packing evergreen tree seedlings (Coles et al., 1992). Environmental samples of moss from the Wisconsin supplier were negative, but *Sporothrix schenckii* was cultured from multiple samples obtained from one of six Pennsylvania tree nurseries, identified as the source for 79 (94%) of cases-associated with handling the 1-to-3 year old stored moss (Coles et al., 1992). Sporotrichosis is also endemic in the sub-Himalayan belt of India, which extends from the northern to the northeast Indian subcontinent (Haldar et al., 2010; Verma et al., 2012). However, attempts to recover *S. schenckii* from soil and other environmental sources to which many patients had been exposed were negative (Haldar et al., 2010; Verma et al., 2012).

Epidemic sporotrichosis is also considered an emerging zoonosis. From 1998 to 2004, 759 humans, 64 dogs and 1503 cats were diagnosed with sporotrichosis in a research center in Brazil (Schubach et al., 2008). Eighty five percent of dogs and 83.4% of patients with sporotrichosis were reported to have had contact with cats with sporotrichosis, and 55.8% of the latter reported cat bites or scratches (Schubach et al., 2008). Sporotrichosis in cats always preceded its occurrence among their owners and their domiciliary canine contacts. The zoonotic potential of cats was further demonstrated by the isolation of *S. schenckii* the fragments of skin lesion in the cats, and from material collected from their nasal and oral cavities (Schubach et al., 2008).

Ecology of Agents of Chromoblastomycosis

Chromoblastoycosis, a chronic subcutaneous mycosis of skin and subcutaneous tissue is caused by several species of dematiaceous (melanized) fungi; the causative agents include *Fonsecaea pedrosoi*, *F. monophora*, *Cladophialophora carrionii*, *Phialophora verrucosa* and *Rhinocladiella aquaspersa*. These fungi occur as filamentous molds with characteristic sporulation in culture and in the environment. In the infected tissues, all agents of chromoblastomycosis occur in an identical form, viz. globose to oval cells with thick brown walls that multiply by septation, often referred to as septate bodies or sclerotic cells (Kwon-Chung and Bennett, 1992). Although the disease has been described worldwide, the incidence is greater in tropical and subtropical regions located between 30° N and 30°S (Kwon-Chung and Bennett, 1992). Countries with the highest number of cases are Madagascar and Brazil (Lopez Martinez and Mendez Tovar, 2007). Several other African, Latin American and Asian countries, such as Gabon, Colombia, Venezuela, Cuba, the Dominican Republic, Mexico, Japan and India also have high prevalence rates (Gugnani et al., 1978; Lopez Martinez and Mendez Tovar, 2007). *P. verrrucosa*, *F. pedrosoi* and *C. carrionii* have been isolated from soil, plant debris and decaying wood (Gezuele et al., 1972; Vincente et al., 2001; Lopez Martinez and Mendez Tovas, 2007). Infection usually occurs following traumatic implantation of the fungus from the environment through penetrating thorn or splinter wounds. Thorns of *Mimosa pudica* have also been implicated as a source of infection (Salgado et al., 2004).

Ecology of *Penicillium marneffei*

Penicillium marneffei, the only dimorphic species of the genus *Penicillium* is the etiological agent of an opportunistic infection, penicilliosis marneffei. This infection occurs among human immunodeficiency virus (HIV)-infected and other immunocompromised patients in several regions of southeast Asia, viz. Thailand, southern China, Taiwan, Hong Kong, Malaysia, Indonesia, Viet Nam, Myanmar (Burma), and parts of northeastern India (Gugnani et al., 2004; Vanittanakom et al., 2006; Sood and Gugnani, 2012). *P. marneffei* is a filamentous mold in culture forming conidiophores with curved chains of conidia. It produces a characteristic red diffusible pigment in the culture medium. In the infected tissue it occurs as a fission yeast (Kwon-Chung and Bennett, 1992).

The ecology of *P. marneffei* is enigmatic. An important issue is whether the human disease, penicilliosis marneffei, occurs as a consequence of zoonotic (animal) or sapronotic (environmental) transmission. In other words, the mode of acquisition of human penicilliosis marneffei is still not clear. A number of studies have firmly established four species of bamboo rats as enzootic reservoirs of infection: *Rhizomys sinensis*, *R. pruinosus*, *R. sumatrensis*, and the reddish-brown subspecies of *Cannomys badius* (Deng et al., 1986; Wei et al., 1987; Ajello et al., 1995). These studies have shown that, within these susceptible species, the prevalence of infection varies widely across Southeast Asia. This finding suggests either that there are regional variations in the endemicity of infection or that there are geographical variations in the predisposition to infection within different species of bamboo rats. A study by Gugnani et al. (2004) comprehensively surveyed six species of sympatric rodents (*Bandicota bengalensis*, *Rattus norvegicus*, *R. rattus*, *R. niditus*, *Mus musculus*, and indigenous reddish-brown *C. badius* [*n* =182]) from bamboo plantations in Manipur state, northeastern India, and found that only *C. badius* harbored infection. As all the rodents in this area presumably have approximately similar exposure rates, this study provides strong evidence that there exist host-specific factors that govern infection. A study by Chariyalertsak et al. (1996) in the Chiang Mai region, Thailand, showed that all 51 animals of the grayish-black subspecies of *C. badius* were negative for *P. marneffei* while three of the 10 rats (30%) in the reddish-brown group were positive, thus demonstrating the intraspecific variation in the prevalence of infection between subspecies of bamboo rats. Taken together, these studies indicate that the distribution and prevalence of prevalence of *P. marneffei* in rodents has a host as well as a geographical component. If these studies are correct and *P. marneffei* infection is in fact sapronosis (rather than a zoonosis), then the fungus must have a reservoir in the environment (Vanittanakom et al., 2006). However, despite extensive efforts, attempts to recover soil isolates of *P. marneffei* have met with only limited success. Vanittanakom et al. (1995) demonstrated 80 to 85% recovery of CFU after three days of incubation from sterilized soil seeded with *P. marneffei*; however, the recovery from non-sterile soil seeded with the fungus was only 6%. Another laboratory study has demonstrated that *P. marneffei* can survive in sterile soil for several weeks but can survive for only a few days in non-sterile soil (Joshi et al., 2003); these studies suggest that survivability in the soil when faced with natural fungal competitors may be limited. Deng et al. (1986) isolated *P. marneffei* from three soil samples collected from the burrows of *Rhizomys pruinosus* (1986), and Chariyalertsak et al. (1996) were able to recover *P. marneffei* from one out of 28 soil samples collected from the burrows of *R. sumatrensis*. However, to date no attempts to recover *P. marneffei* from environments other than those that are intimately associated with bamboo rats have been successful, and definitive proof of an environmental reservoir for *P. marneffei* within the soil, or other substrates, is still lacking.

The history and mode of transmission of the organism remains unclear. The natural habitat of the fungus and its exact route of transmission have not been described. Soil exposure, often during the rainy season, has been suggested to be a serious risk factor (Chariyalertsak et al., 1996). Also, many isolates from bamboo rats and humans were shown to share identical multi-locus genotypes (Vanittanakom et al., 1996; Gugnani et al., 2004). These data show that either transmission of *P. marneffei* can occur from rodents to humans or rodents and humans are co-infected from common environmental sources (Vanittanakom et al., 2006).The main route of infection is thought to be through inhalation of conidia into the lungs, where it can then disseminate via a hematogenous route to other body locations, especially the liver (Vanittanakom et al., 2006).

Conclusion

There have been several reviews on the ecological aspects of individual pathogenic fungi. This chapter has attempted to present a concise update of our knowledge of different dimorphic pathogenic fungi, including the principal agents of chromoblastomycosis. It is hoped that incorporated information would contribute to a better understanding of the epidemiology of human infections caused by dimorphic fungi. Such knowledge should also help in planning strategies for prevention of human infections due to these fungi. Most outbreaks of infection due to systemic dimorphic fungi are associated with activities that disrupt contaminated soil in endemic areas. Currently, no vaccines are available for prevention of these mycoses. Decontaminating infected soil with a 3% formalin solution, with the assistance of local and federal agencies, can prevent aerosolization of conidia (CDC24/7, 2013a). Educating individuals residing or traveling in endemic areas about the risk of exposure, including both leisure and work activities will help in the control of disease. Workers at risk of environmental exposure to *H. capsulatum* might get useful information from skin testing with the antigen, histoplasmin. The results of skin testing would inform each worker of his or her status regarding either susceptibility to infection by *H. capsulatum* (a negative skin test) or partial protection against ill effects if re-infected (a positive skin test) (Lenhart et al., 2003). Advance preparation reduces exposure to contaminated soil, bat and bird dwellings and inoculum. During removal of an accumulation of bat or bird manure from an enclosed area such as an attic, dust control measures should be used. Wearing a NIOSH (National Institute of Occupational Safety and Health)-approved respirator (CDC24/7, 2013b) and other items of personal protective equipment is also recommended to further reduce the risk of *H. capsulatum* exposure. Water sprays should be used during demolition work to decrease dust. A special precaution is required in relation to environmental exposure to *Coccidioides immitis/posadasii*. It is difficult to avoid exposure to *Coccidioides* in areas endemic for this fungus; people who are at higher risk should try to avoid breathing in large amounts of dust if they are in endemic areas by wearing facial masks at dust raising sites or activities (CDC, 2013b).

Acknowledgements

The author is very grateful to Prof. H.S. Randhawa, INSA Emeritus Scientist, Department of Medical Mycology, V.P. Chest Institute University of Delhi, Delhi, India for permitting me to use the image of mycelial form of *B. dermatitidis* from his recent publication, and to Dr. David Ellis, School of Biomedical and Molecular Science, University of Adelaide, Adelaide, Australia for permitting me to use his images of mycelial and yeast forms of *S. schenckii*, and that of the arthroconidia of *C. immitis/C. posadasii*. Some of the images are from websites with free access. Other images are from my own work.

References

Al-Doory, Y. and Kalter, S.S. 1967. The isolation of *Histoplasma duboisii* and keratinophilic fungi from soils of East Africa. Mycopathologia, 31: 289–295.

Ajello, L. 1971. Coccidioidomycosis and histoplasmosis: a review of its epidemiology. Mycopathologia, 45: 221–230.

Ajello, L. 1983. Histoplasmosis—a dual entity: histoplasmosis capsulati and histoplasmosisduboisii. Igiene. Mod., 79: 3–30.

Ajello, L., Padhye, A.A., Sukroongreun, S., Nilakul, C.H. and Tantimavanic, S. 1995. Occurrence of *Penicillium marneffei* infections among wild bamboo rats in Thailand. Mycopathologia, 131: 1–8.

Ajello, L. and Hay, J. (eds.). 1998. Topley and Wison's Microbiology and Microbial Infections. 9th ed. Vol. 4. Medical Mycology, London: Arnold.

Albornoz, M. 1971. Isolation of *Paracoccidioides brasiliensis* from rural soil in Venezuela. Sabouraudia, 9:248–252.

Arentes, T.D., Theodoro, R.C., Da Gracia Marcori, S.A. and Baggali, A. 2013. Detection of *Paracoccidioides* spp. in environmental aerosol samples. Med. Mycol., 51: 83–92.

Bagali, E., Franco, M., Bosco, Hebeler-barbosa, F., Trinca, L.A. and Montenegro, M.R. 2003. High frequency of *Paracoccidioides brasiliensis* infection in armadillos *(Dasypus novemcinctus)*: an ecological study. Med. Mycol., 41: 217–223.

Bagali, E. and Barrozo, S.L. 2005. New strategies and opportunities for the ecoepidemiological study of *Paracoccidioides brasiliensis*: sentinel animal, molecular biology and geoprocessing. Rev. Inst. Med. Trop. S. Paulo, 47(suppl. 14): 16.

Bagagli, E., Theodoro, R.C. and Bosco, S.M.G. 2008. *Paracoccidioides brasiliensis*: phylogenetic and ecological aspects. Mycopathologia, 165: 197–207.

Barker, B.M., Tabor, J.A., Shubitz, L.F., Perril, R. and Orbach, M.J. 2012. Detection and phylogenetic analysis of *Coccidioides posadasii* in Arizona soil samples. Fung. Ecol., 6: 163–176.

Baumgardner, D.J., Knavel, E.M., Steber, D. and Swain, G.R. 2006. Geographic distribution of human blastomycosis cases in Milwaukee, Wisconsin, USA: association of urban watersheds. Mycopathologia, 161: 275–282.

Brandt, M.E. and Warnock, D.W. 2007. Histoplasma, Blastomyces, Coccidioides, and other dimorphic fungi causing systemic mycoses. *In*: Murray, P. (ed.). Manual of Clinical Microbiology, 9th ed., 9: 1857–1865.

Bravo, T.C. 2012. New observations on the epidemiology of sporotrichosis and *Sporothrix schenckii* complex 1. Sources of infection and occupational risks. Rev. Latinoamer. Patol. Clin., 59: 88–100.

Butler, T.M., Gleiser, C.A., Bernal, J.C. and Ajello, L. 1988. Case of disseminated African histoplasmosis in a baboon. J. Med. Primatol., 17: 153–161.

Butler, T.M. and Hubbard, G.B. 1991. An epizootic of histoplasmosis duboisii (African histoplasmosis) in an American baboon colony. Lab. Anim. Sci., 41: 407–410.

CDC 24/7. 2013a. Sources of histoplasmosis and blastomycoses.

CDC 24/7. 2013b. Saving Lives. Protecting people.

Chariyalertsak, S., Vanittanakom, P., Nelson, K.E., Sirisanthana, T. and Vanittanakom, N. 1996. *Rhizomys sumatrensis* and *Cannomys badius*, new natural animal hosts of *Penicillium marneffei*. J. Med. Vet. Mycol., 34: 105–110.

Coles, F.B., Schuchat, A.J.R., Hibbs, S.F., Kondracki, I.F., Salkin, I.F., Dixon, D.M., Chang, H.G., Duncan, R.A., Hurd, N.J. and Morse, D.L. 1992. A Multistate outbreak of sporotrichosis associated with sphagnum moss. Am. J. Epidemiol., 136: 475–478.

Collie, L., Balows, A. and Sussman, M. 1998. Topley & Wilson's Microbiology and Microbial Infections, 9th ed, Vol. 4. Arnold, London, Sydney, Auckland, New York.

Colombo, A.L., Tobón, A., Restrepo, A., Queiroz-Telles, F. and Nucci, M. 2011. Epidemiology of endemic systemic fungal infections in Latin America. Med. Mycol., 49: 785–98.

Conti-Dias, I.A. 2007. On the unknown ecological niche of *Paracoccidioides brasiliensis*. Our hypothesis of 1989: present status and perspectives. Rev. Inst. Med. trop. S. Paulo., 49 no. 2 São Paulo Mar./Apr. 2007.

Cordeiro, R.A., Brilhante, R.S.N., Rocha, M.F.G., Fechine, M.A.B., Camara, L.M.C. and Sidrim, J.J.C. 2006. Phenotypic characterization and ecological features of *Coccidioides* spp. from Northeast Brazil.Med. Mycol., 44: 631–639.

de Hoog, G. S., Guarro, J., Gene, J. and Figueras, M.J. 2000. Atlas of Clinical Fungi, 2nd ed., Centraalbureau voor Schimmelcultures, Utrecht, The Netherlands.

Deng, Z., Yun, M. and Ajello, L. 1986. Human penicilliosis marneffei and its relation to the bamboo rat (*Rhizomys pruinosus*). J. Med. Vet. Mycol., 24: 383–389.

Dias, M.A., Oliveira, R.M., Giudice, M.C., Netto, M.H., Jardao, L.R. et al. 2011. Isolation of *Histoplasma capsulatum* from bats in the urban area of Sao Paulo State, Brazil. Epidemiol. Infect., 139: 1642–1646.

Dierks, F.H. 1965. *Histoplasma capsulatum* in fecal contents and organs of bats in the Canal Zone. Am. J. Trop. Med. Hyg., 14: 433.

DiSalvo, A.F. and Johnson, W.M. 1979. Histoplasmosis in South Carolina: support for microfocus concept. Am. J. Epidemiol., 109: 480–492.

DiSalvo, A.F. 2009. Mycology Chapter 6: blastomycosis (Blastomyces dermatitidis). Microbiology and Immunology. Online <http://pathmicro.med.sc.edu/mycology/mycology-6.htm>.

Dixon, D.M. and Salkin, I.F. 1991. Association between the human pathogen *Sporothrix schenckii* and Sphagnum Moss. pp. 237–249. *In*: Adrews, J.H. and Hirano, S.S. (eds.). Microbial Ecology of Leaves. New York: Springer-Verlag.

Eulálio, K.D., Macêdo, R.L., Cavalcanti, M.A.S., Martins, L.M.S., Lazera, M.S. and Wanke, B. 2000. *Coccidioides immitis* isolated from armadillos (*Dasypus novemcinctus*) in state of Piaui. Annalsof 4° ISHAM World Congress, Buenos Aires.

Eulálio, K.D., Salmito, M.A., Cruz, J.R.M. and Lazera, M.S. 2000. Coccidioidomycosis among armadillo hunters in northeastern Brazil: a new outbreak in the state of Piaui. Annals of 4° ISHAM World Congress, Buenos Aires.

Fisher, M.C., Koenig, G.L., White, T.J. and Taylor, J.W. 2002. Molecular and phenotypic description of *Coccidioides posadasii* sp. nov., previously recognized as the non-California population of *Coccidioides immitis*. Mycologia, 94: 73–84.

Franco, M., Negro D.G., Silva-Lacaz and Restrepo-Moreno, A. 1994. Paracoccidioidomycosis, CRC Press.

Franco, M., Bagagli, E., Scapolio, S. and Lacaz, C.S. 2000. A critical analysis of isolation of *Paracoccidioides brasiliensis* from soil. Med. Mycol., 38: 185–191.

Galgiani, J.N. 2010. Coccidioidomycosis in Principles of Infectious Diseases, ed. Mandel, G.L., Bennett, J.E. and Doli, R. Churchill & Livingstone, Philadelphia, p. 3334.

Gezuela, E., Mackiinon, J.E. and Conti-Diaz, I.A. 1972. The frequent isolation of *Phialophora verrucosa* and *Phialophora pedrosoi* from natural sources. Sabouraudia, 10: 266–273.

Ghosh, A., Chakrabarti, A., Shar, V.K., Singh, K. and Singh, A. 1989. Sporotrichosis in Himachal Pradesh (north India). Trans. R. Soc. Trop. Med. Hyg., 93: 41–45.

Greene, D.R., Koeng, G., Fisher, M.C. and Taylor, J.W. 2000. Soil isolation and molecular identification of *Coccidioides immitis*. Mycologia, 92: 406–410.

Gopalakrishnan, R., Nambi, P.S., Ramasubramanian, V., Ghafur, K. and Parameswaran, A. 2012. Histoplasmosis in India: Truly uncommon or uncommonly recognized J.A.P.I., 60: 25–28.

Grose, E. and Tramsitt, J.R. 1965. *Paracoccidioides brasiliensis* recovered from the intestinal tract of three bats (*Artibeus lituratus*) in Colombia, S.A. Sabouraudia, 4: 124–125.

Gugnani, H.C. 1972. Fungi recovered from lungs of small mammals in India. Mykosen, 15: 479–485.

Gugnani, H.C., Egere, J.U., Suseelan, A.V., Okoro, A.N. and Onguigbo, W.I.B. 1978. Chromomycosis caused by *Phialophora pedrosoi* in Eastern Nigeria. J. Trop. Med. Hyg., 81: 208–210.

Gugnani, H.C., Mutoe-Okafor, F.A., Dupont, B. and Kaufman, L. 1994. A natural focus of *Histoplasma capsulatum* var. *duboisii* in a bat cave. Mycopathologia, 127: 151–157.

Gugnani, H.C. and Muotoe-Okafor, F.A. 1998. African histoplasmosis: a review. Rev. Iberoam. Micol., 14: 155–159.

Gugnani, H.C. 2000. Histoplasmosis in Africa: A review. Indian J. Chest. Dis. & Allied Sci., 42: 271–278.

Gugnani, H.C., Fisher, M.C., Paliwal-Joshi, A., Vanittanakom, N., Singh, I. and Yadav, P.S. 2004. *Cannomys badius* as a natural animal host of *Penicillium marneffei* in India. J. Clin. Microbiol., 42: 5070–5075.

Gugnani, H.C., Paliwal-Joshi, A., Rahman, H., Padhye, A.A., Singh, T.S.K., Das, T.K., Khanal, B., Bajaj, R.S. and Chukhani, R. 2007. Occurrence of pathogenic fungi in soil of burrows of rats and of other sites in bamboo plantations in India and Nepal. Mycoses, 50: 507–511.

Haldar, N., Sharma, M.K. and Gugnani, H.C. 2007. Sporotrichosis in north-east India. Mycoses, 50: 201–204.

Heitman, J., Scott, G.F., Aaron, P.M. and Edwards, J.E., Jr. 2006. Molecular Principles of Fungal Pathogenesis. 1st ed. New York. Heitman, J., Filler, S.G. Jr., Edwards, J.E. and Mirchell, A.P. (eds.). ASM Press, 321–331.

Joshi, A., Gugnani, H.C. and Vijayan, V.K. 2003. Survival of *Penicillium marneffei* in sterile and unsterile soil. J. Mycol. Med., 13: 211–212.

Kauffman, C.A. 2011. Blastomycosis. *In*: Goldman, L. and Schafer, A.I. (eds.). Cecil Medicine, 24th ed. Philadelphia, Pa: Saunders Elsevier; Chap 342.

Khan, Z.U., Randhawa, H.S. and Lulla, M. 1982. Isolation of *Blastomyces dermatitidis* from the lungs of a bat, *Rhinopoma hardwickei hardwickei* Gray, in Delhi. Sabouraudia, 20: 137–144.

Klein, B.S., Vergeront, J.M., Weeks, R.J., Kumar, U.N., Mathai, G., Varkey, B. Kaufman-Bradsher, L.R.W. Stoebig, J.F. and Davis, J.P. 1986. Isolation of *Blastomyces dermatitidis* in soil associated with a large outbreak of blastomycosis in Wisconsin. N. Engl. J. Med., 314: 529–534.

Klein, B.S., Vergeront, J.M., DiSalvo, A.F., Kaufman, L. and Davis, J.P. 1987. Development of long-term specific cellular immunity after acute *Blastomyces dermatitidis* infection: assessments following a large point-source outbreak in Wisconsin. Ann. Rev. Resp. Dis., 136: 1333–1338.

Kolivaras, K.N., Johnson, P.S., Comrie, A.C. and Yoof, S.R. 2001. Environmental variability and coccidioidomycosis (valley fever). Aerobiology, 17: 31–42.

Kwok, H.K., Chan, J.W., Li, I.W., Chu, S.Y. and Lam, C.W. 2009. Coccidioidomycosis as a rare cause of pneumonia in non-endemic areas: a short exposure history should not be ignored. Respirology, 14: 617–620.

Kwon-Chung, K.J. and Bennett, J.E. 1992. Medical Mycology. Philadelphia, Lea Febiger, pp. 248–279, 337–355, 356–396, 464–513, 469–512, 594–619, 707–732, 755–760.

Legendere, A.M. 2006. Blastomycosis. pp. 569–576. *In*: Greene, C.E. (ed.). Infectious Diseases of the Dog and Cat. 3rd ed. St. Louis, Mo: Saunders.

Lenhart, S.W., Shafer, M.R., Singal, M. and Heijeh, R.A. 2003. Histoplasosis, protecting workers at risk. Department of Health and Human Services. CDC. Workplace, Safety and Health.

Lima, R.A., Picanço, Y.V., Cordeiro Rde, A., de Camargo, Z.P., Queiroz, J.A., de Araujo, R.W., de Mesquita, J.R. and Sidrim, J.J. 2012. Coccidioidomycosis in armadillo hunters from the state of Ceará, Brazil.Mem. Inst. Oswaldo Cruz., 107: 813–5.

Long, S.S., Pickering, L.K. and Prober, C.G. 2012. Principles and practice of pediatric infectious diseases, 3rd ed. Philadelphia: Churchill Livingstone, 1213–1217.

Lopez Martinez, R. and Mendez Tovar, L.J. 2007. Chromoblastomycosis. Clin. Dermatol., 25: 188–194.

McMuray, D.N. and Greer, D.L. 1979. Immune responses in bats following intranasal infection with *Histoplasma capsulatum*. Am. J. Trop. Med. Hyg., 28: 1036–1039.

Muotoe-Okafor, F.A., Gugnani, H.C. and Gugnani, A. 1996. Skin and serum reactivity to histoplasmin in the vicinity of a natural focus of *Histoplasma capsulatum* var. *duboisii*. Mycopathologia, 134: 71.

Negroni, P.E. 1966. *Paracoccidioides brasiliensis* vive saprotificamente en el suelo Argentino. Prensa. Med. Argent., 53: 2381–2382.

Pan and Wu, S.X. 1989. Mycologic investigation on *Rhizomys pruinosus senex* in Guangxi as natural carrier with *Penicillium marneffei*. Chung. Hua. I. Hsueh. Tsa. Chih., 102: 477–485.

Perera, P. and Stone, S. 2002. Coccidioidomycosis in workers at an archeologic site—Dinosaur National Monument, Utah, June–July 2001. Ann. Emerg. Med., 39: 566–569.

Petersen, L.R., Marshall, S.L., Barton, C., Hajjeh, R.A., Lindsley, M.D., Warnock, D.W. et al. 2004. Coccidioidomycosis among workers at archeological site, northeastern Utah. Emerg. Infect. Dis., 10: no. 4-AprAvailable from: http://wwwnc.cdc.gov/eid/article/10/4/03-0446.htm.

Randhawa, H.S., Chaturvedi, V.P., Kini, S. and Khan, Z.U. 1985. *Blastomyces dermatitidis* in bats: first report of its isolation from the liver of *Rhinopoma hardwickei hardwickei* Gray. Sabouraudia, 23: 78–86.

Randhawa, H.S., Chowdhary, A., Kathuria, S., Roy, P. Misra, D.S., Jain, S. and Chugh, T.D. 2013. Blastomycosis in India: report of an imported case and current status. Med. Mycol., 51: 185–192.

Roberts, G.D. 1975. The Epidemiology of Sporotrichosis. pp. 227–236. *In*: Al-Doory, Y. (ed.). Epidemiology of Human Mycotic Diseases. Springfield, Ill. Charles C Thomas Pub.

Salgado, C.G., da Silva, J.P., Diniz, J.A.P., da Silva, M.B., da Costa, P.F., Teixeira, C. and Salgado, U.I. 2004. Isolation of *Fonsecaea pedrosoi* from thorns of *Mimosa pudica*, a probable natural source of chromoblastomycosis. Rev. Inst. Med. trop. S. Paulo, Vol. 46 no. 1 São Paulo, Jan./Feb.

Sanyal, M. and Thammayya, A. 1975. *Histoplasma capsulatum* in the soil of Gangetic Plain in India. Indian J. Med. Res., 63: 1020–1028.

Schneider, E., Hajjeh, R.A., Spiegel, R.A., Jibson, R.W., Harp, E.L., Marshall, G.A., Robert, A.G., McNeil, M.M., Pinner, R.W., Baron, R.C., Burger, R.C., Hutwagner, L.C., Leo Kaufman, C.C., Reef, S.E., Feldman, G.M., Pappagianis, D. and Werner, S.B. 1997. A coccidioidomycosis outbreak following the Northridge, Calif, earthquake. J.A.M.A., 277: 904–908.

Schubach, A., Barros, M.B. and Wanke, B. 2008. Epidemic sporotrichosis. Curr. Opin. Infect. Dis., 21: 129–133.

Sifuentes-Osornio, J., Corzo-León, E.D. and Ponce-de-León, L.A. 2012. Epidemiology of invasive fungal infections in Latin America. Curr. Fungal Infect. Rep., 6: 23–34.

Silva-Vergara, M.L., Martinez, R. and Chadu, A., Madeira, M., Freitas-Silva, G. and Leite Maffei, C.M. 1998. Isolation of a *Paracoccidioides brasiliensis* strain from the soil of a coffee plantation in Ibia, State of Minas Gerais, Brazil. Med. Mycol., 36: 37–41.

Silva-Vergara, M.L., Martinez, R., Carmago, Z.P., Malta, M.H.B., Maffe, C.M.L. and Chadu, A. 2000. Isolation of *Paracoccidioides brasiliensis* from armadillos (*Dasypus novemcinctus*) in an area where the fungus was recently isolated from soil. Med. Mycol., 38: 193–199.

Taylor, M.L., Chavez-Tapia, C.B., Vergas-Zanez, R., Rodrigueg-Arrelanis, G., Pena-Sandoval, Toriello, C., Pérez, A. and Reyes-Montes, M.R. 1999. Environmental conditions favoring bat infection with *Histoplasma capsulatum* in Mexican shelters. Am. J. Trop. Med. Hyg., 61: 914–919.

Terçarioli, G.R.E., Bagagli, G.M., Reis, R.C., Theodoro, S.D.M.G., Bosco, S.A.G., Macoris, and Richini-Pereira, V.B. 2007. Ecological study of *Paracoccidioides brasiliensis* in soil: growth ability, conidia production and molecular detection. BMC Microbiology, 7: 92, doi:10.1186/1471-2180-7.

Tesh, R.B. and Schneidau, J.D., Jr. 1967. Experimental infection of bats (*Tadarida brasiliensis*) with *Blastomyces dermatitidis*. J. Infect. Dis., 11: 188–192.

Tewari, R.P., Wheat, L.J. and Ajello, L. 1998. Agents of histoplasmosis. pp. 373–393. *In*: Ajello, L. and Hay, J. (eds.). Topley and Wilson's Microbiology and Microbial Infections. 9th ed. Vol. 4. London: Arnold.

Vanittanakom, N., Mekaprateep, M., Sriburee, P., Vanittanakom, P. and Khanjanasthiti, P. 1995. Efficiency of flotation method in the isolation of *Penicillium marneffei* from seeded soil. J. Med. Vet. Mycol., 33: 271–273.

Vanittanakom, N., Cooper, C.R., Chariyalertsak, S., Youngchim, S., Nelson, K.E. and Sirisanthana, T. 1996. Restriction endonuclease analysis of *Penicillium marneffei*. J. Clin. Microbiol., 34: 1834–1836.

Vanittanakom, N., Cooper, C.R., Jr., Fisher, M.C. and Sirisanthana, T. 2006. *Penicillium marneffei* infection and recent advances in the epidemiology and molecular biology aspects. Clin. Microbiol. Rev., 19: 95–110.

Vicentie, V.A., Atilli-Angelis, D., Pie, M.R., Queiroz-Telles, F., Cruz, L.M.N., Ajafzadeh, M.J., de Hoog, G.S., Zhao, J. and Pizzirani-Kleiner, A. 2008. Environmental isolation of black yeast-like fungi involved in human infection. Stud. Mycol., 61: 137–144.

Wei, X.G., Ling, Y.M., Li, C. and Zhang, F.S. 1987. Study of 179 bamboo rats carrying *Penicillium marneffei*. China J. Zoonoses, 3: 34–35.

Williams, P.L., Sable, D.L. and Smyth, L.T. 1979. Symptomatic coccidioidomycosis following a severe natural dust storm. Chest, 76: 566–70.

Zhang, X. and Andrews, J.H. 1993. Evidence for growth of *Sporothrix schenckii* on dead but not on living sphagnum moss. Mycopathologia, 123: 87–94.

CHAPTER 16

Mycoses Caused by Dematiaceous Fungi with Special Reference to the Recent Fungal Meningitis Epidemic Caused by *Exserohilum rostratum*

Jalpa P. Tewari[1,2,*] and *Sanjiv Tewari*[2]

ABSTRACT

The last few decades have seen a significant rise in the types of opportunistic dematiaceous fungi and human diseases caused by them. The presence of melanin, serving ecological and pathogenesis functions, is one of the features common to all these fungi. It has been estimated that this is a group of more than 150 species belonging to 70 genera of fungi. The diseases caused by these fungi are varied and classified into three types: chromoblastomycosis, eumycetoma and phaeohyphomycosis. The incidence of these fungal diseases, compared

[1] Department of Agricultural, Food & Nutritional Science, Faculty of Agricultural, Life & Environmental Sciences, University of Alberta, Edmonton, Alberta, Canada, T6G 2P5.
[2] Lung, Allergy, Sleep Centers of America, 224 W. Exchange Street, Ste. 380, Akron, Ohio 44302, USA.
* Corresponding author: jalpaptewari@gmail.com

to some others, is relatively low; hence, there is little clinical emphasis and experience in dealing with them. Some of them are prevalent in parts of the world with limited financial resources, which complicates developing appropriate management strategies. Recently, in many parts of the United States of America, there was an epidemic of fungal cerebral meningitis caused by the dematiaceous fungus, *Exserohilum rostratum*, emanating from injections of contaminated methylprednisolone acetate. This disease caused considerable morbidity and mortality, and calls for more stringent quality control measures for the medical products used.

Introduction

Numerous darkly pigmented fungi are causal agents of human and animal diseases. These pathogens are a diverse group of hyphal or pseudohyphal fungi that have brown-colored cell walls due to the presence of melanin. The hyphae are septate and branched. Some of these fungi also have yeast-like cells. They also produce numerous kinds of darkly-colored spores. Such fungi are designated as being melanized or dematiaceous (Etymology; Gk., "deme" meaning "bundle"). The term dematiaceous is a misnomer but is firmly entrenched in mycological and clinical usage to indicate brown-colored fungi (Revankar and Sutton, 2010). The melanin production may vary with conditions of growth of the fungus and among different fungi. In fact, it has been demonstrated in almost all the so-called non-dematiaceous hyaline-looking clinical fungi also such as *Candida albicans*, *Histoplasma capsulatum*, and others (Revankar and Sutton, 2010). Melanin is a polyphenolic compound and is present in many microbes and other groups of organisms. It can be histochemically demonstrated using the Masson-Fontana stain. It is extremely stable and resistant to many destructive processes (Revankar and Sutton, 2010). Melanin is thought to play an important role in evading host defenses (Walsh et al., 2004) and in survival of the pathogen in harsh environments. It also helps protect these fungi from macrophages, neutrophils and from antifungal agents (Santos et al., 2013).

The dematiaceous fungi, causing diseases in humans and animals, are a group of over 150 species belonging to 70 genera. They include both anamorphic (hyphomycetous and coelomycetous) and teleomorphic genera belonging to three main orders, Chaetothyriales, Dothideales and Pleosporales (Revankar and Sutton, 2010; Santos et al., 2013). The frequency with which they cause human diseases has increased in parallel with the increase in chronically immunosuppressed patients (Revankar and Sutton, 2010).

Three clinical syndromes are caused by dematiaceous fungi and include chromoblastomycosis, eumycetoma and phaeohyphomycosis. They are distinguishable based on histopathological evidence (Revankar, 2007). The characteristics of chromoblastomycosis and eumycetoma are detailed and compared by La Hoz and Baddley (2012). Both these diseases are subcutaneous infections common in tropical and subtropical regions and share many features in common. Phaeohyphomycosis is a "catch-all term" which includes the remaining infections caused by the melanized fungi (Revankar, 2007). Mycetomas caused by actinomycetes are called actinomycetomas (Pang et al., 2004) and are not discussed here. Some aspects of the diseases caused by dematiaceous fungi are also discussed in Chapter 15 of this book.

Chromoblastomycosis

This disease was first noted in 1911 in Sao Paulo, Brazil by de Moraes Pedroso and the first North American case was reported from Boston in 1915 by Medlar and Lane (Martínez and Tovar, 2007; Torres-Guerrero et al., 2012). Aptly named, the most frequent etiologic agent is *Fonsecaea pedrosoi* and the diagnostic histological feature is the Medlar body.

Although the disease has been reported from temperate areas also, it is primarily a disease of male agricultural workers in the tropical and subtropical regions (Elgart, 1996). Chromoblastomycosis has been reported from many places from all the continents. Internationally, the highest numbers of cases have been reported from Madagascar, Natal state of South Africa, Brazil and Costa Rica (Martínez and Tovar, 2007). The disease is also frequent in some other regions of central and southern Africa (Martínez and Tovar, 2007). Cases have also been reported from Australia, China, Cuba, Dominican Republic, Gabon, India, Japan, Mexico, Puerto Rico, Sri Lanka, Venezuela and several countries from Europe (Torres-Guerrero et al., 2012; Weedon et al., 2013). The disease is relatively less common in the United States but has been reported from Louisiana, Massachusetts and Texas (La Hoz and Baddley, 2012; Torres-Guerrero et al., 2012).

This disease, which occurs predominantly in men, produces slowly progressing chronic skin and subcutaneous nodular or verrucous mycosis of the lower extremities (legs and feet), also affecting hands, wrists, forearms, trunk, buttocks and the neck (Torres-Guerrero et al., 2012; Weedon et al., 2013). The lesions are nodular, tumoral, cicatricial, verrucose, erythematous plaque and with lymphatic dissemination, but almost never involve bone or muscle (Pang et al., 2004; Torres-Guerrero et al., 2012; Badali et al., 2013). The infection has a low potential for dissemination and does not seem to have any recognizable predisposing factors, however, autoinoculation may occur through scratching (Pang et al., 2004; La Hoz and Baddley, 2012). Systemic infection in lungs, brain, peritoneum and esophagus does occur rarely by *Exophiala jeanselmei* and a species of *Rhinocladiella* (Nucci et al., 2001; Torres-Guerrero et al., 2012). Such infections can cause fungemia with predisposing risk factors such as immunodeficiency, cancer, AIDS, agranulocytosis, use of steroids and broad-spectrum antibiotics, neutropenia, postoperative state, central venous catheter and blood-product transfusion (Nucci et al., 2001; Torres-Guerrero et al., 2012). The lesions contain "black dots" which consist of pigmented sclerotic bodies in clumps (Elgart, 1996). The sclerotic bodies consist of single and clustered thick-walled brown muriform fumagoid cells or Medlar bodies that resemble copper pennies (Pang et al., 2004; Torres-Guerrero et al., 2012). Mycologically, the Medlar bodies resemble the resistant survival thick-walled cells, the chlamydospores, produced singly or in groups by numerous fungi. The Medlar bodies are regarded as being pathognomonic for chromoblastomycosis. They are considered to be the "invasive form" and are also the extremotolerant survival phase of *Cladophialophora yegresii* in the environmental substrate/host which is a cactus plant (de Hoog et al., 2007).

The causal fungi are present in plant debris, decaying wood and soil, and the disease may be initiated as a result of accidental traumatic inoculation of the pathogen through structures such as splinters and thorns (Elgart, 1996; Martínez and Tovar, 2007). It is an occupational disease and most patients are rural agricultural workers who work barefoot (Torres-Guerrero et al., 2012). *Fonsecaea pedrosoi* and *Cladophialophora carrionii* are the two common causal agents of chromoblastomycosis (Martínez and Tovar, 2007). However, some other causal agents include *Fonsecaea compacta*, *Cladophialophora arxii*, *Phialophora verrucosa*, *Rhinocladiella aquaspersa*, *Exophiala spinifera*, *Mangiella dermatitidis*, and *Botryomyces caespitosus* (Pang et al., 2004; Martínez and Tovar, 2007; Torres-Guerrero et al., 2012). These fungi are slow-growing, low in virulence and exhibit relatively high heat tolerance (40–42°C; Torres-Guerrero et al., 2012). Some of these species are distributed preferentially in certain areas. *Fonsecaea pedrosoi* is the commonest etiologic agent in Brazil (Amazon region), Mexico, northern Madagascar and southern China, and *Cladophialophora carrionii* is prevalent in Australia, southern Madagascar, South Africa, northwestern Venezuela and northern China (La Hoz and Baddley, 2012; González et al., 2013; Yang et al., 2013). *Fonsecaea monophora* is a species segregated from *F. pedrosoi* and is also prevalent in southern China (Torres-Guerrero et al., 2012; Lu et al., 2013). *Fonsecaea pedrosoi* has been isolated from the thorns of *Mimosa pudica* in Brazil and from plant debris, soil and other materials in Uruguay (Gezuele et al., 1972; Salgado et al., 2004). Similarly, *Phialophora verrucosa* was also isolated from these substrates in Uruguay (Gezuele et al., 1972). However, despite these observations, demonstration of a direct etiological relationship between isolation of the pathogen from the natural substratum and disease is still lacking. A recently described new species, *Cladophialophora yegresii*, segregated from *C. carrionii*, like other dematiaceous molds, is also low in virulence and is associated with cacti in Venezuela (de Hoog et al., 2007; Torres-Guerrero et al., 2012). It is obvious that at least some species are adapted to particular environmental conditions, for example the climate is humid in the Amazon region of Brazil whereas it is arid and hot in northwestern Venezuela.

Single or clustered Medlar bodies in scrapings or biopsies are diagnostic for chromoblastomycosis (Pang et al., 2004; Torres-Guerrero et al., 2012). Samples for histopathology should be obtained from parts of the lesion where a dark spot is visible and examined in 10–20% potassium hydroxide by light microscopy. This will increase the diagnostic yield for the presence of Medlar bodies where they can be found extracellularly or intracellularly in giant cells or macrophages (La Hoz and Baddlet, 2012). Other diagnostic techniques are antigen test, sequencing of the Internal Transcribed Spacer (ITS) regions of rDNA, and genus-specific Polymerase Chain Reaction (PCR; Badali et al., 2013). Due to many overlapping symptoms with other diseases, special care should be taken in diagnosing chromoblastomycosis. This disease is common in many financially stressed parts of the world; hence this is a neglected mycosis and many treatment options have not been studied and validated. Small initial lesions are treatable but the outcomes are poor in chronic situations. Surgical excision is an option based on the location and degree of lesioning. Some physical methods such as thermotherapy, cryotherapy and photodynamic therapy may be potentially useful. The chemotherapeutic agents intraconazole and terbinafine may be used alone or in combination with the physical methods (La Hoz and Baddlet, 2012; Torres-Guerrero et al., 2012). Intravenous amphotericin B has also been used but with due caution for possible nephrotoxicity (Torres-Guerrero et al., 2012). Thermotherapy may include extended use of pocket warmers at temperatures over 40–42°C to kill the fungus (Torres-Guerrero et al., 2012). This temperature range is just above the tolerance threshold of these fungi. Cryotherapy or cryosurgery uses liquid nitrogen to freeze the diseased tissues. The ultra-low temperatures of liquid nitrogen do not kill the Medlar bodies but induce the healing necrosis and inflammation factors (Queiroz-Telles and de C.L. Santos, 2013). Photodynamic therapy entails topical or systemic administration of a photosensitizer which is photo-activated by light to elicit an oxygen-dependent cytotoxic reaction (Lu et al., 2013). In a study, the lesions applied with 20% methylene blue cream were irradiated with a red light-emitting diode with beneficial effects (Lyon et al., 2011; Queiroz-Telles and de C.L. Santos, 2013). Photodynamic therapy was initially developed for cancer treatment but has lately been also used for the treatment of mucocutaneous candidiasis, pityriasis versicolor, dermatophytosis and onychomycosis (Queiroz-Telles and de C.L. Santos, 2013).

The outcomes of chromoblastomycosis therapies may not be entirely satisfactory perhaps in part due to multiple causal fungi reacting differently to the treatments used. Also, the Medlar bodies may be resistant to penetration by the drugs. Chromoblastomycosis is an occupational disease and sociological and financial constraints of the patients play a role in less than ideal management of the disease (Martínez and Tovar, 2007). Finally, there has been interest in developing a DNA-hsp65 vaccine which will allow the host to restore depressed cellular immunity reducing the effects of antifungal therapies, in addition to other beneficial effects (Siqueira et al., 2013).

Eumycetoma

This is an ancient disease and has the distinction of being one of the oldest infections described. A skeleton from an adult woman dating 300–600 AD from a stone sarcophagus in a burial cave near Bet Guvrin, Israel showed symptoms of a possible mycetoma (McGinnis, 1996). However, the Vedic hymns of the Hindu religious book *Atharva Veda* (at the end of the 2nd millennium BCE) contain the first account of this disease. In this Indian book, mycetoma is described as "pada valmikam" (meaning foot-anthill) referring to pain similar to ant-hill activity (McGinnis, 1996; Revankar and Sutton, 2010).

The eumycetoma is caused by more than 30 species of dematiaceous and hyaline fungi of diverse phylogenetic affiliations (McGinnis, 1996; de Hoog et al., 2004, 2013; Mattioni et al., 2013). The hyaline causal fungi are not addressed in the present chapter. The fungus *Madurella mycetomatis* is the commonest cause of eumycetoma around the world particularly in India and Africa, and *Pseudoallescheria boydii* (*Scedosporium apiospermum*) is found in the United States of America (Pang et al., 2004; Revankar and Sutton, 2010). The morbid manifestation "Madura foot" caused by *Madurella mycetomatis* was first described from Madurai, India in 1842 (Mattioni et al., 2013). Some other causal fungi of eumycetoma include *M. grisea*, *Leptosphaeria senegalensis* and *Pyrenochaeta romeroi* (de Hoog et al., 2004; Pang et al., 2004; Revankar, 2007). Eumycetoma occurs between latitudes 15°S and 30°N but is commonest among male individuals between the Tropics of Cancer and Capricorn in countries such as India, Sudan, Indonesia, Mexico, Pakistan, Senegal, Somalia, Brazil, Colombia, and Venezuela (McGinnis, 1996). It is also reported from the Middle East and Latin America, and migrant patients from the Sub-Saharan Africa and Sri Lanka have been seen in France (Mattioni et al., 2013). Sudan is part of the "mycetoma belt" with disease foci in southern India, Somalia and Senegal (de Hoog et al., 2004). The endemic areas have a hot and dry climate.

Eumycetomata infections are non-contagious and are caused by fungi which are thought to be carried in soil and woody plants. Localized traumas, such as those caused by thorns, wooden splinters, scratches and many other sources are thought to introduce the etiologic fungi in the infection court in the subcutaneous tissue (McGinnis, 1996; de Hoog et al., 2004, 2013; Pattanaprichakul et al., 2013). The penetrating trauma-causing structures also include unusual agents such as a sea urchin spine indicating unique ecological distribution of the eumycetoma-causing agent, *Exophiala jeanselmei* (Pattanaprichakul et al., 2013). It is an occupational disease and most infections occur in the feet and other extremities.

In general, the etiological agents of eumycetoma have been isolated from soil or plant samples indicating that they reside in the environment (McGinnis, 1996; de Hoog et al., 2004). The recovery rate of *Madurella mycetomatis*, the principal causal agent of eumycetoma, from soil was extremely low but it was detectable by polymerase chain reaction (PCR) in culture-negative soil (Thirumalachar and Padhe, 1968; Ahmad et al., 2002; de Hoog et al., 2004). The aforesaid plant samples (two cases) yielded the etiological agents *Leptosphaeria senegalensis* and *Pyrenochaeta romeroi. Madurella mycetomatis*, the primary causal agent of eumycetoma, has remained elusive as to its inoculation source habitat. Also, it has so far never been cultured from thorns (de Hoog et al., 2013). This fungus is reported as a sterile dematiaceous mycelium isolated from diseased human tissues in a recent nesting study within the fungal family Chaetomiaceae using sequences of the universal fungal barcode gene rDNA and the partial LSU gene sequence (de Hoog et al., 2013). This fungal family is mainly coprophilous suggesting that the ubiquitously present cattle dung may play a role in the epidemiology of this disease in rural East Africa (de Hoog et al., 2013). However, in another study, the isolates of *M. mycetomatis* are reported to be frequently sterile, occasionally producing phialides and globose conidia (Revankar and Sutton, 2010). *Exophiala jeanselmei*, another etiological fungus of eumycetoma, has also never been isolated from the environment (Pattanaprichakul et al., 2013). *Madurella mycetomatis* has been assumed to be a saprotroph but de Hoog et al. (2004) doubt this hypothesis. Clearly, more research is needed to clarify the ecology of eumycetoma-causing fungi.

The triad of symptoms, including a painless subcutaneous mass, multiple sinuses, and discharge that contains grains (see below), is pathognomonic for eumycetoma (Suleiman and Fahal, 2013). The disease remains localized in the cutaneous and subcutaneous tissues, fascia and bones, and very slowly develops into a mixed suppurative granulomatous lesion (McGinnis, 1996). The infected tissues contain grains of the pathogen which are white, black or pale in color (de Hoog et al., 2004; Pang et al., 2004; Revankar and Sutton, 2010). Recently, a yellow-grain fungal eumycetoma caused by

Pleurostomophora ochracea has been described (Mhmoud et al., 2012). Yellow-grain eumycetomata had so far been observed only in actinomycetomas (Mhmoud et al., 2012). The color of grains is pathogen-specific. The grains consist of aggregates of the etiological agent surrounded by a dense matrix containing melanin. Melanin protects the pathogen, provides pathogenicity factors, and impedes the penetration of antifungal agents (McGinnis, 1996; Pang et al., 2004; Revankar and Sutton, 2010; Ibrahim et al., 2013). The grains resemble the sclerotia produced by many fungi.

Eumycetomata progress slowly, and are mostly diagnosed through histopathologic analysis of biopsies or tissue culture. Disease progression can be assessed with techniques such as x-ray, ultrasound, Computerized Tomography scan (CT scan), and Magnetic Resonance Imaging (MRI; Suleiman and Fahal, 2013). The eumycetoma requires extended systemic antifungal treatment with azole compounds and echinocandins, in addition to surgery, due to extensive deep tissue involvement and tumefaction of infection (Revankar and Sutton, 2010; Mattioni et al., 2013; Suleiman and Fahal, 2013). The surgical treatment may require local excision, repeated debridement and amputations (Ezaldeen et al., 2013).

Phaeohyphomycosis

The etiological agents of this disease are also dematiacious, and belong to varied morphological and phylogenetic affiliations (Rinaldi, 1996; Revankar and Sutton, 2010; Abu-Elteen and Hamad, 2012). The dematiaceous fungi cause diseases collectively known as phaeohyphomycosis (Etymology; Gk., "Phaeo" meaning "dark") after accounting for all those that cause chromoblastomycosis and eumycetoma (Revankar, 2007; Revankar and Sutton, 2010). Spores are the diagnostic dispersal propagules of these fungi and infection of humans takes place through inhalation and introduction of pathogens through local trauma.

The last few decades have seen a significant rise in the types of fungi and numbers/types of diseases caused by these emerging opportunistic fungi, some of which may be true pathogens also. The diseases caused by these fungi are varied and include allergic diseases, and superficial, deep local, pulmonary, central nervous system and disseminated infections (Revankar and Sutton, 2010). Immunocompetent hosts may generally suffer from different cutaneous and subcutaneous infections whereas both immunocompetent and immunocompromised hosts may get invasive and disseminated infections (Abu-Elteen and Hamad, 2012).

The fungi causing phaeohyphomycosis are ubiquitous, being common in soil and decaying plant debris. Many of them are plant pathogens also. Nosocomial materials can also harbor fungi and infection by *Exserohilum rostratum*, possibly from infested intravenous dressings, causing cutaneous phaeohyphomycosis in a child with leukemia, has been reported (Saint-Jean et al., 2007). Potentially human pathogenic species of the black yeast, *Exophiala*, have also been reported by Biedunkiewicz and Schulz (2012) and others from drinking tap water. *Exophiala* spp. is also known to occur in water supply systems in the United States and Portugal (Biedunkiewicz and Schulz, 2012). Recent information from many countries is indicating preservation and selection of pathogenical fungi in anthropogenic domestic and medical environments where they are repeatedly exposed and selected against polyextreme conditions. Some of these polyextreme anthropogenic emerging domestic and medical ecological niches for pathogenic black yeasts (*Exophiala* spp.) and some other pathogenic fungi include bathrooms, swimming pools, sinks, kitchens, dishwashers, steam baths, drainpipes and chlorine dioxide-treated dental unit waterlines (Gostinčar et al., 2011; Zalar et al., 2011).

Some common etiological fungi of phaeohyphomycosis from different sources are species of *Bipolaris, Curvularia, Exserohilum, Drechslera, Exophiala salmonis, Cladosporium, Phialophora, Wangiella* and *Alternaria* (Castón-Osorio et al., 2008). Several dematiaceous fungi are neurotropic and infect the nervous system. Some such fungi include *Cladophialophora bantiana, Bipolaris spicifera, Exophiala* spp., *Wangiella dermatitidis, Ramichloridium obovoideum, Chaetomium atrobrunneum, C. strumarium* and *Exserohilum rostratum* (Abbott et al., 1995; Revankar et al., 2004; Abu-Elteen and Hamad, 2012). A literature review of 101 clinical cases revealed that *Cladophialophora bantiana* was the most frequent and that *Ramichloridium mackenziei* was present exclusively in patients from the Middle East (Revankar et al., 2004). Infection of the Central Nervous System (CNS) results in a high rate of mortality (Rosow et al., 2011). In 2012 and 2013, there was a multistate outbreak of fungal meningitis in the United States of America after injection with Methylprednisolone Acetate (PF) contaminated with *Exserohilum rostratum* (Kauffman et al., 2013). An overview of this epidemic will be presented separately after this section along with a general account of phaeohyphomycosis caused by this pathogen.

Phaeohyphomycosis is caused by brown-colored fungi containing melanin in their cell walls which can be confirmed by Fontana-Mason stain. Diagnostic procedures for the diseases caused by melanized fungi include direct microscopy and histopathology on surgical biopsies, culturing, radiology, antigen testing, serology and molecular diagnostics (Revankar and Sutton, 2010). Identification of the etiological agent requires expert mycological training and appreciation of fungal growth stages, e.g., *Exophiala dermatitidis* is dimorphic and grows as black yeast in culture and as a mycelial fungus in the host tissue (Walsh et al., 2004). Some isolates of clinical dematiaceous fungi are non-sporulating and require molecular methods for identification (Santos et al., 2013). Proper caution should be exercised when working with melanized clinical fungi as many of them are hazardous and sporulate profusely in culture producing air-dispersed spores. *Cladophialophora bantiana* should be handled using Biosafety Level 3 (http://www.atcc.org/products/all/58035.aspx; Revankar, 2007) whereas *Exophiala* spp. require Biosafety Level 2 (i.e., for potential pathogens; Biedunkiewicz and Schulz, 2012). Caution should be exercised in working with saprobic and plant parasitic fungi many of which are being increasingly recognized as human pathogens, especially for immunosuppressed patients and those with underlying health problems. Many of these fungi are soil inhabitants and superficially appear innocuous. One of the authors of this chapter (J.P. Tewari) in his lifetime has unknowingly casually handled numerous such potentially hazardous dematiaceous fungi; for example he, along with his group, described the new fungus *Achaetomium strumarium* (Rai et al., 1964; now known as *Chaetomium strumarium*) which is now recognized as being neurotropic and a causal agent of CNS phaeohyphomycosis (brain abscess and encephalitis; Abbott et al., 1995; Revankar et al., 2004; Aribandi et al., 2005; Revankar and Sutton, 2010). Treatment of phaeohyphomycosis includes surgical debridement and antifungal drugs such as amphotericin B or azole compounds (Walsh et al., 2004; Revankar, 2007; Castón-Osorio et al., 2008; Revankar and Sutton, 2010).

Phaeohyphomycosis Caused by *Exserohilum rostratum*

The genus *Exserohilum* has three species (*E. longirostratum*, *E. mcginnisii* and *E. rostratum*) which are well documented human pathogens. They have been recognized as emerging agents of phaeohyphomycosis for a few decades (McGinnis et al., 1986; Alder et al., 2006). Recent molecular evidence has indicated that these three species may be conspecific (Revankar and Sutton, 2010; da Cunha et al., 2012), although there are morphological differences among them. *Setosphaeria rostrata* is the teleomorph or the sexual state of *E. rostratum* which is the anamorph or the asexual state (McGinnis et al., 1986). Conidia of *E. rostratum*, borne sympodially on conidiophores, are 6–16 distoseptate, slightly curved or cylindrical to ellipsoidal to rostrate, smooth-walled, with protruding strongly truncate pigmented hila, germinating from one or both end cells and often other cells, and have darkly pigmented basal and distal septa (McGinnis et al., 1986; Padhey et al., 1986). *Exserohilum* is a cosmopolitan human pathogen occurring primarily in the warm, tropical and subtropical regions such as the southern United States, India and Israel (Alder et al., 2006; da Cunha et al., 2012). The genus is primarily terrestrial but occurs in marine habitat also (Alder et al., 2006). It is a pathogen in some domestic animals and marine invertebrates (Padhye et al., 1986; Sappapan et al., 2008). Contaminated intravenous dressings can be a nosocomial source of infection (Saint-Jean et al., 2007). It has been collected from marine environment associated with the cynobacterial mat off the island of Lanai, Hawaii (Tan et al., 2004). Additionally it is common in soil, rotting wood, plant debris and is parasitic on numerous plants such as many grasses and crop plants from many parts of the world (http://www.ars.usda.gov/is/np/pearlmillet/fungelb.htm; Farr and Rossman, 2013). The high parasitic potential of *E. rostratum* on plants is exemplified by the fact that it is virulent and has been used as a biocontrol agent for some grassy weeds (Yamaguchi et al., 2009; Ng et al., 2011).

Exserohilum rostratum is a dematiaceous fungus and contains dihydroxynaphthalene melanin which provides it a fitness advantage in the host (Alder et al., 2006; Kontoyiannis et al., 2013). It is adapted to grow at human body temperatures (Padhye et al., 1986; Kontoyiannis et al., 2013). Infection takes place through inhalation and implantation of the pathogen in the sinuses and lungs, and it can also take place subsequent to local trauma through the skin and eyes (Aquino et al., 1995; Bell et al., 2013). Both in immunocompromised and immunocompetent subjects, infections are commonly initiated through the skin and paranasal sinuses (Aquino et al., 1995). A recent epidemic of *E. rostratum* meningitis in the United States of America was caused through iatrogenic introduction of the pathogen in patients through contaminated epidural or paraspinal glucocorticoid injections (Kainer et al., 2012; Bell et al., 2013; Kontoyiannis et al., 2013). *Exserohilum* species are opportunistic pathogens and are uncommonly encountered causing infection in humans. Infections are usually restricted to allergic sinusitis, keratitis, localized soft-tissue and skin infections, and rarely invasive infection in immunocompromised patients (Saint-Jean et al., 2007; Kauffman et al., 2013). Nevertheless, a few cases of cerebral abscess, osteomyelitis, prosthetic valve endocarditis and disseminated infections have been reported (Saint-Jean et al., 2007). A review of 101 cases of culture-proven primary central nervous system phaeohyphomycosis from English-language literature (1966–2002) did not reveal any infections caused by *Exserohilum* species (Revankar et al., 2004). Hsu and Lee (1993) reported thick-walled brown fungal elements in the intraepidermal microabscesses and in the suppurative and granulomatous infilterate in the dermis. This observation is interesting as these elements, at least superficially, resemble the Medlar bodies produced in chromoblastomycosis. Diagnosis of *Exserohilum* species infections has been made by culture and microscopic evaluations of biopsy samples supported by appropriate diagnostic examinations (such as fiberoptic endoscopy, computerized tomography (CT), etc., as appropriate; Aquino et al., 1995; Alder et al., 2006; Saint-Jean et al., 2007). Treatment has included surgery, amphotericin B and azole compounds (itraconazole, ketoconazole and voriconazole; Aquino et al., 1995; Alder et al., 2006; Saint-Jean et al., 2007). Further methods of diagnosis and treatment have been developed following the recent epidemic of

Exserohilum rostratum meningitis in the United States of America and these are outlined in the following section.

Exserohilum rostratum Meningitis Epidemic in the United States

Chronic low back and lower extremity pains have been increasing in incidence and intensity (Benyamin et al., 2012). This has necessitated the use of glucocorticoid injections for pain management, a process in which complications through contaminated injectables, generally in the form of meningitis and epidural abscesses, have only rarely been observed (Lockhart et al., 2013). The iatrogenic contaminant pathogens in the past have mainly been bacteria and a few fungi (*Exophiala dermatitidis* and *Aspergillus fumigatus*; http://www.cdc.gov/mmwr/preview/mmwrhtml/mm5149a1.htm; Gunaratne et al., 2006, 2007; Kolbe et al., 2007). In contrast, the spectrum of contaminant fungi was different in the *Exserohilum rostratum* meningitis epidemic.

The Centers for Disease Control and Prevention (CDC) was first notified on September 21, 2012 by the Tennessee Department of Health (TDOH) of several cases of fungal meningitis. Following this notification, the CDC sounded an alert regarding the epidural or paraspinal steroid injections of contaminated preservative-free methylprednisolone acetate compounded and distributed by the New England Compounding Center (NECC), Framingham, Massachusetts linked to these fungal infections (Kerkering et al., 2013; Lockhart et al., 2013). A multistate investigation ensued by local, state and federal agencies and health care providers. This emergent public health response was not subject to review by the CDC's Institutional Review Board (Lockhart et al., 2013). The contaminated methylprednisolone acetate consisted of three lots containing 17,675 vials which were distributed in the United States in 23 states to 76 facilities possibly exposing 13,534 persons to potentially contaminated epidural, spinal, paraspinal, peripheral joint or other injections (Smith et al., 2012, 2013). The largest care-associated iatrogenic fungal disease outbreak in the United States ensued with 750 infections and 64 deaths, which were attributable to the injection of contaminated preservative-free methylprednisolone acetate (http://www.cdc.gov/hai/outbreaks/meningitis-map-large.html; Bell and Khabbaz, 2013). Michigan followed by Tennessee has been the most affected state as of September 06, 2013, with total infections/deaths linked to steroid injections as being 264/19 and 153/16, respectively (http://www.cdc.gov/hai/outbreaks/meningitis-map-large.html). In Ohio, the respective counts have been 20/1.

Fungal infections linked to the steroid injections have resulted in meningitis, stroke, paraspinal/spinal infection and peripheral joint infection (http://www.cdc.gov/hai/outbreaks/meningitis-map-large.html). Further pathological complications after this date have been followed by magnetic resonance imaging (Malani et al., 2013). Meningitis was the more frequent disease up to six weeks after the outbreak, however, localized spinal and paraspinal infections became the main manifestations of contaminated steroid injections thereafter (Malani et al., 2013). The precise cause of this epidemic remains unknown, however, glucocorticoid-induced immunosuppression, in a closed space where the cerebrospinal fluid is located, may be a predisposing factor for fungal infection (Lionakis and Kontoyiannis, 2003; Kolbe et al., 2007; Kontoyiannis et al., 2013). Also, fungal growth may have increased upon exposure to steroids as many fungi have steroid receptors (Feldmesser, 2013).

The specimens/cultures from various infections processed by the CDC, consisted of body fluids [cerebrospinal (CSF), epidural, and synovial fluids], biopsy and autopsy tissues, and cultures sent from the state-level facilities (Lockhart et al., 2013). The investigations included cultural/morphological, molecular, histochemical and antifungal susceptibility techniques. Out of the submitted specimens, the majority of patients were diagnosed to have meningitis (220 or 47%) followed by spinal or paraspinal infections (116 or 25%). Others had peripheral joint infections (21 or 4%), and one each had abscess, peripheral joint infection, and stroke (Lockhart et al., 2013). The isolated fungi (# of isolates) submitted from different states and confirmed by the CDC included *Exserohilum rostratum* (48), *Alternaria alternata* (2), *Aspergillus fumigatus* (1), *A. terreus* (3), *A. tubingensis* (1), *Chaetomium* sp. (1), *Cladosporium cladosporioides* (4), *Cladosporium* sp. (1), *Epicoccum nigrum* (1), *Paecilomyces niveus* (1), *Penicillium paneum* (1), *Scopulariopsis brevicaulis* (1), *Stachybotrys chartarum* (1), and a nonspeciated ascomycete (1) (Lockhart et al., 2013). In addition, fungi/bacteria (# of isolates) from unopened vials from the three lots included *Paecilomyces formosus* (1), *Exserohilum rostratum* (3), *Rhodotorula larynges* (2), *Cladosporium cladosporioides* (1), *Bacillus subtilis* (1), and *B. pumilus* (1) (Lockhart et al., 2013). Overall, 150 (32%) patients tested positive or yielded cultures of *E. rostratum* which were most similar in morphology to *E. longirostratum*, a taxon synonymous with the former species (Lockhart et al., 2013).

Infections by dematiaceous fungi are uncommon and those by *Exserohilum rostratum* are relatively still more uncommon, although infections caused by this group as a whole has been increasing in numbers during the last few decades. From the meager literature on diseases caused by *E. rostratum*, it appears that amphotericin B along with an azole compound (e.g., itraconazole, ketoconazole, and voriconazole) are usually the drugs of choice (Aquino et al., 1995; Alder et al., 2006; Saint-Jean et al., 2007). Lockhart et al. (2013) determined the antifungal susceptibility MIC_{50} and MIC_{90} values of *E. rostratum* isolates encountered during the epidemic for voriconazole (1 and 2 µg/ml, respectively), itraconazole (0.5 and 1 µg/ml, respectively), posaconazole (0.5 and 1 μg/ml, respectively), isavuconazole (4 and 4 μg/ml, respectively), and amphotericin B (0.25 and 0.5 μg/ml, respectively). The pros and cons of various drugs with respect to *E. rostratum* have also been investigated by Kontoyiannis et al. (2013) and Pappas et al. (2013). It is indicated that voriconazole is lipophilic and penetrates the CNS highly. Hence, it is recommended as the initial antifungal agent to be used. Amphotericin B has high activity against *E. rostratum* but shows dose-limiting nephrotoxicity. Interim treatment guidelines for the current contaminated steroid episode have been released by the CDC (http://www.cdc.gov/hai/outbreaks/clinicians/guidance_cns.html#monitoring). These guidelines also emphasize the risk of neurotoxicity and extensive drug interactions with voriconazole.

References

Abbott, S.P., Sigler, L., McAleer, R., McDough, D.A., Rinaldi, M.G. and Mizell, G. 1995. Fatal cerebral mycoses caused by the ascomycete *Chaetomium strumarium*. J. Clin. Microbiol., 33: 2692–2698.

Abu-Elteen, K.H. and Hamad, M.A. 2012. Changing epidemiology of classical and emerging human fungal diseases: a review. Jordan J. Biol. Sci., 5: 215–230.

Alder, A., Yaniv, I., Samra, Z., Yacobovich, J., Fisher, S., Avrahami, G. and Levy, I. 2006. *Exserohilum*: an emerging human pathogen. Eur. J. Clin. Microbiol. Infect. Dis., 25: 247–253.

Ahmad, A., Adelmann, D., Fahal, A., Verbrugh, H., van Belkum, A. and de Hoog, G.S. 2002. Environmental occurrence of *Madurella mycetomatis*, major agent of human eumycetoma in Sudan. J. Clin. Microbiol., 40: 1031–1036.

Aquino, V.M., Norvell, J.M., Krisher, K. and Mustafa, M.M. 1995. Fatal disseminated infection due to *Exserohilum rostratum* in a patient with aplastic anemia: case report and review. Clin. Infect. Dis., 20: 176–178.

Aribandi, M., Bazan III, C. and Rinaldi, M.G. 2005. Magnetic resonance imaging findings in fatal primary cerebral infection due to *Chaetomium strumarium*. Australas. Radiol., 49: 166–169.

Badali, H., Fernández-González, M., Mousavi, B., Illnait-Zaragozi, M.T., González-Rodríguez, J.C., de Hoog, G.S. and Meis, J.F. 2013. Chromoblastomycosis due to *Fonsecaea pedrosoi* and *F. monophora* in Cuba. Mycopathologia, 175: 439–444.

Bell, W.R., Dalton, J.B., McCall, C.M., Karram, S., Pearce, D.T., Memon, W., Lee, R., Carroll, K.C., Lyons, J.L., Gireesh, E.D., Trivedi, J.B., Cettomai, D., Smith, B.R., Chang, T., Tochen, L., Ratchford, J.N., Harrison, D.M., Ostrow, L.W., Stevens, R.D., Chen, L. and Zhang, S.X. 2013. Iatrogenic *Exserohilum* infection of the central nervous system: mycological identification and histopathological findings. Mod. Pathol., 26: 166–170.

Bell, B.P. and Khabbaz, R.F. 2013. Responding to the outbreak of invasive fungal infections. JAMA, 309: 883–884.

Benyamin, R.M., Manchikanti, L., Parr, A.T., Diwan, S., Singh, V., Falco, F.J.E., Datta, S., Abdi, S. and Hirsch, J.A. 2012. The effectiveness of lumbar interlaminar epidural injections in managing chronic low back and lower extremity pain. Pain Physician, 15: E363–E404.

Biedunkiewicz, A. and Schulz, Ł. 2012. Fungi of the genus *Exophiala* in tap water-potential etiological factors of phaeohyphomycosis. Mikol. Lek., 19: 23–26.

Castón-Osorio, J.J., Rivero, A. and Torre-Cisneros, J. 2008. Epidemiology of invasive fungal infection. Interntl. J. Antimicrob. Agents, 32(Suppl. 2): S103–S109.

da Cunha, K.C., Sutton, D.A., Gené, J., Cappila, J., Cano, J. and Guarro, J. 2012. Molecular identification and *in vitro* response to antifungal drugs of clinical isolates of *Exserohilum*. Antimicrob. Agents Chemother., 56: 4951–4954.

de Hoog, G.S., Adelmann, D., Ahmad, A.O.A. and van Belkum, A. 2004. Phylogeny and typification of *Madurella mycetomatis*, with a comparison of other agents of eumycetoma. Mycoses, 47: 121–130.

de Hoog, G.S., Ahmad, S.A., Najafzadeh, M.J., Sutton, D.A., Keisari, M.S., Fahal, A.H., Eberhardt, U., Verkleij, G.J., Xin, L., Stielow, B. and van de Sande, W.W.J. 2013. Phylogenetic findings suggest possible new habitat and routes of infection of human eumycetoma. PLOS Neg. Trop. Dis., 7: e2229, 1–7.

de Hoog, G.S., Nishikaku, A.S., Fernandez-Zeppenfeldt, G., Padin-González, C., Burger, E., Richard-Yegres, N. and van den Ende, A.H.G.G. 2007. Molecular analysis and pathogenicity of the *Cladophialophora carrionii* complex, with description of a novel species. Stud. Mycol., 58: 219–234.

Elgart, G.W. 1996. Chromoblastomycosis. Dermatol. Clin., 14: 77–83.

Ezaldeen, E.A., Fahal, A.H. and Osman, A. 2013. Mycetoma herbal treatment: The Mycetoma Research Centre, Sudan experience. PLOS Neg. Trop. Dis., 7: e2400, 1–5.

Farr, D.F. and Rossman, A.Y. 2013. Fungal Databases, Systematic Mycology and Microbiology Laboratory, ARS, USDA. Retrieved October 17, 2013, from http://nt.ars-grin.gov/fungaldatabases/.

Feldmesser, M. 2013. Fungal disease following contaminated steroid injections. Am. J. Pathol., 183: 661–664.

Gezuele, E., Mackinnon, J.E. and Cinti-Díaz, I.A. 1972. The frequent isolation of *Phialophora verrucosa* and *Phialophora pedrosoi* from natural sources. Sabouraudia, 10: 266–273.

González, G.M., Rojas, O.C., Bocanegra-Garcia, V., González, J.G. and Garza- González, E. 2013. Molecular diversity of *Cladophialophora carrionii* in patients with chromoblastomycosis in Venezuela. Med. Mycol., 51: 170–177.

Gostinčar, C., Grube, M. and Gunde-Cimerman, N. 2011. Evolution of fungal pathogens in domestic environments? Fungal Biol., 115: 1008–1018.

Gunaratne, P.S., Wijeyaratne, C.N., Chandrasiri, P., Sivakumaran, S., Sellahewa, K., Fernando, R., Wanigasinghe, J., Jayasinghe, S., Ranawala, R., Riffsy, M.T.M. and Seneviratne, H.R. 2006. An outbreak of *Aspergillus* meningitis following spinal anaesthesia for caesarean section in Sri Lanka: a post-tsunami effect? Ceylon Med. J., 51: 137–142.

Gunaratne, P.S., Wijeyaratne, C.N. and Seneviratne, H.R. 2007. *Aspergillus* meningitis in Sri Lanka-a post sunami effect? N. Engl. J. Med., 356: 754–756.

Hsu, M.M.-L. and Lee, J.Y.-Y. 1993. Cutaneous and subcutaneous phaeohyphomycosis caused by *Exserohilum rostratum*. J. Am. Acad. Dermatol., 28: 340–344.

Ibrahim, A.I., El Hasan, A.M., Fahal, A. and van de Sande, W.W. 2013. A histopathological exploration of the *Madurella mycetomatis* grain. PLoS One, 8, e57774: 1–6.

Kainer, M.A., Reagan, D.R., Nguyen, D.B., Wiese, A.D., Wiese, M.E., Ward, J., Park, B.J., Kanago, M.L., Baumblatt, J., Schaefer, M.K., Berger, B.E., Marder, E.P., Min, J.-Y., Dunn, J.R., Smith, R.M., Dreyzehner, J. and Jones, T.F. 2012. Fungal infections associated with contaminated methylprednisolone in Tennessee. N. Engl. J. Med., 367: 2194–2203.

Kauffman, C.A., Pappas, P.G. and Patterson, T.F. 2013. Fungal infections associated with contaminated methylprednisolone injections. N. Engl. J. Med., 368: 2495–2500.

Kerkering, T.M., Grifasi, M.L., Baffoe-Bonnie, A.W., Bansal, E., Garner, D.C., Smith, J.A., Smith, J.A., Demicco, D.D., Schleupner, C.J., Aldoghaither, R.A. and Savaliya, V.A. 2013. Early clinical observations in prospectively followed patients with fungal meningitis related to contaminated epidural steroid injections. Ann. Intern. Med., 158: 154–161.

Kolbe, A.B.L., McKinney, A.M., Kendi, A.T.K. and Misselt, D. 2007. *Aspergillus* meningitis and discitis from low-back procedures in an immunocompetent patient. Acta Radiol., 48: 687–689.

Kontoyiannis, D.P., Perlin, D.S., Roilides, E. and Walsh, T.J. 2013. What can we learn and what do we need to know amidst the iatrogenic outbreak of *Exserohilum rostratum* meningitis? Clin. Infect. Dis., 57: 853–859.

La Hoz, R.M. and Baddley, J.W. 2012. Subcutaneous fungal infections. Curr. Infect. Dis. Rep., 14: 530–539.

Lionakis, M.S. and Kontoyiannis, D.P. 2003. Glucocorticoids and invasive fungal infections. Lancet, 362: 1828–1838.

Lockhart, S.R., Pham, C.D., Gade, L., Iqbal, N., Scheel, C.M., Cleveland, A.A., Whitney, A.M. Noble-Wang, J., Chiller, T.M., Park, B.J., Litvintseva, A.P. and Brandt, M.E. 2013. Preliminary laboratory report of fungal infections associated with contaminated methylprednisolone injections. J. Clin. Microbiol., 51: 2654–2661.

Lu, S., Lu, C., Zhang, J., Hu, Y., Li, X. and Xi, L. 2013. Chromoblastmycosis in mainland China: a systematic review on clinical characteristics. Mycopathologia, 175: 489–495.

Lyon, J.P., Azevedo, C.M.P.S., Moreira, L.M., Lima, C.J. and Resende, M.A. 2011. Photodynamic antifungal chemotherapy against chromoblastomycosis. Mycopathologia, 172: 293–297.

Malani, A.N., Vandenberg, D.M., Singal, B., Kasotakis, M., Koch, S., Moudgal, V., Jagarlamudi, R., Neelakanta, A., Otto, M.H., Halasyamani, L., Kaakaji, R. and Kauffman, C.A. 2013. Magnetic resonance imaging screening to identify spinal and paraspinal infections associated with injections of contaminated methylprednisolone acetate. JAMA, 309: 2465–2472.

Martínez, R.L. and Tovar, L.J.M. 2007. Chromoblastomycosis. Clin. Dermatol., 25: 188–194.

Mattioni, S., Develoux, M., Brun, S., Martin, A., Jaureguy, F., Naggara, N. and Bouchaud, O. 2013. Management of mycetomas in France. Med. Mal. Infect., 43: 286–294.

McGinnis, M.R. 1996. Mycetoma. Dermatol. Clin., 14: 97–104.

McGinnis, M.R., Rinaldi, M.G. and Winn, R.E. 1986. Emerging agents of phaeohyphomycosis: pathogenic species of *Bipolaris* and *Exserohilum*. J. Clin. Microbiol., 24: 250–259.

Mhmoud, N.A., Ahmed, S.A., Fahal, A.H., de Hoog, G.S. and van den Ende, A.H.G.G. 2012. *Pleurostomophora ochracea*, a novel agent of human eumycetoma with yellow grains. J. Clin. Microbiol., 50: 2987–2994.

Ng, S.C., Kadir, J., Hailmi, M.S. and Rahim, A.A. 2011. Efficacy of *Exserohilum longirostratum* on barnyard grass (*Echinochloa crus-galli* spp. *crusgalli*) under field conditions. Biocontrol Sci. Technol., 21: 449–460.

Nucci, M., Akiti, T., Barreiros, G., Silveira, F., Revankar, S.G., Sutton, D.A. and Patterson, T.F. 2001. Nosocomial fungemia due to *Exophiala jeanselmei* var. *jeanselmei* and a *Rhinocladiella* species: newly described causes of bloodstream infection. J. Clin. Microbiol., 39: 514–518.

Padhey, A.A., Ajello, L., Wieden, M.A. and Steinbronn, K.K. 1986. Phaeohyphomycosis of the nasal sinuses caused by a new species of *Exserohilum*. J. Clin. Microbiol., 24: 245–249.

Pang, K.R., Wu, J.J., Huang, D.B. and Tyring, S.K. 2004. Subcutaneous fungal infections. Dermatol. Ther., 17: 523–531.

Pappas, P.G., Kontoyiannis, D.P., Perfect, J.R. and Chiller, T.M. 2013. Real-time treatment guidelines: considerations during the *Exserohilum rostratum* outbreak in the United States. Antimicrob. Agents Chemother., 57: 1573–1576.

Pattanaprichakul, P., Bunyaratavej, S., Leeyaphan, C., Sitthinamsuwan, P., Sudhadham, M., Muanprasart, C., Feng, P., Badali, H. and de Hoog, G.S. 2013. An unusual case of eumycetoma caused by *Exophiala jeanselmei* after a sea urchin injury. Mycoses, 56: 491–494.

Queiroz-Telles, F. and de C.L. Santos, D.W. 2013. Challenges in the therapy of chromoblastomycosis. Mycopathologia, 175: 477–488.

Rai, J.N., Tewari, J.P. and Mukerji, K.G. 1964. *Achaetomium*, a new genus of Ascomycetes. Can. J. Botany, 42: 693–97.

Revankar, S.G. 2007. Dematiaceous fungi. Mycoses, 50: 91–101.

Revankar, S.G. and Sutton, D.A. 2010. Melanized fungi in human disease. Clin. Microbiol. Rev., 23: 884–928.

Revankar, S.G., Sutton, D.A. and Rinaldi, M.G. 2004. Primary central nervous system phaeohyphomycosis: a review of 101 cases. Clin. Infect. Dis., 38: 206–216.

Rinaldi, M.G. 1996. Phaeohyphomycosis. Dermatol. Clin., 14: 147–153.

Rosow, L., Jiang, J.X., Deuel, T., Lechpammer, M., Zamani, A.A., Milner, D.A., Folkerth, R., Marty, F.M. and Kesari, S. 2011. Cerebral phaeohyphomycosis caused by *Bipolaris spicifera* after heart transplantation. Transpl. Infect. Dis., 13: 419–423.

Saint-Jean, M., St-Germain, G., Laferrière, C. and Tapiero, B. 2007. Hospital-acquired phaeohyphomycosis due to *Exserohilum rostratum* in a child with leukemia. Can. J. Infect. Dis. Med. Microbiol., 18: 200–202.

Salgado, C.G., da Silva, J.P., Diniz, J.A.P., da Silva, M.B., Costa, P.F., Teixeira, C. and Salgado, U.I. 2004. Isolation of *Fonsecaea pedrosoi* from thorns of *Mimosa pudica*, a probable natural source of chromoblastmycosis. Rev. Inst. Med. Trops. S. Paulo, 46: 33–36.

Santos, D.W.C.L., Padovan, A.C.B., Melo, A.S.A., Gonçalves, S.S., Azevedo, V.R., Ogawa, M.M., Freitas, T.V.S. and Colombo, A.L. 2013. Molecular identification of melanised non-sporulating moulds: a useful tool for studying the epidemiology of phaeohyphomycosis. Mycopathologia, 175: 445–454.

Sappapan, R., Sommit, D., Ngamrojanavanich, N., Pengpreecha, S., Wiyakrutta, S., Sriubolmas, N. and Pudhom, K. 2008. 11-hydroxymonocerin from the plant endophytc fungus *Exserohilum rostratum*. J. Nat. Prod., 71: 1657–1659.

Siqueira, I.M., Ribeiro, A.M., de Medeiros Nóbrega, Y.K., Simon, K.S., Souza, A.C.O., Jerônimo, M.S., Neto, F.F.C., Silva, C.L., Felipe, M.S.S. and Bocca, A.L. 2013. DNA-hsp65 vaccine as therapeutic strategy to treat experimental chromoblastomycosis caused by *Fonsecaea pedrosoi*. Mycopathologia, 175: 463–475.

Smith, R.M., Schaefer, M.K., Kainer, M.A., Wise, M., Finks, J., Duwve, J., Fontaine, E., Chu, A., Carothers, B., Reilly, A., Fiedler, J., Wiese, A.D., Feaster, C., Gibson, L., Griese, S., Purfield, A., Cleveland, A.A., Benedict, K., Harris, J.R., Brandt, M.E., Blau, D., Jernigan, J., Weber, T. and Park, B.J. 2012. Fungal infections associated with contaminated methylprednisolone injections-preliminary report. N. Engl. J. Med., DOI: 10.1056/NEJMoa1213978.

Smith, R.M., Schaefer, M.K., Kainer, M.A., Wise, M., Finks, J., Duwve, J., Fontaine, E., Chu, A., Carothers, B., Reilly, A., Fiedler, J., Wiese, A.D., Feaster, C., Gibson, L., Griese, S., Purfield, A., Cleveland, A.A., Benedict, K., Harris, J.R., Brandt, M.E., Blau, D., Jernigan, J., Weber, T. and Park, B.J. 2013. Fungal infections associated with contaminated methylprednisolone injections. N. Engl. J. Med., 369: 1598–1609.

Suleiman, A.M. and Fahal, A.H. 2013. Oral cavity eumycetoma: a rare and unusual condition. Oral Surg. Oral Med. Oral Pathol. Oral Radiol., 115: e23–e25.

Tan, R.X., Jensen, P.R., Williams, P.G. and Fenical, W. 2004. Isolation and structure assignments of rostratins A-D, cytotoxic disulfides produced by the marine-derived fungus *Exserohilum rostratrum*. J. Nat. Prod., 67: 1374–1382.

Thirumalachar, M.J. and Padhye, A.A. 1968. Isolation of *Madurella mycetomi* from soil in India. Hindustan Antibiot. Bull., 10: 314–318.

Torres-Guerrero, E., Isa-Isa, R., Isa, M. and Arenas, R. 2012. Chromoblastomycosis. Clin. Dermatol., 30: 403–408.

Walsh, T.J., Groll, A., Hiemenz, J., Fleming, R., Roilides, E. and Anaissie, E. 2004. Infections due to emerging and uncommon medically important fungal pathogens. Clin. Microbiol. Infect., 10(Suppl. 1): 48–66.

Weedon, D., van Deurse, M., Allison, S. and Rosendahl, C. 2013. Chromoblastomycosis in Australia: an historical perspective. Pathology, 45: 489–491.

Yamaguchi, K.-i., Nagai, K. and Matsumoto, E. 2009. Conidia production of *Exserohilum rostratum*, a biocontrol agent against red sprangletop, by a two-phase system using sponge matrix. Bull. Minamikyushu Univ., 39A: 73–77.

Yang, Y.-P., Li, W., Huang, W.-M., Zhou, Y. and Fan, Y.-M. 2013. Chromoblastomycosis caused by *Fonsecaea*; clinicopathology, susceptibility and molecular identification of seven consecutive cases in southern China. Clin. Microbiol. Inf., 19: 1023–1028.

Zalar, P., Novak, M., de Hoog, G.S. and Gunde-Cimerman, N. 2011. Diswashers—a man-made ecological niche accommodating human opportunistic fungal pathogens. Fungal Biol., 115: 997–1007.

CHAPTER 17

Fungi Associated with Animal Diseases in India

Rishendra Verma, Abhishek* and *Harshit Verma*

ABSTRACT

Fungal diseases in livestock are assuming greater importance as they not only affect the health but also make the country's livestock industry uneconomical as these infections lead to decrease or complete cessation of milk yields, abortion, sterility and deterioration in hide quality. Moreover, fungal diseases are well established as zoonotic diseases and thereby cause a public health problem. Hence, fungal diseases need further attention as no specific medicines are available against most of the diseases and none of them have so far been eliminated (Sikdar, 1998).

The frequency of fungal infections in livestock has risen dramatically in recent years. Filamentous fungi and yeasts infect animals including their keratin, hide, skin and various organs leading to their superficial, subcutaneous or systemic mycoses. Dermatophytes like *Trichophyton, Microsporum* and *Epidermatophyton* attack skin, hair and nails. Mammalian and avian aspergillosis has been recorded as epidemics. Phycomycetous fungi have been isolated from feed and fodder of animals or from other animal sources. Yeast and dimorphic fungi like *Candida* and *Rhodotorula* have also been recovered from the alimentary and reproductive tracts as well as from cases of mastitis. Systemic mycoses such as cryptococcosis, blastomycosis and histoplasmosis have been reported infrequently, while rhinosporidosis has been well

Indian Veterinary Research Institute, Izatnagar-243 122 (Uttar Pradesh).
* Corresponding author

documented mostly from the southern part of India. This chapter embodies a consolidated information of those fungi which have been directly or indirectly associated with mycoses in different animals in India.

Introduction

Fungi, both filamentous and yeasts, infect animals including their keratin, hide, skin and various organs leading to their superficial, subcutaneous or systemic mycoses. Unfortunately in India, these problems especially dermatophytoses have not received their due attention because of its lesser severity, lack of adequate diagnostic and clinical facilities, availability and feasibility of treatment of such infections, particularly in small and large ruminants, non-availability of vaccines and most importantly ignorance of the animal keepers. The information that we have for these ailments (mycotic infections) of animals in India pertains to the isolation of causal organisms, their identification and demonstration in histopathological preparations. Although, the economic losses due to fungal infections in animals, has not been estimated, nevertheless, mycotic infections impede their growth, affect hide productivity and in some cases even lead to chronic mastitis and can even cause abortion. These conditions certainly attribute to the economics of livestock and poultry industry as well as trade. This chapter overviews all important types of mycoses involved in superficial and systemic infections as well as fungi associated with reproductive infections and chronic mastitis in animals in India.

Dermatophytosis

Dermatophytosis is an integumentary disease caused by the fungi belonging to a group referred to as dermatophytes. It is popularly known as ring worm and it affects the keratinized tissues of the body, e.g., skin, nail, hoof and hairs. The disease leads to a discoloration of skin followed by the erythema, exudation, heat and alopecia. Ringworm has been reported in all species of domesticated animals irrespective of age and sex. *Trichophyton verrucosum* is the most common dermatophyte affecting bovines in India whereas in dogs and cats, *Microsporum canis* is the causative agent.

Epidemiologically, the dermatophytes are categorized into anthropophilic (*Epidermatophyton floccosum, Trichophyton mentagrophytes,* and *T. rubrum*), zoophilic (*Microsporum canis, T. equinum,* and *T. nanum*) and geophilic (*M. gypseum, T. ajelloi,* and *T. terrestre*). The difference in the chemical composition of various keratins, which the dermatophytes have to digest with their enzymes, may well account for this selective parasitism in human and animals. There are several instances where human infections by zoophilic dermatophytes are almost invariably acquired through contact with infected animals. About 10–20% (urban area) and 70–80% (rural area) of human ringworms are of animal origin. It would be evident from the list given below that a number of anthropophilic dermatophytes have been isolated from animals.

Trichophyton Group		
Cattle	*T. verrucosum*	(Klokke, 1964; Klokke and Durairaj, 1967; Singh and Singh, 1970b; Gugnani, 1972; Tewari, 1972; Chatterjee and SenGupta, 1979; Sharma et al., 1979; Pal and Singh, 1983; Sarkar et al., 1985)
	T. rubrum	(Chakrabarty et al., 1954; Singh and Singh, 1970b; Chatterjeee et al., 1978; Chatterjee and SenGupta, 1979; Sharma et al., 1979; Thakur et al., 1983b; Sarkar et al., 1985)
	T. mentagrophytes	(Gupta et al., 1968; Monga et al., 1974; Sharma et al., 1979; Chatterjee and SenGupta, 1979; Pal and Singh, 1983; Thakur et al., 1983b; Sarkar et al., 1985; Sharma et al., 2010)
	T. violaceum	(Gupta et al., 1968; Gupta et al., 1970; Tewari, 1972)
Buffaloes	*T. verrucosum*	(Singh and Singh, 1970b; Pal and Singh, 1983; Thakur et al., 1983b)
	T. mentagrophytes	(Chatterjee and SenGupta, 1979; Pal and Singh, 1983; Thakur and Verma, 1984)
	T. ajelloi	(Saxena, 1972)
	T. terrestre	(Gupta et al., 1970)
Calves	*T. verrucosum*	(Padhye et al., 1966; Tewari, 1972; Saxena and Mehra, 1973; Khanna et al., 1974; Monga et al., 1974; Sharma, 1979)
	T. mentagrophytes	(Padhye et al., 1966; Sharma et al., 2010)

Sheep	*T. verrucosum*	(Thakur et al., 1983; Sarkar et al., 1985)
	T. mentagrophytes	(Misra, 1971; Tewari, 1972)
Goat	*T. verrucosum*	(Chatterjee and SenGupta, 1979; Pal and Singh, 1983; Sarkar et al., 1985)
	T. mentagrophytes	(Chatterjee and SenGupta, 1979; Pal and Singh, 1983; Sarkar et al., 1985)
Dog	*T. verrucosum*	(Singh and Singh, 1970b)
	T. mentagrophytes	(Saxena, 1972; Tewari, 1972, Das and Sambamurti, 1973; Chatterjee and SenGupta, 1979; Sharma et al., 1979; Sharam et al., 2010; Gangil et al., 2012)
	T. simii	(Kulkarni et al., 1969; Singh and Singh, 1970a; Gugnani, 1972; Das and Sambamurti, 1973)
	T. rubrum	(Chakrabarty et al., 1954; Padhye et al., 1966; Tewari, 1972)
	T. violaceum	(Singh and Singh, 1970a)
Horse	*T. equinum*	(Padhye et al., 1966; Singh, 1981)
	T. tonsurans var. *sulfureum*	(Kulkarni et al., 1969)
	T. mentagrophytes	(Chatterjee and SenGupta, 1979)
Mules	*T. verrucosum*	(Das et al., 1984)
	T. mentagrophytes	(Das et al., 1984)
Pig	*T. mentagrophytes*	(Chatterjee and SenGupta, 1979)
	T. rubrum	(Chaudhary et al., 1968)
Camel	*T. schoenleni*	(Chatterjee et al., 1978)
Poultry	*T. mentagrophytes*	(Singh and Singh, 1970c; Tewari, 1972)
	T. simii	(Mohapatra and Mahajan, 1970; Singh and Singh, 1970c; Gupta et al., 1970; Tewari, 1972; Gugnani and Randhawa, 1973)
	T. violaceum	(Singh, 1981)

Microsporum Group		
Cattle	*M. gypseum*	(Padhye et al., 1966; Gupta et al., 1970; Saxena, 1972; Chatterjee and SenGupta, 1979)
	M. canis	(Padhye et al., 1966)
Sheep	*M. gypseum*	(Thakur et al., 1983a)
Goat	*M. gypseum*	(Thakur and Verma, 1984; Sarkar et al., 1985)
Horse	*M. gypseum*	(Singh and Singh, 1970c; Gupta et al., 1970; Saxena and Mehra, 1973; Pal et al., 1994)
Pig	*M. nanum*	(Gupta et al., 1970; Chatterjee and SenGupta, 1979; Das et al., 1980; Sarkar et al., 1985)
Dog	*M. canis*	(Khandhari and Sethi, 1964; Gupta et al., 1968; Singh and Singh, 1970a; Saxena, 1972; Navjeevan et al., 2005)
	M. gypseum	(Saxena and Mehra, 1973; Gugnani et al., 1977; Chatterjee and SenGupta, 1979; Navjeevan et al., 2005; Sharma et al., 2009, 2010; Gangil et al., 2012)
	M. distortum	(Chatterjee et al., 1980)
	T. mentagrophytes	(Navjeevan et al., 2005)

Aspergillosis

Aspergillosis is primarily a disease of the respiratory tract characterized by inflammatory granulomatous lesions and caused by *Aspergillus fumigatus, Aspergillus flavus* and other *Aspergillus* species. Hematogenous infection to other organs may occur. The disease is relatively rare in domestic and pet animals but occurs frequently in avian species. In poultry, it occurs in two forms. Firstly, acute outbreaks with high morbidity and high mortality in younger birds and secondarily, chronic infection affecting individual adult birds. Aspergillosis is more important in turkeys but may also affect chickens. The acute form of this disease is usually called as "Brooder Pneumonia".

Mammalian Aspergillosis

Aspergillosis in bovine has also been reported in India. Association of *Aspergillus glaucus* with bovine haematuria was first reported along with demonstration of *Aspergillus* sp. in the kidney and bladder by Datta (1984), and in calves affecting skin by Nag and Mallick (1961). Later on Naryana et al. (1964) demonstrated a fungus in sections of tissues of lung nodule with pulmonary aspergillosis in Murrah she-buffalo. Then after, a number of reports established the association of different species of Aspergilli in a variety of pulmonary ailments of animals—bull, buffalo, pig, goats and sheep (Damodaran, 1971; Sadana and Kalra, 1974; Chauhan and Dwivedi, 1974; Gill et al., 1977; Sharma and Dwivedi, 1977; Singh et al., 1977; Uppal et al., 1978; Baruah et al., 1984; Ponnuswamy, 2010).

Haematogenous spread of *A. oryzae* in 6–8 months old desi lamb was demonstrated by the presence of branched, septate hyphae in the area of necrosis and thrombus within blood vessels. The fungus was isolated in culture (Krishna et al., 1981). *A. fumigatus* isolated and identified from corneal ulcer of a buffalo was found pathogenic to mice (Pal, 1983). *A. fumigatus* pathogenic to mice was isolated from nasal exudate of one of 23 mules and one of the seven camel dromedaries. It was also isolated from the soil of a dromedary shed and a mule shed, and also from corneal ulcer of a buffalo (Pal, 1983; Pal and Mehrotra, 1984).

Eye swabs of mules, dogs, fowls, cattle, buffaloes and camels with various ophthalmological problems were examined and *A. fumigatus* was isolated from the swabs of dog, bull, mule and fowl (Pal and Mehrotra, 1986). *A. terreus* was found to be associated with cancerous tissue of the horn core in a weak and old debilitated cow (Pal, 1989b). Pal et al. (2011) reported canine otitis due to *Aspergillus niger* in two out of 36 dogs.

Avian Aspergillosis

A fungus resembling *A. fumigatus* has been isolated from the eggs of infected hens (Asthana, 1944). Lesions, in the affected birds, were found in the lungs, bronchi, trachea, intestine, caecum, kidney, ovaries, testes and heart. Later on, fungi like *A. niger* (Dhawedkar and Dhansear, 1971; Singh and Malhotra, 1970), *Aspergillus* sp. (Dhawedkar and Dhansear, 1971) and *A. flavus* (Singh and Malhotra, 1970) were isolated from dead-in-shell eggs.

Cutaneous aspergillosis was reported in poultry (Kalapesi and Purohit, 1956; Naryana, 1961). *A. fumigatus* was isolated and demonstrated histopathologically in the lung, heart, trachea, air sacs and intestine in poultry (chicks, duck, pigeon, etc.; Prasad and Srivastav, 1964; Banerjee and Bhattacharya, 1965; Babbras and Radhakrishnan, 1967; Sharma et al., 1979; Rao and Choudhary, 1980; Pal and Dabiya, 1984; Rao and Acharjyo, 1988;

Pal et al., 1989; Pal, 1992, 2003; Singh et al., 2009; Lakshman et al., 2009). Besides other Aspergilli, *A. nidulans* has also been found to be responsible for pulmonary aspergillosis and a variety of histological changes have been recorded by different workers (Dhodapkar and Awadhiya, 1969; Shah and Mohanty, 1974; Deka and Rao, 1988).

Other fungal genera recovered from birds, having mild or no respiratory tract lesions or from healthy ones, are *Aspergillus* (different species), species of *Cladosporium, Curvularia, Fusarium, Mucor, Penicillium, Rhizopus, Sepedonium, Scopulariopsis, Trichosporon,* and *Tritirachium* (Garg and Sethi, 1969; Dhopapkar and Awadhiya, 1969; Sharma and Sethi Negi, 1971; Parihar and Singh, 1971; Jand and Dhillon, 1973; Christopher, 1977; Sharma et al., 1979; Rao et al., 1982; Pal and Dabiya, 1984; Rao and Acharjyo, 1988; Pal et al., 1989; Pal, 1992).

Maduramycosis or Madura Foot

Maduramycosis is a chronic granulomatous infection characterized by the presence of multiple discharging sinuses in the affected part. It occurs in both animals and man and caused by the number of soil inhabiting fungi, which get implanted in to the tissue usually through trauma on the foot. The disease was first reported in India in the province of Madurai. Maduramycosis is endemic in certain parts of the country. It is predominant in Tamil Nadu, Kerala, Rajasthan and West Bengal. A great variety of fungi have been incriminated as etiologic agents in animals like *Curvularia geniculata* and *Helminthosporium speciferum*. The infection of phycomycosis was for the first time, reported in India in a mare. It was confined to the skin and subcutaneous tissues of the brisket region and the fungus was identified as *Entomophthora coronata* (Chauhan et al., 1973). Mycotic gastritis in piglets was reported and non-septate freely branching hyphae were seen on the surface of ulcers and in the underlying tissues (Chauhan and Sadana, 1973; Mataney et al., 1975; Sadana and Kalra, 1973; Damodaran and Sundararaj, 1976).

Mycotic ruminitis and mycotic stritis due to species of *Syncephalastrum* and *Mucor* in an eight month old calf of non-descript breed was reported (Nayak et al., 1975; Damodaran et al., 1976). Granulomatous lesions in lungs and lymph nodes of buffaloes at various abattoirs were observed and phycomycotic hyphae were demonstrated in two cases (Gill and Singh, 1976). Vanamaya et al. (1983) also observed phycomycotic pneumonia in a buffalo (*Bubalus bubalis*).

Candidiasis

Candidiasis is a sporadic disease of the alimentary canal of poultry caused by the *Candida albicans* and is a common endogenous infection of animals. Sometimes this disease leads to a high mortality rate in a flock. Predisposing factors are important for establishing the infection. The disease may cause mastitis or abortion in cattle. Swine are occasionally affected and the disease is rare in dogs, cats and horses. It becomes more severe during long term treatment with corticosteroids or oral antibiotics. Candidiasis in poultry is popularly known as "thrush". Many species of *Candida* have been reported from the domesticated animals which are responsible for clinical spectrum of diseases like *Candida tropicalis, Candida krusei* and *Candida parapsilosis.*

Isolation of *Candida albicans, C. krusei, C. parakrusei, C. stellatoidea* and *C. tropicalis* has been reported from the crops of fowls (Pathak and Singh, 1962). Singh and Singh (1972) isolated 14 strains of *C. parakrusei* from a nodular lesion in the intestine of goat but they failed to demonstrate the organism in PAS stained sections of the tissue. Shah et al. (1982) reported cutaneous candidiasis in 5–30 weeks old Japanese quails (*Coturnix coturnix japonica*). Histopathologically, granulomatous reaction in the dermis and hyper keratosis in foot pad lesions were seen. *C. albicans* was isolated and demonstrated histologically in skin scrapings of one month old male lion cub (*Panther leo*). The sections stained with PAS showed septate branched filaments together with yeast-like structures in both degenerated epidermis and in dermal layers (Verma et al., 1983). *C. albicans* and *C. guilliermondi* were isolated from pneumonic lungs obtained from an abattoir (Baruah et al., 1984). The mycological survey of the alimentary tract of chicken and quail yielded isolation of *C. albicans, C. tropicalis, C. pseudotropicalis* and *C. stellatoidea* (Verma, 1984). Sikdar and Upal (1985) isolated *C. guillermondi* from suspected case of bovine lymphagnitis. Pal and Ragi (1989) reported *C. tropicalis* from a lymph node of buffalo slaughtered for food. *C. albicans,* from 11.4% stomatitis cases of dogs, was reported by Jadav and Pal (2006).

Cryptococcosis

Cryptococcosis is a sub-acute or chronic disease of animals caused by the yeast like fungus *Cryptococcus neoformans*. The disease in dogs and cats is characterized by pulmonary and central nervous system involvement and/or by localized lesions of the oral and nasal mucosa. The infection in cattle, usually involves the mammary tissue and adjacent lymph nodes. In horse, manifestations are nasal discharges and respiratory distress associated with nasal granuloma. It is serious and infections of *Cryptococcus neoformans* are known to occur from pigeon droppings and soil. The fungus was also recovered from the sputum of an attendant of the pigeon house who presented chest pain, low grade fever, mild cough and mucoid expectoration (Gugnani et al., 1972), but not from the samples of pigeon viscera (Gugnani et al., 1976).

C. neoformans has also been isolated from various materials associated with different birds including their droppings. A zoo aviary, which was inhabited by 4 spp. of munia birds, Red Munia (*Amandava amandava*), Green Munia (*Estrilda* [A.] *formosa*), Black Headed Munia (*Lonchura malacca*) and Crested Bunting (*Melophus lathami*), also showed the presence of *C. neoformans* var. *neoformans* (Pal, 1989b).

Blastomycosis

Blastomycosis is a granulomatous and suppurative disease of various animals and caused by the dimorphic fungus *Blastomyces dermatitidis.* It starts from a primary pulmonary focus and may disseminate to the body organs or skin. Dogs are the most susceptible animals. It is also known as "Gilchrist" disease and North American Blastomycosis. Greatly enlarged mediastinal and bronchial lymph nodes, which appear on the radiograph as a dense mass at the bifurcation of the trachea is a common feature of the disease in dogs. Mainly young dogs are affected. Initially this disease has been reported from the North America but, few cases have also been reported from India. There are reports on the isolation of *Blastomyces dermatitidis* from the lungs of an insectivorous bat (*Rhinopoma hardwickei*) trapped in Delhi (Khan et al., 1982) and on demonstration of characteristic single buckling cells of *Blastomyces* in lung sections of an aged male pariah dog stained with methanamine silver (Iyer, 1983). Tufted pockhard (*Aythya fuligata*), a migratory bird, has also been found to have *Blastomyces dermatitidis* (Rawal et al., 1988). Organisms resembling *Blastomyces* sp. were demonstrated in fetal contents of aborted fetus stomach (Sulochana et al., 1970). A yeast-like fungus showing single and multiple buddings in pulmonary tissue sections, morphologically resembling *Blastomyces* and forming caseo-calcified mycotic granuloma in a sheep and a goat, has also been observed (Sharma and Dwivedi, 1977).

Rhinosporidiosis

Rhinosporidiosis is caused by the fungus *Rhinosporidium seeberi.* Rhinosporidiosis is a chronic subcutaneous infection of the mucous membrane of the nasal cavity, characterized by polypoid growth. This disease has been reported in cattle, horses, mules, dogs, goats, geese and ducks (Jungerman and Schwartzman, 1990). The disease is endemic in certain parts of India like Tamil Nadu, Andhra Pradesh, Orissa and West Bengal. Both animal and human cases have been reported. The disease in man is more severe than in animals and disseminated lesions have been reported in man only and not in animals. The primary lesion of rhinosporidiosis is a polyp, which may be pedunculated or sessile, pink in color and usually not larger than 3 cm in diameter. The growth is often soft and friable, bleeds easily, and has a lobulated surface which creates a cauliflower-like appearance. Rhinosporidiosis was for the first time reported in cattle (Ayyar, 1929) followed by its occurrence in equines (Ayyar, 1932; Rao, 1938; Sahai, 1938; Rao et al., 1951) and goats (Singh, 1941). Incidence was found to be more in male bullocks because of their extensive use for ploughing and more so because of the puncturing of the nasal septum at the age of 3 or 4 years for the nose ring (Rao, 1938). Bilateral nasal infection by *Rhinosporidium seeberi* showing no gross lesions was reported on the basis of histological demonstration of sporangia of *R. seeberi* involving left nostril of a water buffalo. Attempts to isolate the fungus *in vivo* were unsuccessful (Rao, 1951).

Histoplasmosis

Histoplasmosis is a systemic fungal infection of animals and man caused by the dimorphic fungus *Histoplasma capsulatum*. It is primarily a disease of the reticulo-endothelial system which may progressively become a fatal disease. Symptoms of the disease vary greatly, but it affects primarily lungs and sometime other organs may also be affected. It is also known as "Darling's" disease. Dog is the most susceptible animal and the disseminated infection is rare in animals other than dog. Initially, *Histoplasma farciminosus* was isolated, identified and demonstrated in sections of intestinal ulcers and associated lymph nodes of a mule (Iyer, 1936). Ocular involvement of eye lids with local and palpebral lesions and button like ulcers have since been reported (Kapur, 1952; Singh, 1965; Singh, 1966). Disease simulating epizootic lymphangitis appeared in donkeys, mules and ponies involving mainly the lymphatics of the fore or hind limbs. Eye lesions were not observed.

Studies on wound smears and tissue sections of natural and experimental cases of epizootic lymphangitis revealed that the organism possessed a double membranous wall and not a capsule. It was found that *H. farciminosum*, after colonization usually in a traumatized area, spreads initially by way or lymphatics and later via the vascular system. Tissue reactions to *H. farciminosum* infection were principally of reticuloendothelial hyperplasia and fall into four distinct stages: (a) polymorphonuclear leucocytic infiltration followed by (b) lymphocytic infiltration, (c) macrophage formation and (d) giant cells formation. Proliferation of arterial tunica media was observed in early stages (Singh et al., 1965).

Flies (*Musca* and *Stomoxys*) have been reported to play a vital role in the transmission of infection (Singh, 1965). Incubation period for the disease varied from 4 to 72 days with an average of 34.89 days (Singh, 1965). Horse, mule, donkey, camel and pig act as hosts of the disease. Mice and rabbits have been infected experimentally (Singh, 1966; Singh and Varmani, 1966). Some of the serological tests were evaluated for the diagnosis of infection (Sikdar, 1966).

Mycotic Mastitis

Mastitis is a disease complex of different etiology and different degrees of intensity along with variations in duration and residual effects. The causative agents of mastitis may contaminate the milk from affected cows and many render it unsuitable for human consumption. The various agents that cause clinical and subclinical mastitis may be hazardous to human health. In India, Singh and Singh (1968), Monga and Kalra (1971), Sharma et al. (1977), Misra and Panda (1986), Mehrotra and Rawat (1989) and Singh et al. (1992) have reported various types of fungi as the etiological agents of mastitis in dairy cattle.

Varying incidence of mycotic mastitis has been reported due to yeast like fungi (Dhanda and Sethi, 1962). Clinically affected udders of cows, buffaloes and goats have shown mycotic mastitis (Monga et al., 1970; Monga and Kalra, 1971, 1972; Jand and Dhillon, 1975a).

Jand and Dhillon (1975b) induced experimental mastitis in healthy goats by infusing isolates of *Candida albicans*, *C. guilliermondii*, *C. parapsilosis*, *C. stellatoidea*, *C. tropicalis* and *Rhodotorula glutinis* into the udder. Recently, Pauchari et al. (2013) reported that the species of *Candida* and Aspergilli are the main fungi involved in bovine mastitis cases in Mathura region of Uttar Pradesh.

Fungi isolated associated with cases of mastitis in animals are listed below:

Cryptococcus sp.	(Dhanda and Sethi, 1962; Sharma, 1983; Pal, 1989a)
Cryptococcus neoformans	(Monga et al., 1970; Monga and Kalra, 1972; Pal, 1989a)
Candida sp.	(Verma, 1982; Shah et al., 1986; Simaria and Dholakia, 1986; Mahapatra et al., 1996)
Candida albicans	(Sharma et al., 1977; Verma, 1982; Sharma, 1983; Rahman and Baxi, 1983; Misra and Panda, 1986; Simaria and Dholakia, 1986; Singh et al., 1998)

Candida tropicalis	(Singh and Singh, 1968; Sharma, 1983; Shah et al., 1986; Simaria and Dholakia, 1986; Verma, 1988)
C. krusei	(Monga and Kalra, 1971; Verma, 1982; Sharma, 1983; Shah et al., 1986; Simaria and Dholakia, 1986)
C. guilliermondii	(Singh and Singh, 1968; Jand and Dhillon, 1975b; Sharma et al., 1980)
C. parapsilosis	(Prasad and Prasad, 1966; Monga et al., 1970; Jand and Dhillon, 1975b; Shah et al., 1986)
Rhodotorula rubra	(Simaria and Dholakia, 1986)
Rhodotorula glutinis	(Jand and Dhillon, 1975b)
Rhodotorula sp.	(Shah et al., 1986)
Geotrichum candidum	(Jand and Dhillon, 1975a; Misra and Panda, 1986)
Geotrichum sp.	(Verma, 1988)
Saccharomyces sp.	(Monga et al., 1970; Verma, 1988)
Aspergillus fumigatus	(Jand and Dhillon, 1975a; Shah et al., 1986; Verma, 1988)
A. flavus	(Sharma et al., 1980; Sharma, 1983; Misra and Panda, 1986; Shah et al., 1986)
A. terreus	(Singh and Singh, 1968; Sharma, 1983)
A. niger	(Misra and Panda, 1986)
A. sydowii	(Misra and Panda, 1986; Shah et al., 1986)
A. chevalieri	(Misra and Panda, 1986)
A. amestelodami	(Misra and Panda, 1986)
A. candidus	(Shah et al., 1986)
Absidia corymbifera	(Shah et al., 1986)
Rhizopus sp.	(Dhanda and Sethi, 1962; Sharma et al., 1980; Sharma, 1983)
Cladosporium olivaceum	(Sharma et al., 1980; Sharma, 1983)
Trichosporon cutaneum	(Sharma, 1983)
Nigrospora oryzae	(Shah et al., 1986)
Penicillium sp.	(Jand and Dhillon, 1975a; Sharma, 1983; Misra and Panda, 1986)

Alternaria sp.	(Jand and Dhillon, 1975a)
Sepedonium sp.	(Sharma et al., 1980)
Fusarium acuminatum	(Shah et al., 1986)
Chyrosporium sp.	(Sharma et al., 1980; Shah et al., 1986)
Curvularia verruculosa	(Shah et al., 1986)
Pithomyces sacchari	(Shah et al., 1986)
Syncephalastrum racemosum	(Shah et al., 1986)

Fungi in the Stomach and Reproductive Tract

A variety of fungal forms have been seen/isolated/cultured from the content of stomach of various animals (Ahmed et al., 1971; Sikdar et al., 1972; Murthi, 1982; Singh and Uppal, 1986a). The fungi were: *A. flavus, A. fumigatus, G. candidum, Mucor* sp., *Scopulariopsis* sp. and *Blastomyces* sp., *Candida pseudotropicalis, C. tropicalis, C. albicans,* and *C. laurentii.* Fungi such as species of *Aspergillus, Mucor, Rhizopus, Fusarium, Absidia, Curvularia, Penicillium, Candida, Geotrichum* and *Paecilomyces* were isolated from the fetal organs and placenta of aborted cows, uterine mucus and semen of bulls (Misra et al., 1984).

Bacterial and fungal seminal vesiculitis (27% incidence) is reported in Indian bulls (Saxena, 1972) of which 11% showed orchitis, with pathological changes in the semen, accesory glands and gonads. *A. terreus, Aspergillus* sp. and *Hormodendrum* sp. were isolated from the seminal vesicles of a slaughtered bull. *A. fumigatus* was isolated from uterine mucus of a Haryana cow having a history of metritis (Pathak and Mittal, 1966).

Fungi have been isolated from preputial washings (*Rhizopus arrhizus*, dematiaceous sp., *Aspergillus terreus*, Aspergillus sp., and a "contaminant fungus"), semen sample (*A. terreus*), seminal vesicles (*Aspergillus* sp., *Hormodendrum* [*Cladosporium*] sp.) and uterine washings (*Aspergillus niger, Aspergillus* sp., *Alternaria* sp., and some unidentified). Some details noted for a bull were the inflammation of seminal vesicle, prostate and testes, from which *A. terreus* was isolated in pure culture (Saxena and Pathak, 1972). Species of *Curvularia, Chyrosporium* and dematiaceous fungi have also been isolated from the uterine discharge of repeat breeder cows and buffaloes (Saxena, 1974).

Abnormal uteri whites/discharges, intra-uterine scrapings of cows and buffaloes yielded species of fungi like *Aspergillus, Trichothecium, Scopulariopsis, Geotrichum* and *Sporotrichum, Rhizopus, Penicillium, Candida,*

Absidia, Rhizopus, Fusarium, Cladosporium, Penicillium, Curvularia, Alternaria, and *Syncephalastrum* (Singh and Uppal, 1982; Patgiri and Uppal, 1983; Singh and Uppal, 1986a,b). A study conducted to isolate fungi from 100 genital and uterine discharge samples from repeat breeder cows and buffaloes and 13 healthy animals in Madhya Pradesh showed 87 samples positive for *A. fumigatus, A. niger, A. flavus, Penicillium, Alternaria* sp., *Candida albicans, C. tropicalis, C. pseudotropicalis* [*Atelosaccharomyces pseudotropicalis*], and *Cryptococcus* sp. whereas *Geotrichum* spp., *Candida albicans, A. fumigatus* and *Cryptococcus* sp. were isolated from healthy animals (Bahekar et al., 2010).

Aspergillus fumigatus, A. flavus, Alternaria sp., *Mucor* sp., *A. niger* and *Mortierella wolfii* were isolated from the reproductive tract of mares for the first time in 1983 (Singh and Uppal, 1986a). Cervical mucus of 22 repeat breeding cows yielded *Aspergillus fumigatus*, species of *Absidia*, and *Rhizopus*, while uterine mucus of 12 buffaloes with the history of repeat breeding yielded *Fusarium* sp., *A. niger, Aspergillus* sp., *Penicillium* sp. and *Rhizopus* sp. (Sulochana et al., 1970).

Candida tropicalis and *C. steillatoidea* were isolated from the semen of buffalo bull with infection in prepuceal sac (Kodagali, 1979). Species of *Aspergillus, Penicillium, Rhizopus* and *Alternaria* were isolated from buffalo bull semen (Singh and Uppal, 1982; Singh and Uppal, 1986a,b). *A. fumigatus, A. flavus* and unidentified yeast cells have also been isolated from the bovine semen samples. Further, mycological examination of 21 neat semen samples yielded species of *Penicillium, Candida albicans, Rhizopus* sp. and *Absidia* whereas from 18 deep frozen semen samples, *A. niger, A. fumigatus,* and species of *Mortierella* and *Absidia* were recovered (Sulochana et al., 1970).

According to an investigation done by John and his colleagues on the vaginal microflora in infertile dogs revealed that out of 17 vaginal discharge samples subjected to culturing, growth of *Candida* (7.7%) was seen in samples (John et al., 2011).

Miscellaneous Mycoses

Chronic granulomatous lesions of the skin of buffaloes in India caused by an unknown fungus were described by Iyer (Iyer, 1968).

Fungal species like *Penicillium vinaceum, Pseudourotium* sp., *Pseudoarachniotus* sp., *Allescheria* sp. and mycelia sterilia were isolated from skin lesions of camels (Sinha, 1987). *A. guadrilineatus* was isolated from mycotic dermatitis in sheep (Singh and Singh, 1968c). Even the skin of healthy calves when examined revealed the presence of *A. fumigatus, A. niger, A. nidulans, Paecilomyces* sp. and *Rhizopus* sp. (Singh and Singh, 1968c). A case of pulmonary infection in a local non-descript, 4 years ewe due to *Rhodotorula rubra* was reported on the basis of isolation and histopathological demonstration (Mongo and Garg, 1979). Mycotic pneumonia in sheep due to an unidentified fungus as seen in the affected lungs was reported (Rahman and Iyer, 1979). *Petriellidium boydii, A. fumigatus, Candida albicans, C. tropicalis, C. guilliermondii* and *Geotrichum candidum* were recovered from the intestinal contents of pigeons (Gugnani et al., 1976). Lymphadenitis in a sheep due to a phycomycetous fungus was reported (Christopher, 1977).

Keratinophilic fungi including *Chrysosporium tropicum, Rollandina hyalinospora,* and *Malbranchea* sp. were isolated from the feathers of 10 *Gallus domesticus* (Banani and Ghosh, 1980). *Phialophora repens* was isolated from the superficial lesions comprising of hyperkeratotic crusts on the lower lip of a poultry attendant (32 years old) (Verma et al., 1986). Disseminated cryptococcosis was diagnosed in a 26 month old WLH layer chicken during routine histopathological investigation based on the morphology of the encapsulated yeasts found in tissue sections of lungs, kidneys, liver and proventriculus (Singh and Dash, 2008).

Thermophilic fungi have also been recovered from many substrates including bird nest materials, vermicompost, cow dung and poultry litter. Among the thermophiles *Humicola lanuginosus* was present nearly in all substrates and *A. fumigatus* was found as a thermotolerant fungus in all the substrates (Rajavaram et al., 2010). Soils of 10 poultry farms from Namakkal and 12 feather-dumping sites from Chennai studied yielded keratinophilic fungi belonging to 34 species and 19 genera, and one non-sporulating fungus. Dermatophytes and closely related fungi were represented by six species belonging to five genera. Fungal species commonly found in the soil samples included *Chrysosporium keratinophilum* (73%), *T. mentagrophytes* (68.2%), *M. gypseum* (64%), *Myceliopthora vellerea* (32%), *Chrysosporium* state of *Arthroderma tuberculatum* (27.3%) and *Geomyces pannorum* (23%). Non-dermatophyte fungi were represented by 28 species belonging to 14 genera and one non-sporulating fungus (Periasamy et al., 2004). Out of 53 soil samples collected from various sites in the vicinity of Vedanthangal Water Bird Sanctuary, seven species of keratinophilic fungi related to five genera were isolated viz., *Auxarthron conjugatum* (1.89%), *Chrysosporium fluviale* (3.77%), *Chrysosporium indicum* (20.75%), *Chrysosporium tropicum* (7.55%), *Chrysosporium* state of *Ctenomyces serratus* (5.66%), *Gymnoascus petalosporus* (1.89%) and *M. gypseum* complex (11.32%). The study showed that migratory birds harbor a variety of keratinophiles and may be a potential source of transfer of these fungi from one location to another (Deshmukh and Verekar, 2011).

Priya and Mini (2008) isolated *Cladosporium* spp. from the pus sample from a 6-year-old crossbred cow.

Jain et al. (2011) reported *Fusarium* sp. as one of the emerging causes of opportunistic mycoses in humans and animals. Approximately, 15 species have been reported to cause human and animal diseases. Common species include: *F. solani* (commonest), *F. oxysporum*, *F. verticoides*, *F. proliferatum* and *F. anthophilum*. Clinical examination of a Labrador retriever bitch and laboratory investigations of 10% KOH digested deep skin scrapings, revealed concurrent *Sarcoptes scabiei* var. *canis* and *M. gypseum* infestations in the bitch suffering from generalized dermatitis under discriminate therapy inclusive of immune-suppressive drugs (Godara and Sharma, 2010).

A total of 134 clinical samples (hair plucks and skin scrapings) collected from dogs with cutaneous lesions of alopecia, hyper-pigmentation, and scales directly examined under a microscope revealed the involvement of various fungi including dermatophytes such as *Trichophytan* spp. that were frequently isolated along with *Microsporum* sp. and *Epidermophytan* sp. Other fungi were *A. niger*, frequently isolated, followed by *A. fumigatus*, phycomycetous species, *Penicillium* sp., *Curvularia* sp. and *Alternaria* sp. and yeast such as *Malassezia pachydermatis* and *C. albicans* (Kumar et al., 2011). Srikala et al. (2010) reported mixed infestation of malasseziosis and demodicosis in dogs. The isolation of *Chryosporium tropicum* from comb lesions in two different breeds of chickens were shown to be pathogenic when inoculated onto prepared skin of guinea pigs. It was concluded that *C. tropicum* can be considered as a pathogenic fungus and a probable cause of dermatomycosis in chickens (Saidi et al., 1994).

From the rumen of Nilgai and Barbari goats, *Piromyces* sp. has been isolated (Chaudhary et al., 2008). Pawaiya et al. (2011) implicated *Candida albicans* to be a causative agent of the invasive mycotic gastritis in a wild Indian crocodile (*Crocodylus palustris*).

References

Ahmed, Z., Pandurang, G. and Rao, V. 1971. Abortion associated with mycotic infections in a cow in Hyderabad. Indian Vet. J., 48: 446–449.

Arya, S.C., Shah, R.L., Panisup, A.S. and Verma, R. Mycotic pneumonitis in quails (unpublished data).

Asthana, R.P. 1944. Aspergillosis in fowls. Proc. Indian Acad. Sci. (Sec. B), 20: 43–47.

Ayyar, V.K. 1929. Rhinosporidiosis in cattle. Trans. Far- Eastern Trop. Med., 7th Congress held in India (cited by Ayyar. 1932).

Ayyar, V.K. 1932. Rhinosporidiosis in equines. Indian Vet. J. Vet. Sci. & AH., 2: 49–52.

Babbras, M.A. and Radhakrishnan. 1967. Aspergillosis in chicks and trial of Hamycin in an outbreak. Hindustan Antibio. Bull., 9: 244.

Bahekar, V.S., Joseph, E., Agrawal, R.G. and Hirpurkar, S.D. 2010. Fungal isolates from repeat breeding cows and buffaloes. Indian Vet. J., 87: 1087–1088.

Balasubramaniam, A. and Sukumar, S. 2007. An overview on outbreaks of candidiasis in poultry. Tamilnadu J. Vet. Anim. Sci., 3: 121–123.

Banani, Sur and Ghosh, Gouri Rani. 1980. Keratinophilic fungi from Orissa, India II: Isolation from feathers of wild birds and domestic fowl. Sabouradia, 18: 275–280.

Banerjee, T.P. and Bhattacharya, M.M. 1965. Aspergillosis in a poultry and in a duck in West Bengal. Indian J. Anim. Hlth., 4: 23.

Baruah, T.P., Datta, B.M. and Rahman, T. 1984. Studies on pathology of porcine pneumonia I Mycotic pneumonia. Indian Vet. J., 61: 9–12.

Chakrabarty, A.N., Ghosh, S. and Blank, F. 1954. Isolation of *Trichophyton rubrum* (Castellani) Sab. 1911, from animals. Conod. J. Comp. Med., 18: 436–438.

Charan, K., Gill, B., Parihar, N.S., Uninithan, R.R. and Mohamed, Shaffi. 1976. Pneumomycosis in a cow having cholangio cellular carcinoma. Indian Vet. J., 53: 252–254.

Chatterjee, A. and Sengupta, D.N. 1979. Ringworm in domestic animals. Indian J. Anim. Hlth., 18: 34–46.

Chatterjee, A., Banerjee, P.G. and Sengupta, D.N. 1980. Isolation of *M. distortum* from ringworm lesions in a dog. Indian J. Anim. Hlth., 19: 59–61.

Chatterjee, A., Chakraborty, P., Chattopadhyay, D. and Sengupta, D.N. 1978. Isolation of *T. schocnleinii* from a camel. Indian J. Anim. Hlth., 17: 167–168.

Chaudhary, U.B., Singh, Monika and Gupta, Arti. 2008. Fibrolytic enzyme activities of anaerobic fungus isolated from Nilgai and goats. Indian Vet. J., 85: 1179–1181.

Chaudhry, P.G., Kulkarni, V.B., Deshpande, C.K. and Bhagat, S.V. 1968. An usual outbreak of rubromycosis in pigs in Bombay. Indian J. Microbiol., 8: 37–40.

Chauhan, H.V.S. and Dwivedi, O.P. 1974. Pneumomycosis in sheep & goats. Vet. Rec., 95: 58–59.

Chauhan, H.V.S. and Sadana, J.R. 1973. Gastric phycomycosis in a piglet. Indian Vet. J., 50: 1155–1156.

Chauhan, H.V.S., Sharma, G.L., Kalra, D.S., Malhotra, F.C. and Kapur, M.P. 1973. A fatal cutaneous granuloma due to *Entomophthora coronata* in a mare. Vet. Rec., 92: 425–427.

Christopher, J. 1977. Hepantic asperigition in a hen—a case report. Indian J. Anim. Res., 11: 57–58.

Damodaran, S. 1971. Mycotic pneumonitis in a bull. Indian Vet. J., 48: 464–465.

Damodaran, S. and Sundararaj, A. 1976. Mycotic gastritis in neonatal pig. Indian Vet. J., 53: 45–46.

Damodaran, S., Ramachandran, P.V., Thainkachalam, M. and Mahalingam, P. 1976. Mycotic gastritis in cattle. Indian Vet. J., 43: 848–851.

Das, J.H. and Sambamurli, B. 1973. Zoophilic, geophylic and authropophilic dermatophytes in Tripati. Indian Vet. J., 50: 1171–1173.

Das, T.K., Zachariah, K. and Mattada , R.R. 1980. Isolation of *Microsporum nanum* from north eastern region of India. Indian J. Microbiol., 20: 326.

Das, T.K., Zachariah, K. and Mattada, R.R. 1984. A note on the dermatomycosis in mules and their possible mode of transmission. J. Remount Vet. Corps., 23: 24–27.

Datta, S.C.A. 1984. Chronic bovine heamaturia. Rep. 1st Imp. Vet. Conf. Lond. Webridge: Imperial Bureau of Animal Health, pp. 78–79.

De, S., Sanyal, P.K., Pan, D., Bera, A.K., Bandyopadhyay, S., Pal, S., Mandal, S.C., Sarkar, A.K., Patel, N.K., Bhattacharya, D. and Das, S.K. 2010. Assessment of genetic relation of Indian isolates of four predatory fungi through RAPD markers. Indian J. Anim. Sci., 80: 711–714.

Deka, P.N. and Rao, A.T. 1988. Aspergillosis caused by *A. nidulans* in ducklings. Cherion., 17: 232–234.

Deshmukh, S.K. and Verekar, S.A. 2011. Incidence of keratinophilic fungi from the soils of Vedanthangal Water Bird Sanctuary (India). Mycoses, 54: 487–490.

Dhanda, M.R. and Sethi, M.S. 1962. Investigation of mastitis in India Res. Ser. 35. Indian Council of Agricultural Research, New Delhi.

Dhawedkar, R.G. and Dhansear, W. 1971. Microflora of dead in shell eggs. Indian Vet., 48: 233–234.

Dhopapkar, B.S. and Awadhiya, R.P. 1969. Studies on the pathology of the lesions of avian pulmonary aspergillosis. JNKVV Res. J., 3: 89–92.

Garg, D.N. and Sethi, M.S. 1969. Fungal flora of avian respiratory tract. Poul. Sci., 48: 339–341.

Gangil, R., Dutta, P., Tripathi, R., Singathia, R. and Lakhotia, R.L. 2012. Incidence of dermatophytosis in canine cases presented at Apollo Veterinary College, Rajashtan, India. Vet. World, 5: 682–684.

Gill, B.S., Singh, B. and Gupta, P.P. 1977. Pulmonary aspergillosis in Indian buffaloes. Mykosen, 20: 65–70.

Gill, B.S. and Singh, B. 1976. Phycomycotic pneumonia in buffaloes: a note. Zentrabl. Veterinarmed., 238: 782–784.

Godara, R. and Sharma, R.L. 2010. Concurrent acarine and dermatophyte infestations in a bitch. J. Vet. Parasitol., 24: 181–183.

Gugnani, H.C., Gupta, N.P. and Shrivastava, J.B. 1972. Prevalence of *Cryptococcus neoformans* in Delhi Zoological Park and its recovery from the sputum of an employee. Indian J. Med. Res., 60: 182–185.

Gugnani, H.C., Randhawa, H.S. and Shrivastava, J.B. 1977. Isolation of dermatophytes and other keratinophilic fungi from apparently healthy skin of domestic animals. Indian J. Med. Res., 59: 1699–1702.

Gugnani, H.C. 1972. *Trichophyton verrucosum* infection in cattle in India. Mykosen, 15: 285–288.

Gugnani, H.C. and Randhawa, H.S. 1973. An epizootic of dermatophytosis caused by *Trichophyton simii* in poultry. Sabourodia, 11: 1–3.

Gugnani, H.C., Sandhu, R.S. and Shome, S.K. 1976. Prevalence of *Cryptococcus neoformans* in avian excreta in India. Mykosen, 19: 183–187.

Gupta, P.K. and Singh, R.P. 1969. Effect of some therapeutics on dermatophytosis (ringworm) in animals. Indian Vet. J., 46: 1001–1007.

Gupta, P.K., Singh, R.P. and Singh, I.P. 1968. Dermatophytes from man, dog and pigs with special reference to *Trichophyton simii* and *Microsporum nanum*. Indian J. Anim. Hlth., 11: 1–3.

Gupta, P.K., Singh, R.P. and Singh, I.P. 1970. A study of dermatomycoses (ringworm) in domestic animals and fowls. Indian J. Anim. Hlth., 9: 85–89.

Iyer, P.K.R. 1936. An unusual case of epizootic lymphangitis in a mule. Indian J. Vet. Sci., 6: 251.

Iyer, P.K.R. 1968. Mycetoma caused by an unknown fungus in Indian buffaloes (*Bos bubalis*). Ceylon Vet. J., 16: 77.

Iyer, P.K.R. 1983. Pulmonary blastomycosis in a dog in India. Indian J. Vet. Path., 7: 60–62.

Jadhav, V.J. and Mahendrapal. 2006. Canine mycotic dermatitis due to *Candida albicans*. Rev. Iberoam Micol., 23: 233–34.

Jain, P.K., Gupta, V.K., Misra, A.K., Gaur, R., Bajpai, V. and Issar, S. 2011. Current status of *Fusarium* infection in human and animal. Asian J. Anim. Vet. Adv., 6: 201–227.

Jairam, R. and Das, A.M. 2005. Cutaneous trichosporonosis (white piedra) in a dog. J. Bombay Vet. College, 13: 114–115.

Jand, S.K. and Dhillon, S.S. 1975. Mastitis caused by fungi. Indian Vet. J., 52: 125–128.

Jand, S.K. and Dhillon, S.S. 1973. Fungi associated with respiratory diseases of poultry. Indian Vet. J., 50: 211.

Jand, S.K. and Dhillon, S.S. 1975. Mycotic mastitis produced experimentally in goats. Mykosen, 18: 363–366.

John, A., Nair, D.R., Sreejith, J.R., Deepthi, L., Praseeda, R. and Ajitkumar, G. 2011. Vaginal microflora and its antibiogram in infertile bitches. Vet-Scan., 6: Article 79.

Kalapesi, R.M. and Purohit, B.L. 1956. Unusual lesion in the subcutis of a fowl due to the fungus *Aspergillus* sp. Curr. Sci., 25: 400.

Kapur, J.R. 1952. Epizootic lymphangitis. Indian Vet. J., 28: 341–343.

Khan, Z.U., Randhawa, H.S. and Lulla, M. 1982. Isolation of *Blastomyces dermitidis* from the lungs of a bat *Rhinopoma hardwickei* gray in Delhi. Sabouradia, 20: 137–144.

Khandhari, K.C. and Sethi, K.K. 1964. Dermatophytosis in Delhi area. Indian J. Med. Res., 42: 324–326.

Khanna, B.M., Datta, S.C., Monga, D.P. and Ger, K.I. 1974. The therapeutic effect of Thiabendazole on ringworm in calves. Indian Vet. J., 51: 562–565.

Kharole, M.U., Gupta, P.P., Singh, B., Mandal, P.C. and Hothi, D.S. 1976. Phycomycotic gastritis in buffalo calves (*Bubalis bubalis*). Vet. Path., 13: 409–413.

Klokke, A.H. 1964. Fungus diseases of skin in North India. Derm. Trap., 3: 108–110.

Klokke, A.H. and Durairaj, P. 1967. The casual agents of superficial mycoses isolated in rural areas of South India. Sabouradia, 5: 153–158.

Kodagali, S.B. 1979. Prevalence of fungi in buffalo semen. Indian Vet. J., 56: 807–809.

Krishna, I., Kulshreshtha, S.B. and Paliwal, O.P. 1981. Systematic aspergillosis in lambs. Indian Vet. Med. J., 5: 70–73.

Kulkarni, V.B., Chaudhri, P.G., Kulkarni, M.P. and Sasane, H.D. 1969. Equine ringworm caused by *Trichophyton tonsurans* var. *sulfureum*. Indian Vet. J., 46: 215–218.

Kulshreshtha, R.C., Verma, Rishendra, Chandiramani, N.K. and Singh, J. 1981. A note on abortion cases in livestock-a ten year record. Dairy Guide, 4: 9–10.

Kumar, K.S., Selvaraj, P., Vairamuthu, S., Nagarajan, B., Nambi, A.P. and Prathaban, S. 2011. Survey of fungal isolates from canine mycotic dermatitis in Chennai. Tamilnadu J. Vet. Anim. Sci., 7: 48–52.

Lakshman, M., Nagamalleswari, Y. and Balachandran, C. 2009. Cutaneous mycotic granuloma (dermatomycosis) in fowl: a pathomorphological observations. Indian J. Vet. Path., 233: 204–206.

Larone, D.H. 1995. Medically Important Fungi. A Guide to Identification. 3rd ed., ASM Press, USA.

Mahapatra, S., Kar, B.C. and Misra, P.R. 1996. Occurrence of mycotic mastitis in buffaloes of Orissa. Indian Vet. J., 73: 1021–1023.

Mandal, P.C. and Gupta, P.P. 1993. Experimental aspergillosis in goats: clinical, haematological and mycological studies. J. Vet. Med.-Series-B, 40: 283–286.

Mandal, P.C., Gill, B.S. and Grewal, G.S. 1981. Spontaneous pneumonia in a she-buffalo caused by *A. corymbifera*. J. Res. PAU., 18: 452–453.

Mataney, C.F., Gupta, S.C. and Iyer, P.K.R. 1975. A note on mycormycosis in a pig. Indian J. Anim. Sci., 45: 96–98.

Misra, P.R. and Panda, S.K. 1986. Some observations on the occurrence of mycotic mastitis in cows in Orissa. Indian Vet. J., 63: 886–888.

Misra, P.K., Mishra, A. and Panda, S. 1984. Fungal agents associated with abortion and repeat breeding in crossbred cattle of Orissa. Indian J. Anim. Hlth., 23: 115–119.

Misra, S.K. 1971. Studies on the incidence of dermatophytosis in man and animals, zoonosis. Experimental infection and chemicals for their antifungal activity. Ph.D. thesis submitted to the Orissa Univ. of Agri. & Tech., Bhubaneshwar.

Mohan, R.N., Sharma, K.N. and Agrawal, V.C. 1986. A note on outbreak of epizootic lymphangitis in equines. Indian Vet. J., 43: 338–339.

Mohapatra, L.N. and Mahajan, V.M. 1970. *Trichophyton simii* infection in man and animals. Mycopath. mycol. Appl., 41: 357.

Mohiuddin, S.M. 1972. Fungal hepatitis in poultry—A note. Indian J. Anim. Sci., 42: 602–624.

Monga, D.P. and Kalra, D.S. 1971. Prevalence of mastitis among animals in Haryana. Indian J. Anim. Sci., 41: 813.

Monga, D.P. and Garg, D.N. 1979. Ovine Pulmonary Infection caused by *Rhodotorula rubra*. Mykosen, 23: 208–211.

Monga, D.P. and Kalra, D.S. 1972. Observations on experimental mycotic mastitis in goats. Mykosen, 15: 199–205.

Monga, D.P., Mohapatra, L.N. and Kalra, D.S. 1970. Bovine mastitis caused by *Cryptococcous neoformans*. Indian J. Med. Res., 58: 1203–1205.

Monga, D.P., Singh, S.U., Satija, K.C. and Khanna, R.S. 1974. Dermatophytosis due to *Trichophyton mentagrophytes*. Haryana Vet., 13: 111–114.

Murray, P.A., Rosebthal, K.S. and Pfaller, M.A. 1999. Manual of Clinical Microbiology. 7th ed., ASM Press, USA.

Murthi, K.K. 1982. Abortions in buffaloes due to *Aspergillus* species. J. Remount Vet. Corps., 21: 75–77.

Nag, P.C. and Mallick, B.S. 1961. Aspergillosis in a calf. Cand. Vet., 2: 30–32.

Naryana, J.V. 1961. A note on a case of Aspergillosis in a domestic fowl. Indian Vet. J., 38: 599–600.

Naryana, J.V.P., Ramarao, J.C. and Sastry, G.A. 1964. Pulmonary Aspergillosis in a Murrah buffalo. Indian Vet. J., 41: 523–525.

Navjeevan, Baksi, S., Lodh, C. and Chakrabarti, A. 2005. Studies on infectious alopecia in dogs. Indian Vet. J., 82: 719–720.

Nayak, B.C., Rao, A.G., Ray, K. and Chandra, S.K. 1975. Mycotic ruminitis in a calf. Indian Vet. J., 53: 56–57.

Nayar, P.R., Sethna, S.B. and Thirumalachar, M.J. 1974. *Trichophytosis* in sheep. Hindustan Antibio. Bull., 16: 190–192.

Pachauri, S., Varshney, P., Dash, S.K. and Gupta, M.K. 2013. Involvement of fungal species in bovine mastitis in and around Mathura, India. doi:10.5455/vetworld.393-395.

Padhye, A.A., Radhakrishnan, C.V. and Thirumalachar, M.J. 1966. Studies on dermatomycosis in Poona. Hindustan Antibio. Bull., 9: 23–26.

Pal, M. 1983. Keratomycosis in a buffalo calf (*Bublus bubalis*) caused by *Aspergillus fumigatus*. Vet. Rec., 113: 67.

Pal, M. 1989a. *Cryptococcus neoformans* var. *neoformans* and munia birds. Mycoses, 32: 250–252.

Pal, M. 1989b. Colonization of a squamous cell carcinoma in the bovine horn core by *Aspergillus terreus*. Mycoses, 32: 197–199.

Pal, M. 1992. Disseminated *Aspergillus terreus* infection in a caged pigeon. Mycopathologia, 119: 137–139.

Pal, M. 2003. Mycotic pneumonia in a captive pigeon due to *Aspergillus fumigatus*. Zoos'—Print-Journal, 18: 1185–1186.

Pal, M. and Dabiya, S.M. 1984. Studies on avian dermatitis caused by *Aspergillus fumigatus*. Indian Vet., 61: 188.

Pal, M., Gangopadhyay, R.M. and Prajapati, K.S. 1989. Systemic aspergillosis in chicks due to *Aspergillus flavus*. Indian J. Anim. Sci., 59: 1074–1075.

Pal, M. and Mehrotra, B.S. 1984. Association of *Aspergillus fumigatus* with rhinitis. Vet. Rec., 115: 167.

Pal, M. and Mehrotra, B.S. 1986. Studies on the association of *Aspergillus fumigatus* with ocular infections in animals. Vet. Rec., 118: 42–44.

Pal, M. and Singh, D.K. 1983. Studies on dermatomycoses in dairy animals. Mykosen, 26: 317–318.

Pal, M. and Verma, J.D. 2003. Canine dermatitis caused by *Geotrichum candidum*. Livestock-Intal., 7: 21–22.

Pal, M., Patil, D.B., Parikh, P.V. and Jhala, S.K. 2011. Canine Otitis due to *Aspergillus niger*—Diagnosis and Chemotherapy. Indian Pet Journal Online Issue: June Volume-12.

Pal, M., Matsusaka, N. and Lee, C.W. 1994. Clinical and mycological observations on equine ringworm due to *Microsporum gypseum*. Korean J. Vet. Clin. Med., 11: 5–8.

Pal, M. and Ragi, A.S. 1989. Fungi isolated from lymphnodes of buffaloes. Mycoses, 32: 578–580.

Pal, M., Salkin, I.F. and Dixon, D.M. 1991. Colonization of a bovine burn wound by *Trichosporon beigelii*. Mycoses, 34(11-12): 513–514.

Parihar, N.S. and Singh, C.M. 1971. Studies on the incidence of avian respiratory mycoses. Indian J. Anim. Sci., 41: 712–720.

Patgiri, G.U. and Uppal, P.K. 1983. Mycoflora of bovine female genital tract affected with various reproductive disorders. Indian J. Comp. Microbiol. Immunol. & Inf. Dis., 4: 19–22.
Pathak, R.C. and Mittal, K.R. 1966. Isolation of *Aspergillus fumigatus* from the cervical mucus of a cow having history of metritis. Curr. Sci., 35: 312.
Pathak, R.C. and Singh, C.M. 1962. Occurrence of *Candida* sp. in crops of fowls. Indian J. Microbiol., 2: 89–90.
Pawaiya, R.V.S., Sharma, A.K., Swarup, D. and Somvanshi, R. 2011. Pathology of mycotic gastritis in a wild Indian freshwater/marsh crocodile (Mugger; *Crocodylus palustris*): a case report. Veterinarni-Medicina, 56: 135–139.
Periasamy-Anbu, A. Hilda and Gopinath, S.C.B. 2004. Keratinophilic fungi of poultry farm and feather dumping soil in Tamil Nadu, India. Mycopathologia, 158: 303–309.
Ponnuswamy, K.K., Arthanarieswaran, M., Selvaraj, P. and Subramanian, M. 2010. Bronchoscopic diagnosis of pulmonary aspergillosis in a buffalo. Indian Vet. J., 87: 383–384.
Prasad, G., Kumar, A. and Sharma, V.D. 1987. Treatment of equine dermatophytosis by garlic. Indian Vet. Med. J., 11: 108–110.
Prasad, H. and Srivastav, C.P. 1964. Aspergillosis in poultry in Bihar. Indian Vet. J., 41: 18.
Prasad, L.B.M. and Prasad, S. 1966. Bovine mastitis caused by a yeast in India. Vet. Rec., 79: 809–810.
Priya, P.M. and Mini, M. 2008. A note on the isolation of *Cladosporium carrionii*. Tamilnadu J. Vet. Anim. Sci., 4: 162.
Rahman, H. and Baxi, K.K. 1983. Prevalence of *Candida albicans* in bovine mastitis. Indian J. Comp. Microbiol. Immunol. & Inf. Dis., 4: 49–50.
Rahman, T. and Iyer, P.K.R. 1979. Studies on pathology of ovine pneumonias. Indian Vet. J., 56: 455–461.
Rajan, A. and Sivadas, C.G. 1972. A study of the distribution of fungal spores in brooder houses. Kerala J. Vet. Sci., 3: 186–187.
Rajavaram, R.K., Sreelatha-Bathini, Sivadevuni-Girisham and Reddy, S.M. 2010. Incidence of thermophilic fungi from different substrates in Andhra Pradesh (India). Intl. J. Pharma Bio-Sci., 1: BS33.
Rao, A.N., Paliwal, O.P. and Ramakrishna, A.C. 1992. A note on gastric phycomycosis in pig. Indian Vet. Med. J., 16: 218–219.
Rao, A.T. and Acharjyo, L.N. 1988. Aspergillosis in some captive birds at Nandankanan Zoo. Indian J. Poult. Sci., 23: 254–257.
Rao, K.S. and Sambamurty, B. 1972. *Candida albicans* gastric ulcers of country swine. Indian Vet. J., 9: 457.
Rao, M.A.N. 1938. Rhinosporidiosis in bovines in Madras Presidency with a discussion of the probable modes of infection. Indian J. Vet. Sci. & AH., 8: 187–188.
Rao, M.R. and Choudary, C. 1980. Aspergillosis in ducks. Poult. Advi., 13: 59–60.
Rao, M.R., Choudhry, C. and Khan, D.I. 1982. An outbreak of acute aspergillosis in chicks. Indian Vet. J., 59: 341–342.
Rao, P.V.R., Jain, S.N. and Rao, T.V.H. 1951. Animal rhinosporidiosis in India with case reports. Ann. Soc. Belge. Med. Trop., 55: 119–124.
Rao, S.R. 1951. Rhinosporidiosis in equines in Bombay State. Bombay Vet. College Magazine, 68–72.
Rawal, I.J., RaoBhan, L.N. and Dutta, P.C. 1988. *B. dermitidis* in tufted pockhard *Aytha fuligata*. Indian Vet. J., 65: 965–996.
Sadana, J.R. and Kalra, D.S. 1973. A note on pulmonary phycomycosis in a pig. Indian J. Anim. Sci., 43: 961–964.
Sadana, J.R. and Kalra, D.S. 1974. A note on aspergillosis in a pig. Indian J. Anim. Sci., 44: 218–219.
Sahai, L. 1938. Rhinosporidiosis in equines. Indian Vet. Sci., 8: 221–223.

Saidi, S.A., Bhatt, S., Richard, J.L., Sikdar, A. and Ghosh, G.R. 1994. *Chrysosporium tropicum* as a probable cause of mycosis of poultry in India. Mycopathologia, 125: 143–147.
Sarkar, S. and Roy, S. 1988. Studies on pathogenesis and histopathology of experimental dermatophytosis in cattle, buffalo and goat. Indian J. Anim. Hlth., 27: 127–131.
Sarkar, S., Sinha, P.P. and Thakur, D.K. 1985. Epidemiology of dermatophytes in domestic animals and its impact on human health. Indian Vet., 62: 1017–1022.
Saxena, S.C. 1972. Mycotic infections of genitalia of domestic animals. Paper presented in the Summer Institute, Deptt. of Bacteriology, Vet. College, Mathura.
Saxena, S.C. and Pathak, R.C. 1972. Isolation of fungi from bovine genitalia. Agra Univ. J. Res. Sci., 21: 21–25.
Saxena, S.P. 1972. Incidence of dermatomycosis in dogs and cats. JNKVV Res. J., 6: 106–108.
Saxena, S.P. and Mehra, K.N. 1973. A note on the occurrence of dermatophyte infection in domestic animals. Indian J. Anim. Sci., 43: 347–350.
Shah, N.M., Dholakia, P.M. and Simaria, B.M. 1986. *In vitro* drug sensitivity trials against yeasts and moulds isolated from bovine udder. Indian Vet. J., 63: 58.
Shah, R.L. and Mohanty, G.C. 1974. A note on mycotic meningio-encephalitis in ducklings caused by *Aspergillus* sp. Indian J. Anim. Sci., 44: 211–213.
Shah, R.L., Mall, M.P. and Mohanty, G.C. 1982. Cutaneous candidiasis in Japanese quail (*Coturnix cotunrnix japonica*). Mycopathologia, 80: 33–37.
Sharma, D.K., Joshi, G., Singathia, R. and Lakhotia, L. 2010. Fungal infections in cattle Gaushala at Jaipur. Haryana Vet., 49: 62–63.
Sharma, D.K., Joshi, Gurudutt, Singathia, R. and Lakhotia, R.L. 2009. Zooanthroponosis of *Microsporum gypseum* infection. Haryana Vet., 48: 108–109.
Sharma, D.N. and Dwivedi, J.N. 1977. Pulmonary mycosis of sheep and goats in India. J. Anim. Sci., 47: 808–813.
Sharma, G., Boro, B.R. and Goswami, J.N. 1980. A note on bovine mastitis due to *C. albicans*. Indian J. Anim. Hlth., 79: 77–78.
Sharma, S.D. 1983. Studies on bovine mastitis with special reference to mycotic infections of udder. Vet. Res. J., 6: 105–106.
Sharma, S.D., Rai, P. and Saxena, S.C. 1977. Solvay of mycotic infections of udder in clinical and subclinical cases of mastitis in cows and buffaloes. Indian Vet. J., 54: 284.
Sharma, V.D. and Sethi Negi, S.K. 1971. Fungal flora of respiratory tract of fowls. Poul. Sci., 50: 1041–1044.
Sharma, V.D., Saxena, K.D.P. and Singh, S. 1979. Prevalence of dermatophytes in man and animals in the vicinity of Pant Nagar. Indian J. Anim. Res., 13: 61–62.
Sharma,V.D., Sethi, M.S. and Joshi, H.C. 1979. Acute Aspergillosis in chicks. Indian Vet. J., 56: 151–152.
Sikdar, A. 1966. Observation on the utility of serological test in the study of epizootic lymphangitis. M.VSc. Thesis, IVRI, Kumaon (UP).
Sikdar, A., Singh, G., Banerjee, M.C. and Sharma, R.N. 1972. Isolation of *Candida pseudotropicalis* from case of abortion among mares: A note. Indian J. Anim. Sci., 42: 737–738.
Sikdar, A. and Uppal, P.K. 1985. Isolation of *Candida guillermodni* from suspected bovine lympangitis cases. Curr. Sci., 54: 151.
Simaria, B.M. and Dholakia, P.M. 1986. Incidence and diagnosis of mycotic mastitis in cattle. Indian J. Vet. Sci., 56: 995–1000.
Singh, B., Chawla, R.C. and Sanota, P.S. 1976. Phycomycotic pneumonia in a pig. Indian Vet. J., 53: 818.
Singh, Balbir. 1941. Rhinosporidiosis, administration report of the disease investigation officer, Central province and Berar, 12: 41–42.
Singh, G.M., Kapur, M.P. and Monga, D.P. 1971. Aspergillosis in chicks embryos. Indian J. Anim. Hlth., 10: 239–240.
Singh, J. and Malhotra, F.C. 1970. Experimental studies on Aspergillosis in chicks. Haryana Vet., 17: 23–28.

Singh, K.P. and Uppal, P.K. 1982. Mycological examination of the uterus and semen of buffaloes. Indian Vet. J., 59: 837–841.

Singh, K.P. and Uppal, P.K. 1986a. Isolation of *Alternaria* sp. from genital tract of buffalo. J. Rec. Adv. Appl. Sci., 1: 89–90.

Singh, K.P. and Uppal, P.K. 1986b. Studies on mycological examination of buffalo semen. J. Rec. Adv. Appl. Sci., 1: 86–88.

Singh, M.P. and Singh, C.M. 1968a. Fungi isolated from clinical cases of bovine mastitis in India. Indian J. Anim. Hlth., 7: 259–263.

Singh, M.P. and Singh, C.M. 1968b. Mycotic dermatitis in Camels. Paper presented in National Seminar in Zoonosis in India. National Institute of Communicable Disease, Delhi.

Singh, M.P. and Singh, C.M. 1968c. Fungi associated with superficial mycosis of cattle and sheep in India. Paper presented in National Seminar in Zoonosis in India. National Institute of Communicable Disease, Delhi.

Singh, M.P. and Singh, C.M. 1970a. Fungi associated with ringworm in dogs. Indian J. Anim. Hlth., 9: 109–110.

Singh, M.P. and Singh, C.M. 1970b. Fungi associated with superficial mycoses of cattle and sheep in India. Indian J. Anim. Hlth., 9: 75–77.

Singh, M.P. and Singh, C.M. 1970c. Aspergilli associated with respiratory tract of poultry, buffaloes, sheep and goats. Indian Vet., 47: 624–627.

Singh, M.P. and Singh, C.M. 1972a. *Candida parakrusei* isolation from goats. Indian J. Anim. Hlth., 11: 111–112.

Singh, M.P. and Singh, C.M. 1972b. *Trichophyton simii* infection in poultry. Vet. Rec., 90: 218.

Singh, N.B., Singh, C.D.N. and Sinha, B.K. 1977. Some observations on the pathology of bovine pulmonary mycosis. Haryana Vet., 16: 78–80.

Singh, N., Yadav, J.S., Singh, A.P. and Sharma, S.N. 1997. Clinico-epidemiological studies on bovine dermatophytosis in and around Bikaner. Indian J. Anim. Sci., 67: 845–848.

Singh, Pritipal, Sood, N., Gupta, P.P., Jand, S.K. and Banga, H.S. 1998. Experimental candidal mastitis in goats: Clinical, haematological, biochemical and sequential pathological studies. Mycopathologia, 140: 89–97.

Singh, R.P. 1981. Equine ringworm—a case study. J. Remount Vet. Corps., 20: 91–93.

Singh, S. 1966. Equine Cryptococosis (epizootic lymphangitis). Indian Vet. J., 32: 260–270.

Singh, S.D. and Dash, B.B. 2008. Sponataneous lesions of cryptococcosis in White Leghorn chicken. Indian J. Vet. Path., 32: 68–69.

Singh, S., Borah, M.K., Sharma, K.K., Joshi, G.D. and Gogoi, R. 2009. Aspergillosis in turkey poults. Indian J. Vet. Path., 33: 220–221.

Singh, T. 1965. Studies on epizootic lymphangitis, modes of infections and transmission of equine histoplasmosis (EL). Indian J. Vet. Sci., 35: 102–110.

Singh, T. 1966. Studies on epizootic lymphangitis. Study of clinical cases and experimental transmission. Indian J. Vet. Sci., 36: 45–49.

Singh, T. and Varmani, B.M.L. 1966. Studies on epizootic lymphangitis. A note on pathogenicity of *H. farciminosum* (Rivolta) for laboratory animals. Indian J. Vet. Sci., 36: 164–167.

Singh, T. and Varmani, B.M.L. 1967. Some observations on experimental infection with *H. farciminosum* (Rivolta) and the morphology of the organism. Indian J. Vet. Sci., 37: 45–57.

Singh, T., Varmani, B.M.L. and Bhalla, N.P. 1965. Studies on epizootic lymphangitis II. Pathogenesis and histopathology of equine histoplasmosis. Indian J. Vet. Sci., 35: 111–120.

Sinha, B.K. 1987. Studies on the occurrence of fungi on apparently healthy skin of calves. Indian J. Comp. Microbiol. Immunol. Inf. Dis., 7: 35–37.

Sivaseelan, S., Srinivasan, P., Balachandran, P., Balasubramaniam, G.A. and Arulmozhi, A. 2009. Pathology of respiratory Aspergillosis in desi chicken. Indian Vet. Path., 33: 218–219.

Srikala, D., Kumar, K.S., Kumar, V.V.V.A., Nagaraj, P., Ayodhya, S. and Rao, D.S.T. 2010. Management of mixed infestation of malasseziosis and demodicosis in dogs. Intas-Polivet., 11: 74–76.

Sulochana, S., Abdulla, P.K. and Punnose, K.T. 1970. Association of mycotic agents with bovine abortions in Kerala. Kerala J. Vet. Sci., 1: 5–12.

Sunitha-Karunakaran, Nair,- G.-K., Nair, N.-D. and Mini,-M. 2010. Systemic Aspergillosis in Emu chicks in an organised farm in Kerala. Vet. World, 3: 453–455.

Tewari, R.P. 1972. Studied on some aspects of mycoses. M.VSc. Thesis submitted to Agra University, Agra.

Thakur, D.K., Misra, S.K. and Chaudhri, P.C. 1983a. Prevalence of dermatomycosis in sheep in Northern India. Mykosen, 26: 271–272.

Thakur, D.K., Singh, N., Misra, S.K. and Singh, K.B. 1985. Histopathology of dermatomycosis in cattle and buffaloes. Indian J. Anim. Sci., 55: 762–765.

Thakur, D.S., Misra, S.K. and Singh, K.B. 1983b. A preliminary study on dermatomycosis in water buffalo in Punjab. Indian J. Vet. Med., 3: 57–60.

Thakur, S.K. and Verma, B.B. 1984. Dermatomycosis due to *M. gypseum* in goats. Indian Vet. J., 61: 1083.

Thomas, M., Abin-Varghese, Samuel, K. and Punnen-Kurian, A. 2011. Zygomycosis in captive Asian elephant (*Elephas maximus*). Zoos'-Print., 26: 27.

Uppal, P.K., Thakur, K.K. and Oberoi, M.S. 1978. Serodiagnosis of mycotic abortions in bovine caused by *Aspergillus fumigatus*. Aspects Allergy & Applied Immunology. Vol XI. ed.

Vanamayya, P.R., Mouli, A.C. and Parihar, N.S. 1983. A note on phycomycotic pneumonia in a buffalo (*Bubalus bubalis*). Indian Vet. J., 60: 490.

Verma, R., Srivastava, D.P., Shukla, P.K. and Iyer, P.K.R. 1986a. Isolation and antifungal-biogram of *Phialophora repens*. Indian J. Comp. Microbiol. Immunol. Inf. Dis., 7: 79–80.

Verma, Rishendra, Srivastava, O.P., Shukla, P.K. and Iyer, P.K.R.I. 1986b. Isolation and antifungal-biogram of *Phialophora repens*. Indian J. Comp. Microbiol. Immunol. Inf. Dis., 7: 79–80.

Verma, Rishendra. 1982. Annual Report. Division of Bact. & Mycology, IVRI, Izatnagar.

Verma, Rishendra. 1988. Studies on clinical and subclinical mastitis bovine mastitis. Indian J. Comp. Microbiol. Immunol. Inf. Dis., 9: 29–33.

Index

T

Color Plate Section

Chapter 3

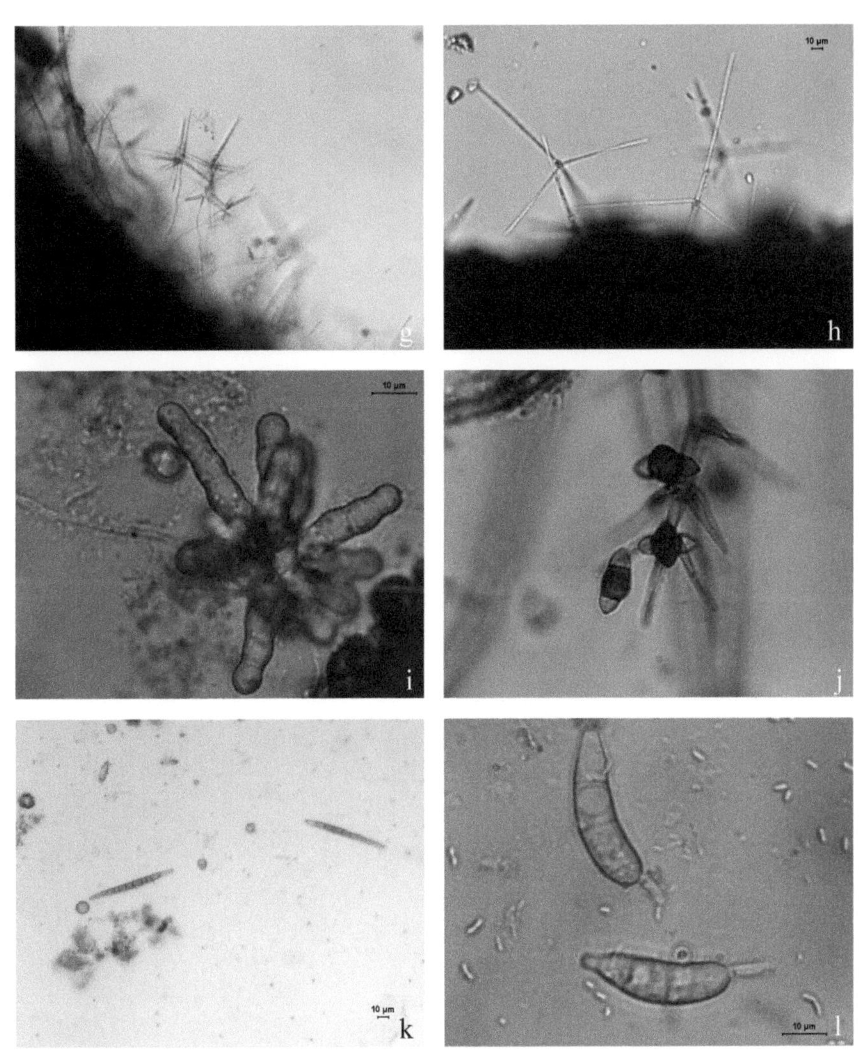

Figure 3.2. Overview of genera of aquatic Hyphomycetes in Brazilian waters: (g, h) *Triscelophorus*; (i) *Pyramidospora*; (j) *Uberispora*; (k) *Camposporium*; (l) *Margaritispora*. Images by Carolina Gasch Moreira.

Chapter 7

Figure 7.4. A small stone fragment from the St. Genis Cathedral in France with evident black clusters from were *Coniosporium uncinatus* was isolated. In the circle, the *Coniosporium* strain seen under phase contrast (magnification 400x).

Chapter 11

Figure 11.3. Stages and structures of the stem rust life cycle: (a) teliospore, (b) basidia and basidiospore, (c) *Berberis* plant (alternate host), (d) pycnia on barberry, (e) aecia on the lower part of the barberry leaf.

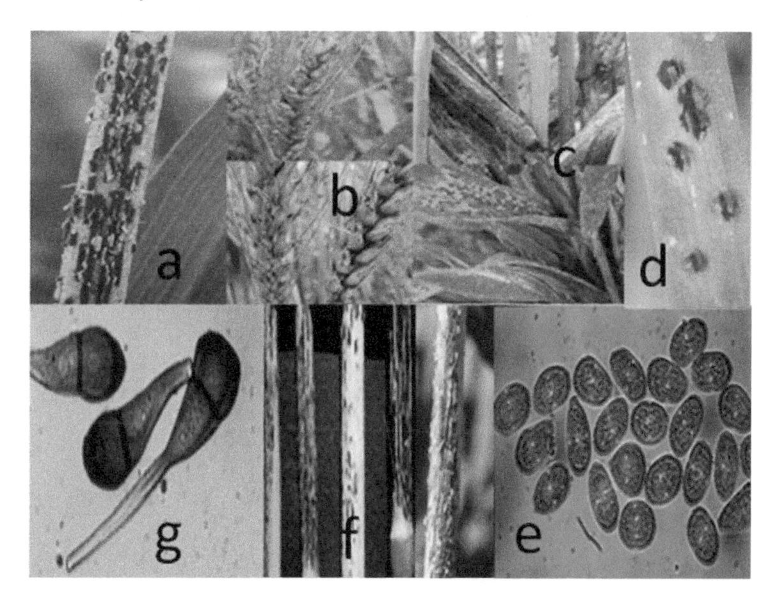

Figure 11.4. Structures present on different parts of the wheat plant: (a) uredinia on stem, (b) uredinia on the head; (c) uredinia on leaf and sheath, (d) uredinia on seedling leaf, (e) urediniospores, (f) telia, and (g) teliospore.

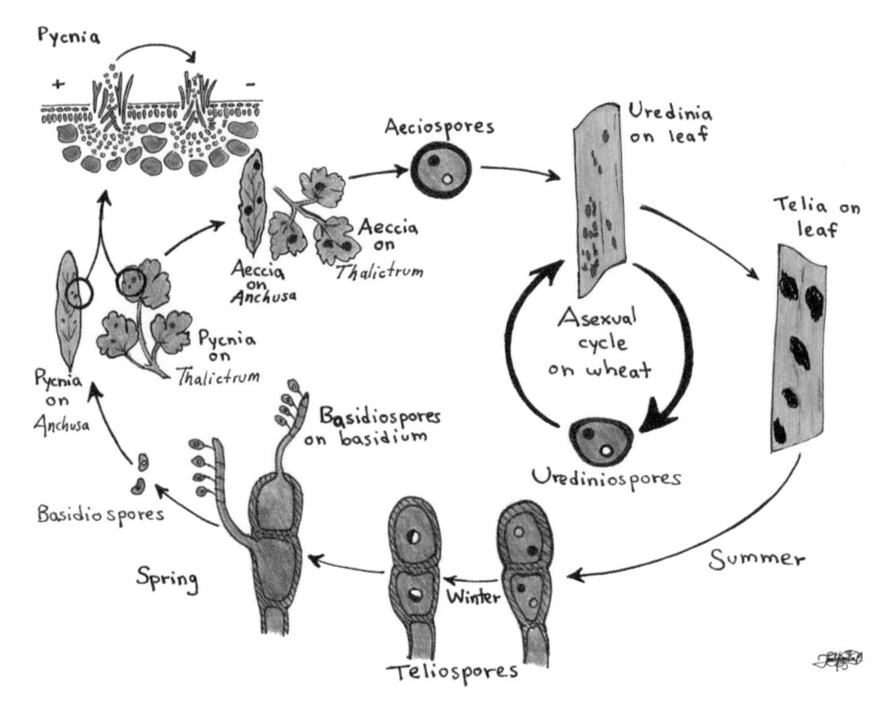

Figure 11.5. Illustrated life cycle of *P. triticina* and *P. tritici-duri.*

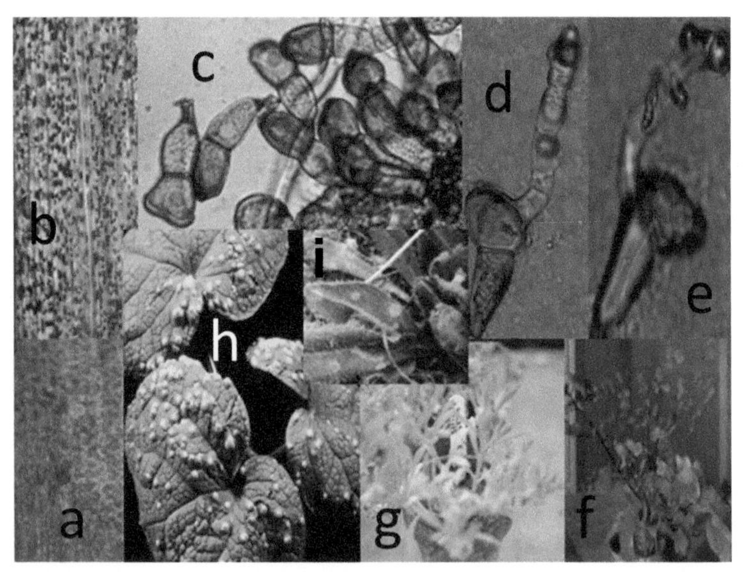

Figure 11.6. Structures and alternate host plants of the life cycle of *P. triticina* and *P. tritici-duri*: (a) Uredinia on wheat leaf, (b) telia, (c) Teliospore, (d) basidia, (e) basidiospore, (f) *Thalictrum* (alternate host), (g) *Anchusa* (alternate host), (h) pycnia on *Thalictrum*, (i) pycnia on *Anchusa.*

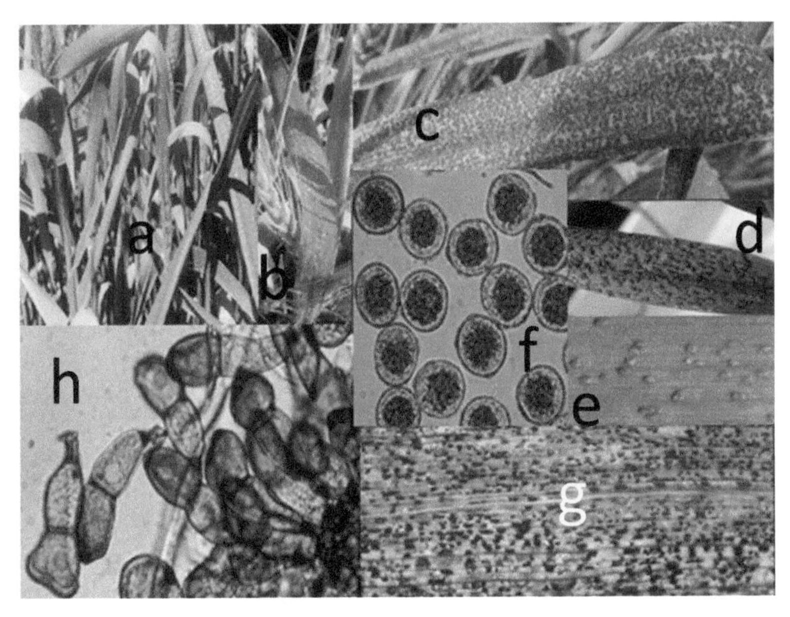

Figure 11.7. Structures on the wheat plant from the life cycle of *P. triticina* and *P. tritici-duri*: (a, b), and (c) Uredinia on flag leaves (field infection), (d,e) uredinia on flag leaf (greenhouse) , (f) urediniospores, (g) telia, (h) teliospores.

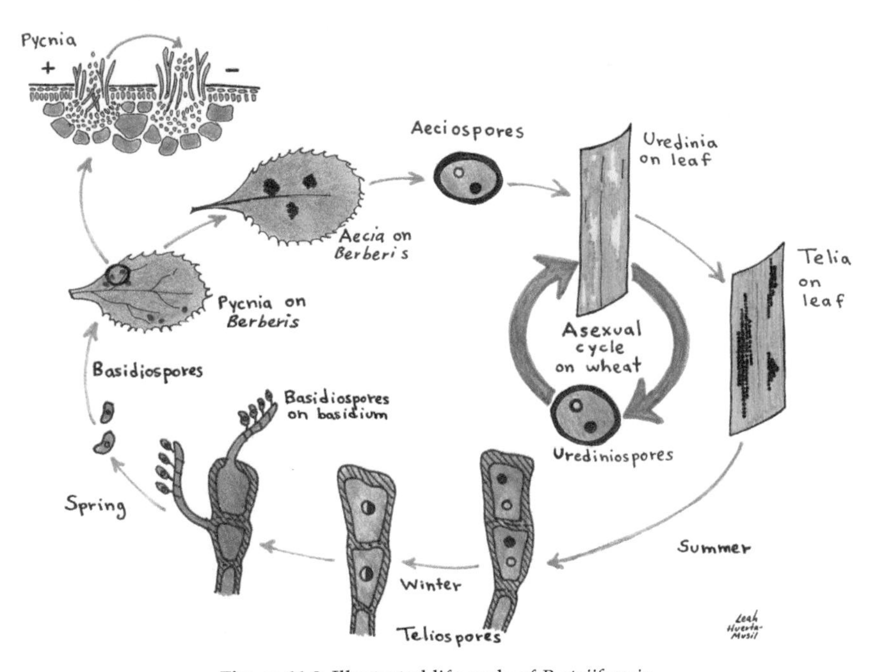

Figure 11.8. Illustrated life cycle of *P. striiformis*.

Figure 11.9. Structures on the wheat plant from the life cycle of *P. striiformis*: (a, b, d, e and f) Uredinia on flag leaves under field conditions, (c) urediniospores, (g) head infection, (h) spikelet infection, (i) telia on leaf, (j) teliospore.

Figure 11.10. Structures and alternate host plants of the life cycle of *P. striiformis*: (a) Uredinia on wheat leaf, (b) telia, (c) Teliospore, (d) basidia and basidiospore, (e) *Berberis* plant (alternate host), (f) pycnia on barberry leaf, (g) aecia on the lower part of the barberry leaf.

Figure 11.11. Bulk collection of urediniospores and field infection on the spreader row in a breeding stripe rust nursery.

Figure 11.12. Steps in the collection of urediniospores and inoculation in the greenhouse.

Figure 11.13. Multiplication of *P. graminis*, RTR race (Singh, 1991) in the greenhouse.

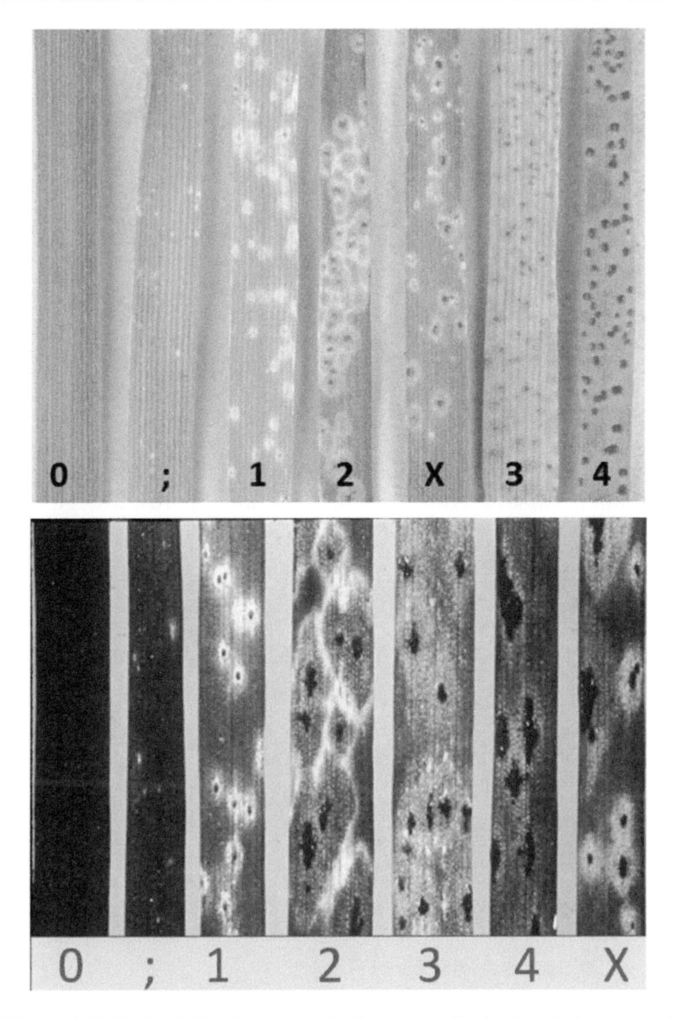

Figures 11.18 and **11.19.** Scale 0–4 to score infection types for leaf and stem rust at seedling stage respectively under greenhouse conditions (Roelfs et al., 1992).

Figure 11.20. Scale 0–9 to score seedling infection types for stripe rust in the greenhouse, After McNeal et al. (1971).

Chapter 12

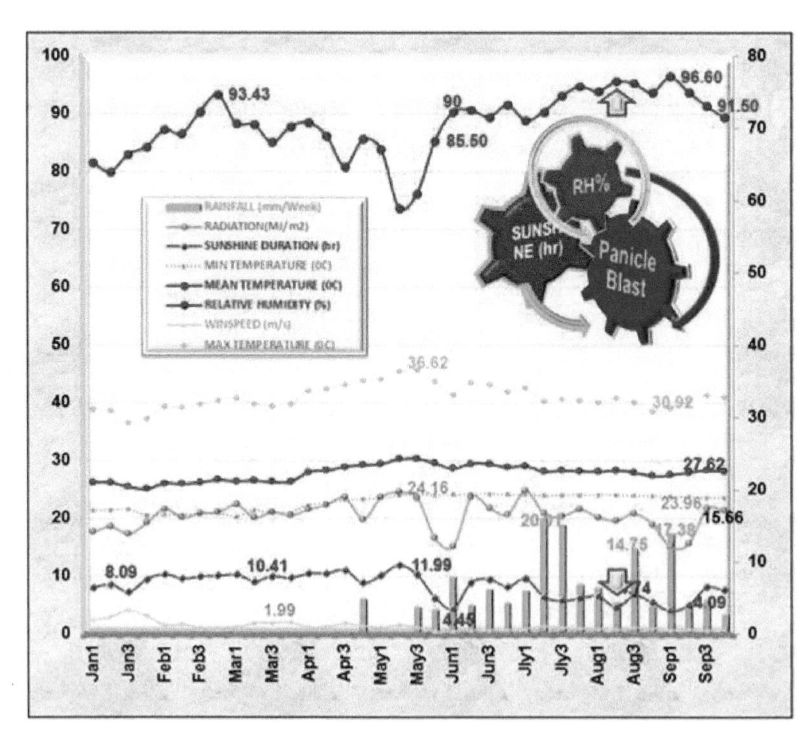

Figure 12.1. Prevalent weather parameters during July to September, 2008–10 were conducive to rice blast endemic in rainfed lowland rice locations, Central Luzon, Philippines.

Figure 12.2. Leaf blast lesions initially appear as brown specks smaller than 1 mm in diameter without sporulation (a), turn roundish to elliptical about 1–3 mm with gray center surrounded by brown margins and/or yellow halo covered with conidia (b); and typical spindle-shaped lesions >3 mm in diameter with necrotic gray centers and reddish round margins which may coalescence (c).

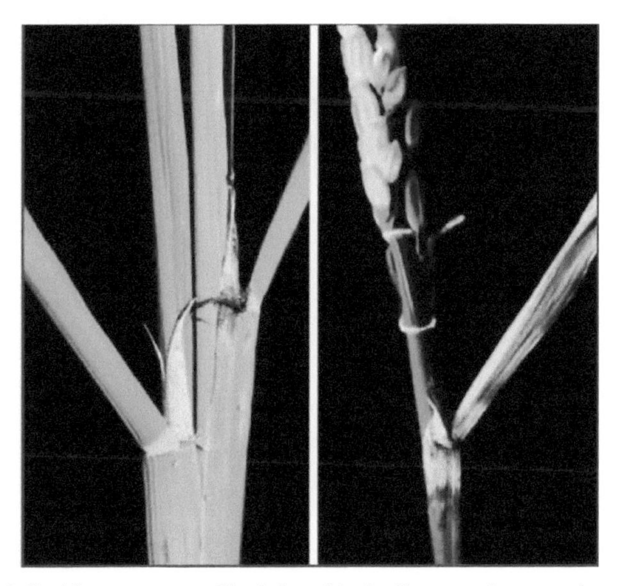

Figure 12.3. Collar blast symptoms. The infected leaf collar turned green to brownish which may expand to the leaf sheath and leaf base and cause collapsed or flabby leaves.

Figure 12.4. Neck blast which causes death and brown internodes under the panicle and at one site of flag leaf. Under favorable weather, the fungus spreads onto the spikelets and produces white powder.

Figure 12.5. Typical symptom of sheath blight on the leaves (left), sclerotial bodies formed at the reproductive crop stage (center), and severe infection of panicles (right).

Figure 12.6. A traditional upland rice variety from Northeastern Central Luzon, Philippines found susceptible to brown leaf spot.

Figure 12.7. Partially emerged panicles due to sheath rot infection showing rotten, red-brown to dark brown spikelets and unfilled, shriveled grains.

Figure 12.8. Foot rot or bakanae induces yellowish green leaves (left) and abnormally elongated tillers on infected PSB Rc82 rice variety (right). Four DNA pathogen profiles detected within the province surveyed for bakanae(insert).

Figure 12.9. Narrow brown leaf spot and leaf scald oftenly occur in poor soil with limited water supply in rain-fed lowland and upland in the humid tropics.

Figure 12.10. Leaf scald lesion initiates along leaf margin and turns gray with alternating narrow reddish-brown bands owing to intermittent rains and warm days in rain-fed and upland areas (left). Worsen microclimatic conditions can rot the spikelets (right).

Figure 12.11. Initial infection of stem rot occurs above the water line and appears as black lesion at maximum tillering stage (left, first two photos). Small black sclerotial bodies are formed inside the plant tissues after the plant dried up (right two photos).

Figure 12.12. Orange-yellow smut balls are formed at early milking stage then turn dark green to almost black with cracked surface.

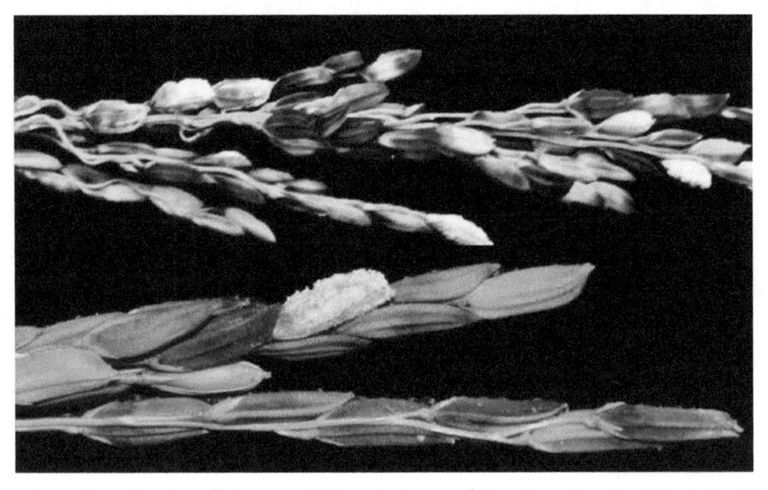

Figure 12.13. *Fusarium graminiarum* and *F. semitectum* are often found on the infected spikelets during hot and humid days in irrigated and rain-fed lowland rice.